NUCLEAR

ENERGY

Principles, Practices, and Prospects

NUCLEAR —

ENERGY

Principles, Practices,
— and Prospects

David Bodansky

University of Washington
Seattle, Washington

American Institute of Physics   **Woodbury, New York**

©1996 by American Institute of Physics.
All rights reserved.
Printed in the United States of America.

AIP Press
American Institute of Physics
500 Sunnyside Boulevard
Woodbury, NY 11797-2999

Library of Congress Cataloging-in-Publication Data
Bodansky, David.
 Nuclear energy : principles, practices, and prospects / David Bodansky.
  p. cm.
 Includes bibliographical references and index.
 ISBN 1-56396-244-6
 1. Nuclear engineering. 2. Nuclear power plants. I. Title.
TK9145. B54 1996   96-148
333.792'4--dc20   CIP

10 9 8 7 6 5 4 3 2 1

Contents

I. INTRODUCTION

II. TECHNICAL BACKGROUND

III. NUCLEAR PRACTICES AND ISSUES

Preface

This book has evolved from notes prepared for students in a physics course designed to cover the major aspects of energy production and consumption. About one-third of the course dealt with nuclear energy, and the notes for that segment were revised and expanded for the present book.

The course assumed that the students had at least one year of college-level physics, thus permitting the inclusion of some technical discussions. The present book, in its occasional use of equations and technical terminology, somewhat reflects the nature of that original audience. Readers with relatively little background in physics and engineering may find it useful to refer to the Appendix on "Elementary Aspects of Nuclear Physics," and to the Glossary.

I have sometimes been asked: "For whom is the book written?" One difficulty in addressing this question has already been touched on. Some of the technical discussions include equations, which is not customary in a book for a "lay audience." Other parts are more elementary than would be the case were this a textbook on nuclear engineering. Nonetheless, most of the key issues can be constructively discussed using little or no mathematical terminology, and I therefore hope that the book will be useful to readers with a wide variety of backgrounds who have an interest in nuclear energy matters.

A more fundamental difficulty lies in the fact that such interest is now at a low ebb. In fact, it is often believed that the era of nuclear fission energy has passed, or is passing. While most informed people are aware that France is highly dependent on nuclear energy, this is ignored as an aberration, holding little broader significance. It is not widely realized that nuclear energy, despite its stagnancy in the United States and most of Europe, is expanding rapidly in Asia. Further, many people who are otherwise well-informed on issues of public policy are surprised to learn that the United States now obtains more than 20% of its electricity from nuclear power.

This book has been written in the belief that it is premature and probably incorrect to assume that there is to be only one era of nuclear power and that this era has passed. The future pattern of nuclear energy use will depend upon developments in a variety of energy technologies and upon public attitudes in differing countries. There can be little certainty as to how these developments will unfold. However, the demands of a growing world economy and the pressures of declining availability of oil will inevitably force a realignment and reassessment of energy options. The goal of this book is to provide basic information to those who want to gain, or refresh, an introductory familiarity with nuclear power, even before broad new reassessments of energy policy are made in the United States and elsewhere.

The preparation of the book has been aided by contributions from many individuals. Among these, I would like especially to acknowledge three. Since I first became interested in energy issues some twenty years ago and continuing until his death in 1991, my understanding of these issues and of nuclear energy in particular benefited greatly from discussions and collaborative writing with my colleague Fred Schmidt. Over the years, I have also gained much from the wisdom of Alvin Weinberg, who has made unique contributions to nuclear energy and its literature and, most recently, has very kindly read and commented upon much of this manuscript. I am also grateful to Peter Zimmerman who served the publisher as an anonymous reviewer of a preliminary draft of this book and who subsequently, anonymity discarded, has been a very constructive critic of a revised draft.

In addition, I am heavily indebted to many other individuals at the University of Washington, in government agencies, in industry, and elsewhere. Some have been generous in aiding with information and insights, some have commented on various chapters as the book has evolved, and some have done both. Without attempting to distinguish among these varied contributions, I particularly wish to thank Mark Abhold, Thomas Bjerdstedt, Robert Budnitz, Thomas Buscheck, J. Gregory Dash, Kermit Garlid, Ronald Geballe, Marc Gervais, Emil Glueckler, Lawrence Goldmuntz, Isaac Halpern, Charles Hyde-Wright, William Kreuter, Jerrold Leitch, Norman McCormick, Thomas Murley, James Quinn, Maurice Robkin, Margaret Royan, Mark Savage, Jean Savy, Fred Silady, Bernard Spinrad, Ronald Vijuk, and Gene Woodruff.

This list is far from exhaustive and I extend my thanks and apologies to the many others whom I have failed to mention. I am also grateful to the University of Washington and the Department of Physics for making it possible for me to teach the courses and devote the time necessary for the development of this book. Finally, I must express my appreciation to my wife, Beverly, for her support and encouragement as the book progressed.

Chapter 1

Nuclear Power Development

1.1 EARLY HISTORY OF NUCLEAR POWER

1.1.1 The Hopes and the Reality

General perceptions of nuclear energy, among both the public and policy makers, have undergone dramatic shifts in the past 50 years. As nuclear energy emerged in 1945 from scientific obscurity and military secrecy, it began to be talked of in speculative terms as an eventual power source. Within a decade an enthusiastic vision developed of a future in which nuclear power would provide a virtually unlimited solution for the world's energy needs. It was not difficult to picture nuclear power as the ideal energy source. With the use of breeder reactors, it would be ample in supply. As experience was gained in reactor construction, it would become economical. And because a nuclear reactor would emit virtually no pollutants, it would be clean, especially in contrast to coal.

There was also a negative side, as some doubters pointed out from the first. Very large amounts of radioactivity would be produced. In principle, practically none need escape, but the possibility of mishaps could not be totally excluded. Further, benign nuclear energy had a malign sibling in nuclear weapons. While it was realized that a reactor itself could not explode like a bomb, there were fears that in some way controlled nuclear energy might go out of control.

The optimists prevailed for two decades, into the early 1970s, and many nuclear reactors were designed, built, and put into operation in the United States and Europe. Part of the motivation for this development was the desire of countries to reduce their heavy dependence on oil, which, they realized, would eventually be in short supply. The first oil crisis came in 1973, even sooner than had been anticipated. Just as the nuclear buildup was gaining momentum, an oil embargo was imposed by the oil-rich Persian Gulf countries as a sequel to the October 1973 war between Egypt and Israel.

An immediate impulse was to rely even more on nuclear power as a substitute for oil. This was especially true in the United States, where nuclear energy had already appeared to many, including the federal government, as an important key to "energy self-sufficiency."[1] But the embargo had unanticipated effects. It focused new attention on the possibility of reducing *all* energy consumption, and rising prices slowed the pace of economic growth. These factors sharply reduced the demand for electricity and therefore the pressure to add new nuclear power plants. At the same time, the costs of nuclear power and fears about nuclear power both began to grow. The Three Mile Island accident in 1979 and the Chernobyl accident in 1986 hit a world becoming more attuned to believing the worst about nuclear power.[2]

Nuclear energy development was stopped or brought to a crawl in all but a few countries during the 1980s and early 1990s. Contributing factors to this decline included a gradual reduction in oil and gas prices, rising nuclear costs, the sluggishness of the growth in energy demand, general fears of nuclear power, and in some countries determined campaigns against it. By 1994 it was easy to think that the age of nuclear power was coming to an end. The apparent rise and fall of nuclear power came quickly: Virtually unheard of in 1944, it was the panacea of 1974 and the pariah of 1994.

But this is an incomplete picture. There was always less unanimity than this description suggests. Not only were there scientists who were aware of nuclear power's potential in 1944, but, more significantly, there were skeptics in 1974, and there remain enthusiasts in 1994. The final verdict has not been given. Nuclear power may seem to be dormant in Germany and dead in Italy, but across the borders in France, it is thriving. The actual picture varies from country to country, with considerable complexity and uncertainty in most industrialized countries.

In the succeeding chapters we will pay only passing attention to the political and psychological factors that have influenced the changing assessments of nuclear power. Instead, the emphasis will be on describing the functioning of nuclear reactors, the extent to which they are used, and the problems surrounding their use. Issues of history and public attitudes will not be totally ignored, but the physical functioning of nuclear power, rather than its social functioning, will be the primary focus.

1.1.2 Developments Prior to the End of World War II

Speculations Before the Discovery of Fission

Atomic energy, now usually called nuclear energy,[3] became a gleam in the eye of scientists in the early part of the 20th century. The possibility that the atom held a vast reservoir of energy was suggested by the large kinetic energy of the particles emitted in radioactive decay and the resultant large production of heat. In 1911, 15 years after the discovery of radioactivity, the British nuclear pioneer Ernest Rutherford called attention to the heat produced in the decay of radium:

---

[1]Just prior to the embargo, an Atomic Energy Commission energy policy statement was formulated in the report *The Nation's Energy Future*, based on studies carried out during the summer of 1973 and submitted to President Nixon on December 1, 1973 [1]. To eliminate the need for oil imports, it called for major contributions from coal, nuclear power, conservation, and domestic oil and gas.

[2]The antinuclear movie *The China Syndrome* had reached theaters slightly before the Three Mile Island accident, indicative of the growing negative image of nuclear power in popular culture.

[3]Although the two terms mean the same thing, there has been some shift in usage with time. Originally, "atomic energy" was the more common designation, but it has been largely replaced by "nuclear energy." For example, the *Atomic* Energy Commission was established in 1946 and the *Nuclear* Regulatory Commission in 1975.

This evolution of heat is enormous, compared with that emitted in any known chemical reaction. . . . The atoms of matter must consequently be regarded as containing enormous stores of energy which are only released by the disintegration of the atom [2].

At this time, however, Rutherford had no concrete idea as to the source of this energy. The magnitudes of the energies involved were gradually put in more quantitative form as information accumulated on the masses of atoms, but until the discovery of fission in 1938, there could be no real understanding of how this energy might be extracted. In the interim, however, important progress was made in understanding the basic structure of nuclei. Major steps included the discovery of the nuclear atom by Rutherford in 1911, the discovery of the neutron as a constituent of the nucleus in 1932, and a series of experiments undertaken in the 1930s by Enrico Fermi and his group in Rome on the interactions between neutrons and matter, which eventually led to the discovery of fission.[4]

As these developments unfolded, there was general speculation about atomic energy. One of the earliest scientists to have thought seriously about the possibilities was Leo Szilard, who was later active in efforts to initiate the U.S. atomic bomb program. He attributes his first interest in the extraction of atomic energy to reading in 1932 a book by the British novelist H. G. Wells. Writing in 1913, Wells had predicted that artificially induced radioactivity would be discovered in 1933 (he guessed the actual year of discovery correctly!) and also predicted the production of atomic energy for both industrial and military purposes. In Szilard's account, he at first "didn't regard it as anything but fiction." A year later, however, two things caused him to turn to this possibility more seriously: (a) he learned that Rutherford had warned that hopes of power from atomic transmutations were "moonshine,"[5] and (b) the French physicist Frederic Joliot discovered artificial radioactivity as predicted by Wells [6, pp. 16–17].

Szilard then hit upon a "practical" scheme of obtaining nuclear energy. At the time it was thought that the beryllium 9 (^{9}Be) nucleus was unstable and could decay into two alpha particles and a neutron. This was a misconception, based on an incorrect value of the mass of the alpha particle; actually the neutron in ^{9}Be is bound, if only relatively weakly. In any event, Szilard thought that it might be possible to "tickle" the breakup of ^{9}Be with a neutron and then use the extra neutron released in the breakup to initiate another ^{9}Be reaction. In each stage, there is one neutron in and two neutrons out. This is the basic idea of a chain reaction. This particular chain reaction cannot work, as was soon realized, because too high a neutron energy is required to cause the breakup of ^{9}Be, and even then there is a net loss of energy in the process, not a gain.

Szilard tried to find other ways to obtain a chain reaction, but his efforts failed. Nonetheless, in the interim he went so far as to have a patent on neutron-induced chain reactions entrusted to the British Admiralty for secret safekeeping, the military potential of nuclear energy being important in his thinking [5, p. 225]. However, the key to a realizable chain reaction—fission of heavy elements—eluded Szilard and all others.

In these early speculations, there was an awareness of the potential of atomic energy for both military and peaceful applications, and the latter loomed large in the thinking of some scientists. For example, in a document dated July 1934, Szilard explained planned experiments that, if successful, would lead to

power production . . . on such a large scale and probably with so little cost that a sort of industrial

---

[4]Historically oriented accounts of these developments and the discovery of fission are given in Refs. [3], [4], and [5]. A discussion of basic nuclear physics concepts and terminology is presented in Appendix A.

[5]It is not clear from the Szilard reference whether the word "moonshine" was Rutherford's own, or whether it was a paraphrase appearing in *Nature* in a summary of Rutherford's talk.

revolution could be expected; it appears doubtful, for instance, whether coal mining or oil production could survive after a couple of years [6, p. 39].

Along the same lines, Joliot prophesied in his 1935 Nobel Prize acceptance speech:[6]

scientists, disintegrating or constructing atoms at will, will succeed in obtaining explosive nuclear chain reactions. If such transmutations could propagate in matter, one can conceive of the enormous useful energy that will be liberated [7, p. 46].

Fission and the First Reactors

Fission of uranium was discovered—or more precisely, recognized for what it was—in 1938.[7] Scientists quickly recognized that there was now in principle a path to a chain reaction and, in addition, that large amounts of energy are released in fission. By early 1939 it was verified that neutrons are emitted in fission, and it soon became apparent that enough neutrons were emitted to sustain a chain reaction in a properly arranged "pile" of uranium and graphite.[8] It took several more years to demonstrate the practicality of achieving a chain reaction. This work was led by Fermi, who had left Italy for the United States, and it culminated in the development and demonstration of the first operating nuclear reactor on December 2, 1942, at an improvised facility in Chicago.

The discovery and preliminary understanding of fission came at a time when the prospect of war was much on people's minds. The start of World War II in Europe in August 1939 ensured that military, rather than civilian, applications of atomic energy would take primacy, and the early work was heavily focused on the military side, in both thinking and accomplishments. A major goal of the nuclear program was the production of plutonium 239 (^{239}Pu), which was recognized to be an effective material for a fission bomb. The ^{239}Pu was to be produced in a reactor, by neutron capture on uranium 238 (^{238}U) and subsequent radioactive decay.

The first reactor in Chicago was very small, running with a total power output of 200 watts. But even before the successful demonstration of a chain reaction in this reactor, plans had started towards construction of the much larger reactors required to produce the desired amounts of plutonium. A pilot plant, designed to produce one megawatt, was completed and put into operation at Oak Ridge, Tennessee, in November 1943 [8, p. 392]. A full-size, 200-megawatt reactor began operating at the Hanford Reservation in Washington state in September 1944—a millionfold increase in power output in less than two years.

The laboratories at Oak Ridge and Hanford were new wartime installations. The Hanford site was not selected until early 1943, and construction on the first reactor began in June 1943. Workers completed the reactor within about 15 months, despite the absence of any directly relevant prior experience.[9] The speed with which the program was pursued is breathtaking by present standards, but it can be understood in the context of the exigencies of World War II.

The pressures of wartime bomb development pushed work on peaceful applications largely into the background, but there was still considerable thinking about future civilian

---

[6]Joliot shared with his wife, Irene Joliot-Curie, the 1935 Nobel Prize in Chemistry for their discovery in 1933 of artificial radioactivity. They were the son-in-law and daughter of Marie and Pierre Curie, recipients (with Antoine-Henri Becquerel) of the 1903 Nobel Prize in Physics, awarded for the discovery of radioactivity

[7]See Sections 4.1 and 4.2 for a brief description of fission and its discovery.

[8]The uranium was the fuel. The graphite served as a "moderator," to slow the neutrons down to energies where they were most effective in producing fission in uranium. See Section 5.2 for a discussion of moderators.

[9]An early summary of this chronology is given in Ref. [8], Ch. XIV.

uses. An official report on the development of the atomic bomb was prepared by Henry Smyth, a Princeton physicist. It was published in 1945, shortly after the end of the war, to inform the public about the bomb project. In a closing section, entitled "Prognostication," Smyth pointed out that

> The possible uses of nuclear energy are not all destructive, and the second direction in which technical development can be expected is along the paths of peace. In the fall of 1944 General [Leslie] Groves appointed a committee to look into these possibilities as well as those of military significance. This committee . . . received a multitude of suggestions from men on the various projects, principally along the lines of the use of nuclear energy for power and the use of radioactive by-products for scientific, medical, and industrial purposes [9, p. 224].

With or without such a committee, it was inevitable that imaginative scientists would consider ways of using nuclear energy for electricity generation. The possibility, for example, of power production from a reactor that used water at high pressure for both cooling the reactor and moderating the neutron energies was suggested as early as September 1944, in a memorandum written by Alvin Weinberg, who was then working closely with Eugene Wigner on reactor designs [10, p. 43]. This is the basic principle of the dominant reactor in the world today, the *pressurized light water reactor* (PWR).[10]

1.1.3 Post-War Developments in the United States

Progress Towards Commercial Nuclear Power

In the years immediately following World War II, the main activities of the American nuclear authorities continued to be directed towards further military developments, but increased attention was turned to electricity generation. A somewhat guarded assessment of the future was presented in the Smyth report:

> While there was general agreement that a great industry might eventually arise . . . there was disagreement as to how rapidly such an industry would grow; the consensus was that the growth would be slow over a period of many years. At least there is no immediate prospect of running cars with nuclear power or lighting houses with radioactive lamps although there is a good probability that nuclear power for special purposes could be developed within ten years [9, p. 225].

This turned out to be rather close to the mark, although the words "slow" and "immediate" may have had different connotations in 1945, in the wake of the rapid pace of wartime development, than they do today.

The development of nuclear power in the United States was in the first instance the responsibility of the Atomic Energy Commission (AEC), an agency established in 1946 to oversee both military and civilian applications of nuclear energy. Although the desirability of moving ahead with nuclear energy was widely accepted, the AEC undertook in 1949 to sponsor a more analytic study, described as

> a study of the maximum plausible world demands for energy over the next 50 to 100 years. The study was envisaged as background for the Commission's consideration of the economic and

---

[10]*Light water reactor* (LWR) refers to a reactor cooled and moderated with ordinary water, "light" being used to differentiate these reactors from those cooled or moderated with heavy water, i.e., water in which the hydrogen is primarily in the form of deuterium (^{2}H). The PWR is one version of the LWR. Characteristics of LWRs are discussed at length in later chapters.

public policy problems related to the development and use of machines for deriving electrical power from nuclear fuels [11].

The study was carried out by a consulting engineer with broad interests, Palmer C. Putnam.[11] The results of Putnam's study appeared in 1953 in the book *Energy in the Future* [11]. In retrospect, the book is a prophetic masterpiece. It started with the consideration of future increases in population, in demand for energy, and in the efficiency of delivering energy. Putnam then addressed the issues of fossil fuel reserves, concluding that we could not live "much longer" off fossil fuels, which he termed "capital energy." He also pointed out the possible dangers of climate change from carbon dioxide produced in the combustion of fossil fuels [11, p. 170].

Putnam next turned to the potential of what is now called renewable energy, which he termed "income energy." This is primarily solar energy, in all its forms. He concluded that the world could not expect to obtain "more than 7 to 15 per cent of the maximum plausible demands for energy from 'income' sources at costs no greater than 2 times present costs" [11, p. 204].

This led Putnam to the conclusion that a new "capital" source of energy would be required, i.e., nuclear energy. With breeder reactors, he indicated that world uranium supplies would suffice for "many centuries" [11, p. 250]. However, he pointed out that nuclear energy could only make a decisive contribution if transportation and home heating were electrified to a much greater extent than was the case in the early 1950s.

In summary, Putnam urged the prompt development of nuclear power, the exploration of nuclear fusion, and "as our ultimate anchor to windward," exploration of ways to obtain solar energy "in more useful forms and at lower costs than now appear possible" [11, p. 255]. This was a prescient book, and it provided a reasoned basis for proceeding with nuclear power. However, so thoughtful an analysis was not really a prerequisite for this move. Nuclear energy was favored by a much more casual, almost romantic, image as a source of abundant, clean energy, by the possible technological imperative to move ahead because it was possible to do so, and by the correct, even if not well quantified, recognition of an eventual limit to fossil fuels.

There were, however, impediments to quick progress. For one, fossil fuels were plentiful in the 1950s, and so no immediate urgency was felt. Nuclear facilities and technical knowledge were under the tight control of the AEC, with many aspects kept secret because of the military connections. Further, there was indecision as to the relative roles to be played by the government and private utilities in the development of nuclear power. Finally, it was not clear which type or types of reactor should be built.

The U.S. Navy made a decision first and, under the leadership of Hyman Rickover, built and began tests of pressurized light water reactors by the first part of 1953 [12, p. 188]. These reactors became the foundation of the U.S. nuclear submarine fleet. After some hesitation and consideration of alternative designs, the AEC announced in autumn 1953 that it would build a 60-MWe[12] power plant [12, p. 194]. Participation by utilities was sought. The general

---

[11]In selecting Putnam, the AEC does not appear to have attempted to stack the deck in favor of nuclear power. He had previously written books with titles such as *Chemical Relations in the Mineral Kingdom, Power From the Wind*, and *Solar Energy* and was one of the designers of the giant Smith–Putnam windmill built at Grandpa's Knob in Vermont in the 1940s.

[12]The maximum electrical power output of a nuclear reactor is commonly indicated by its capacity, in megawatts of electricity produced. This is denoted by the abbreviation MWe, where the e is added to emphasize that reference is being made to the *electrical* power out, rather than the *thermal* power produced in the consumption of the fuel. The latter is designated by MWt. For a reactor operating at an efficiency of 33%, the capacity in MWt is three times the capacity in MWe. Alternatively, capacities can be specified in gigawatts, denoted by GWe and GWt.

reactor configuration was to be the same as that used by the navy, namely, a pressurized light water reactor. A Pennsylvania utility won the competition to participate in this project, contributing the land and buildings and undertaking to run the facility when completed. The reactor was built at Shippingport, Pennsylvania, and was put into operation at the end of 1957 [12, pp. 419–423].

Early Enthusiasm: "Too Cheap to Meter"?

Despite the considerable caution in initiating the United States nuclear power program, many euphoric statements about the future of nuclear power were made in the 1940s and 1950s. Such statements even go back as far as H. G. Wells in 1913. Nuclear power was to be abundant, clean, and cheap. Of course, thoughtful observers modulated and qualified even their optimistic prophecies, but some of the optimism was quite unbridled. For example, an article in *Business Week* in 1947 stated "Commercial production of electric power from atomic engines is only about five years away. . . . There are highly respected scientists who predict privately that within 20 years substantially all central power will be drawn from atomic sources" [13].

However, among all the enthusiastic quotations from that era, one in particular has come to haunt nuclear advocates. In the 1980s, as nuclear power became more expensive than electricity from coal, an earlier phrase, "too cheap to meter," was thrown back in the faces of proponents of nuclear power as illustrating a history of false promises and overweening and foolish optimism.

The phrase originated with Lewis L. Strauss, chairman of the AEC in the 1950s. Speaking at a science writers meeting, he stated: "It is not too much to expect that our children will enjoy electrical energy in their homes too cheap to meter." The phrase was used in a *New York Times* headline on September 17, 1954, the day after the speech.[13] A fuller version of Strauss' remarks as they appeared in his prepared text indicated a broadly euphoric technological optimism about nuclear energy and its applications [14, p. 2].

It is doubtful that many professionals shared this euphoria at the time. A more official version of the AEC position, expressed in Congressional testimony in June 1954, held out the hope that

> [nuclear power] costs can be brought down—in an established nuclear power industry—until the cost of electricity from nuclear fuel is about the same as the cost of electricity from conventional fuels, and this within a decade or two [14, p. 4].

Overall, the history appears to have been one of hesitation and examination, followed by a conviction developing in the 1960s that nuclear power would indeed be cheaper than the fossil fuel alternatives. For a period, in the 1970s, this expectation was fulfilled in the United States, and it is still fulfilled in many countries. The failure of this expectation in the past decade in the United States was a surprise and disappointment to nuclear proponents as well as many neutral analysts. There was a serious misjudgment, but not as egregious a folly as connoted by the "too cheap to meter" phrase.

---

[13]An account of the history of the remark is given in a brief report prepared by the Atomic Industrial Forum (AIF), a nuclear advocacy organization [14]. There is a good chance that Strauss was thinking of fusion power, not fission power, although he could not be explicit because the practicalities of fusion were secret in 1954, with the development of the hydrogen bomb only recently started. The AIF report quotes Lewis H. Strauss, the son of the AEC Chairman and himself a physicist: "I would say my father was referring to fusion energy. I know this because I became my father's eyes and ears as I travelled around the country for him."

1.2 NUCLEAR POWER GROWTH IN THE UNITED STATES

1.2.1 History of Reactor Orders and Construction

The First Reactors

The Shippingport reactor, the first to provide commercial electricity in the United States, was a unique case. Although used for commercial electricity supply, it was largely financed by the federal government and built under navy leadership. Following the order of the Shippingport reactor in 1953, there was a fitful pattern of occasional orders over the next ten years. The only sizable reactors (>100 MWe) ordered before 1962 were the 265-MWe Indian Point 1 (New York), the 207-MWe Dresden 1 (Illinois), and the 175-MWe Yankee Rowe (Massachusetts) reactors, which were ordered in 1955 and 1956 [15].[14] The first of these to go into operation was Dresden 1, in 1960.

The early period of reactor development was characterized by extensive exploration, with a wide variety of reactors being developed for military and research applications and for electricity generation. For the latter, a total of 14 reactors were ordered in the period from 1953 through 1960.[15] With the three exceptions cited above, all had capacities under 100 MWe. They included nine light water reactors (LWRs), not identical by any means, plus five other reactors with a wide variety of coolants and moderators.[16]

Growth Until the Mid-1970s

The exploratory period ended quickly. There was a brief lull in reactor orders after 1960, with only five more orders until 1965, and then a period of rapid expansion. The dominance of LWRs in reactor orders was complete after 1960, the only exception being the gas-cooled Fort St. Vrain reactor, ordered in 1965.

None of the reactors ordered before 1962 had a capacity of as much as 300 MWe. After that, there was a substantial escalation in reactor size, in an effort to gain from expected economies of scale. Some critics believe that the growth in size was too fast to permit adequate learning from experience. The mean size of reactors ordered in 1965 was about 660 MWe, and by 1970 the mean size exceeded 1000 MWe, with some above 1200 MWe. The largest reactors completed and licensed to date in the U.S. have a (net) capacity of 1250 MWe.

Developments Since the Mid-1970s

The history of the deployment of nuclear reactors from 1953 through 1993 is shown in Fig. 1.1, which gives the cumulative capacities of reactors ordered and of reactors licensed to operate.[17] There was a period of rapid growth in the number of orders from 1965 to 1975. At first, reactors were completed within about six years, and by the early 1970s nuclear power had begun to assume significant proportions. At the end of 1974 there were about 50 reactors in operation, with a total capacity of well over 30 GWe, providing 6% of U.S. electricity [16].

---

[14]These reactors have all been shut down, most recently Yankee Rowe in 1991 (see Table 1.2).

[15]These data are from Ref. [15].

[16]These were a fast breeder reactor, a sodium graphite reactor, a high-temperature, gas-cooled reactor, an organic moderator reactor, and a heavy water reactor. Many of these types have since been abandoned worldwide. Indian Point 1, Dresden 1, and Yankee Rowe were all LWRs.

[17]This figure is based on data from Ref. [15].

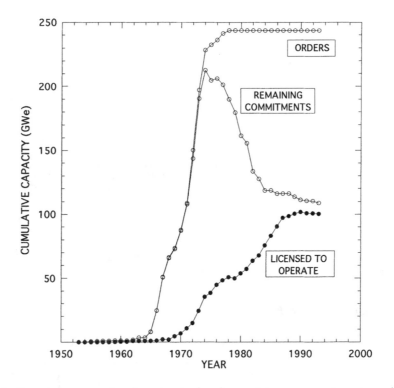

Figure 1.1. Cumulative capacity of reactors ordered, remaining as commitments, and holding operating licenses (including low-power licenses).

After 1974 the rate of putting reactors into operation slowed, due to a drop in the growth in electricity demand. There was a sharp further slowdown following the Three Mile Island accident in 1979 and then a resumption in the early 1980s with the completion of reactors ordered earlier. At the end of 1994 there were 109 operating reactors in the U.S., with a capacity of about 99 GWe.[18] All are light water reactors. Net generation in 1994 was 640 billion kilowatt-hours, or 73 gigawatt-years (GWyr) [16]. The fraction of electricity provided by nuclear power has risen to over 20% in recent years, amounting to 22% in 1994.[19] Through 1990 the driving force in the increased nuclear generation was the addition of new reactors. Since 1990 there has been an increase due to improved operation of existing reactors, with little net change in capacity.[20]

In addition, there were six other reactors in the pipeline as of the end of 1994. Many of these face uncertain fates and may never be completed, although in all cases construction was at least 45% completed and in most cases more than 60% completed [17]. Thus, the number of additional reactors brought on line in the 1990s will be very small, and nuclear power in the United States has reached a plateau. There may be some further improvement

---

[18]Two of the Browns Ferry reactors operated by the Tennessee Valley Authority have been undergoing very extensive modifications and as of the end of 1994 had not generated any power since being shut down for repairs and modifications in the mid-1980s.

[19]This fraction is for the generation by utilities; if generation by nonutility sources is included, the fraction is reduced to about 19% or 20%.

[20]The shutdowns listed in Table 1.2 were partially balanced by the addition of the 1150-MWe Comanche Peak reactor in 1993, resulting in only a slight net decrease in capacity.

Table 1.1. *Cumulative record of nuclear power in the United States, 1953 to 1994.*

Status	Number of Reactors	Net Capacity[a] (GWe)
Cumulative orders	253	243
Orders cancelled	118	129
Ordered and not cancelled	135	114
Shut down before 12/31/94	20	6
Remaining commitments, 12/31/94	115	107
Completed, 12/31/94[b]	109	99
Not completed, 12/31/94	6	7

Sources: Orders and shutdown data are from Ref. [15]. Data on remaining commitments are from Ref. [17].
[a]Capacities from Ref. [15] do not include subsequent capacity reratings of individual reactors; the reratings are reflected in the data of Ref. [17].
[b]Reactors that are completed and have received operating licenses.

in operating performance, but the few additional reactors and the possible improvements could well be balanced or outweighed by additional shutdowns of reactors.

Reactor Orders and Cancellations

A striking feature of the data shown in Fig. 1.1 is the large number of cancellations of reactors that had been ordered. These cancellations are reflected in the drop in the remaining "commitments." Commitments are here defined as the sum of the number of operating reactors and the number of reactors that are at least nominally still under construction or awaiting an operating license. They equal the total number of original reactor orders, less the number of cancelled orders and the number of reactors that have been shut down. Overall, the rapid drop in commitments after 1974 reflects the cancellations, a dearth of new orders, and a few shutdowns. This history is summarized in Table 1.1, which gives the cumulative record of nuclear reactor orders and construction, from the 1950s to the end of 1994.

As seen in Table 1.1, more than half of the reactors that were ordered since 1953 have been cancelled. In some cases, this occurred after construction had started. The large number of cancellations was due in part to unrealistic projections for the growth in electricity use. After 1974, with some slowing of economic growth and a new emphasis on conservation, electricity sales grew at a much slower rate than in previous decades. Utilities that had placed earlier orders found themselves with a surplus of planned capacity.[21] This plus the substantial opposition to nuclear power that arose in the 1970s and intensified after the Three Mile Island accident in 1979 led to the many cancellations.

Quantitative detail on the history since 1953 is shown in Fig. 1.2, which indicates the numbers of reactors ordered in each year and the number of these orders that were not cancelled, i.e., that were built or remain under construction. Most reactors ordered by 1970 were eventually built. After 1970 there was a large surge in orders, but most of these were later cancelled, including all orders placed in 1974 and later. New orders fell abruptly in 1974 and 1975, and no orders have been placed since 1978. Overall, almost all reactors now in operation were ordered in the period from 1965 through 1973. This is a strikingly compressed interval, and one that did not allow much opportunity for the manufacturers and utilities to learn from experience.

---

[21]Electricity sales grew at a rate of 7.5% per year in the decade from 1963 to 1973. Sales actually dropped in 1974, and the average annual increase was only 2.3% from 1973 to 1983 and 2.9% from 1983 to 1993.

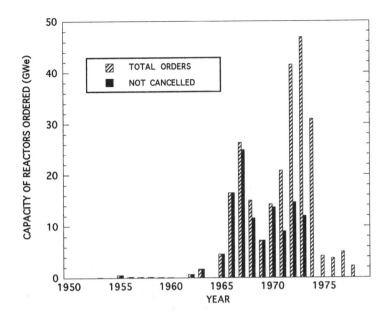

Figure 1.2. Reactor orders in the United States, 1953–1978 (in GWe of total capacity). Annual figures are given for all orders and for those that were not subsequently cancelled. The "not cancelled" category includes operating reactors, reactors that have been shut down, and partially completed reactors.

1.2.2 Capacity Factors

The cessation of reactor orders and the absence of any new orders on the horizon represent a discouraging picture for U.S. nuclear power. A bright spot, however, has been the recent improvement in operating performance. An overall measure of how well a reactor is operating is given by its *capacity factor*. The capacity factor in any period is the ratio of the total electrical output of the reactor during that period to the output if the reactor had run continuously at full power. Capacity factors are usually expressed in percent, the ideal being 100%. Although capacity factors above 90% in a given year are not uncommon,[22] routine maintenance and variations in demand generally keep the capacity factors below 90% over a more extended period. A long-term capacity factor greater than 80% is considered very good.

A summary of the record of the average capacity factor for U.S. reactors is given in Fig. 1.3 [16]. The capacity factor generally rose in the period from 1973 to 1978 but dropped following the Three Mile Island accident, which precipitated a period of precautionary repairs and modifications, some entailing long periods during which the reactor was out of operation. With the completion of most of these modifications and the effects of a concerted industry effort to improve the reliability of reactor operation, there has been a significant increase in the annual average capacity factor since 1989.

A study made in early 1995 examined trends in capacity factors, considering both changes with time and differences among different types of reactors [19]. Capacity factors were taken as averages over three-year periods, providing values that are less perturbed by maintenance

---

[22]The 90% level was exceeded by 29 U.S. reactors in 1994 [18].

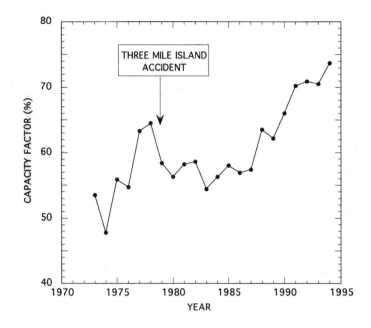

Figure 1.3. Mean capacity factor of U.S. reactors in a given year, 1973–1994.

and refueling interruptions than are one-year values. There were 108 reactors licensed for the full period from 1992 through 1994. At the extremes, 39 reactors had capacity factors over 80% for this period, while 24 other reactors had capacity factors below 60%. The median was 77%. Average capacity factors have risen during the past decade and the fraction of reactors with capacity factors above 70% was more than three times as great for 1992–1994 as for 1982–1984. On the other hand, a number of poor performers remain, including two reactors operated by the Tennessee Valley Authority (Browns Ferry 1 and 3) that have been shut down since 1985.

Reactors were compared in different categories including reactor type, size, and age. The median capacity factors were somewhat higher for PWRs than for BWRs and also slightly higher for newer reactors (less than eighteen years old) than for older reactors. However, there are very substantial overlaps in the distributions. Smaller reactors (≤700 MWe) had a slightly higher median capacity factor than larger reactors (≥1020 MWe), although the highest capacity factor over the three years was for a 1070-MWe reactor (91.5% for San Onofre 2).

Overall, the capacity factor comparisons indicate an improvement with time in the construction and operation of reactors. There was a plateau in performance in the years 1991 to 1993 but, as seen in Fig. 1.3, a rise again in 1994.

An increase in the capacity factor from, say, 60% to 80% corresponds to an increase in output of 33%, equivalent to the addition of roughly 30 reactors. The capacity factor has a large impact on the cost of electricity. The total annual expenses of a nuclear utility do not change much as the total electricity output rises or falls. The cost of electricity from a given reactor is therefore reduced by about 25% if the capacity factor is increased from 60% to 80%.

Table 1.2. *History of reactors shut down in the United States, omitting reactors with capacity under 100 MWe.*

Reactor	Capacity (MWe)	Year of Shutdown	Year of Initial Operation
Trojan	1130	1992	1976
San Onofre 1	436	1992	1968
Yankee Rowe	175	1991	1961
Shoreham[a]	809	1989	-
Fort St. Vrain	330	1989	1979
Rancho Seco	918	1989	1975
Three Mile Island 2	900	1979	1978
Dresdan 1	207	1978	1960
Indian Point 1	265	1974	1963

Sources: Capacity data are from Ref. [15]. Years of shutdown and initial operation are from Ref. [17].
[a]The Shoreham reactor received a low-power license but was never put into full-scale operation.

1.2.3 Permanent Reactor Shutdowns

One of the threats thought to face the nuclear industry is the prospect of the shutdown of reactors before the expiration of their operating licenses (commonly issued for 40 years). Most of the 20 reactors that were shut down before the end of 1993 were small, built in the early days of nuclear power development. Although 14 reactors ordered in the period from 1953 through 1962 have been shut down, the total capacity of these was only slightly more than 1 GWe. In addition, six reactors of later vintage were shut down by 1993; these are of greater concern. The history of reactor closures is summarized in Table 1.2, omitting reactors with a capacity of under 100 MWe.

These reactors fall into several categories:

- Three relatively small, older reactors (Indian Point 1, Dresden 1, and Yankee Rowe).
- Three Mile Island 2, which went into operation in 1978 and was severely damaged in 1979, in one of the world's two major reactor accidents (the other being the much worse accident at Chernobyl). The damage to the reactor and resulting contamination to the building made it impossible to put the reactor back into operation.
- Four reactors with a history of poor performance and low capacity factors, which weakened their economic and political viability (Rancho Seco, Fort St. Vrain, San Onofre-1, and Trojan).
- Shoreham in Long Island, New York, which operated briefly at very low power after issuance of a low-power license in 1985. However, local and state opposition prevented the issuance of a license to operate at full power, and after lengthy negotiations, the reactor was abandoned, with the date of "shutdown" not well defined.

Each of these categories, except the first, carriers its own cautionary note. In the case of Three Mile Island 2, it is obvious: A serious accident must be avoided. The weak performers in the third group are a reminder that a reactor is vulnerable, for both economic and political reasons, if its performance is poor. The fourth, the Shoreham reactor, is probably the most worrisome to utilities, because it is a case where a reactor was built, but public hostility prevented its use.

The set of shutdowns, particularly those of 1991 and 1992, was interpreted by some as the forerunner of a cascade of closures. However, following these four closures in two years, there were no additional cases in the years 1993 through 1995.

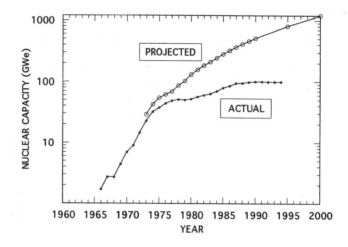

Figure 1.4. Comparison of U.S. nuclear capacity, projected in 1972 and actual.

1.2.4 Failures of Prediction

The changing fortunes of nuclear power in the United States can be seen by comparing early projections for its growth with the actual subsequent developments. In 1972 the Forecasting Branch of the AEC Office of Planning and Analysis made a projection of future growth in nuclear capacity, based on past trends in energy use and increases in electricity capacity [20]. Its forecasts for the "most likely" case are shown in Fig. 1.4, together with the actual history. The projected capacities for 1990 and 2000 were 508 GWe and 1200 GWe, respectively. The actual capacity for 1990 was 100 GWe, and it cannot be much greater for the year 2000.[23]

This was a spectacular failure of prediction, but one which was in tune with the conventional wisdom of the time. It is natural to speculate as to the implications for today. Alternatively, we can congratulate ourselves on our present ability to avoid comparable errors, or we can wonder what new predictive errors are now being committed.

1.3 WORLDWIDE GROWTH OF NUCLEAR POWER

The discussion above has emphasized the U.S. nuclear program. But the United States was not alone in having an early interest in nuclear energy. Other countries had similar interests, although their development lagged because they lacked the head start provided by the World War II atomic bomb program and they had smaller technological and industrial bases.

For countries that wanted nuclear weapons, the priority was the same as that of the United States. The bomb came first and peaceful nuclear energy later. Thus, the construction and testing of nuclear weapons was achieved by the USSR in 1949, Britain in 1952, France in 1960, and China in 1964. Commercial nuclear electricity followed: the USSR started with several 100-MWe reactors in 1958;[24] Britain with two 50-MWe reactors at Calder Hall in

---

[23]For example, a capacity of 103 GWe in 2000 is projected in *Annual Energy Outlook 1994*, published by the Department of Energy, a successor to the AEC [21].

[24]A 5-MWe electric plant was put into operation in Obinsk in June 1954, which has been cited by some Soviet authors as being the "first nuclear power plant in the world" (see Ref. [22]).

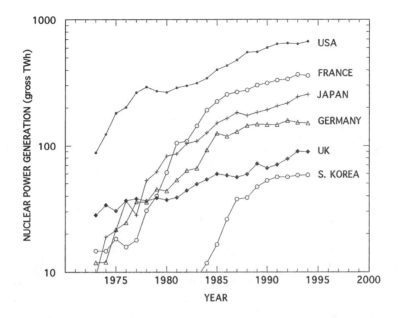

Figure 1.5. Growth of annual nuclear power generation in selected countries, 1973–1994.

1956 (preceding the U.S. reactor at Shippingport); France with a 70-MWe prototype reactor at Chinon in 1964; and China with a 300-MWe reactor that was essentially finished in 1991 but was not listed as being in commercial operation until 1994 [17].

Additional countries had no intention of building nuclear weapons and went directly to nuclear reactors. These included other countries of Western Europe beyond France and Britain, as well as Japan. As in the United States, the reactors for electricity generation put into operation in the late 1950s were relatively small in size and number. Large-scale worldwide exploitation of nuclear power did not begin until the 1960s and remained on a relatively modest scale until the 1970s. The major expansion came after 1973, with world nuclear power generation rising by more than a factor of ten between 1973 and 1992.

Changes in nuclear generation from 1973 to 1994 are shown in Fig. 1.5 for several countries [16]. France, Japan, and Germany began major programs after the United States but have had larger fractional growth rates. In France and Japan growth has been particularly strong and, unlike the case for Germany, is continuing. South Korea, the latest major entry, has had rapid growth in the past decade, but its overall program is still relatively small. The United Kingdom, one of the original leaders in nuclear power, has lagged substantially.

By 1993 nuclear power provided over 6% of the world's commercial primary energy and about 18% of the world's electric power [24]. Summary data on the role of nuclear power and its growth from 1973 to 1993 are shown in Table 1.3 for the world, the countries in the Organization for Economic Cooperation and Development (OECD),[25] the United States, and the most successful nuclear country, France.

Some of these data might seem to suggest a vigorous and thriving enterprise. This is not the case in most countries. World nuclear power generation rose from 191 TWh in 1973 to

---

[25]The OECD as of 1993 was composed of the countries of Western Europe, plus Australia, Canada, Japan, New Zealand, Turkey, and the United States.

Table 1.3. *International growth of nuclear power, 1973 to 1993.*

	World	OECD	U.S.	France
Net generation, 1973 (TWh)	191	177	83	11
Net generation, 1993 (TWh)	2080	1707	610	350
Ratio of generation: 1993 to 1973	11	10	7	≈30
Fraction of electricity, 1993 (%)[a]	18	26	21	78

Sources: 1973 net generation data are from Ref. [23], 1993 net generation data are from Ref. [24], world fraction of electricity is calculated from Ref. [24], OECD fraction of electricity is calculated from Ref. [25], and U.S. and French fractions of electricity are from Ref. [26].
[a]The nuclear fraction is specified here for utility sources only; the fraction is slightly less if non-utility generation is included.

2080 TWh in 1993, corresponding to an average increase of 13% per year or a doubling time of under six years.[26] But this rapid increase has ended. From 1988 to 1993 the average annual increase was only about 3%, and the increases can be expected to be even smaller for the remainder of the 1990s, due to the small number of new reactors in the construction pipeline. Few new reactors are going into operation, and the shutdown of old units has meant a very slow increase in total nuclear capacity, averaging well under 2% per year from 1988 to 1993.[27] Nuclear power's share of total world electricity generation stayed roughly constant from 1988 to 1993, rising only from 17% to 18% in this period [24], following two decades of rapid increases.

Data for all countries with nuclear power are presented in Table 1.4, which gives for each country its nuclear generation and nuclear power's fraction of its total electricity generation. As of the end of 1994 there were 424 operating reactors in the world, distributed among 30 countries, with a combined net generating capacity of 338 GWe [17].[28] Nuclear power accounted for roughly 18% of world electricity generation in 1994 and for much larger fractions in many countries, headed by France, which for several years has obtained more than 70% of its electricity from nuclear power.

The number of new reactors coming on line in recent years has been partly balanced by cancellations. Thus the comparable figures for the end of 1989 were 416 reactors with a capacity of 311 GWe [28]. Among the larger changes in the following five years were the shutdown after German unification of five Soviet-manufactured LWRs in the former East Germany and the startup of eleven new reactors in Japan.

1.4 PROBLEMS FACING NUCLEAR ENERGY

The decline in the growth of nuclear power discussed in the preceding sections can be attributed both to a less favorable economic environment and to concerns about the safety of nuclear power. The nuclear decline is not universal, with France, Japan, and South Korea standing as conspicuous exceptions. Nevertheless, for the world as a whole, there has been a dramatic difference between the expectations of 1970 and the reality of today.

---

[26]1 terawatt-hour (TWh)=1 billion kilowatt-hours (kWh).
[27]Generation of nuclear electricity has increased somewhat more rapidly than nuclear capacity in recent years, because the operating reactors are being used more efficiently, with a tendency to phase out some of the least efficient.
[28]There are some small differences among references in defining the number of operating reactors; part of the difference stems from alternative definitions of the start of operation and of the status of some reactors that may be shut down.

Table 1.4. *World nuclear status, 1994.*

Country	Number of units	Capacity (GWe)	Net Generation (TWh)	Net Generation (GWyr)	Percent Nuclear[a]
United States	109	99.2	639	73.0	22
France	55	57.4	342	39.0	75
Japan	49	38.9	258	29.5	31
Germany	21	22.7	143	16.6	29
Canada	22	15.4	102	11.6	19
Russia	25	19.8	98	11.2	11
United Kingdom	34	11.5	79	9.1	26
Sweden	12	10.1	70	8.0	51
Ukraine	14	12.1	69	7.9	34
South Korea	9	7.2	56	6.4	35
Spain	9	7.1	53	6.0	35
Belgium	7	5.5	38	4.4	56
Taiwan	6	4.9	33	3.8	32
Switzerland	5	3.0	23	2.6	37
Finland	4	2.3	18	2.1	30
Bulgaria	6	3.4	15	1.8	46
China	3	2.1	14	1.5	1.5
Hungary	4	1.7	13	1.5	44
Slovak Republic	4	1.6	12	1.4	49
Czech Republic	4	1.6	12	1.4	28
South Africa	2	1.8	10	1.1	5.7
Argentina	2	0.9	7.7	0.9	14
Lithuania	2	2.8	6.6	0.8	76
Slovenia	1	0.6	4.4	0.5	38
India	9	1.6	4.3	0.5	1.4
Mexico	1	0.7	4.3	0.5	3.2
Netherlands	2	0.5	3.7	0.4	4.9
Pakistan	1	0.1	0.5	0.1	1.0
Kazakhstan	1	0.1	0.4	0.0	0.6
Brazil	1	0.6	0.0	0.0	0.0
WORLD TOTAL	424	338	2130	243	18[b]

Sources: Number of units and capacity data, as of the end of 1994, are from Ref. [17]. The generation and percent nuclear data from Ref. [27], based on IAEA figures.
[a]The nuclear fraction is here specified for utility sources only; for some countries, the fraction is slightly less if non-utility generation is included.
[b]This is a 1993 value [24], estimated to be little changed for 1994.

Some of the reasons have been economic. A slowing in the overall growth in energy consumption has lessened the strain on fossil fuel supplies. That strain was further relieved by additional oil and gas production in many parts of the world and the partial replacement of fossil fuels by nuclear power. In this changed balance between supply and demand, fossil fuel prices have dropped since the early 1980s, even when expressed in current dollars.[29] In the meantime, nuclear power costs have risen. Thus, in many countries, including the United States, nuclear power has lost its cost advantage over coal and even natural gas.

[29]Correcting for inflation (i.e., making the comparison in constant dollars), the price of fossil fuels used by U.S. utilities in 1994 was less than one-half the price in 1981 [16, pp. 13, 123].

But economic factors are only part of the story. There has also been widespread concern over the possibility of reactor accidents, the disposal of the radioactive wastes from nuclear power, and the possibility that nuclear fuel or nuclear facilities might be diverted from reactors and used to produce nuclear bombs. Part of the increase in costs stems from measures taken in response to these concerns.

In the remainder of the book we will discuss basic aspects of nuclear power. The discussion starts with several chapters giving background information on nuclear technology. These are followed by chapters on the operation and safety of nuclear reactors, the nuclear fuel cycle (which extends from obtaining uranium to the disposal of nuclear wastes), questions of nuclear weapons proliferation, and nuclear costs. In a final chapter, we will briefly consider the prospects for nuclear power in the United States and other countries.

REFERENCES

1. U.S. Atomic Energy Commission, *The Nation's Energy Future, A Report to Richard M. Nixon, President of the United States*, submitted by Dixy Lee Ray, December 1, 1973.
2. Ernest Rutherford, "Radioactivity," in *Encyclopaedia Britannica*, 11th ed.
3. Charles Weiner, "1932—Moving Into the New Physics," in *History of Physics*, edited by Spencer R. Weart and M. Phillips, 332–339 (New York: American Institute of Physics, 1985).
4. Emilio Segrè, *From X-Rays to Quarks: Modern Physicists and Their Discoveries* (San Francisco: W. H. Freeman, 1980).
5. Richard Rhodes, *The Making of the Atomic Bomb* (New York: Simon and Schuster, 1986).
6. Spencer R. Weart and Gertrud W. Szilard, eds., *Leo Szilard: His Version of the Facts* (Cambridge: MIT Press, 1978).
7. Bertrand Goldschmidt, *Atomic Rivals*, translated by George M. Temmer (New Brunswick, N.J.: Rutgers University Press, 1990).
8. Samuel Glasstone, *Sourcebook on Atomic Energy* (New York: Van Nostrand, 1950).
9. Henry D. Smyth, *Atomic Energy for Military Purposes* (Princeton: Princeton University Press, 1945).
10. A. M. Weinberg to R. L. Doan, memorandum, September 18, 1944. In Alvin M. Weinberg, *The First Nuclear Era: The Life and Times of a Technological Fixer* (New York: AIP Press, 1994).
11. Palmer C. Putnam, *Energy in the Future* (New York: D. Van Nostrand, 1953).
12. Richard G. Hewlett and Jack M. Holl, *Atoms for Peace and War: 1953–1961* (Berkeley: University of California Press, 1989).
13. "What Is the Atom's Industrial Future," *Business Week*, March 8, 1947, pp. 21–22. Cited in *The American Atom*, edited by Robert C. Williams and Philip L. Cantelon (Philadelphia: University of Pennsylvania Press, 1984).
14. Atomic Industrial Forum, *'Too Cheap to Meter?' Anatomy of a Cliché*, May 1980.
15. Nuclear Energy Institute, "Historical Profile of U.S. Nuclear Power Development, 1994 Edition," *USCEA Energy Data* (Washington, D.C.: 1994).
16. U.S. Department of Energy, *Monthly Energy Review, September 1995*, Energy Information Administration report DOE/EIA-0035(95/09) (Washington, D.C.: U.S. DOE, 1995).
17. "World List of Nuclear Power Plants," *Nuclear News* 38, no. 3 (March 1995): 27–42.
18. Nuclear Energy Institute, "U. S. Nuclear Plants Top 75-Percent Capacity," *Nuclear Energy Insight* (March 1995): 8.
19. E. Michael Blake, "U.S. Capacity Factors: Still higher, from necessity," *Nuclear News* 38, no. 7 (May 1995): 28–34.
20. U.S. Atomic Energy Commission, *Nuclear Power 1973–2000*, report Wash-1139(72) (Washington, D.C.: 1972).
21. U.S. Department of Energy, *Annual Energy Outlook 1994, with Projections to 2010*, Energy Information Administration Report DOE/EIA-0383(94) (Washington, D.C.: U.S. DOE, 1994).
22. A. M. Petrosyants, *From Scientific Search to Atomic Industry*, translated from the 1972 Russian edition (Danville, Ill. Interstate Printers & Publishers, 1975).
23. U.S. Department of Energy, *International Energy Annual 1983*, report DOE/EIA-0219(83) (Washington, D.C.: Energy Information Administration, 1984).
24. U.S. Department of Energy, *International Energy Annual 1993*, report DOE/EIA-0219(93) (Washington, D.C.: Energy Information Administration, 1995).

25. Organization for Economic Co-operation and Development, *Energy Balances of OECD Countries 1992–1993* (Paris: OECD, 1995).
26. "Nuclear power contributions in 1993," *Nuclear News* 37, no. 9 (July 1994): 47.
27. "Nuclear power contributions in 1994," *Nuclear News* 38, no. 8 (June 1995): 48.
28. "World List of Nuclear Power Plants," *Nuclear News* 36 (February 1990): 63–82.

Chapter 2

Natural Radioactivity and Radiation Exposures

2.1 BRIEF HISTORY

Radioactivity and the associated radiation exposures are sometimes thought of as environmental problems that have been created by modern science and technology. However, substantial amounts of radioactivity exist in nature and have existed on Earth since its original formation. All species evolved on Earth, for better or worse, in this radioactive environment. Radioactivity could be plausibly termed the oldest "pollutant," if one chooses to so describe an integral part of the natural world.

With the advent of controlled nuclear fission, we have obtained the ability to create concentrations of radioactivity that far exceed those encountered in the natural environment. An effort is made to isolate or shield the radioactive material to minimize human exposure, but it is to be expected that some exposure will nonetheless occur. To gain perspective on the seriousness of the potential resulting problems, we consider in this chapter the sources and amounts of natural radioactivity, along with the health impacts of radiation. Natural radioactivity here provides a qualitative reference scale or benchmark.

Human awareness of the existence of ionizing radiation dates only to the turn of the century. Wilhelm Roentgen discovered x rays in 1895, and within the next five years Henri Becquerel and Marie and Pierre Curie discovered the previously unsuspected ionizing radiations from uranium ore. These were alpha particles, beta particles, and gamma rays from the radioactive decay of uranium and its associated products. The nature of these radiations and of radioactivity was quickly elucidated by investigators in France, Great Britain, and elsewhere.

The benefits of x rays for diagnostic purposes were recognized almost immediately, and the purported benefits of radium, extracted from uranium ore, were proclaimed soon after. The belief that radium had curative properties continued into the 1920s, with horrors such as the sale of "medicines" laced with radium [1]. More remarkably, patronage of underground sites that featured enhanced exposure to radon persisted in the United States into the 1990s (see e.g., Ref. [2]). At the same time, radium was used for more justifiable purposes, including cancer therapy and industrial applications.

Along with a growing use of x rays and natural radioactive materials for a wide variety of purposes, there arose a belated recognition of the health hazards from x rays and other ionizing radiations. These were recognized following excessive exposures of radiologists, x-ray technicians, and, in a particular tragedy, radium watch dial painters. The inherent dangers were first widely recognized in the 1920s, and serious attempts to establish safety standards date from the late 1920s, starting with occupational hazards. Since the 1950s the setting of standards has been expanded to include those for the general population, usually at levels substantially stricter than those for occupational exposures.

2.2 MEASURES OF RADIOACTIVE DECAY

2.2.1 Half-Life and Mean Life

The "activity" of a radioactive sample is equal to the number of decay events in the sample per unit time. As discussed in more detail in the Appendix, the activity of the sample decreases exponentially with time, proportional to $e^{-\lambda t}$, where λ is the *decay constant* and t is the time. The rate of decay can be characterized in terms of λ, the mean life (τ), or the *half-life* (T). The designation in terms of half-life is the most frequently used in general discussions of radioactivity. The relationship between these parameters is

$$T = \ln 2/\lambda = \tau \ln 2. \tag{2.1}$$

The half-life is the time interval in which the rate of radioactive decay decreases by a factor of 2. For example, in ten half-lives, the activity drops by a factor of $2^{10} = 1024$. Different radionuclides are characterized by different half-lives, which vary from a small fraction of a second to billions of years.

2.2.2 Units of Radioactivity

The original unit for measuring the amount of radioactivity was the *curie* (Ci), with derivative units in common use ranging from the picocurie (pCi) to the megacurie (MCi).[1] The curie was defined in 1910 to be the amount of radon (^{222}Rn) in equilibrium with 1 gram of radium (^{226}Ra) [3, p. 448], but the usage evolved to become the amount of any radionuclide that underwent radioactive decay at the same rate as 1 gram of ^{226}Ra. Using current numbers for the half-life ($T = 1600$ years) and atomic mass ($M = 226.03$ u) of ^{226}Ra, this rate is 3.66×10^{10} disintegrations/second.

---

[1]The derivative units use the standard prefixes: *femto* (f)$=10^{-15}$, *pico* (p)$=10^{-12}$, *nano* (n)$=10^{-9}$, *micro* (μ)$=10^{-6}$, *milli* (m)$=10^{-3}$, *kilo* (k)$=10^3$, *mega* (M)$=10^6$, *giga* (G)$=10^9$, and *tera* (T)$=10^{12}$.

Table 2.1. *Specific activities S of selected radionuclides.*

Nuclide	Isotopic Abund (%)	Atomic Mass (u)	Half-life (years)	S (Bq/g)	S (μCi/g)
^{239}Pu	—	239.05	2.411×10^4	2.30×10^9	6.20×10^4
^{238}U	99.27	238.05	4.468×10^9	1.24×10^4	0.336
^{235}U	0.72	235.04	0.704×10^9	8.00×10^4	2.16
^{232}Th	100.00	232.04	14.05×10^9	4.06×10^3	0.110
^{226}Ra	—	226.03	1600	3.66×10^{10}	9.89×10^5
^{87}Rb	27.83	86.91	47.5×10^9	3.21×10^3	0.087
^{40}K	0.0117	39.96	1.277×10^9	2.59×10^5	7.01

Source: Data are from Ref. 6.

With this definition, the numerical value of the curie would have to be modified whenever a more precise value for T (or M) is found. The curie was subsequently redefined in 1950 as an exact unit, namely, "the quantity of any radioactive nuclide in which the number of disintegrations per second is 3.700×10^{10}" [4, p. 472]:

$$1 \text{ curie} = 3.7 \times 10^{10} \text{ disintegrations per second.}^2$$

In international usage, and gradually in much of U.S. usage, the curie is being supplanted by the becquerel (Bq), introduced along with other SI units in 1975. The becquerel is defined as the activity corresponding to one disintegration per second:

$$1 \text{ becquerel} = 1 \text{ disintegration per second} = 2.703 \times 10^{-11} \text{ Ci.}$$

The magnitude of the activity (in Bq) is λN, where λ is the decay constant (in sec^{-1}) of the radionuclide and N is the number of nuclei in the sample.

It is helpful to obtain some sense of scale: What is a "large" amount of radioactivity? A typical room contains slightly more than 1 pCi (0.037 Bq) of ^{222}Rn per liter of air. A 70-kg person has a potassium 40 (^{40}K) body content of about 0.1 μCi (3.7×10^3 Bq). Intense sources used for cancer therapy have ranged in size up to about 10,000 Ci (3.7×10^{14} Bq). The core of a typical commercial nuclear reactor has an activity of about 15 billion Ci (5.6×10^{20} Bq) just after being turned off following a period of prolonged operation.

2.2.3 Specific Activity

A radioactive substance can be characterized in terms of its *specific activity S* defined as the activity per unit mass. The specific activity of a single radionuclide (in Bq/g) is

$$S = \lambda N = \frac{\ln 2}{T} \times \frac{N_A}{M}, \tag{2.2}$$

where N is the number of nuclei per gram, M is the atomic mass of the radionuclide (in atomic mass units, u), T is its half-life (in seconds), and N_A is Avogadro's number. Specific activities for selected radionuclides are given in Table 2.1.

---

^{2}There may be a minor semantic problem here, in that the curie is used to denote both an amount of material and a disintegration rate (see, e.g., Ref. [5]). However, this rarely, if ever, leads to an ambiguity.

As often is the case in discussing natural radioactivity, if the measured mass is the mass of the element, then Eq. 2.2 must be modified to take into account the isotopic abundance of the radionuclide:

$$S=\frac{\ln 2}{T}\times\frac{f_i N_A}{M_E},$$

(2.3)

where f_i is the fractional isotopic abundance (by number of atoms) of the radionuclide and M_E is the atomic mass of the element.

2.3 NATURAL RADIOACTIVITY

2.3.1 Origin of Natural Radioactivity

Nucleosynthesis

The existence of natural radioactivity, i.e., the formation of the radionuclides, cannot be understood in isolation, apart from the understanding of the formation of the stable nuclei. They were produced in the same processes of nucleosynthesis, primarily in the evolution and explosion of stars. Only matters of relatively fine nuclear detail make some species stable and some unstable. A sketch showing the relationship between the atomic number and atomic mass number for the nuclides found in nature is shown in Fig. 2.1. These are predominantly

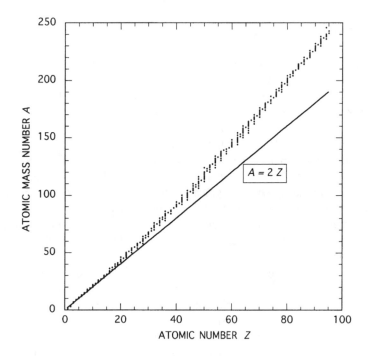

Figure 2.1. Nuclides in nature: the trajectory of atomic mass numbers and atomic numbers for nuclides that are stable against beta-particle emission. For reference, the path $A=2Z$ is shown in the dotted line.

nuclei that are stable against beta-particle decay. Many other nuclides can be formed in cosmic rays, accelerators, and nuclear reactors, but these have relatively short half-lives.

The first phases of nucleosynthesis, which build the elements up to the general region of atomic mass number $A = 60$, proceed through a series of reactions initiated primarily by protons and alpha particles. Some of the nuclides produced are stable, and some are radioactive. The latter are almost all beta-particle emitters, with half-lives that are short compared to the age of the Earth. In consequence, there are very few natural radioactive isotopes in this mass region. The one important exception is ^{40}K.

Beyond the $A \approx 60$ region, nucleosynthesis proceeds primarily by neutron capture. The neutron-rich products of neutron capture decay by β^- emission to more stable isobars, and the buildup to heavy elements continues by a sequence of neutron capture and β^- decay events. Again, most of the radioactive nuclei that were formed in this fashion have had ample time to decay since they were formed. The most prominent exception is rubidium 87 (^{87}Rb).

As the buildup by neutron capture proceeds up to and past uranium (atomic number $Z = 92$), the energy corresponding to the repulsive coulomb force grows more rapidly than the energy corresponding to the attractive nuclear force, and alpha-particle decay and fission become possible. In several important cases, nuclei are formed that are stable against β^- decay but that can decay by alpha-particle emission with half-lives comparable to the age of the Earth. These nuclei are responsible for the radioactive decay series discussed in the next section. For nuclei of very high mass, fission terminates the neutron-capture chain and adds to the abundances of nuclei of intermediate mass (in the $A = 80$ to $A = 160$ region). Buildup can begin again in a new neutron-capture chain, with these fission fragments as starting points.

Radioactive Products of Nucleosynthesis

The material of the solar system was formed by several generations of such synthesis processes, possibly including the explosion of a supernova near the protosolar nebula shortly before the formation of the Sun and the condensation of the planets. In the 4.6 billion years since the formation of the Earth, any nucleosynthesis products with short half-lives have disappeared. The only naturally radioactive relics of nucleosynthesis are nuclei that have very long half-lives or are decay products of such nuclei.

One group of natural radioactive nuclides are the intermediate-mass beta emitters. As mentioned above, a very large number of β^--unstable nuclei were formed in nucleosynthesis, but few remain today because typical lifetimes for β^- emission range from seconds to days. The existence of some long-lived exceptions is due to special circumstances in the nuclear properties of these nuclides and their decay products. With many β^--unstable nuclides having been formed, the chance is enhanced that a few will have the requisite unusual properties.

Two of these β^--emitting nuclides are of particular environmental relevance: ^{40}K ($T = 1.28 \times 10^9$ years) and ^{87}Rb ($T = 47.5 \times 10^9$ years). These nuclides produce nontrivial radiation exposures to the human body because of the potassium and rubidium naturally present in the tissues. The calculated exposure due to the ^{40}K is considerably greater than that due to the ^{87}Rb, so sometimes only ^{40}K is considered.

At higher atomic masses, the important radionuclides are the members of three radioactive decay series, headed by long-lived parent nuclei. These are discussed in greater detail in the following section.

In addition to the nucleosynthesis products, some radioactive nuclei are being continually produced by cosmic rays, but the amounts are so small that they have little practical significance except as historical markers for determining ages of samples in which they are found. The most important of these is carbon 14 (^{14}C), produced in the interaction of cosmic-ray neutrons with nitrogen 14 (^{14}N) in the atmosphere.

2.3.2 Radioactive Series in Nature

The Occurrence of Three Series

The radioactive decay series are groups of radioactive nuclei that arise from the production of long-lived alpha-particle emitters in nucleosynthesis. The "head" of each series is either a direct product of nucleosynthesis or a remnant of the synthesis of short-lived parents. As it decays, it feeds a set of radioactive progeny, which decay either by alpha-particle emission or β^- emission. In alpha-particle decay, the atomic mass number A is reduced by 4. In β^- emission, A is unchanged.

This means that there are in principle four possible series of radioactive nuclei. They can be classified by the magnitude of the parameter δ, if the atomic mass number is expressed as $A = 4n + \delta$. Here, n is an integer that decreases by unity in alpha-particle decay, and δ is a fixed constant for each series, with possible magnitudes of 0, 1, 2, or 3. The observed series in nature are given in Table 2.2.

The three series listed in Table 2.2 are headed by nuclei with half-lives of the order of one to ten billion years. If the half-life were much longer, the decay would be so slow as to be undetectable. If the half-life were much less, the series would have disappeared. This is the case for the $A = 4n + 1$ series, which is not found in nature because no member of the series has a sufficiently large half-life. For example, the longest-lived $A = 237$ isobar is ^{237}Np, with a half-life of 2.1×10^6 years, and there exists no alternative long-lived parent at higher or lower A for a $4n + 1$ series.

The Uranium Series

The uranium series, headed by ^{238}U, includes two radionuclides of some special importance, in addition to ^{238}U itself. One of the progeny, ^{226}Ra, was the source of the intense activity found in uranium ore in the early studies carried out by the Curies. It appears as a strong source, i.e., one with a high specific activity, because of its intermediate half-life of 1600 years. A longer half-life would reduce the specific activity. A much shorter half-life would have made it harder to notice in slow chemical separations (as practiced by the Curies), as well as of little practical use.

Table 2.2. *Natural radioactive decay series.*

Series Type	Series Name	Parent Nucleus	Half-life (10^9 yr)	Stable Nucleus	Number of Decays (N)[a]	
					α	β^-
$A=4n$	Thorium	^{232}Th	14.05	^{208}Pb	6	4
$A=4n+2$	Uranium	^{238}U	4.47	^{206}Pb	8	6
$A=4n+3$	Actinium	^{235}U	0.704	^{207}Pb	7	4

[a]$N_\alpha = (A_i - A_f)/4$; $N_{\beta^-} = 2N_\alpha - (Z_i - Z_f) = (A_i/2 - Z_i) - (A_f/2 - Z_f)$, where i and f denote *initial* (parent) and *final* (stable), respectively.

No other member of any of the natural radioactive series has a half-life lying within a factor of 20 of the ^{226}Ra half-life. Its intermediate half-life, coupled with the relatively high abundance of the parent, ^{238}U, made ^{226}Ra unique for practical applications of radioactivity, until an array of "artificial" radioisotopes became available as products of nuclear reactions. Radium was long used for purposes as varied as cancer therapy and luminescent watch dials.

^{226}Ra decays by alpha-particle emission to ^{222}Rn ($T=3.8$ days). This is the nuclide responsible for the radiation exposures arising from indoor radon. Radon is a noble gas and reacts little with other substances. Therefore, it itself is not trapped in the lungs when inhaled. However, ^{222}Rn has a number of alpha-particle emitting progeny with short half-lives. These are present in the air as decay products of ^{222}Rn and are to a considerable extent retained in the lungs when inhaled. They therefore lead to significant radiation doses to the lungs (see Table 2.5).

Steady-State Relation in Radioactive Decay Series

In a radioactive decay series, each decay product is itself radioactive until the end of the series is reached. It decays at a rate determined by its half-life. In each of the series, the longest half-life is that of the original parent, or head of the series, and the series reaches a steady state.

If the radioactive series is in an isolated volume of material, then in the steady state each member of the series decays at a virtually constant rate. We say "virtually" constant because there is a very slow change governed by the very slow decay of the head of the series. In the steady state, the number of decays per unit time is the same for each member of the series:

$$\lambda_1 N_1 = \lambda_2 N_2 = \lambda_3 N_3 = \dots \ . \tag{2.4}$$

The steady-state condition for a radioactive decay series is also known as *secular equilibrium*. The secular equilibrium expression of Eq. 2.4 can be derived rigorously from the equations that describe a series of decaying nuclei, for the limit that the head of the series has a decay constant λ small compared to that of the other members of the series or, equivalently, a half-life T large compared to that of the other members.

The result can also be seen to be qualitatively reasonable. Consider the relationship between the first (long-lived) and second (shorter-lived) member of the series. If the condition of Eq. 2.4 is violated so that, for example, $\lambda_2 N_2 > \lambda_1 N_1$, then the loss from the "pool" of species 2 exceeds the flow into it, and N_2 will decrease until the inequality is removed. A similar restoration to the condition of Eq. 2.4 occurs if $\lambda_2 N_2 < \lambda_1 N_1$. By an extension of the same argument, secular equilibrium is established throughout the radioactive series.

The condition described above applies to isolated material. It does not apply exactly to the natural radioactive series in all environments, due to chemical and physical processes that transfer different elements from one locale to another at different rates. Fractionation can be important in the dissolution and removal of radionuclides by water flowing through rocks and in the transfer of radionuclides from the ocean to ocean sediments.

2.3.3 Concentrations of Radionuclides in the Environment

Radionuclides in the Earth's Crust

There are substantial variations in the concentrations of chemical elements and hence of radionuclides in different parts of the Earth's crust. For example, for the crust as a whole, the

Table 2.3. *Average abundances of natural radionuclides in the upper continental crust, the oceans, and ocean sediments.*

	^{40}K	^{87}Rb	^{232}Th	^{238}U
Upper continental crust				
Elemental abundance (ppm)	28 000	112	10.7	2.8
Activity^a (Bq/kg)	870	102	43	35
Activity (nCi/kg)	23	2.7	1.2	0.9
Activity^b (kCi/km^3)	66	8	3.3	2.6
Oceans				
Elemental concentration (mg/liter)	399	0.12	1×10^{-7}	0.0032
Activity^a (Bq/l)	12	0.11	4×10^{-7}	0.040
Activity (nCi/l)	0.33	0.003	1×10^{-8}	0.0011
Ocean sediments				
Isotopic abundances (mg/kg)	≈2		5.0	1.0
Elemental abundance^c (ppm)	17 000		5.0	1.0
Activity^b (Bq/kg)	500		20	12
Activity (nCi/kg)	14		0.5	0.3

Sources: Elemental abundance data for the continental crust are from Ref. [9], p. 46. Concentrations of K and Rb in the oceans are from Ref. [9], p. 15; those for Th and U are from Ref. [10]. Isotopic abundances in ocean sediments are from Ref. [11], pp. 23–24.
^{a}Calculated from elemental abundances, using Eq. 2.3 with parameters of Table 2.1.
^{b}Calculated for a density of 2.8 g/cm^3.
^{c}Calculated from isotopic abundances.

uranium abundance by mass is about 3 parts per million (ppm) (i.e., 0.0003%), with common rocks having uranium concentrations ranging from 0.5 to 5 ppm [7, p. 61]. Concentrations in uranium ores commonly exceed several hundred ppm, and there are deposits that are reported to have concentrations as high as 65% of U_3O_8, equivalent to 550 000 ppm of U [8, p. 409].

For a detailed picture of radionuclides in nature, it is useful to separate rocks by rock type. It is found, for example, that the abundance of radionuclides is greater in granite than in carbonate rocks such as limestone [7, p. 61]. In some cases, these differences are of crucial importance, as for uranium ores. However, for present purposes of providing perspective, it is of interest to consider the broad average for the continental crust. (Average values for the upper continental crust have been presented by a number of authors; we follow Ref. [7] in adopting those presented by Taylor and McLennan [9, p. 46] for use in Table 2.3.)

Table 2.3 shows that the magnitude of the activity for the β^- emitters, ^{40}K and ^{87}Rb, is significantly greater than that for the listed alpha-particle emitters. However, the disparity in overall effect is much less than might appear because uranium 238 (^{238}U) and thorium 232 (^{232}Th) each head long series of radionuclides, and the activities of the progeny should be added. In particular (see Table 2.2), each ^{238}U decay is followed by seven additional alpha-particle decays and six β^- decays.[3] Further, alpha-particle emitters in general release substantially more energy per disintegration than do β^- particle emitters.

We can get a sense of the magnitudes involved by estimating the total activity in the Earth's crust down to some appropriate depth. The average density of the Earth's continental crust is 2.8 g/cm^3, and the average depth is approximately 17 km [12, p. 11]. Arbitrarily taking 1 km as a depth of interest, it being within relatively easy accessibility, the mass per

---

[3]At any particular location the magnitudes of the activities of different species in a decay series may not be the same, due to differences in chemical and physical properties, which cause different movements of these elements. Globally, however, the series are in secular equilibrium, and the total activities are the same for each member.

square kilometer of surface area is 2.8×10^{12} kg. For this depth, the ^{238}U and ^{232}Th activities are therefore each roughly 3×10^3 Ci per km^2 of surface (see Table 2.3). The land area of the continental United States (48 contiguous states) is about 8×10^6 km^2, and therefore the activities of ^{238}U and ^{232}Th are each about 2×10^{10} Ci, down to the 1 km depth. The activity represented by the entire series is about a factor of ten greater, i.e., roughly 2×10^{11} Ci for each series. Adding these activities and the decay of ^{40}K and ^{87}Rb gives a total of roughly 10^{12} Ci over the area of the United States, again down to a depth of 1 km. Such numbers help provide a sense of scale, when one considers the additions to the radioactivity in the ground projected in plans for nuclear waste disposal (see Chapter 8).

Radionuclides in the Oceans

Over the billions of years since the Earth was formed, there has been a substantial interchange of material between the land and oceans. For example, flowing water erodes rocks and carries material from the rocks into the oceans. It therefore is to be expected that the same elements that are present in the Earth's crust will also be present, to one degree or another, in the oceans.

However, a high concentration of an element in the Earth's crust does not necessarily mean a high concentration in the ocean. Table 2.3 gives data on the concentrations of some natural radionuclides in the oceans and ocean sediments. It is seen, for example, that the ratio of the concentration (in parts per million by mass) in the oceans to the concentration in the Earth's crust is of the order of 10^{-2} for potassium and about 10^{-8} for thorium. These differences are in part due to differences in the solubility of the compounds in which the elements are normally found in the crust and in their rates of removal by erosion. They are due also to large differences in the average residence time of elements in the ocean, before transfer to ocean sediments in the ocean floor. For example, the residence time of potassium is about ten million years, while the residence time of thorium is less than ten years [9, pp. 15–16].

The volume of the oceans is 1.4×10^{21} liters [11, p. 22]. This means that the total activity of ^{40}K in the oceans is about 4×10^{11} Ci and the total activity of ^{238}U in the oceans is about 1×10^9 Ci. Thus, in the oceans ^{40}K is the dominant species in terms of total activity. Its ratio to ^{238}U (and, much more dramatically, ^{232}Th) is greater in the ocean than in the Earth's crust. Fractionation processes, such as rapid removal of ^{230}Th, reduce the abundances of many of the ^{238}U series progeny in the oceans to levels well below secular equilibrium with ^{238}U (see, e.g., Ref. [11]).

Radionuclides in the Human Body

The chief source of radioactivity in the human body is ^{40}K. In the 70-kg "reference man," considered a prototype by the International Commission on Radiological Protection (ICRP), there are 140 g of potassium ($M = 39.10$ u) [7, p. 109]. This corresponds to an activity of 4340 Bq or 0.12 μCi. Other radionuclides in the body, present in lesser amounts, include members of the ^{238}U and ^{232}Th series, ^{14}C, and ^{87}Rb. For example, in a typical U.S. diet, ^{238}U, ^{226}Ra, and ^{210}Po are each ingested at a rate in the neighborhood of 1 pCi/day [7, p. 110].

2.4 RADIATION EXPOSURES

2.4.1 Radiation Dose Units

Radiation Exposure and Radiation Dose

The terms *radiation exposure* and *radiation dose* are often treated in casual usage as being almost interchangeable. However, they are not identical, and in fact their definitions are quite different. The dose is a quantitative measure of the impact of radiation, closely related to the energy deposited by incident radiation per unit mass. Exposure is now used in two senses: (a) in a specialized sense in connection with the roentgen unit (see below) and (b) as a general qualitative term to indicate "the incidence of *radiation* on living or inanimate matter" [5, p. 46]. Aside from the special case of the roentgen, *dose* is the appropriate term for quantitative descriptions while *exposure* describes a general qualitative situation. Thus, for example, reports of the National Council on Radiation Protection and Measurements (NCRP) carry descriptive titles such as *Ionizing Radiation Exposure of the Population of the United States* [13], while the quantitative results presented are given in terms of the doses resulting from the exposures.

Ionization Measure of Exposure: The Roentgen

Historically, radiation exposures were first expressed in terms of the *roentgen* (R). The definition was based on the ionization produced by x rays or gamma rays in air, with one R equivalent to the deposition of 87 ergs per gram of air.[4]

The roentgen has more recently been redefined in terms of SI units to be the amount of x-ray or gamma-ray radiation that produces 2.58×10^{-4} coulomb of charge in one kilogram of air. This redefinition leads to the same numerical value for the roentgen as it had previously. Historically, the roentgen was a convenient unit to introduce because early work with radiation was concerned primarily with x rays, and radiation was commonly detected with instruments that measured the amount of ionization produced in air.

Absorbed Dose: Rad and Gray

In radiation dosimetry the energy deposited by the radiation, usually in human or animal tissue, is a more pertinent quantity than ionization in air, and the roentgen has been largely replaced by a more general dose unit based on energy deposition. This is the *absorbed dose*, sometimes also known as the *physical dose*. The units of absorbed dose are the *rad* and, more recently in the SI system, the *gray* (Gy).[5] The rad and the gray apply to any absorber, with the absorbers of greatest interest being human tissues. The definitions of the units essentially define the meaning of the absorbed dose:

---

[4]More specifically, the roentgen was defined as the amount of x-ray or gamma-ray radiation that produces one electrostatic unit of charge (esu) in one cm³ of air (at standard temperature and pressure). The corresponding energy deposition is $E = I/\rho e = 87$ ergs/g (in air), where the average energy expenditure per ion pair is $I = 33.8$ eV $= 5.41 \times 10^{-11}$ ergs [14, p. 18], the electron charge is $e = 4.80 \times 10^{-10}$ esu, and the density of air is $\rho = 0.001293$ g/cm³.

[5]The situation for dose units is similar to that for other SI units competing with older units. The SI dose units, the gray and the sievert (see below), are used in most foreign countries and in much of the U.S. scientific literature. The older units, the rad and the rem, nonetheless remain in common usage in the United States.

One gray is the absorbed dose corresponding to the deposition of 1 joule/kilogram in the absorber.

One rad is the absorbed dose corresponding to the deposition of 100 ergs/g in the absorber.

From these definitions, it follows that

$$1 \text{ gray} = 100 \text{ rad}.$$

The rad and the gray have the advantage over the roentgen in not being restricted to photons, of being defined in terms of exact numbers, which will not require revision if constants are better determined (e.g., the average ionization energy of electrons in air), and of being directly applicable to any tissue. A gamma-ray flux that delivers an exposure of 1 roentgen in air would deliver a dose of approximately 0.96 rad to human tissue.[6] Thus the roentgen and rad are numerically roughly equivalent.

Biological Damage: Deterministic and Stochastic

In describing biological effects of radiation, a distinction is made between *deterministic* and *stochastic* effects.[7] Deterministic effects depend upon the killing of many cells over a relatively short period of time. They are induced by intense exposures, and the outcome of this exposure is reasonably well defined. The magnitude of the dose determines the *intensity* of the effect. The most conspicuous deterministic outcome is the death of the victim within a short time, up to several months, after the exposure.

Stochastic effects are by definition of a random nature, with the likelihood of one or another outcome a matter of statistical probability. The modification of a limited number of cells by radiation can lead either to the induction of cancer in the exposed person or to genetic damage, which affects later generations. The magnitude of the dose determines the *probability* of the effect.

The impacts of radiation exposure from nuclear power are primarily stochastic. Even in the accident at Chernobyl, only a relatively limited number of workers received high enough acute radiation doses to cause either death within a few months or other deterministic effects. Much larger populations were exposed to lower doses. The expected appearance of cancer in these populations will be delayed, typically by more than ten years after the time of exposure, and it is not possible to identify specific individual victims.

Dose Equivalent: Rem and Sievert

Biological damage is not determined solely by the total energy deposition per unit mass. It is found to depend also upon the density of the ionization. The energy loss per unit distance traversed is much greater for alpha particles than for electrons, and the disturbance to the tissue is more localized in space.[8] As a result, the deposition of 100 ergs by alpha particles in a given tissue mass is more injurious than the deposition by electrons of the same amount of energy.

In radiobiology the rate of energy loss is termed the *linear energy transfer* (LET) rate, and

---

[6]This follows from the fact that the mass absorption coefficient of gamma rays is about 10% greater in tissue than in air.

[7]See, e.g., Refs. [15], p. 4, and [16], p. 8.

[8]A corollary is that the distance penetrated in matter by alpha particles is much less than the distance for electrons of roughly comparable energy.

Table 2.4. *Average quality factors and the absorbed dose corresponding to a dose equivalent of unity for selected radiations.*

Type of Radiation	Quality Factor[a] Q	Absorbed Dose[b] D (for $H=1$)
X rays and gamma rays	1	1
Electrons (including beta particles)	1	1
Neutrons (depending on energy)	5–20	0.2–0.05
Alpha particles and fission fragments	20	0.05

[a]From Ref. [15], p. 6, where this quantity is termed the "radiation weighting factor."
[b]For H in rem or Sv, D is in rad or gray, respectively.

radiations are classified as low LET and high LET. The low-LET radiations are x rays, gamma rays and beta particles. X rays and gamma rays, being neutral, are not themselves ionizing particles, but they transfer their energy to electrons. In all these cases, the ionizing particle is an electron for which the rate of energy loss per unit path length is small. The high-LET radiations are alpha particles and neutrons. Neutrons themselves are not ionizing particles, but their interactions with nuclei of the material produce ionizing particles, for example, in elastic scattering with hydrogen nuclei (protons).

The *dose equivalent* is introduced to take into account the *relative biological effectiveness* of different types of radiation. It is defined so that equal dose equivalents have an equal chance of inducing stochastic effects, independent of the type of radiation. The dose equivalent H is related to the absorbed dose D by a dimensionless parameter commonly known as the *quality factor Q*, which takes on different values for different radiations:

$$H=QD. \qquad (2.5)$$

The unit of dose equivalent corresponding to the rad is the *rem*, an abbreviation for roentgen-equivalent-man. The analogous unit corresponding to the gray is the *sievert* (Sv). Thus, 1 sievert=100 rem.

Strictly speaking, the relative biological effectiveness of different radiations depends upon the tissue exposed, the type of effect in question, the energy of the radiation, and the intensity of the exposure. However, there is usually insufficient information about biological effects to make such precise distinctions. Instead, Q is commonly specified as an approximate average conversion factor—loosely speaking, an average relative biological effectiveness.[9]

Average values of the quality factor for radiations of interest are given in Table 2.4. These correspond to the standard values recommended in the International Commission on Radiological Protection (ICRP) Publication 60 [15, p. 6] and are widely accepted by other bodies (see, e.g., Ref. [17], p. 375). As seen from these data, an absorbed dose of 1 rad from exposure to alpha particles has the same overall biological impact, in terms of cancer induction, as an absorbed dose of 20 rad from gamma rays.

A word of caution is in order here. In the example cited, for a dose equivalent of 200 mSv, the potential impact is a stochastic one, and the use of the dose equivalent is appropriate. However, at considerably higher dose levels, when deterministic effects dominate, use of the quality factors of Table 2.4 for neutrons and alpha particles would overestimate the effect [15, p. 15]. It is therefore common practice to describe high doses in terms of the absorbed dose, in rad or gray.

---

[9]In recognition of the imprecision in usage, the International Commission on Radiological Protection has recently recommended that the parameter of Eq. 2.5 be termed the *radiation weighting factor* rather than the quality factor [15, p. 5], but we will here use quality factor as it is still commonly employed in the literature.

Effective Dose Equivalent and Tissue Weighting Factors

It is of course more serious if the entire body receives a given exposure than if only a single organ receives that exposure. Using appropriate weighting factors, it is possible to translate an organ dose into the whole-body dose that would produce the same overall risk of a fatal cancer. Such a translated dose is called the *effective dose equivalent*.[10] When doses are described, it is important to know whether the reference is to the dose equivalent over a limited region of the body or to the effective dose equivalent.

The effective dose equivalent H_E is related to the individual tissue doses H_T and D_T by tissue weighting factors w_T and a summation over tissues:

$$H_E = \Sigma w_T H_T = \Sigma w_T Q D_T. \tag{2.6}$$

Standard values for the weighting factors w_T are 0.20 for the gonads; 0.12 for bone marrow, colon, lung, and stomach (each); 0.05 for the bladder, breast, liver, oesophagus, and thyroid (each); 0.01 for the skin and bone surface (each); and 0.05 for the remainder of the body [15, p. 8]. The sum of these weighting factors is unity. Therefore, when the dose is uniform over the entire body, the effective dose equivalent and the "whole-body dose" are the same.

In judging the stochastic consequences of a given exposure, the effective dose equivalent (or the effective dose, in the ICRP usage) is the relevant quantity. It is an appropriate dose for both describing exposures and setting radiation standards.[11]

Collective Dose

The *collective dose* to a population is the sum of the doses to the individuals in that population. If the impact of the exposure is a stochastic one, where the outcome for any particular individual is a matter of statistical chance, there is no way of identifying which individuals will be, or have been, harmed by the exposure. However, the collective dose is an indication of the impact on the population as a whole. If one has confidence in the model relating health effects to dose, and in particular the linearity hypothesis (see Section 2.5.3), the collective dose can be used, for example, to calculate the number of expected fatal cancers in the population. The dose of greatest interest is the collective effective dose equivalent (in person-Sv), which in the terminology suggested in ICRP Publication 60 can be shortened to collective effective dose.

Dose Commitment

The concept of the *dose commitment* is particularly relevant to the intake of radionuclides into the body. If the radionuclide has a long residence time in the body, there will be an exposure lasting for many years. Thus, specifying the dose in a given year gives an incomplete picture. The dose commitment is the cumulative dose over the relevant future time (perhaps 50 years on average), allowing for a decrease in annual dose at a rate dependent

---

[10]The ICRP has changed its usage and now replaced the effective dose equivalent by the "effective dose" (in sieverts), denoted by the symbol E [15, p. 7]. This shortens the terminology. As discussed in NCRP Report No. 116, there are subtle differences between the definitions of the effective dose E and the effective dose equivalent H_E, but we will ignore these here [16, p. 21].

[11]The terminology may be somewhat in transition, with the U.S. Nuclear Regulatory Commission standards expressed in terms of the effective dose equivalent [17, §20.1003] and NCRP recommendations for standards expressed in terms of the effective dose [16]. The difference, however, is more one of terminology than substance.

upon the radioactive half-life for decay and the biological half-life for residence in the body. In this spirit, it is possible to consider both an individual dose commitment and a collective dose commitment.

2.4.2 Sources of Radiation Exposure

Natural Sources of Radiation

Radiation doses from the natural environment arise from inhalation of radon, from radionuclides in the Earth's crust, from cosmic rays, and from radionuclides in the body. The magnitudes of typical radiation doses in the United States are listed in Table 2.5. (The numbers are for 1980–1982, but there has been little change since.) It is seen that the natural sources are the largest contributors.

Indoor radon, more particularly the exposure to the lung due to inhaled alpha-particle emitting progeny of ^{222}Rn, is the largest single contributor. Radon is a gas. It enters the house from the soil, where it is produced in the alpha-particle decay of ^{226}Ra, a member of the ^{238}U decay chain. Uranium, and with it radium and radon, is present in all soil. Radon doses vary greatly from house to house, depending upon the radium content of the underlying soil, the soil porosity, and the house construction. A rough average radon dose is 2 mSv per year.

The next largest natural contribution, of about 0.39 mSv/yr, is from radionuclides in the body, most importantly ^{40}K, which is unavoidably present due to the large concentration of potassium in human tissue. Terrestrial radiation, arising from radionuclides in the ground, and cosmic rays each give annual doses of about 0.3 mSv, averaged over the country as a whole. In aggregate the dose from natural sources, excluding radon, averages about 1

Table 2.5. *Typical radiation exposures in the United States, 1980–1982 (annual effective dose equivalent).*

Radiation Source	Effective dose equivalent	
	(mSv/yr)	(mrem/yr)
Natural sources		
Indoor radon	2.0	200
Terrestrial radiation	0.28	28
Cosmic rays	0.27	27
Radionuclides in body	0.39	39
Total of natural sources (rounded)	3.0	300
Medical sources[a]		
Diagnostic x rays	0.39	39
Medical treatments (nuclear)	0.14	14
Other		
Consumer products[b]	0.1	≈10
Nuclear fuel cycle[c]	0.0005	0.05
TOTAL (rounded)	3.6	360

Source: Data based on Ref. [13], pp. 13, 15, and 53.
[a]Medical exposures vary widely among individuals. The quoted numbers are averages over the entire population.
[b]Major sources include building materials and the domestic water supply.
[c]Based on calculated exposure of people within 50 miles of nuclear facilities and averaged over entire population.

mSv/yr. Until the early 1980s many analyses ignored radon, and the total natural dose was taken to be about 1 mSv/yr. The actual average for the United States, with radon, is about 3 mSv/yr.

There are wide variations about this average, apart from differences in doses from radon. These variations stem from differences in the concentrations of gamma-ray emitting nuclei in the ground and differences in altitude. In parts of Colorado, to take an extreme U.S. case, the terrestrial radiation dose and the cosmic ray dose each exceed 1 mSv/yr, about triple the mean. Even greater excursions occur for radon exposures, with indoor radon levels in some houses more than 100 times the U.S. average.

Anthropogenic Sources of Radiation

The largest exposures from radiation due to human activities are those produced medically, primarily from diagnostic x rays and fluoroscopies and secondarily from radiation treatments. The received doses vary widely from individual to individual. A rough population average is 0.5 mSv/yr. Miscellaneous other sources of exposure, grouped under "consumer products" in Table 2.5, are estimated to cause a dose of about 0.1 mSv/yr.

The entire civilian nuclear fuel cycle, involving mining, fuel fabrication, and reactor operation, contributes a negligible dose, averaging about 0.0005 mSv/yr.[12] The largest contributors to this dose are from uranium mining and processing operations, including the release of radon from the mill tailings, which are the unused residue of the processes by which uranium for the reactors is extracted for use as fuel.

While to date the exposures from the nuclear fuel cycle have been, on average, negligibly small, the picture is incomplete. Fuel consumed in the reactors has been stored at the individual reactor sites, and thus there has been no exposure due to transportation, reprocessing, or disposal of spent fuel. If these steps are carried out without mishap, there will be no significant dose increments, but observers differ on the likelihood that major mishaps can be avoided. Further, no reactor accidents in the U.S. to date have released large amounts of activity. A Chernobyl-type accident, should it occur, would substantially increase the average exposures from nuclear operations over those cited here.[13]

2.5 EFFECTS OF RADIATION

2.5.1 Sources of Information

Data on the effects of radiation come from studies of Hiroshima and Nagasaki atomic bomb survivors, individuals receiving high exposures in the course of medical treatments, and uranium miners and other miners, as well as from experiences in isolated accidents and animal experiments. Further information may eventually come from the study of the effects of the Chernobyl nuclear reactor accident. Aside from the Hiroshima and Nagasaki studies,

---

[12]The collective effective dose equivalent for all the operations associated with a large reactor (with a capacity of 1 GWe and operating with a capacity factor of 80%) is estimated by the NCRP to be 1.36 person-Sv/yr [18, p. 160]. For a U.S. capacity of about 100 GWe, this would imply a national collective dose of 136 person-Sv/yr, or an average individual dose of about 5×10^{-7} Sv/yr.

[13]There were some releases from above-ground atomic bomb tests and intentional releases of ^{131}I from plants at the Hanford reservation. However, above-ground bomb tests were terminated by the United States, the USSR, and the United Kingdom in 1963, and the large Hanford releases of ^{131}I ended after the early 1950s. These releases are of historical interest, and extensive efforts are underway to assess their consequences, but they have no relevance to current or anticipated future exposures.

the data base on cancer mortality in humans is small, and much of these data are pertinent only to a single organ—for example, lung cancer in uranium miners.[14]

Extensive efforts by many individuals and groups have gone into the attempt to extract as much information as possible from the limited observational data. Official advice as to radiation protection is provided internationally by the International Commission on Radiological Protection (ICRP) and in the United States by the National Council on Radiation Protection and Measurements (NCRP). Each issues a series of major publications or reports, summarizing the status of knowledge on particular topics and sometimes embodying specific recommendations. Other influential reports are the so-called BEIR Reports, which are prepared by the Committee on the Biological Effects of Ionizing Radiations of the National Research Council (an arm of the National Academy of Sciences and associated academies), and the reviews published at roughly four-year intervals by the United Nations Scientific Committee on the Effects of Atomic Radiation (the UNSCEAR reports).

U.S. radiation protection regulations, with force of law, are formulated by the Nuclear Regulatory Commission (NRC) and the Environmental Protection Agency (EPA). The regulations are contained in Titles 10 and 40, respectively, of the Code of Federal Regulations (CFR).

The evaluations, recommendations, and regulatory limits are under continuous but slowly evolving review. Their scientific basis rests on very extensive research activities carried out throughout the world. In the United States, these were supported by the Atomic Energy Commission after World War II and more recently by its successor organization, the Department of Energy, along with other federal agencies. Much has been learned from the long array of studies and analyses, but as discussed below, some crucial questions remain inadequately resolved.

2.5.2 Effects of High Radiation Doses

Deterministic Effects

It is well established that high radiation doses are fatal. For a dose in the neighborhood of 3 to 5 Gy received over a short period of time, there is approximately a 50% chance of death within 60 days, although the probability of death is influenced by the prior health of the individual and the treatment administered.[15] For Hiroshima victims, 50% lethality was produced at about 3 Gy. For Chernobyl victims receiving doses between about 4 and 6 Gy, 7 of 23 died. Doses above 6 Gy were lethal in 21 of 22 cases (see Section 12.3.4).

For doses between 1 and 4 Gy, there are clinical symptoms, included in the phenomenon of "radiation sickness," with some chance of fatalities at the top end of the range. For doses below 1 Gy, there are usually no clinical symptoms.

Observational Evidence for Cancer at High Doses

The increased cancer risk is the most serious long-term stochastic consequence of high radiation exposures. The evidence is unambiguous that cancer can be caused by high but

---

[14]Even the Hiroshima–Nagasaki data base is quite limited. For the roughly 76 000 survivors studied, there were 5936 cancer fatalities in the period 1950 to 1985, of which 344 were "excess"—that is, attributable to exposures from the bombs rather than to natural causes (as gauged from the cancer rates in unexposed populations) [19, p. 8].

[15]See Refs. [20], pp. 565–574, 595 and [15], p. 105.

sublethal radiation doses. The quantitative risk factor for cancer induction has been estimated by a number of scientific groups, in the United States most importantly by the BEIR Committee. The most recent studies are based largely on a 1986 analysis of the Hiroshima–Nagasaki data. These analyses are carried out by correlating the health histories of survivors with the magnitudes of the exposures they received.

The conclusions of the latest published BEIR study, the 1990 BEIR V report [21], differed substantially from those of the 1980 BEIR III report [22]. Two factors caused a major reevaluation of the effects of radiation between the times of the two studies. During the more prolonged observation period, extending to 40 years after the bombing, additional excess cancers were observed, raising the cumulative number of cancer fatalities. In addition, a reexamination of evidence on the exposures received by bomb victims led to the conclusion that the average neutron exposures had been considerably overestimated in earlier work.[16] Both factors had the effect of raising the extracted value for number of cancer fatalities per unit exposure. This led to substantially higher risk estimates in BEIR V than in BEIR III.

The BEIR V study concluded that for an instantaneous whole-body exposure of 100 mSv (10 rem), the expected excess cancer mortality is 800 deaths per 100 000 exposed individuals [21, p. 175]. Put another way, an "average" individual receiving a dose of 100 mSv is estimated to have an 0.8% chance of suffering a fatal cancer as a result of this exposure.

More recently, new studies of the neutron fluxes have suggested that the DS86 dosimetry, on which the results used in BEIR V were based, may have underestimated the exposures at Hiroshima from neutrons [24, 25]. This conclusion comes from measurements of chlorine 36 (^{36}Cl) in concrete cores from buildings near the bomb locations. ^{36}Cl is produced by slow neutron capture on ^{35}Cl and thus provides a measure of the thermal neutron flux. If these results are confirmed and can be shown to apply to fast neutrons as well, the conclusions of BEIR V may have to be modified back again in the direction of the BEIR III estimates.

According to ICRP Publication 60, the Hiroshima–Nagasaki data give clearcut (i.e., at 95% confidence level) indications of cancer induction only at doses above about 200 mSv, although there is weaker positive evidence down to about 50 mSv [15, p. 16]. It is also important to note that the Hiroshima–Nagasaki data refer to exposures received over a very brief period, i.e., to very high dose rates per unit time.

2.5.3 Effects of Low Doses

Observational Evidence for Cancer at Low Doses

Virtually all our evidence on the harmful stochastic effects of radiation, for studies in both animals and man, comes from observations at high doses or high dose rates. However, most exposures of concern, such as those from nuclear weapon tests, nuclear energy (including even the Chernobyl accident), and indoor radon, involve doses and dose rates much below those of Hiroshima and Nagasaki.

There have been extensive efforts to determine from direct observation the rates of cancer induction in populations exposed to radiation at doses that are only moderately higher than normal. These include studies of populations living in regions of above-average natural radiation background, of workers in the nuclear industry, and of groups that received moderate exposures from nuclear weapons tests. In no case is there unambiguous evidence

---

[16]Calculations and measurements contributing to the newer dosimetry, revised in 1986 and called the DS86 dosimetry, included calculations of neutron and gammas ray fluxes from the bombs, measurements of gamma-ray induced luminescence, and measurements of activation products of neutron capture [23].

of a positive effect, although often the results are controversial.

There are major statistical uncertainties in trying to identify a small number of additional cancers in the presence of the background of the many cancers that occur without excess radiation. The statistical difficulties can be compounded by the subdivision of the data. For example, if a population is divided into six age groups, and one considers ten cancer types, there are a total of 60 categories. On average, there should be three cases of a departure of two standard deviations or more from the norm, even if there is no true effect. When a two-standard-deviation anomaly is seen under such circumstances, is the effect "significant"? Beyond the statistical problems, there are difficulties in determining whether an observed higher (or lower) than average cancer rate in a given population could be caused by some unidentified confounding factor, rather than by higher radiation doses.

As a result of these difficulties, there is no consensus as to what has been learned from the studies of cancer incidence in populations exposed to small radiation doses. In a summary made in ICRP Publication 60, it is judged:

> Overall, studies at low dose, while potentially highly relevant to the radiation protection problem, have contributed little to quantitative estimates of risk.[15, p.17].

The BEIR V report also attempted a review of the available epidemiological studies at low doses but was unable to draw any crisp conclusions. The authors found reassurance that the extrapolations from results at higher doses did not underestimate the risk at low doses, but they could not exclude the possibility of an overestimate:

> Studies of the imputed effects of irradiation at low doses and low dose rates fulfill an important function even though they do not provide sufficient information for calculating numerical estimates of radiation risks. They are the only means available now for determining that risk estimates based on data accumulated at higher doses and higher dose rates do not underestimate the effects of low-level radiation on human health [21, p. 371].

The Linearity Hypothesis

In the absence of direct evidence on the rate of cancer induction at low doses, estimates are made by extrapolation from observations at high doses. The *linearity hypothesis* provides a simple assumption: The cancer risk (above the normal rate) is proportional to the excess radiation dose. Under this hypothesis, if a dose of 1 Sv in a brief period corresponds to an 8% excess chance of an eventual fatal cancer, then a dose of 1 mSv (gradual or sudden) corresponds to an 0.008% excess chance.

Many alternatives to the linearity assumption have been proposed; some of these are illustrated in Fig. 2.2:

1. A dose response curve that corresponds to greater effects at low doses than implied by linearity; this is termed "supralinearity" (curve A).

2. A linear–quadratic dose response in which the risk is dependent upon the sum of linear and quadratic terms. Depending upon the relative importance of the two terms, the extrapolated risk may be substantially depressed at low doses (curve C).

3. A dose-response curve that takes on negative values at low doses, corresponding to a beneficial effect of small radiation doses; this is termed "hormesis" (curve D).

4. The existence of a threshold, below which there is no appreciable rate of cancer induction (not shown).

5. A dose or dose-rate effectiveness factor, to reduce the risk-per-unit dose at low doses or low-dose rates (or both) below the level observed at high doses and high-dose rates (not shown).

The linearity assumption corresponds to curve B in Fig. 2.2.

All of the above alternatives with the exception of (1) correspond to reduced effects at low doses. Of these, the only one that has been adopted in major recent studies is (5), the inclusion of a dose or dose-rate effectiveness factor (DREF). At an earlier time, as reflected in the 1980 BEIR III study, the linear–quadratic model (2) had been favored [22]. Alternatives (1) and (3) are outside the mainstream of standard assessments and can be considered maverick opinions. This does not in itself prove they are wrong.

The validity of the linearity hypothesis and the alternatives suggested above could in principle be resolved from either conclusive observational data at low radiation doses or a full understanding of the damage mechanisms. But, as discussed above and implied by the persistence of alternatives, neither is available.

Importance of Low Doses

It is unfortunate from a policy standpoint that the available information on the effects of low doses is as inadequate as it is, because most exposures of concern are for low doses and low dose rates. For example, most of the calculated fatal cancers from the Chernobyl nuclear power accident are the result of cumulative doses in the neighborhood of 1 or 2 mSv over a lifetime (see Section 12.3.5). Even accepting the linearity assumption, the individual risk from such exposures is small (in the neighborhood of 1 in 10^4). But there still might be a large total number of fatalities in a population of hundreds of millions. Similarly, the doses from indoor radon, averaging about 2 mSv per year, are the largest source of collective population exposure in the United States. Should they be a matter of concern? In the absence

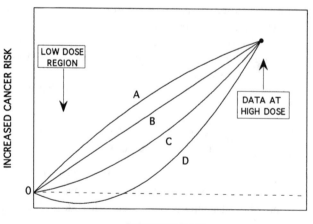

Figure 2.2. Schematic representation of several alternative assumptions for the extrapolation of the cancer risk vs radiation dose to low-dose levels, given a known risk at a high dose: supralinear (A), linear (B), linear–quadratic (C), and hormesis (D).

of firm knowledge about the effects of low-level radiation, there are no definite answers to these crucial questions in radiation protection.

2.5.4 Conclusions of Advisory Bodies on Low-Dose Effects

BEIR V: Linearity with no DREF

In its 1990 report (BEIR V) [21], the BEIR Committee adopted the linearity assumption for cancers other than leukemia, with no dose or dose-rate reduction factor.[17] The risk of excess cancer mortality is taken to be 0.08 per Sv over a wide range of dose levels. The use of this assumption can be illustrated by applying it to the case of a constant low-level lifetime exposure. Consider a population of 100,000, with a representative distribution by age and sex. The "normal" cancer mortality in such a population is about 19 000 [21, p. 172]. If this population sustains a 1 mSv/yr dose for a period of 70 years, the cumulative exposure is 70 mSv per person, or 7000 person-Sv, and the calculated excess cancer mortality is 560.[18] This would represent an increase of about 3% in the total cancer rate.

However, the BEIR V report expresses concern about this sort of extrapolation to doses and dose rates that are far lower than those on which the fundamental risk estimate is based, especially the Hiroshima–Nagasaki data. Thus, while adopting the linear extrapolation as a basis for predicting the risk at low doses, the BEIR V report also indicates the possibility that there is no risk from exposures in the neighborhood of several mSv per year:

> departure from linearity cannot be excluded at low doses below the range of observation. Such departures could be in the direction of either an increased or decreased risk. Moreover, epidemiological data cannot rigorously exclude the existence of a threshold in the millisievert dose range. Thus the possibility that there may be no risks from exposures comparable to external natural background radiation cannot be ruled out. At such low doses and dose rates, it must be acknowledged that the lower limit of the range of uncertainty in the risk estimates extends to zero [21, p. 181].

ICRP and NCRP Reports: Linearity Modified by a DREF

Although no dose or dose-rate effectiveness factor was adopted in BEIR V, despite the expressed misgivings, such a factor has been adopted in the recent ICRP, UNSCEAR, and NCRP reports. ICRP Publication 60 recommends using a reduction factor of about 2 [15, p. 18] at doses below 0.2 Gy and at higher doses if the dose rate is less than 0.1 Gy/h. The recommendation is limited to low-LET radiation, and thus applies to gamma-ray and beta-particle exposures, but not to neutron or alpha-particle exposures. The UNSCEAR report recommends the same factor under the same circumstances but limits its use to dose rates of 0.006 Gy/h or less [26, p. 682]. The NCRP report concludes that a dose-rate effectiveness factor should be applied for "persons exposed to low doses and at low-dose rates and the factor could range between two and three" [19, p. 112].

It should be noted that the "low" dose designations of the ICRP and UNSCEAR recommendations are high doses by any normal standard. A dose of 0.2 Gy, which translates to 0.2 Sv or 200 mSv for low-LET radiation, is more than 50 times the U.S. average annual dose

---

[17]Leukemia is responsible for roughly 4% of normal cancer deaths, 12% of excess fatalities for a single exposure of 0.1 Sv, and slightly less than 12% of excess fatalities for a lifetime exposure of 1 mSv/yr [21, p. 172].

[18]The actual quoted numbers in BEIR V are 520 per 100 000 for men and 600 per 100 000 for women [21, p. 172].

of 3.6 mSv (see Table 2.5). Similarly, a dose rate of 0.006 Gy/h (equivalent to 6 mSv/h) means a larger dose in one hour than normally received in one year.

Summary of Estimates Relating to Cancer Induction at Low Doses

In summary, if the BEIR V linear assumption is adopted for low-level exposures, then the relation between the risk of fatal cancer and the effective dose equivalent from radiation exposure is

Risk of eventual fatal cancer: 0.08 per Sv (0.0008 per rem).

The ICRP report starts with a slightly higher overall population risk factor at high doses but applies a reduction factor for low doses and dose rates [15, p. 20]. The ICRP result for low-LET radiation is

Risk of eventual fatal cancer: 0.05 per Sv (0.0005 per rem).

The NCRP report, which took into account the earlier reports, arrived at the same result. It recommended that for setting radiation protection guidelines, a dose-rate effectiveness factor of two should be used, and it found a risk factor for the general population in agreement with the ICRP, namely 0.05 per Sv [19, p. 4]. For a worker population, it recommended a slightly lower risk factor, namely 0.04 per Sv. The difference stems from the difference in the age distributions of the two populations.

Regulatory Guidelines and Scientific Assessments

These risk estimates, for low doses and low dose rates, do not provide well-established numbers based on a firm scientific foundation. They instead represent estimates that, as appropriate for guidelines to be used in radiation protection, probably "err on the side of caution." If the effects of low-level radiation are not well known, it is appropriate for an advisory body that will influence regulations to take a conservative approach. This does not, however, necessarily represent the best scientific estimate.

The dichotomy is illustrated by two passages from NCRP Report No. 116, *Limitation of Exposure to Ionizing Radiation* [16]. In a section on the objectives of radiation protection, the report states:

> the Council assumes that, *for radiation-protection purposes, the risk of stochastic effects is proportional to dose without threshold, throughout the range of dose and dose rates of importance in routine radiation protection* [p. 10].

Slightly later, the report points out the uncertainties in the extrapolation from high doses to low doses and dose rates:

> This uncertainty is estimated to be about an additional factor of two or more since, at very low doses, the possibility that there is no risk cannot be excluded [p. 13].

This second quotation was in the context of a discussion of occupational dose limits, where conservatism was somewhat tempered by comparisons to the general level of industrial hazards. But obviously the possibility of no risk at very low doses applies to the general population as well.

For the purposes of this book, we will in general use the ICRP and NCRP risk factor of 0.05 per Sv for fatal cancer induction, although sometimes cognizance will be taken of the uncertainties involved.

2.5.5 Genetic Effects

There is no evidence that directly demonstrates genetic damage in the offspring of people exposed to radiation, although through the 1950s there was considerable emphasis on potential genetic effects of radiation, based on animal studies. Since then, in the words of NCRP Report No. 116, "the genetic risks were found to be smaller and cancer risks larger than were thought at the time" [p. 12].

One might expect that there would be evidence from the exhaustively studied atomic bomb victims. But here the evidence is negative: on comparing offspring of heavily irradiated parents with offspring of parents who received little or no radiation, "no statistically significant effects of radiation have been demonstrated to date" [21, p. 95]. In these studies, stillbirths, abnormalities at birth and early deaths among children were all compared, with no overall statistically significant positive effects.[19]

Inferences as to genetic effects come therefore from animal experiments, primarily with fruit flies and mice. The atomic bomb results serve to verify that the deduced effects in humans are not thereby substantially underestimated. In fact, they suggest (but do not establish at a 95% confidence level) that extrapolating from animal data overestimates the genetic effects of radiation in humans.

It has been common in discussing genetic effects to introduce the concept of the doubling dose, namely, the dose at which the normal incidence of genetic defects is doubled. A recent evaluation of the overall evidence on radiation-induced genetic effects is contained in NCRP Report No. 115:

> The doubling dose (DD) for genetic diseases that cause morbidity or mortality in humans is now estimated to be about 1.7–2.2 Sv or about 3.4–4.4 Sv for exposures at low-dose rates. . . . Because of the large uncertainties in both estimates and the desire to provide adequate protection, a risk of severe hereditary effects for the general population (for all generations) of 1×10^{-2} Sv^{-1} should provide a reasonable basis for dose limitation [19, p. 3].

It may be noted that this risk factor is one-fifth the risk factor for fatal cancers.

Because cancer effects are both larger and better documented than genetic effects, it is rather common to couch the effects of radiation exposures in terms of cancer incidence alone. Thus, for example, in the discussion of the Chernobyl accident in Chapter 12, the focus is on cancer incidence rather than genetic or other effects such as mental retardation. These are not unimportant, but whatever their importance, it appears to be less than that of cancer.

---

[19]It is interesting to speculate why it is the common impression that there was great genetic damage from the atomic bombs. Perhaps the misconception arose in part from the fact that there *was* substantial damage to children of pregnant mothers at Hiroshima and Nagasaki, and in part because responsible people did not wish to risk underestimating the horrors of the bombs. Horrors there were, but the genetic ones, if any, are not great enough to be identified.

2.6 RADIATION STANDARDS

2.6.1 Standards for the General Public

From about 1960 to 1990 the standard established nationally and internationally was that the average additional exposure to the population should not exceed 1.7 mSv/yr, and for no individual (excluding radiation workers) should it exceed 5 mSv/yr. This was the maximum "permitted" dose, over and above the dose received from natural and medical sources. This average incremental exposure limit was roughly equal to the basic natural and medical exposure (omitting radon, which had been more or less ignored!). Coupled with these limits was the further proviso that the radiation doses be kept "as low as practicable," in recognition of the fact that it is prudent to incur no excess radiation unless there are compensating benefits.

A reduction of the dose limit for the general population to 1 mSv/yr was recommended by the ICRP in 1990 [15, p. 45], and was adopted by the U.S. Nuclear Regulatory Commission effective January 1, 1994 [17, §20.1301]. This limit refers to exposures from a single facility licensed by the NRC, excluding medical facilities.

The reduction will not have a practical impact, as far as its stipulations for the general population are concerned. Most of the nuclear fuel cycle, including reactor operation, already falls under an EPA whole-body dose limit of 0.25 mSv/yr for total exposures from *all* covered facilities [27, §190.10]. In addition, nuclear waste disposal facilities covered by the EPA are to satisfy a 0.15 mSv/yr limit, under provisions adopted at the end of 1993 [28].[20]

Actual exposures to the public from nuclear power operations and facilities are much lower than the regulatory limit of 0.25 mSv/yr, and the limit is therefore not presently constraining (see Table 2.5). It would be neither violated nor even approached except in the case of an accident, in which case the existence of regulations would be moot. The regulatory limits provide a safeguard, however, against negligent operation of existing facilities. Without a history of external pressures from regulatory agencies, radiation exposures from nuclear facilities might not be as low as they now are.

Under the terms of the Safe Water Drinking Act, the EPA has established stringent limits on the contamination of drinking water from a wide variety of pollutants, including radionuclides. For beta-particle and gamma-ray emitters, the radionuclide concentrations in community water systems must not produce a dose above 4 mrem (0.04 mSv) per year for a person drinking 2 liters per day from this source [27, §141.16]. The limit for alpha-particle emitters is stated in terms of maximum concentrations (in picocuries per liter), rather than in terms of a radiation dose [27, §141.15], but the intent is to achieve the same limit of 4 mrem/year.

Uranium and radon are explicitly excluded from control in these regulations. The chief ingestion danger for uranium is from its chemical properties, not its radioactive emissions, because the half-life of ^{238}U is large and its rate of decay slow. For radon, there is a special problem, still not resolved, because many water supplies cause a radon dose in excess of a 4 mrem/year limit, but this dose is still small compared to the radon inhalation dose in indoor (or even outdoor) air.

The EPA has taken an advisory, but not regulatory, position on radon exposures in air. Regulation is much more difficult here, because radon is a natural product and limits would have to be imposed house by house. For radon, the EPA suggests that remedial action be

---

[20]The regulation of nuclear waste disposal facilities is discussed further in Section 10.1.4. At present, the EPA stipulation applies only to the WIPP facility in New Mexico.

Table 2.6. *EPA and NRC standards for U.S. radiation protection, expressed in terms of annual effective dose equivalent.*

Source of Exposure	Dose Limit (mSv)	Agency
Occupational	50	NRC
General public		
Any licensed facility	1	NRC
Nuclear power facility	0.25	EPA
High-level waste repository	0.15	EPA
Public water supplies	0.04	EPA
Indoor radon (general public)[a]	8	EPA

[a]This is not a limit but rather a level at which the EPA recommends remedial action.

taken if the indoor concentration exceeds 4 pCi of ^{222}Rn per liter of air. This concentration corresponds to an annual dose of approximately 8 mSv/yr.[21]

A summary of these standards and recommendations, along with those for occupational exposure, is given in Table 2.6. It is seen that the limits established by the EPA on nuclear facilities are much lower than the level at which action is recommended to reduce indoor radon levels.

2.6.2 Occupational Exposures

For many years the limit for occupational exposures was 50 mSv/yr. The International Council on Radiation Protection and Measurements has recently reduced its recommended limits. The new ICRP recommended limits are 100 mSv (10 rem) for occupational exposures over a 5-year period, equivalent to an average of 20 mSv per year. The limit of 50 mSv in any one year is retained. In addition, for pregnant women, there is a limit of 5 mSv for the remainder of the pregnancy (after it is known) [15].

The recent changes in U.S. Nuclear Regulatory Commission regulations [17] also include changes for occupational exposures. The overall occupational limit is basically unchanged, remaining at 50 mSv/yr, but there is some modification of the way in which the 50 mSv is calculated. In addition, the general principle that radiation doses should be kept "as low as is reasonably achievable" (ALARA) has been incorporated [17, §20.1101].

Were a worker exposed to the maximum dose limit over a working lifetime of 40 years, the cumulative total would amount to 2 Sv. Adopting a risk factor of 0.04 per Sv, this would imply an 8% chance of a fatal cancer due to occupational exposure. Average occupational exposures have been far below the limit, except for uranium miners, but nonetheless there are pressures for a formal reduction in the NRC occupational limit.

2.6.3 Annual Limits on Intake

The radiation dose per unit of activity, for ingestion or inhalation of radionuclides, depends upon the time the radionuclide remains in the body, the tissues in which it is located, the rate of radioactive decay, the type of emitted radiation, and the energy deposited. These considerations go into the calculation of annual limits on intake (ALI) for the radionuclides.

---

[21]The effective dose equivalent in mSv is here established using the following approximate equivalences: 1 pCi/l≈0.2 working level months per year≈2 mSv effective dose equivalent per year.

Table 2.7. *Annual limits on intake (ALI) for occupational exposure, for selected radionuclides.*

Radionuclide	Half-life (years)	ALI for Ingestion (μCi)	ALI for Ingestion (mg)	ALI for Inhalation (μCi)	ALI for Inhalation (mg)
^{3}H	1.23×10^1	80 000	8×10^{-3}	80 000	8×10^{-3}
^{14}C	5.73×10^3	2000	4×10^{-1}	2000	4×10^{-1}
^{40}K	1.28×10^9	300	4×10^4	400	6×10^4
^{60}Co	5.27×10^0	200	2×10^{-4}	30	3×10^{-5}
^{90}Sr	2.88×10^1	30	2×10^{-4}	4	3×10^{-5}
^{99}Tc	2.11×10^5	4000	2×10^2	700	4×10^1
^{131}I	2.20×10^{-2}	30	2×10^{-7}	50	4×10^{-7}
^{137}Cs	3.01×10^1	100	1×10^{-3}	200	2×10^{-3}
^{210}Pb	2.23×10^1	0.6	8×10^{-6}	0.2	3×10^{-6}
^{226}Ra	1.60×10^3	2	2×10^{-3}	0.6	6×10^{-4}
^{232}Th	1.41×10^{10}	0.7	6×10^3	0.001	9×10^0
^{235}U	7.04×10^8	10	5×10^3	0.04	2×10^1
^{238}U	4.47×10^9	10	3×10^4	0.04	1×10^2
^{237}Np	2.14×10^6	0.5	7×10^{-1}	0.004	6×10^{-3}
^{239}Pu	2.41×10^4	0.8	1×10^{-2}	0.006	1×10^{-4}
^{241}Am	4.32×10^2	0.8	2×10^{-4}	0.006	2×10^{-6}

Source: Data are from Ref. [17], Appendix B.

Table 2.7 gives, for selected radionuclides, the ALI values adopted by the NRC for occupational exposure. These are extracted from a comprehensive table in the Code of Federal Regulations [17, App. B to §20.1001–20.2402]. The values in Table 2.7 correspond to the more restrictive of two criteria for a given radionuclide: (1) a committed effective dose equivalent of 50 mSv or (2) a committed dose equivalent to an individual organ of 500 mSv. The dose can depend upon the form in which the radionuclide enters the body and on its residence time in the body. In such cases the limit in Table 2.7 is the more restrictive of the possibilities given in Ref. 17.

The ALI values are given in Table 2.7 in both microcuries and milligrams. The former is related to the radiation dose and the latter to the amount likely to be ingested or inhaled. Comparing ^{239}Pu to ^{226}Ra, for example, it is seen that ^{239}Pu is the more hazardous for ingestion on a per curie basis, while it is less hazardous on a per gram basis. It also is to be noted that ^{239}Pu is much more hazardous for inhalation than for ingestion, because most ingested plutonium is soon excreted.

REFERENCES

1. Roger M. Macklis, "The Great Radium Scandal," *Scientific American* 269, no. 2 (August 1993): 94–99.
2. Eric Morgenthaler, "For a Healthy Glow Some Old Folks Try a Dose of Radon," *Wall Street Journal*, October 12, 1990: 1.
3. Samuel Glasstone, *Sourcebook on Atomic Energy* (New York: Van Nostrand, 1950).
4. Robley D. Evans, *The Atomic Nucleus* (New York: McGraw-Hill, 1955).
5. American Nuclear Society, *Glossary of Terms in Nuclear Science and Technology* (La Grange Park, Illinois, American Nuclear Society, 1986).
6. Jagdish K. Tuli, *Nuclear Wallet Cards* (Upton, New York: Brookhaven National Laboratory, 1995).
7. National Council on Radiation Protection and Measurements, *Exposure of the Population in the United States and Canada from Natural Background Radiation*, NCRP report no. 94 (Washington, D.C.: NCRP, 1987).
8. DeVerle P. Harris, "World Uranium Resources," *Annual Review of Energy* 4 (1979): 403–432.
9. Stuart Ross Taylor and Scott M. McClennan, *The Continental Crust: its Composition and Evolution* (Oxford: Blackwell Scientific Publications, 1985).

10. J. H. Chen, R. Lawrence Edwards, and G. J. Wasserburg, "^{238}U, ^{234}U, and ^{232}Th in Seawater," *Earth and Planetary Science Letters* 80 (1986): 242–251.

11. P. Kilho Park, Dana R. Kester, Iver W. Duedall, and Bostwick H. Ketchum, "Radioactive Wastes and The Ocean: An Overview," in *Wastes in the Ocean*, vol. 3 (New York: John Wiley, 1983), 3–46.

12. P. A. Cox, *The Elements: Their Origin, Abundance and Distribution* (Oxford: Oxford University Press, 1989).

13. National Council on Radiation Protection and Measurements, *Ionizing Radiation Exposure of the Population of the United States*, NCRP report no. 93 (Washington, D.C.: NCRP, 1987).

14. Gudrun A. Carlsson, "Theoretical Basis for Dosimetry," in *The Dosimetry of Ionizing Radiation*, edited by K. R. Kase, B. E. Bjärngard, and F. H. Attix (Orlando: Academic Press, 1985).

15. International Commission on Radiological Protection, "1990 Recommendations of the International Commission on Radiological Protection," ICRP Publication 60, *Annals of the ICRP* 21, nos. 1–3 (Oxford: Pergamon Press, 1991).

16. National Council on Radiation Protection and Measurements, *Limitation of Exposure to Ionizing Radiation*, NCRP report no. 116 (Washington, D.C.: NCRP, 1993).

17. *Energy, U.S. Code of Federal Regulations*, title 10, part 20 (1993).

18. National Council on Radiation Protection and Measurements, *Public Radiation Exposure from Nuclear Power Generation in the United States*, NCRP report no. 92 (Washington, D.C.: NCRP, 1987).

19. National Council on Radiation Protection and Measurements, *Risk Estimates for Radiation Protection*, NCRP report no. 115 (Washington, D.C.: NCRP, 1993).

20. United Nations Scientific Committee on the Effects of Atomic Radiation, *Sources, Effects, and Risks of Ionizing Radiation* (New York: United Nations, 1988).

21. National Research Council, *Health Effects of Exposure to Low Levels of Ionizing Radiation, BEIR V*, report of the Committee on the Biological Effects of Ionizing Radiations (Washington, D.C.: National Academy Press, 1990).

22. National Research Council, *The Effects on Populations of Exposure to Low Levels of Ionizing Radiation: 1980*, report of the Committee on the Biological Effects of Ionizing Radiations (Washington, D.C.: National Academy Press, 1980).

23. William C. Roesch, ed. *U.S.–Japan Joint Reassessment of Atomic Bomb Radiation Dosimetry in Hiroshima and Nagasaki*, final report, vol. 1 (Hiroshima: Radiation Effects Research Foundation, 1987).

24. T. Straume *et al.*, "Neutron Discrepancies in the DS86 Hiroshima Dosimetry System," *Health Physics Society's Newsletter* 63 (1992): 421–426.

25. T. Straume, L. J. Harris, A. A. Marchetti, and S. D. Egbert, "Neutrons Confirmed in Nagasaki and at the Army Pulsed Radiation Facility: Implications for Hiroshima," *Radiation Research* 138 (1994): 193–200.

26. United Nations Scientific Committee on the Effects of Atomic Radiation, *Sources and Effects of Ionizing Radiation* (New York: United Nations, 1993).

27. *Protection of Environment, U.S. Code of Federal Regulations*, title 40 (1993).

28. "Environmental Radiation Protection Standards for the Management and Disposal of Spent Nuclear Fuel, High-Level and Transuranic Radioactive Wastes," *Federal Register* 58 (December 20, 1993): 66398.

Chapter 3

Neutron Reactions

3.1 OVERVIEW OF NUCLEAR REACTIONS

3.1.1 Neutron Reactions of Importance in Reactors

The term *nuclear reaction* is used very broadly to describe any of a wide array of interactions involving nuclei. Innumerable types of nuclear reactions can occur in the laboratory or in stars, but in considering energy from nuclear fission, interest is limited almost entirely to reactions initiated by neutrons. Here, the important reactions are those that occur at relatively low energies, several MeV or less, characteristic of neutrons produced in nuclear fission. These reactions are elastic scattering, inelastic scattering, neutron capture, and fission.[1]

Elastic Scattering

In elastic scattering, a neutron and nucleus collide with no change in the structure of the target nucleus (or of the neutron). An example, for the elastic scattering of neutrons on carbon 12 (^{12}C), is

$$n + {}^{12}C \rightarrow n + {}^{12}C .$$

---

[1]This is not a fully exhaustive list. Other reactions are sometimes possible, for example, those in which an alpha particle is produced, but the exceptions only rarely are of interest in nuclear reactors.

Although the structure of the ^{12}C nucleus is unchanged, the neutron changes direction and speed, and the ^{12}C nucleus recoils. The total kinetic energy of the system is unchanged, but some of the neutron's energy is transferred to the ^{12}C target nucleus.

Elastic neutron scattering can occur with any target nucleus, but in reactors it is of greatest importance when the target nucleus is relatively light and the loss of kinetic energy of the neutron is therefore relatively large. In such cases, elastic scattering is an effective means of reducing the energy of the neutrons without depleting their number. This is the process of *moderation* (see Section 5.2).

Inelastic Scattering

Inelastic scattering differs from elastic scattering in that the target nucleus is left in an excited state. It decays, usually very quickly, to the ground state, with the emission of one or more gamma rays. An example is

$$n+{}^{12}C\rightarrow n+{}^{12}C^*\rightarrow n+{}^{12}C+\gamma\text{'s},$$

where the asterisk indicates an excited state of ^{12}C. The total kinetic energy of the neutron and carbon nucleus after the scattering is less than that of the incident neutron before the scattering, the difference equaling the energy of the gamma rays. In nuclear reactors, inelastic scattering is usually unimportant at the low neutron energies of interest, and it will be ignored in much of the discussion below.[2]

Neutron Capture

In the first stage of many reactions, the neutron combines with the target nucleus to form an excited *compound nucleus*. The term neutron capture is usually restricted to those cases where the excited compound nucleus decays by the emission of gamma rays. An example is

$$\text{formation:}\quad n+{}^{238}U\rightarrow{}^{239}U^*$$

$$\text{de-excitation:}\quad {}^{239}U^*\rightarrow{}^{239}U+\gamma\text{'s}.$$

Here, the compound nucleus is $^{239}U^*$, where the asterisk again indicates an excited state. The number of gamma rays emitted in the de-excitation of $^{239}U^*$ may vary from one to many, but the gamma rays themselves are not the product of chief interest. The important effect here is the transformation of uranium 238 (^{238}U) into uranium 239 (^{239}U).

Neutron capture can occur for almost any target nucleus, although at widely differing rates. Neutron-capture reactions play two general roles in nuclear reactors: (a) they consume neutrons that might otherwise initiate fission, and (b) they transform nuclei produced in fission into other nuclei. In addition, the specific reaction indicated above is of special interest because it is the first step in the formation of plutonium 239 (^{239}Pu) in reactors. The ^{239}Pu results from the β^- decay of ^{239}U ($T=23.5$ minutes) to neptunium 239 (^{239}Np), followed by the β^- decay of ^{239}Np ($T=2.355$ days) to ^{239}Pu. ^{239}Pu has a half-life of 2.411×10^4 years and therefore does not decay appreciably while in the reactor.

---

[2]Inelastic scattering is impossible if the excitation energy of the lowest excited state of the target nucleus is greater than the incident neutron energy.

Nuclear Fission

Energy production in a nuclear reactor derives from fission. In a typical fission reaction, the excited compound nucleus divides into two main fragments plus several neutrons. The energy released in fission comes from the large kinetic energy of the fission fragments. A typical fission reaction, here illustrated for a ^{235}U target, is of the general form

$$n + {}^{235}U \rightarrow {}^{236}U^* \rightarrow {}^{144}Ba + {}^{89}Kr + 3n .$$

The products in this example are barium 144 (atomic number $Z=56$), krypton 89 ($Z=36$), and 3 neutrons. Many other outcomes are also possible, always subject to the condition that the sums of the atomic numbers and atomic mass numbers of the products are the same as those of the initial system, 92 and 236, respectively.

Both the ^{144}Ba and ^{89}Kr nuclei are radioactive. For each, its formation is followed by a series of beta decays that continues with the successive emission of β^- particles until a stable isobar is reached. In addition, gamma rays are emitted in the de-excitation of the two fission fragments (^{144}Ba and ^{89}Kr), assuming they are formed in excited states, as well as in the de-excitation of the products of the successive β^- decays.

Fission is possible for only a very few target nuclei, the most important cases being isotopes of uranium and plutonium.

3.1.2 Reaction Cross Sections

Definition of Cross Section

The neutrons in a reactor may interact with a variety of target nuclei. Possible reactions are elastic scattering, capture, and in some cases, fission or inelastic scattering. Which of these reactions occurs for an individual neutron is a matter of chance. The probability of each outcome is commonly couched in terms of the reaction *cross section* σ for the event.

The term cross section suggests an area, and the reaction cross section for a given reaction can be thought of as the effective cross-sectional or projected area of a nucleus as a target for that reaction. The so-called *geometric cross section* of a nucleus in fact corresponds to a physical area. It is defined as the area of the disc presented to incident particles, namely πR^2, where R is the effective radius of the nucleus.[3]

Although a nucleus does not have a sharp boundary, its density is rather uniform over its volume and falls off fairly rapidly at the exterior. Thus, it is meaningful to speak in terms of a nuclear radius. Nuclear densities are approximately the same for all nuclei, independent of nuclear mass, and the nuclear radius is therefore proportional to the cube root of the nuclear mass. For heavy nuclei such as ^{238}U, the nuclear radius is roughly 9×10^{-13} cm, and the geometric cross section is therefore in the neighborhood of 2.5×10^{-24} cm^2. In common notation, this is expressed as 2.5 barns (b), where $1 \; b = 10^{-24} \; cm^2 = 10^{-28} \; m^2$.

Actual cross sections for specific reactions may be much larger or much smaller than the geometric cross section. Larger cross sections are difficult to understand if one thinks in terms of a classical geometric picture. They are a consequence of the wave properties of moving particles.[4] These properties are most important for neutrons at low energies, and they

---

[3]The small but finite radius of the neutron can be taken into account in assigning a numerical magnitude to the effective radius.

[4]It would carry the discussion too far afield to discuss the relationship between wave and particle aspects of matter. An elementary introduction is given, for example, in Ref. [1].

explain the large reaction cross sections for very low-energy neutrons.

Cross sections are appreciably smaller than the geometric cross section when the particular reaction is relatively improbable. If the reaction cannot occur at all—for example, if the energy in the initial system is too low to create the final system—then the cross section is zero.

The cross section for a given reaction can be defined by relating it to the probability that the reaction occurs. For neutrons traversing a short distance δx, the probability of a reaction is given by

$$\delta P = N \sigma \delta x , \tag{3.1}$$

where N is the number of nuclei per unit volume. For a homogenous material, $N = \rho N_A / M$, where M is the atomic mass of the material, ρ is the density, and N_A is Avogadro's number.

The probability δP is an experimentally determined number for a given path length δx. It is the ratio of the number of neutrons undergoing the given reaction to the number of incident neutrons. The number of nuclei per unit volume N is in principle known or measurable. Therefore, Eq. 3.1 defines the cross section σ for that reaction.

Mean Free Path

Neutrons do not travel a well-defined distance through material before undergoing interactions. They are removed exponentially, at separate rates for each of the possible reactions.[5] The average distance traversed by a neutron before undergoing a reaction of the type specified is the *mean free path* λ for the reaction. Numerically, it is equal to the reciprocal of $\delta P / \delta x$ in Eq. 3.1, namely,

$$\lambda = \frac{1}{N \sigma}. \tag{3.2}$$

Thus, the mean free path is inversely proportional to the magnitude of the cross section. While the mean free path can be defined for individual types of reactions, the concept is most suggestive when the cross section in Eq. 3.2 is the total cross section σ_T. If the medium has more than one nuclide, then the overall mean free path is found by replacing $N \sigma$ in Eq. 3.2 by the summation $\Sigma N_i \sigma_i$ over different species.

Total Cross Section

The *total cross section* σ_T for neutron reactions is the sum of the individual cross sections. If we limit consideration to the reactions considered above, then

$$\sigma_T = \sigma_{el} + \sigma_a = \sigma_{el} + \sigma_{in} + \sigma_\gamma + \sigma_f , \tag{3.3}$$

where σ_T is the sum of the elastic scattering cross section (σ_{el}) and the *absorption cross section* (σ_a), and the absorption cross section is the sum of the cross sections for inelastic scattering (σ_{in}), capture (σ_γ), and fission (σ_f).

---

[5]This is analogous to radioactive decay, as is seen more explicitly by rewriting Eq. 3.1 in the equivalent form: $-(dn/n) = N \sigma dx$, where n is here the number of neutrons in a neutron beam.

Competition Between Capture and Fission

For neutrons in a nuclear fuel, such as uranium or plutonium, the interesting reactions are fission and capture. (Elastic scattering changes the neutron energy only slightly, due to the high mass of the target, and has little importance.) If the goal is fission, as it is with ^{235}U or ^{239}Pu, capture has the effect of wasting neutrons. The ratio of the capture cross section to the fission cross section is therefore an important parameter. It is commonly denoted by the symbol α, defined as

$$\alpha(E_n) = \frac{\sigma_\gamma(E_n)}{\sigma_f(E_n)}, \tag{3.4}$$

where the dependence of the cross sections and of α on neutron energy, E_n, is shown explicitly.

3.1.3 Neutron Reactions in Different Energy Regions

As discussed in more detail in Section 5.1.1, a nuclear chain reaction is sustained by the emission of neutrons from fissioning nuclei. The neutrons emitted in fission have a broad energy spectrum, with the peak lying between several hundred keV and several MeV and with a typical central energy in the neighborhood of 1 MeV. Reactions take place at all neutron energies E_n, from several MeV down to a small fraction of 1 eV. We are interested in the main features of the reaction cross sections over this entire region.

Such information is obtained from experimental measurements, and extensive measurements have been carried out. We here will consider results for neutron reactions with ^{235}U, partly because of the importance of ^{235}U reactions in nuclear reactors and partly to illustrate general aspects of neutron-induced reactions. The illustrative graphs of cross sections vs. energy in Figs. 3.1, 3.3, and 3.4 are taken from a Brookhaven National Laboratory report [2].[6]

Experimentally measured cross sections for nuclear fission are shown in these figures for three somewhat distinct energy regions, in which the variation with energy of the cross sections differ significantly. There is no universal terminology to characterize the regions. We will here term them the *continuum* region, the *resonance* region, and the *low-energy* region. Although there are no clear boundaries, the continuum region corresponds to the higher neutron energies, from about 0.01 MeV to 25 MeV. The resonance region extends from about 1 eV to about 0.01 MeV. The low-energy region extends from zero energy up to about 1 eV.[7]

3.2 CROSS SECTIONS IN THE RESONANCE REGION

3.2.1 Observed Cross Sections

We start with a discussion of the intermediate-energy region, the resonance region, because phenomena observed here are pertinent to understanding characteristics of the cross sections at higher and lower energies. Fig. 3.1 shows fission cross sections in a section of this region

---

[6]These figures are copied from the 1965 second edition of report BNL-325. While not the latest available data, they bring out in clear fashion the key aspects of the cross sections.

[7]In other usage, what is here termed the continuum region is called the *fast neutron* region; what is here termed the low-energy region is divided into the *epithermal* and *thermal* regions.

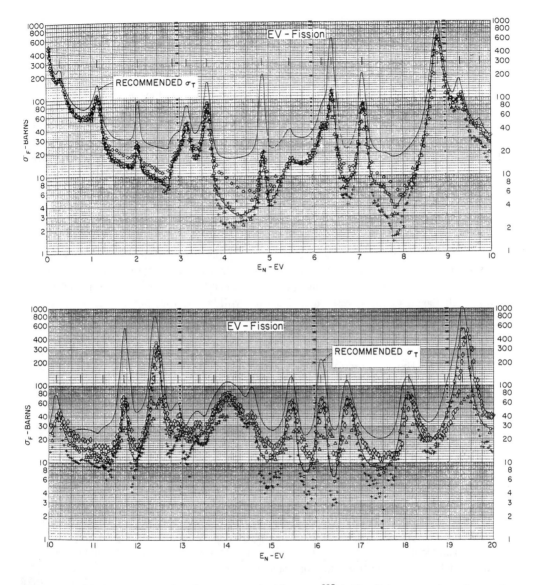

Figure 3.1. Fission cross section for neutrons incident on ^{235}U. The light curve gives the total cross section σ_T. (From Ref. [2], p. 92-235-40.)

for neutrons incident on ^{235}U. The cross section for neutron capture in ^{238}U is qualitatively similar in this energy region. The striking aspect of Fig. 3.1 is the rapid variations in cross section as a function of energy. For example, σ_f falls from 500 b at 19 eV to less than 10 b at 17.5 eV.

The high peaks are due to resonances in the neutron cross section. Resonances in neutron cross sections are analogous to the resonance absorption of light in a gas at wavelengths characteristic of the gas.[8] If the neutron energy matches the energy difference between the

---

[8]This is the phenomenon that produces the so-called Fraunhofer dark lines in the solar spectrum.

ground state of, say, ^{235}U and an excited state of ^{236}U, then there can be a large cross section for absorption.[9] The result of this absorption is a compound nucleus of ^{236}U. The number of possible resonance lines depends upon the number of available states in the compound nucleus. If the number is not too great, there will be separated, discrete resonance lines, as seen in Fig. 3.1.

In Fig. 3.1 the total cross section σ_T is plotted along with σ_f. In the energy region covered, the only important reactions are fission, neutron capture, and elastic scattering. Thus, $\sigma_T = \sigma_f + \sigma_\gamma + \sigma_{el}$. Over this region, σ_{el} is fairly constant, at about 10 b, while the magnitudes of σ_γ and σ_f and their ratio vary considerably. To give approximate cross sections for the resonance peak at $E_n = 18.05$ eV: $\sigma_f = 70$ b, $\sigma_\gamma = 50$ b, and $\sigma_{el} = 10$ b, for a total cross section $\sigma_T = 130$ b.

3.2.2 Shape of the Resonance Peak

If measurements are made of cross section vs. energy (as displayed in Fig. 3.1), peaks in the cross section do not get narrower as the neutron energy is more tightly defined. There is a certain inherent width to the resonance peak, which can be parametrized in terms of a *level width* Γ, characteristic of the resonance. Neutrons will be absorbed not only at some central energy E_0, but over a band of energies, with the absorption strongest in the region from $E_0 - \Gamma/2$ to $E_0 + \Gamma/2$.

For a single isolated resonance, the shape of this absorption peak is given by the *Breit–Wigner formula*. The cross section for formation of the compound nucleus σ_c, which we take here to be approximately equal to the absorption cross section σ_a, can be written in a simplified form as[10]

$$\sigma_a = \frac{C/\mathrm{v}}{(E-E_0)^2 + \Gamma^2/4}, \tag{3.5}$$

where E = energy of the incident neutron; E_0 = neutron energy at resonance; v = velocity of the incident neutron; C = a constant, for a given resonance; and Γ = the width for the level, different for each level.

If the neutron kinetic energy E is very close to the resonance energy E_0, the reaction cross section may be very large, up to hundreds of barns, and these neutrons are all very likely to be absorbed. Neutrons that differ in energy by several tenths of an eV, on the other hand, can have a much larger mean free path and escape absorption.

3.2.3 Level Widths and Doppler Broadening

The level width Γ, as seen from Eq. 3.5, is the full-width at half-maximum of the resonance peak. That is, when $E - E_0 = \Gamma/2$, the magnitude of the cross section is one-half of its maximum value. There is a minimum width for any level, the *natural* width Γ_n. It is related to the mean lifetime τ for the decay of the compound nucleus by the expression $\Gamma_n = h/2\pi\tau$, where h is Planck's constant. A typical width for an isolated resonance for neutrons interacting with ^{235}U or ^{238}U nuclei is about 0.1 eV.

---

[9]For both light and neutrons, strong absorption is not assured if the energies match, but it is made possible. In addition, other criteria must be satisfied, for example, those relating to the angular momenta of the states.

[10]See, e.g., Ref. [3] for simple forms of the Breit–Wigner formula as well as a more complete treatment.

The actual level width is increased over the natural width by the thermal motion of the uranium nuclei in the fuel at nonzero temperatures. This is called *Doppler broadening* because the broadening is due to the motion of the nuclei, analogous to the change in the observed frequency of sound or light due to the motion of the source or receiver. The Doppler broadening increases as the temperature of the medium increases, as illustrated schematically in Fig. 3.2.

The effect of Doppler broadening is to decrease the neutron-absorption cross section at the center of a resonance peak and to increase the cross section in the wings of the peak. At the center of a resonance E_0, the cross section is sufficiently large that almost all the neutrons with energy E_0 are absorbed. For example, if the cross section for neutron absorption in ^{238}U is 500 b, a value that is exceeded for a considerable number of resonances, the mean free path of a neutron in UO_2 fuel is under 0.1 cm. If the cross section is somewhat reduced at E_0, as happens with increased Doppler broadening, most of these neutrons are still absorbed.

However, in the wings of the resonance, at either side of the peak, the cross sections are low enough that many neutrons are not absorbed. At these energies, the increase in the cross section with increasing temperature (see Fig. 3.2), means a greater chance of neutron absorption. Summed over all neutron energies, the overall effect of the increased Doppler broadening at higher temperatures is to increase the total resonance absorption in the fuel. This has important implications for reactor safety (see Section 11.2.2).

3.3 CROSS SECTIONS IN THE CONTINUUM REGION

For neutrons with energies in excess of tens of kilovolts, the accessible states in the compound nucleus are so numerous that their spacing is small compared to their width. Thus the levels overlap. A compound nucleus is still formed, but it involves a continuum of overlapping states. The cross section varies relatively smoothly with energy, without the rapid changes characteristic of the resonance region. The cross section for fission in ^{235}U is displayed for this region in Fig. 3.3.

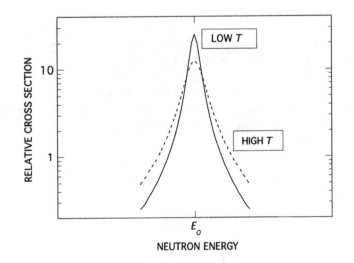

Figure 3.2. Schematic representation of Doppler broadening of Breit–Wigner resonance at high temperature: Cross section as a function of neutron energy at high and low temperatures.

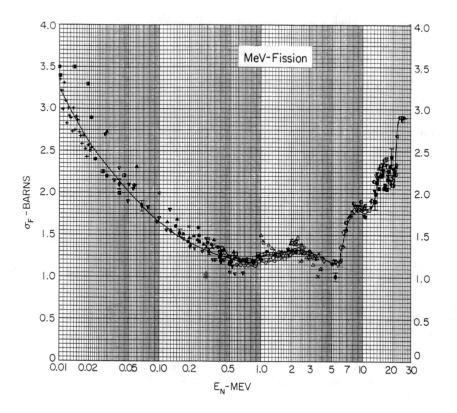

Figure 3.3. Fission cross section for neutrons incident on ^{235}U. (From Ref. [2], p. 92-235-47.)

For 1-MeV neutrons incident upon ^{235}U, the main decay modes of the ^{236}U compound nucleus are fission, emission of a neutron to an excited state of ^{235}U (inelastic scattering), and γ-ray emission (capture). The cross sections for these reactions at 1 MeV are approximately σ_f=1.2 b, σ_{in}=1.1 b, and σ_γ=0.11 b, respectively [2]. The sum of the cross sections for the possible reactions (excluding elastic scattering) is 2.4 b, close to the geometric cross section cited in Section 3.1.2 for the neighboring nucleus, ^{238}U.

At somewhat lower energies, the relative importance of the different decay modes changes. For example, at E_n=0.01 MeV, σ_f=3.2 b and σ_γ=1.15 b. The cross section for inelastic scattering is very small, in part because of the lack of suitable low-lying excited states of ^{235}U. The total absorption cross section $\sigma_f+\sigma_\gamma$ is here 4.4 b, well above the geometric cross section.

3.4 THE LOW-ENERGY REGION

3.4.1 Low-Energy Region and the 1/v Law

For slow neutrons, below 1 eV, one is interested in variations of cross section with energy over a very small energy region. For example, from 0.1 eV to 0.025 eV, the neutron velocity drops by a factor of two, but the energy change may be small compared to the width of a

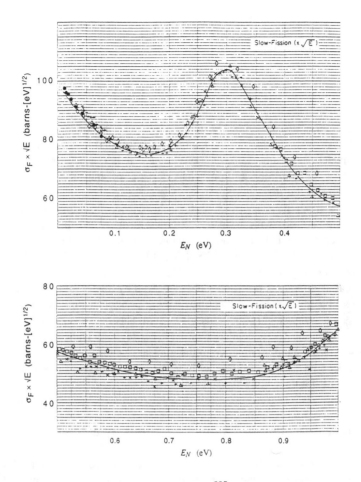

Figure 3.4. Fission cross section for neutrons incident on ^{235}U, 0.01 eV to 1.0 eV. The ordinate is the product $\sigma_f E^{0.5}$, where E is in eV. (From Ref. [2], p. 92-235-32.)

resonance, typically of the order of 0.1 eV.[11] The behavior of the cross section over so narrow a region is often dominated by the tail of a single resonance. If the denominator in Eq. 3.5 does not change appreciably over this interval, the cross section is proportional to $1/v$, where v is the neutron velocity.

The $1/v$ dependence is contingent upon the presence of an appropriately located resonance (or resonances), and it characterizes many, although not all, neutron-absorption cross sections at very low energies. It is exhibited in Fig. 3.4, which shows the fission cross section for ^{235}U below 1 eV. Were the cross section σ itself plotted, it would exhibit a steady rise as E_n falls, proportional to $1/v$. Instead, the curve in Fig. 3.4 is plotted with the ordinate as $\sigma_f \times E^{0.5}$. There is not much more than a factor of two difference between the highest and lowest value of the ordinate in Fig. 3.4, although the energy varies by a factor of almost 100 and the velocity by a factor of almost 10. Thus, the $1/v$ dependence of σ holds reasonably well. If it held perfectly, the curves in Fig. 3.4 would be horizontal lines.

---

[11]While the discussion has been couched in terms of a single resonance, a $1/v$ dependence can also occur with several resonances contributing to the cross section.

The cross section itself can be found from Fig. 3.4 by dividing the observed ordinate by $\sqrt{E_n}$, where E_n is the neutron energy in electron volts. At 1 eV, σ_f=65 b, while at 0.0253 eV, $\sigma_f = 92/\sqrt{0.0253} \doteq 580$ b.

3.4.2 Thermal Neutrons

In many reactors, a succession of elastic collisions with the nuclei of the so-called moderator (see Section 5.2) reduces the neutron energies to lower and lower values, until the motion of the target nuclei of the moderator cannot be neglected. The neutron energies then approach a Maxwell–Boltzmann distribution, characteristic of the temperature of the fuel and moderator. This is the *thermal* region.[12]

So-called thermal neutron cross sections have been measured in the laboratory using monoenergetic neutrons with a velocity v of 2200 m/sec. At this velocity, E_n=0.0253 eV. This energy corresponds to the most probable velocity of a Maxwell–Boltzmann velocity distribution, where $T=E_n/k=0.0253\times(1.602\times10^{-19})/1.381\times10^{-23}=293$ K. Thus, v here characterizes thermal neutrons, at T=20 °C.

The actual neutrons in a reactor are characterized by higher temperatures, roughly 300 °C, and are distributed in energy rather than having a single value. Nonetheless, use of these nominal thermal-energy cross sections gives quite accurate results. This follows, somewhat fortuitously, from the $1/v$ dependence of the cross section. Thus, in subsequent chapters, we will follow what is standard practice in the nuclear reactor literature, namely, to characterize thermal neutron interactions in the fuel in terms of the cross sections at a single energy. The justification for this procedure is sketched in Section 3.4.3.

3.4.3 Characterization of Thermal Neutrons by a Single Energy

We here demonstrate that the total reaction rate for neutrons with an arbitrary energy (or velocity) distribution equals the reaction rate for monoenergetic neutrons, if the neutron cross section is inversely proportional to the neutron velocity: $\sigma \propto 1/v$.[13] Consider neutrons with a velocity distribution $n(v)$, i.e., $n(v)dv$ is the number of neutrons per unit volume with a velocity in the interval from v to $v+dv$. The rate of reactions of neutrons with nuclei in a homogeneous target is[14]

$$r=\int_0^\infty Nvn(v)\sigma(v)dv, \tag{3.6}$$

where N is the number of target nuclei per unit volume and $\sigma(v)$ can be expressed as

---

[12]Thermalization of the neutrons is not complete because absorption of neutrons removes some of them before equilibrium is reached. Thus, there is an excess high-energy tail in the actual velocity or energy distribution, lying above the exact Maxwell–Boltzmann spectrum (and a suppression of low-energy neutrons) [4, p. 332 ff.].

[13]The remainder of this section is a rather technical justification of the approximation of a $1/v$ dependence of the cross section, and is not necessary for the understanding of subsequent chapters.

[14]We are here speaking as if the target nuclei are at rest. This is not true in thermal equilibrium, where all nuclei are in motion. However, if the target has high atomic mass M, such as for U or Pu, it is a reasonably good description, because at fixed temperature the average velocity of nuclei varies as $M^{-0.5}$. For any targets of any mass, the present analysis is valid if v is interpreted to be the *relative* velocity of the neutron and target.

[15]For a more detailed discussion of the interplay between the velocity dependence of the cross section and the shape of the thermal spectrum, see Ref. [5], pp. 45–52. This reference uses the notation $\hat{v}$ and (in the absence of Westcott parameter corrections) $\hat{\sigma}$ for what are called v_0 and $\sigma(v_0)$ above.

$$\sigma(v) = \frac{v_0}{v}\,\sigma(v_0)\,. \tag{3.7}$$

As discussed above, in practice σ_0 is found at $v_0 = 2200$ m/sec. Substituting the expression for $\sigma(v)$ into Eq. 3.6, the $1/v$ dependence cancels v in the integrand, and the reaction rate becomes

$$r = N v_0 \sigma(v_0) \int_0^\infty n(v)\,dv = n N v_0 \sigma(v_0)\,, \tag{3.8}$$

where n is the number of neutrons per unit volume. If n and N are expressed in m^{-3}, v_0 in m/sec, and σ in m^2, then r will be in units of m^{-3}-sec^{-1}.

The expression on the right-hand side of Eq. 3.8 is the reaction rate for monoenergetic neutrons of velocity v_0. Thus, the use of the nominal thermal values measured at $v_0 = 2200$ m/sec ($E = 0.0253$ eV) gives the rate of neutron reactions for any energy (or velocity) distribution, as long as the $1/v$ dependence of cross sections holds over the energies of the distribution. It does not matter that the actual temperature in the reactor is about twice the temperature at the nominal thermal energy or that there is a broad distribution in neutron energies.[15]

Departures from the $1/v$ dependence of neutron cross section and from a thermal distribution of neutron energies cannot be neglected entirely if accurate calculations are desired. Correction factors to take into account these departures from the idealized behavior are known as Westcott parameters, and tables of these parameters are found in standard nuclear engineering references (see, e.g., Refs. [5] and [6]). However, these corrections are not large in the cases of greatest interest and will be ignored in the qualitative discussions below. In nonthermal reactors, where the $1/v$ dependence does not describe the neutrons, none of these simplifications is relevant, and it is necessary to examine $\sigma(E)$ in detail.

REFERENCES

1. David Halliday, Robert Resnick, and Kenneth S. Krane, *Physics, Volume Two, Extended Version*, Fourth Edition (New York: John Wiley, 1992).
2. J. R. Stehn, M. D. Goldberg, R. Wiener-Chasman, S. F. Mughabghab, B. A. Magurno, and V. M. May, *Neutron Cross Sections, Volume III, Z=88 to 98*, report no. BNL-325, 2nd ed., supplement no. 2 (Upton, N.Y.: Brookhaven National Laboratory, 1965).
3. Emilio Segrè, *Nuclei and Particles*, 2nd ed. (Reading, Mass.: W. A. Benjamin, 1977).
4. Alvin M. Weinberg and Eugene P. Wigner, *The Physical Theory of Neutron Chain Reactors* (Chicago: University of Chicago Press, 1958).
5. Manson Benedict, Thomas H. Pigford, and Hans Wolfgang Levi, *Nuclear Chemical Engineering*, 2nd ed. (New York: McGraw-Hill, 1981).
6. John R. Lamarsh, *Introduction to Nuclear Engineering*, 2nd ed. (Reading, Mass.: Addison-Wesley, 1983).

Chapter 4

Nuclear Fission

4.1 DISCOVERY OF FISSION

Prior to the actual discovery of the neutron in 1932, it had long been suspected, in part due to a suggestion by Ernest Rutherford in 1920, that a heavy neutral particle existed as a constituent of the atomic nucleus.[1] The discovery of the neutron, as with many of these early discoveries, was accidental. In the bombardment of beryllium by alpha particles from a naturally radioactive source, a very penetrating radiation had been observed in experiments begun in 1928. In view of its penetrating power, the radiation was first thought to be gamma radiation, but the radiation was found to eject energetic protons from paraffin, which would not have been possible with gamma rays.

While in retrospect the explanation is obvious, it was not until early 1932 that James Chadwick demonstrated that the radiation was made up of particles (already recognized to be neutral) with a mass close to that of a proton. The ejected protons were the recoil products of elastic scattering between these particles and hydrogen nuclei in the paraffin. Thus, the previously suspected neutron had been found. The reaction responsible for the neutron production was

$$^4\text{He} + {}^9\text{Be} \rightarrow {}^{12}\text{C} + \text{n}.$$

With a source of neutrons readily available, it was natural to explore what reactions the neutrons might induce in collisions with nuclei of various elements. There was particular interest in doing this because, while protons and alpha particles of the energies then obtainable had too little kinetic energy to overcome the coulomb repulsion of the nuclei of heavy elements, there would be no coulomb force acting upon the neutrons. Enrico Fermi in Italy quickly became the leader in this work, undertaking a systematic program of bombardment

---

[1] For a brief account of this history, see, e.g., Refs. [1] and [2].

by neutrons of nuclei throughout the periodic table.

The bombardment of uranium, undertaken in 1934, proved to be particularly interesting because many new radioactive products were discovered. Fermi at first believed that this was due to the formation of a transuranic element (atomic mass number $A = 239$) through neutron capture in ^{238}U, followed by a chain of radioactive decays. In fact, when the decision was made to award the 1938 Nobel Prize in physics to Fermi for his discoveries of radioactivity produced with neutrons, it was still thought that the most interesting of these were transuranic elements and their radioactive progeny.[2] Efforts were made to isolate the hypothesized new elements chemically, but it proved impossible to demonstrate properties that differed from those of known lighter elements.

Finally, in 1938 Otto Hahn and Fritz Strassmann in Germany demonstrated that a residue of neutron bombardment of uranium that had previously been thought to be radium (atomic number $Z = 88$), a presumed decay product of the radioactive chain, was in fact barium ($Z = 56$). This result was communicated by them to Lise Meitner, a former colleague of Hahn who was then a refugee in Sweden, and she and her nephew Otto Frisch formulated the first overall picture of the fission process. From known nuclear binding energies, it was clear that there should be a large accompanying energy release, and Frisch quickly demonstrated this release experimentally. Thus, fission was hypothesized and demonstrated in late 1938 and early 1939.

The potential practical implications of fission were immediately recognized, and work on fission quickly intensified in Europe and the United States. The first detailed model of the fission process was developed by Niels Bohr and John Wheeler, during a visit by Bohr to Princeton University in early 1939, the so-called liquid drop model. The essential features of this model are still accepted.

Among the early insights from this work was the explanation of the puzzling fact that fission was copiously produced in natural uranium by both very slow neutrons (under about 0.1 eV in energy) and moderately fast neutrons (above about 1 MeV), but not to any appreciable extent by neutrons of intermediate energy (say, 0.5 MeV). Bohr recognized, in what Wheeler suggests was a rather sudden inspiration, that the slow neutrons cause fission in the relatively rare isotope ^{235}U, for which the fission cross section is high.[3] Fast neutrons (above one MeV) are required to produce fission in the abundant isotope ^{238}U.

4.2 SIMPLE PICTURE OF FISSION

4.2.1 Coulomb and Nuclear Forces

The energy available for fission (or other processes) within the nucleus is the result of the combined effects of the coulomb force between the protons, which is repulsive, and the nuclear force between nucleons (neutrons or protons), which is attractive. In a stable nucleus, the nuclear force is sufficiently strong to hold the nucleus together. In alpha-particle emitters, the coulomb repulsion causes eventual decay, but on a slow time scale. Fission represents a more extreme "victory" for the coulomb force.

The fact that fission occurs for heavy nuclei follows from the dependence of the forces and the corresponding potential energies on nuclear charge and mass. For the repulsive coulomb force, the potential energy is positive and increases as Z^2/R, where the nuclear radius R is

---

[2]The suggestion by the German chemist Ida Noddack that lighter elements were being produced was largely ignored because she lacked the eminence and the unambiguous evidence to gain credence for her radical suggestion.

[3]A brief account of this history is given in Ref. [3].

proportional to $A^{1/3}$. The atomic number Z is roughly proportional to the atomic mass number A, and the coulomb potential energy therefore rises roughly as $A^{5/3}$.

For the attractive nuclear force, the potential energy is negative. In light of the saturation of nuclear forces, which limits the number of nucleons with which each nucleon interacts, the magnitude of the nuclear potential energy is proportional to A. Thus the ratio of the magnitudes of the coulomb and nuclear potential energies rises with increasing A, and eventually the coulomb force dominates. This limits the maximum mass of stable nuclei.

The existence of ^{235}U and ^{238}U in nature is evidence that the balance between their nuclear and coulomb forces is not such as to cause instability, beyond alpha-particle decay at a low rate. However, the fact that no elements of higher atomic number are found in nature suggests that these uranium isotopes are "vulnerable," and that fission is possible if they are sufficiently excited. As described in the liquid drop model, the excitation can take the form of a shape oscillation, in which the nucleus deforms into a dumbbell shape (where the two ends of the dumbbell are usually not of equal size). The coulomb repulsion between these two parts may be sufficient to overcome the remaining attraction from the nuclear force, and the parts move further apart, culminating in fission.

4.2.2 Separation Energies and Fissionability

The striking qualitative fact that thermal neutrons can produce fission in ^{235}U but not in ^{238}U is explained by rather small differences in the excitation energies of the nuclei ^{236}U and ^{239}U, produced when neutrons are absorbed in ^{235}U and ^{238}U, respectively. Consider, for example, the neutron-capture process in which the excited ^{236}U nucleus is formed, followed by gamma-ray emission to bring ^{236}U to its ground state:

$$n + {}^{235}U \rightarrow {}^{236}U^* \rightarrow {}^{236}U + \gamma\text{'s}.$$

Here, the energy carried off by the gamma rays is equal to the excitation energy E of the excited compound nucleus ^{236}U*. It is given by the difference in the masses of the initial and final constituents.[4] Assuming the kinetic energies to be zero (or negligible), the excitation energy E equals the energy required to remove a neutron from ^{236}U to form ^{235}U+n, the so-called *separation energy S*.

In terms of the atomic masses M and the mass of the neutron m_n, these energies are

$$E = S = [M(^{235}U) + m_n - M(^{236}U)]c^2,$$

where c is the velocity of light. Expressing masses in atomic mass units (u) and noting that 1 u corresponds to 931.50 MeV:

$$E = S = (235.043925 + 1.008665 - 236.045563) \times 931.5 = 6.55 \text{ MeV}. \tag{4.1}$$

The meaning of Eq. 4.1 can be stated as follows: (a) The ^{236}U excitation energy E following the capture of a thermal neutron in ^{235}U is 6.55 MeV; or (b) To separate a neutron from a ^{236}U nucleus (originally in its ground state) and form n+^{235}U, an energy must be supplied that is at least as great as the separation energy S of 6.55 MeV.

For the capture reaction described above, the excitation energy is carried off by gamma rays. However, the absorption of thermal neutrons can in some cases lead to fission. For ^{235}U, in particular, fission is the dominant process. In general, thermal neutron absorption leads to

---

[4] In the standard nuclear terminology, this is the *Q value* of the reaction.

fission if the excitation energy is sufficiently high. Relevant excitation energies are listed in Table 4.1 for thermal neutron absorption in a number of nuclei of interest.

If an appreciable fission yield is produced by thermal neutrons incident upon a target nucleus, that nuclear species is termed *fissile*. It is seen in Table 4.1 that the fissile nuclei are all odd-A nuclei, and the resulting compound nuclei all have excitation energies exceeding 6.3 MeV. For the nonfissile nuclei, the excitation energies are about 5.0 MeV. However, some of the nonfissile nuclei can play an important role in nuclear reactors. Neutron capture in ^{232}Th and ^{238}U leads, with intervening β^- decays, to the production of the fissile nuclei ^{233}U and ^{239}Pu. Therefore, ^{232}Th and ^{238}U are termed *fertile*.

The question arises, How much excitation energy is required for fission to occur? An alternate means of exciting nuclei is by bombardment with gamma rays. If fission then occurs, it is called *photofission*, to distinguish the process from the more common neutron-induced fission. It was early found that the thresholds for photofission in uranium and thorium isotopes lie in the range from 5.1 to 5.4 MeV [4, p. 492]. This means that an excitation energy between about 5.1 and 5.4 MeV is required for photofission. For all fissile nuclei, the excitation energies are higher than the photofission threshold. For the nonfissile nuclei, they are below this threshold.

The association between high separation energies and odd-A target nuclei (i.e., even-A compound nuclei) results from a special property of nuclear forces, the pairing energy. In nuclei where both Z and A are even (for example, ^{234}U, ^{236}U, and ^{240}Pu), there are even numbers of both protons and neutrons. There is a stronger binding for members of a neutron pair than for unpaired neutrons (as in ^{239}U, for example). Thus, the neutron separation energy is high for these even–even nuclei, and their excitation energy is high when they are formed by neutron capture. The fact that all fissile target nuclei are odd-A nuclei is not an accident, but rather a manifestation of pairing interactions in nuclei.

4.2.3 Fission Cross Sections with Fast and Thermal Neutrons

For orientation purposes, the cross sections for fission in ^{235}U, ^{238}U, and ^{239}Pu are plotted in Fig. 4.1. (The data are extracted from detailed plots given in Ref. [5].) In the thermal region, the cross sections in ^{235}U and ^{239}Pu are hundreds of barns. In general, they then fall with increasing energy, with very rapid fluctuations in the resonance region. In the 1 to 10 MeV region, the fission cross sections are in the vicinity of 1 or 2 b for these nuclei.

The cross section for ^{238}U fission is negligible at low neutron energies. For neutrons with energies of several eV, σ_f is typically in the neighborhood of 10^{-5} to 10^{-4} b [5], and it is still under 0.02 b at $E_n = 1.0$ MeV. However, it then rises rapidly to a plateau of about 0.6 b,

Table 4.1. *Excitation energies E^* for thermal neutron capture.*

Target Nucleus	Compound Nucleus	Excitation Energy (MeV)	Fissile Target?
^{232}Th	^{233}Th	4.79	No
^{233}U	^{234}U	6.84	Yes
^{235}U	^{236}U	6.55	Yes
^{238}U	^{239}U	4.81	No
^{239}Pu	^{240}Pu	6.53	Yes
^{240}Pu	^{241}Pu	5.24	No
^{241}Pu	^{242}Pu	6.31	Yes

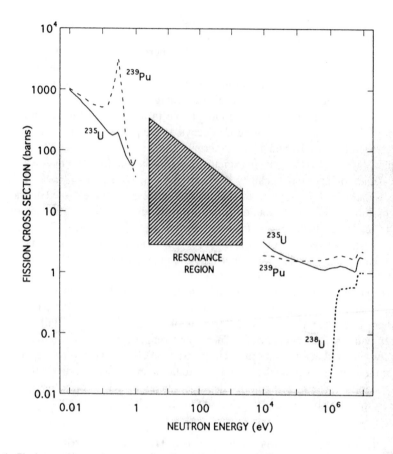

Figure 4.1. Fission cross sections as a function of neutron energy E_n for neutrons on ^{235}U, ^{238}U, and ^{239}Pu. The shaded region is a schematic of the resonance region where the cross section varies rapidly; the boundaries of the shaded region are *not* limits on the peaks and valleys of the resonance cross sections.

which extends from 2 to 6 MeV. In this region, σ_f in ^{238}U is about one-half the magnitude of σ_f in ^{235}U. Thus, while ^{238}U is not "fissile" in the sense that the term is used—namely, having a significant fission cross section for thermal neutrons—the ^{238}U fission cross sections become appreciable for neutron energies above about 1 MeV.

4.3 PRODUCTS OF FISSION

4.3.1 Mass Distribution of Fission Fragments

The fission reaction, again taking the example of ^{235}U, can be written in the somewhat general form[5]

$$n + {}^{235}U \rightarrow {}^{236}U^* \rightarrow (A_1, Z_1) + (A_2, Z_2) + Nn + \gamma\text{'s}.$$

---

[5]We here ignore relatively rare additional particles, most notably alpha particles.

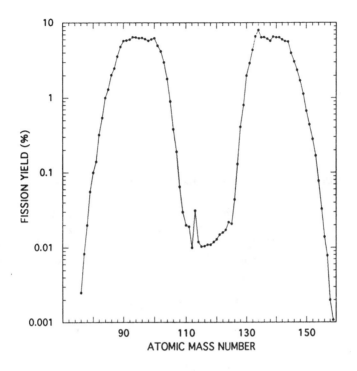

Figure 4.2. Yield of fission fragments as a function of atomic mass number A for thermal fission of ^{235}U (in percent per fission).

This expression represents the fission of ^{236}U into two fragments (with atomic mass numbers A_1 and A_2) and N neutrons. From conservation of charge and of number of nucleons, $Z_1+Z_2=92$ and $A_1+A_2+N=236$. N is a small number, typically 2 or 3. If $A_1=A_2$, or nearly so, the fission process is called *symmetric fission*. Otherwise, it is *asymmetric fission*. The relative yields of fission fragment nuclei of different atomic mass number are plotted in Fig. 4.2.[6]

As seen in Fig. 4.2, thermal fission of ^{235}U leads overwhelmingly to asymmetric fission.[7] The fission yield is dominated by cases where one fragment has a mass number A between about 89 and 101, and the other has a mass number between about 133 and 144. The fission yields, namely the fraction of the events that produce nuclei of a given A, lie between about 5% and 7% in this region, with a single anomalously high value at $A=134$. For symmetric fission (A of about 116), the fission yield at a given value of A is only about 0.01%. The integral of the area under the curve of Fig. 4.2 is 200%, as there are two fragments per fission.

Although in most cases the massive products from the breakup of the fissioning nucleus are the two fission fragments and neutrons, there is also a small possibility of emission of other nuclei, especially alpha particles. Alpha-particle emission, with energies up to 20 MeV, occurs in 0.2% of fission events [7, p. 590]. While of some interest in its own light, this

---

[6]Data for Fig. 4.2 are taken from Table 2.9 of Ref. [6].

[7]This fission process for neutrons incident on ^{235}U is commonly referred to as "fission of ^{235}U," although of course the fissioning nucleus is ^{236}U. Usually this does not lead to ambiguity, because the meaning is clear from the context.

contribution is too small to be of any importance in considering nuclear reactors.

The fission fragment yields for other fissile targets, in particular ^{233}U and ^{239}Pu, are similar to those in Fig. 4.2, although the minima are not quite as deep and there are small displacements in A due to the differences in nuclear mass number. Fast fission yields also have the same general character, for both fissile and fertile targets [6]. Again, asymmetric fission dominates, with peak yields primarily in the range between 5% and 7%, although there are some exceptions and differences in detail.

4.3.2 Neutron Emission

Number of Neutrons as a Function of Incident Neutron Energy

The average number of neutrons emitted in fission ν is a crucial parameter in establishing the practicality of a chain reaction. The number of neutrons varies from event to event, ranging from 0 to about 6 [7, p. 587]. Typically, the number is 2 or 3. Because of its importance, the magnitude of ν has been determined with high precision. It increases with neutron energy and can be described approximately by the relations [8, p. 61]

$$^{235}\text{U}: \nu = 2.432 + 0.066E, \quad \text{for } E < 1 \text{ MeV}$$

$$^{235}\text{U}: \nu = 2.349 + 0.15E, \quad \text{for } E > 1 \text{ MeV}$$

$$^{239}\text{Pu}: \nu = 2.874 + 0.138E, \quad \text{for all } E,$$

where the incident neutron energy E is in MeV. It is seen that ν is significantly higher for ^{239}Pu than for ^{235}U; e.g., 3.0 vs. 2.5, at 1 MeV.

Energy Spectrum of Neutrons

The neutrons are emitted with a distribution of energies, with a relatively broad peak centered below 1 MeV. Several approximate semiempirical expressions are used to represent the actual neutron spectrum. One, which is in a form that lends itself easily to calculation, is (see, e.g., Ref. [7], p. 586)

$$f(E) = 0.770\sqrt{E}e^{-0.775E}, \tag{4.2}$$

where E is the neutron energy in MeV, and $f(E)dE$ is the fraction of neutrons emitted with energies in the interval E to $E+dE$. An alternative expression for $f(E)$ has the slightly different form (see, e.g., Ref. [9], p. 75)

$$f(E) = 0.453e^{-1.036E} \sinh[(2.29E)^{0.5}], \tag{4.3}$$

where again E is in MeV.

The spectra corresponding to Eqs. 4.2 and 4.3 are plotted in Fig. 4.3. It is seen that there is little difference between the curves for the two expressions. In both cases, there is a peak at roughly 0.7 MeV. At $E=1$ MeV, $f(E)$ is only slightly below its peak value, and at $E=2$ MeV it is still above 60% of the peak value. The median neutron energy is about 1.5 MeV. There is a substantial tail extending to higher energies, and the "average" neutron energy is commonly taken to be about 2 MeV.

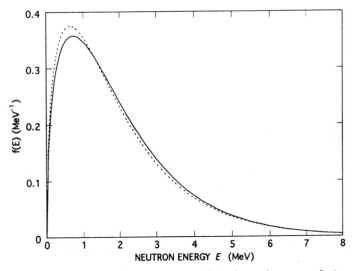

Figure 4.3. Energy spectrum $f(E)$ of neutrons emitted in thermal neutron fission of ^{235}U. The dotted curve corresponds to Eq. 4.2 and the solid curve to Eq. 4.3.

Delayed Neutrons

Most of the neutrons produced in fission are emitted promptly, essentially simultaneously with the fission itself. However, some neutrons are emitted from nuclei produced in the β^- decay of the original fission fragments. These are called *delayed neutrons*. The delay arises from the beta-decay process and has a time scale determined by the β^--decay half-life of the precursor to the neutron emitter. The neutron emission itself follows virtually immediately after the β^- decay.

An example of delayed neutron emission occurs at $A=87$. Bromine 87 (^{87}Br, $Z=35$) is formed in about 2% of ^{235}U fission events. Its β^- decay, with a half-life of 56 sec, proceeds to a number of different states of krypton 87 (^{87}Kr, $Z=36$) In 2.3% of the cases, the resulting ^{87}Kr state decays by neutron emission to ^{86}Kr. In aggregate, for every 100 ^{235}U fission events, there are 0.05 delayed neutrons with a ^{87}Br precursor. Considering all delayed neutron paths, with other precursors, there are 1.58 delayed neutrons per 100 fissions [7, p. 589]. Given an average of 2.42 prompt neutrons per fission, this means that the fraction of fission neutrons that are delayed, conventionally designated as β, is about 0.0158/2.42. Thus, for ^{235}U, about 0.65% of fission neutrons are delayed neutrons. The precursors for these neutrons have half-lives ranging from 0.2 to 56 sec, with the largest group having a half-life of about 2 sec.

This might seem to be a small point—either a small detail introduced for completeness or a minor nuisance. It is neither. As discussed in Section 5.3.3, the existence of these delayed neutrons is crucial for the stable operation of nuclear reactors.

4.3.3 Decay of Fission Fragments

Gamma-Ray Emission from Excited Nuclei

If the initial fission fragments are formed in excited states, as is usually the case, they will decay by gamma-ray emission to the ground state before beginning a β^--decay chain.

Although the half-lives for β^- decay are typically seconds or longer, gamma decay usually occurs within about 10^{-14} seconds of the fission event. The gamma rays have an average energy of about 1 MeV, and overall about 7 MeV are emitted in the form of gamma rays [7, p. 590]. In a reactor, this energy is almost entirely recovered because the gamma rays are absorbed in a distance small compared to the dimensions of the reactor.

Beta Decay

For the stable nuclei that exist in nature with atomic mass number up to $A=40$, the ratio Z/A is close to 1/2. Beyond this, as one goes to heavier nuclei, Z/A gradually decreases. In fission, aside from the few emitted neutrons, the average value of Z/A of the initial fragments must be the same as that of the fissioning nucleus. Thus, for their mass region, the fission fragments will be proton-poor, or equivalently, neutron-rich. For example, if we consider the nucleus at $A=234/2=117$, the stable nucleus is tin 137 (^{117}Sn, $Z=50$), while the fragments of symmetric fission of ^{235}U (with the emission of 2 neutrons) are two palladium 117 (^{117}Pd, $Z=46$) nuclei. Each of the ^{117}Pd fragments undergoes four β^- decays to reach the stable ^{117}Sn nucleus.

As a more typical example, we can consider the fission event used previously as an illustrative case:

$$n + {}^{235}U \rightarrow {}^{144}_{56}Ba + {}^{89}_{36}Kr + 3n.$$

This is followed by the β^--decay chains (half-lives in parentheses):[8]

$$A=144: \quad Ba \ (11 \ s) \rightarrow La \ (41 \ s) \rightarrow Ce \ (285 \ d) \rightarrow Pr \ (17 \ m) \rightarrow Nd$$

$$A=89: \quad Kr \ (3 \ m) \rightarrow Rb \ (15 \ m) \rightarrow Sr \ (51 \ d) \rightarrow Y.$$

The "final" products are the (virtually) stable nuclei $^{144}_{60}$Nd and $^{89}_{39}$Y.

Some qualifications should be made on the designations "final" and "stable." Neodymium 144 (^{144}Nd) is not truly stable. It has a half-life of 2.3×10^{15} years. Of course, this is stable enough for any practical purposes. A much more significant point is illustrated by the $A=144$ chain, in particular involving cerium 144 (^{144}Ce), with a half-life of 285 days. Only part of the ^{144}Ce decays before the fuel is discharged from the reactor, which on average takes place one or two years after the fission event. Thus only part of the ^{144}Ce decay energy should be included in considering the energy budget of the reactor. Further, the remaining activity will appear in the radioactivity inventory of the spent fuel. More generally, although most of the radioactive fission products decay quickly enough to have disappeared before the spent fuel is removed, the longer-lived species remain. This is the origin of the nuclear waste disposal problem.[9]

---

[8]We here use the common abbreviations s=seconds, m=minutes, and d=days.
[9]Actually, nuclides with half-lives of less than a year, such as ^{144}Ce, are not the major problem, because they are essentially gone after a decade or two. It is the longer-lived nuclides, with half-lives of many years, that create the demand for repositories that will be safe for thousands of years.

4.4 ENERGY RELEASE IN FISSION

4.4.1 Energy of Fission Fragments

The energy release in fission can be calculated in two ways: (a) from the mass differences and (b) from the coulomb potential. Method (a) gives the more accurate result. Method (b) is the more physically instructive in terms of mechanism. We shall perform the calculation, using both methods, for the same case that we have been citing:

$$n + {}^{235}U \rightarrow {}^{144}_{56}Ba + {}^{89}_{36}Kr + 3n.$$

In terms of the atomic masses, the energy release is

$$Q = [M({}^{235}U) - M({}^{144}Ba) - M({}^{89}Kr) - 2m_n]c^2 = 173 \text{ MeV},$$

where the atomic masses M of ${}^{235}U$, ${}^{144}Ba$, ${}^{89}Kr$, and the neutron are 235.0439 u, 143.9228 u, 88.9177 u, and 1.0087 u, respectively.

Thus, on the basis of mass differences, the energy carried by the fission fragments and neutrons is 173 MeV. For the fission fragments alone, it is about 167 MeV. This is for only one particular fission channel, but it is reasonably representative. The calculation tacitly assumes that the fragments are formed in their ground state. However, they are commonly formed in excited states, and some of the available energy is carried off by de-excitation gamma rays rather than particle kinetic energy. But either way, the energy is part of the total energy budget.

A rough alternate calculation of the energy release can be made from consideration of the coulomb repulsion, with the aid of some gross simplifications, including the approximation that the entire energy release is given by the final kinetic energy of the two fragments. We assume that at some stage in the fission process the two fragments have become spheres that are just touching, and we neglect the energy changes in getting to this point. We further assume that at this point, the nuclear force between the fragments will have become vanishingly small and that one only need consider the coulomb repulsion between them. If we take the nuclear radius to be given by $r = r_0 A^{1/3}$, with $r_0 = 1.4 \times 10^{-13}$ cm, then the radii of the ${}^{144}Ba$ and ${}^{89}Kr$ nuclei are 7.34×10^{-13} cm and 6.25×10^{-13} cm, respectively, for a total separation between the centers of these fragments of $R = 13.6 \times 10^{-13}$ cm.

The potential energy for two uniformly charged spheres, carrying charges Q_1 and Q_2 and with centers separated by a distance R, is

$$PE = \frac{1}{4\pi\epsilon_0} \frac{Q_1 Q_2}{R}, \tag{4.4}$$

where $\epsilon_0 = 8.85 \times 10^{-12}$ (in SI units). Given that $Q = Ze$, Eq. 4.4 gives the result that $PE = 3.42 \times 10^{-11}$ J $= 214$ MeV. All of this potential energy is converted into kinetic energy of the fission fragments, and thus the fission fragment energy is about 214 MeV.

This result is about 25% greater than the much more accurate result from mass excesses. Nonetheless, there is a rough agreement between the results of the two estimates. It is impressive that so simple a model captures enough of the essential features of the fission process to give an approximately correct result.

4.4.2 Total Energy Budget

The total energy release in the fission of ^{235}U can be divided into two components. We list these below and give energies associated with each contribution, averaged over the various possible outcomes of the fission process (from Ref. [7], p. 591):

1. *The energy released in the initial fission process (180.5 MeV).* As discussed above, this energy can be determined from the mass differences. Contributions include

- The kinetic energy of the fission fragments (168.2 MeV)
- The kinetic energy of the prompt neutrons (4.8 MeV)
- The energy carried off by prompt gamma rays (7.5 MeV)

2. *The energy released in the decay of the fragments to more stable nuclei (14.6 MeV).* This component includes

- The kinetic energy of the emitted β^- rays (7.8 MeV), which is almost all recovered in the reactor
- The energy of the neutrinos emitted in β^- decay (12 MeV), which is lost to the reactor. It is somewhat greater than the beta-particle energy because, by conservation of momentum, the neutrino typically carries away more than half of the available energy in beta decay.
- The energy of the gamma rays accompanying β^- emission (6.8 MeV), which is almost all recovered in the reactor
- The energy of the delayed neutrons, which is negligible because delayed neutrons represent only a small fraction of all neutrons

For ^{235}U, the sum of the contributions above gives an average total energy recovery of 195 MeV per fission event. For ^{239}Pu, the corresponding energy is 202 MeV [7, p. 591].

This is not quite a complete accounting of the energy budget in fission. In an actual reactor, not all neutrons produce fission. Some are captured with the release of energy in the form of gamma rays. For each fission event, there are enough capture events to add another 5 MeV [10, p. 46]. Thus, on average the energy delivered to the reactor is 200 MeV per ^{235}U fission event, where the accompanying capture events are included. For ^{239}Pu, the corresponding energy is about 207 MeV per fission event.

REFERENCES

1. Charles Weiner, "1932—Moving Into The New Physics," in *History of Physics*, edited by Spencer R. Weart and Melba Phillips (New York: American Institute of Physics, 1985).
2. Emilio Segrè, *From X-Rays to Quarks: Modern Physicists and Their Discoveries* (San Francisco: W. H. Freeman, 1980).
3. John A. Wheeler, "Fission in 1939: The Puzzle and the Promise," *Annual Review of Nuclear and Particle Science* 39 (1989): xiii–xxviii.
4. Irving Kaplan, *Nuclear Physics* (Cambridge: Addison-Wesley, 1954).
5. Victoria McLane, Charles L. Dunford, and Philip F. Rose, *Neutron Cross Section Curves*, Vol. 2 of *Neutron Cross Sections* (New York: Academic Press, 1988).
6. Manson Benedict, Thomas H. Pigford, and Hans Wolfgang Levi, *Nuclear Chemical Engineering*, 2nd ed. (New York: McGraw-Hill, 1981).
7. Emilio Segrè, *Nuclei and Particles*, 2nd ed. (Reading, Mass.: W. A. Benjamin, 1977).
8. James J. Duderstadt and Louis J. Hamilton, *Nuclear Reactor Analysis* (New York: John Wiley, 1976).
9. John R. Lamarsh, *Introduction to Nuclear Engineering*, 2nd ed. (Reading, Mass.: Addison-Wesley, 1983).
10. Ronald Allen Knief, *Nuclear Engineering: Theory and Technology of Commercial Nuclear Power*, 2nd ed. (Washington, D.C.: Hemisphere Publishing, 1992).

Chapter 5

Chain Reactions and Nuclear Reactors

5.1 CRITICALITY AND THE MULTIPLICATION FACTOR

5.1.1 General Considerations

Effective Multiplication Factor

Release of significant amounts of energy from nuclear fission requires a *chain reaction*. For the buildup of a chain reaction, each generation must have more fission events than the preceding one. This means that the average number of neutrons produced in a fission event must be significantly greater than unity. This excess is necessary because some neutrons will be lost, instead of inducing fission in the next generation. For example, some neutrons will be lost to capture reactions, and others will exit the region in which the chain reaction is to be established.

The condition for establishing a chain reaction in a nuclear reactor is commonly expressed as the achievement of *criticality*. This can be quantified in terms of the *criticality factor* or *effective multiplication factor* k, defined as the ratio of the number of neutrons produced by fission in one generation to the number in the preceding generation.[1] A system is termed *critical* if $k=1$. It is *subcritical* if $k<1$ and *supercritical* if $k>1$.

---

[1] In some notations, the effective multiplication factor is denoted k_{eff}.

Number of Neutrons per Fission Event, ν

A quantity of key importance in establishing the possibility of fission is the magnitude of ν, defined as the average number of neutrons produced per fission event. Before a nuclear chain reaction could be thought to be practical, it was necessary to establish that ν was large enough to make the idea feasible. Depending upon the magnitude of ν, achievement of criticality might be easy, impossible, or possible but difficult.

Measurements to establish ν for uranium were undertaken almost immediately after the discovery of fission. In 1939 an erroneously high value of 3.5 was reported by a French group headed by Frederic Joliot [1, p. 48], and a more accurate value, about 2.5, by Enrico Fermi's group at Columbia University [2]. This was high enough to make the achievement of a chain reaction plausible. (See Section 4.3.2 and Table 5.1 for more recent values of ν.)

Chain Reactions in Uranium and the Use of Moderators

A typical fission neutron has an energy in the neighborhood of 1 or 2 MeV. For ^{238}U, the fission cross section σ_f is small below 2 MeV (see Fig. 4.1), and over much of the relevant neutron energy region the capture cross section σ_γ is appreciable [3]. More specifically, below 1.3 MeV, $\sigma_\gamma/\sigma_f > 1$ and below 1 MeV, $\sigma_\gamma/\sigma_f \gg 1$. With these cross sections, a fast-neutron chain reaction cannot be sustained in ^{238}U.

For ^{235}U, the situation is more favorable. At 1 MeV, $\sigma_f = 1.2$ b and σ_γ is only one-tenth as great. However, the isotopic abundance of ^{235}U in natural uranium is 0.72%, while the abundance of ^{238}U is 99.27%. Thus, natural uranium is not much better than pure ^{238}U in the ability to sustain a chain reaction with fast neutrons.

There are several possible solutions, if one is restricted to uranium: (a) increase the fraction of ^{235}U, (b) reduce the energy of the neutrons to a region where the cross sections are more favorable, or (c) both. Solution (a) was the solution for a bomb. Solution (b) was used by Fermi for the first operating chain reactor. Solution (c) is the approach adopted for most present-day reactors used to generate electricity.

The "favorable" energy is the thermal energy region. At the nominal thermal energy of 0.0253 eV, the ratio of the cross section for fission in ^{235}U to capture in ^{238}U is greater than 200. Under these circumstances, a chain reaction is possible, even in natural uranium. In so-called thermal reactors, which have been and remain the dominant type of nuclear reactor, the neutron energy is reduced from the MeV region to the thermal region by successive elastic collisions with light nuclei. This is the process of moderation, and the material with low atomic mass is the moderator. These will be discussed at greater length in Section 5.2.

5.1.2 Formalism for Describing the Multiplication Factor

Number of Neutrons per Absorption Event, η

In this section we consider chain reactions in thermal reactors. As mentioned earlier, the term "thermal" is commonly used to refer specifically to room temperature, $T=293$ °K and $kT=0.0253$ eV, because cross sections are normally measured in the laboratory with the sample at this temperature. When neutrons are thermalized, they come into equilibrium with the local medium. The thermalization occurs primarily in the moderator, and therefore the moderator temperature is the relevant one. It is higher than room temperature, but kT will still be well under 0.1 eV.[2] In this section we assume that a means is found to reduce the neutron energy from the 1 MeV region to below 0.1 eV, and in that context we will discuss the formalism used to describe the multiplication factor.

The multiplication factor can be considered in terms of the fate of a neutron emitted in fission. The neutron faces a number of conceptual branch points. At each branch point, a "yes" means that fission remains possible, while a "no" means that the neutron is lost to fission. The overall probability of fission is the product of the individual probabilities of a yes answer at each of the branch points. We first define a new parameter η

η=average number of fission neutrons per thermal neutron *absorbed* in fuel.

The absorption cross section σ_a is the sum of the fission cross section σ_f and the capture cross section σ_γ. The parameters η and ν differ in that η refers to the number of neutrons per absorption event, while ν refers to the number of neutrons per fission event. For a fissile nuclide (e.g., ^{235}U), it follows that

$$\frac{\eta}{\nu}=\frac{\sigma_f}{\sigma_a}=\frac{1}{1+\alpha},$$ (5.1)

where $\alpha=\sigma_\gamma/\sigma_f$, as in Eq. 3.4, and $\sigma_a=\sigma_f+\sigma_\gamma$.

For a mixture of isotopes, the fission and capture cross sections must be summed for the various components. For the case of uranium as the fuel,[3] if the isotopic fraction (by number of atoms) of ^{235}U is x

$$\frac{\eta_F}{\nu}=\frac{x\sigma_f(235)}{x\sigma_a(235)+(1-x)\sigma_a(238)},$$ (5.2)

where η_F refers to the mixture of uranium isotopes in reactor fuel, and η refers to pure ^{235}U. For ^{238}U, $\sigma_f\dot{=}0$ and $\sigma_a=\sigma_\gamma$. The ratio of η_F to η follows from Eqs. 5.1 and 5.2:

$$\frac{\eta_F}{\eta}=\frac{x\sigma_a(235)}{x\sigma_a(235)+(1-x)\sigma_a(238)}.$$ (5.3)

For $x=1$, as in pure ^{235}U, Eq. 5.2 reduces to Eq. 5.1 and $\eta_F=\eta$.

Thus, despite the emphasis on the quantity ν (the average number of neutrons produced per fission event in ^{235}U), the key quantity in determining the progress from generation to generation in a chain reaction is η_F. This is the average number of neutrons produced per absorption event in uranium. To summarize, η_F is less than ν for two reasons: (1) some of the thermal neutron absorptions in ^{235}U result in capture, not fission; and (2) some of the thermal neutrons absorbed in uranium are absorbed in ^{238}U, not ^{235}U.

Five-Factor and Four-Factor Formulae

The effective multiplication factor can be expressed as a product of factors:

$$k=\eta_F f p \epsilon P_L,$$ (5.4)

---

[2]For example, the moderator temperature, namely, the water temperature, in a pressurized water reactor is about 300 °C (573 °K), corresponding to $kT=0.05$ eV [4, p. 81].

[3]For simplicity, we ignore in Eq. 5.2 the presence of plutonium isotopes. In practice, as uranium fuel is consumed in a reactor, the ^{235}U fraction (x) decreases and ^{239}Pu and other plutonium isotopes are produced, as taken into account in more complete treatments.

where η_F=number of fission neutrons per thermal neutron absorbed in the fuel; p=fraction of fission neutrons that avoid capture and reach thermal energies; f=fraction of thermal neutrons captured in the fuel; ϵ=enhancement factor due to contribution from fast neutrons; and P_L=probability that the neutron will not leak out of the reactor.

Eq. 5.4 is based on the conceptual picture that fission occurs at thermal energies and therefore only in ^{235}U (in a uranium reactor). This is a good approximation, but not a perfect one, because fast neutrons can also initiate fission, albeit with a smaller cross section. The enhancement factor ϵ compensates for that oversimplification. Typically, ϵ is about 1.02.

A neutron faces two main hurdles before it can initiate fission. First, as quantified in the factor p, it must be thermalized before capture occurs. If the reactor fuel is primarily ^{238}U, then capture will occur mostly when the neutron energy matches resonance energies for neutron capture in ^{238}U. The factor p is commonly termed the *resonance escape probability*.

Second, as quantified in the factor f, once the neutron is thermalized it must be absorbed in the fuel, rather than in something else. All the other materials of the reactor offer competition for absorption of thermal neutrons. These materials include the moderator, the reactor coolant (if different from the moderator), and structural materials. The factor f is commonly termed the *thermal utilization factor*.

The loss factor P_L is sometimes subdivided into loss of fast neutrons and loss of thermal neutrons. If the system is too small, the losses will be large, and a chain reaction becomes impossible. Often the other extreme is considered, at least for conceptual purposes, and an infinite system is considered. In that case, P_L=1, and Eq. 5.4 reduces to the so-called *four-factor formula* (Eq. 5.5). The resulting multiplication factor is commonly designated as k_∞ (where $k=k_\infty P_L$):[4]

$$k_\infty = \eta_F f p \epsilon. \tag{5.5}$$

5.1.3 Numerical Values of Thermal Reactor Parameters

Table 5.1 lists cross sections and other pertinent parameters for important nuclear fuel constituents. The *fissile* nuclei are those with high fission cross sections at thermal energies. Several of the remainder, most importantly ^{232}Th and ^{238}U, are termed *fertile* nuclides. Although they have negligible cross sections for fission at thermal energies, they capture neutrons leading to the formation of the fissile nuclides, ^{233}U and ^{239}Pu.

The magnitudes of η given in Table 5.1 are for pure isotopes. For natural uranium, x=0.0072. Then, from Eq. 5.3, η_F/η=0.648 and η_F=1.35, using the parameters of Table 5.1. For uranium enriched to x=0.03, η_F=1.84.

As the reactor fuel is consumed, the ^{235}U fraction in the fuel will decrease, and it is to be expected that η_F will be lower than the 1.84 calculated for a 3% ^{235}U enrichment. This is illustrated in typical quoted values of the parameters in Eq. 5.4 for a thermal reactor: η_F=1.65, p=0.87, f=0.71, ϵ=1.02, and P_L=0.96 [7, p. 83]. These give the product k=1.00.

It is clear from the preceding discussion that greater enrichment in ^{235}U will increase η_F. It will also increase the thermal utilization factor f, because the large cross section for fission in ^{235}U increases the relative probability of absorption in the fuel, rather than in other materials.

---

[4]Strictly speaking, the magnitudes of the parameters on the right-hand side of Eq. 5.5 differ from those in Eq. 5.4 because of differences in the conditions in an infinite reactor and in a finite one [5]. However, these small differences will be ignored.

Table 5.1. *Parameters at thermal energies (0.0253 eV) for selected actinides (cross sections are in barns).*

Nuclide	σ_f	σ_γ	σ_a	α	ν	η
Fissile						
^{233}U	529	45.5	575	0.0861	2.493	2.296
^{235}U	583	98.3	681	0.1687	2.425	2.075
^{239}Pu	748	269	1017	0.360	2.877	2.115
^{241}Pu	1011	358	1369	0.354	2.937	2.169
Fertile						
^{232}Th	$<3\times10^{-6}$	7.37	7.37			
^{238}U	4×10^{-6}	2.68	2.68			
^{240}Pu	0.06	290	290			

Source: Data are from Ref. [6], part B.

5.2 THERMALIZATION OF NEUTRONS

5.2.1 Role of Moderators

Kinematic Relations

As explained in Section 5.1.1, fission in natural uranium will not result in a chain reaction if the neutrons interact at energies close to those at which they are emitted, in the vicinity of 1 or 2 MeV. The neutrons are reduced in energy from this region to the more favorable region below 1 eV region by elastic collisions with the nuclei of the *moderator*.

The energy transfer in an elastic collision depends on the angle at which the incident neutron is scattered. The energy of the scattered neutron E_s can be calculated from the kinematics of "billiard ball" collisions, i.e., by the application of conservation of energy and momentum. If a neutron of initial energy E_i is scattered at 0°—namely, it continues in the forward direction—then the energy of the scattered neutron will be $E_s = E_i$. In brief, nothing is changed. If the neutron is scattered at 180°, then the energy of the scattered neutron can be shown to be

$$E_s = \left(\frac{M-m_n}{M+m_n}\right)^2 E_i = \left(\frac{A-1}{A+1}\right)^2 E_i, \tag{5.6}$$

where M is the atomic mass of the moderator, m_n is the mass of the neutron, and the approximation has been made that $M = A$ and $m_n = 1$ (both in atomic mass units, u). Eq. 5.6 describes the condition for minimum neutron energy or maximum energy loss. If we define δ to be the ratio of *average* scattered neutron energy to E_i, and assume this average to be the mean of the 0° and 180° cases, it follows that

$$\delta = \frac{1}{2}\left[1 + \left(\frac{A-1}{A+1}\right)^2\right] = \frac{A^2+1}{A^2+2A+1}. \tag{5.7}$$

For example, $\delta = 0.50$ for hydrogen ($A = 1$), $\delta = 0.86$ for carbon ($A = 12$), and $\delta = 0.99$ for uranium ($A = 238$).

Obviously, Eq. 5.7 could not give a correct average if the scattering events were either predominantly forward or predominantly backward. Less obviously, it gives an exactly correct result if the scattering is isotropic, namely, an equal number of neutrons scattered in all directions.[5] This isotropy condition is well satisfied for neutrons incident on light elements such as hydrogen and carbon at the energies relevant to fission, and therefore Eq. 5.7 is a good approximation.

An effective moderator—i.e., one that reduces the neutron energy with relatively few collisions—requires a low value of parameter δ. As seen from Eq. 5.7 and illustrated by the examples above, this means that the moderator must have a relatively low atomic mass number A. This rules out using uranium as its own moderator. Instead, hydrogen in water (H_2O), deuterium in heavy water (D_2O), or carbon in graphite (C) are commonly used as moderators.

Number of Collisions for Thermalization

The factor δ can be used to calculate the number of collisions n required to reduce the *mean* neutron energy by any given factor. For example, to reduce the mean neutron energy from 2 MeV to 1 eV, a factor of 2×10^6, $\delta^n = 5 \times 10^{-7}$. For hydrogen ($\delta = 1/2$), this gives $n = 21$. However, the distribution of neutron energies becomes highly skewed in successive collisions, with many neutrons having very small energies and a few neutrons having high energies. These relatively few high energy neutrons elevate the mean energy to a level considerably higher than the more representative *median* energy. Thus, the decrease in mean energy provides only a crude characterization of the extent to which the neutrons have lost energy, and δ, while simple to calculate, is not widely used.

It is more common to characterize the evolution of the neutron energy distribution in terms of a quantity which is less perturbed by a few remaining high-energy neutrons, namely the logarithm of the energies. The mean *logarithmic energy decrement* ξ is defined as[6]

$$\xi = \left\langle \ln\left(\frac{E_i}{E_s}\right) \right\rangle.$$

(5.8)

Here, ξ is the mean of the logarithm of the ratio of the incident and scattered neutron energies in successive generations. Based on the distribution of the logarithm of the energy as the reference, the "average" neutron energy changes by a factor of $e^{-\xi}$ in successive generations. It can be shown that for an isotropic distribution in scattering angle (see, e.g., Ref. [8], p. 624)

$$\xi = 1 - \frac{(A-1)^2}{2A} \ln\left(\frac{A+1}{A-1}\right),$$

(5.9)

where A is the atomic mass number, and the small difference between A and the atomic mass M is ignored. To reduce the neutron energy by a factor of 2×10^6, as discussed above, requires n generations, where

$$e^{n\xi} = 2 \times 10^6.$$

(5.10)

It is desirable that thermalization be accomplished in as few collisions as possible, to minimize the loss of neutrons by absorption in ^{238}U and other materials of the reactor.

---

[5]Here, isotropy must be in the center-of-mass system, i.e., a system moving at the velocity of the center of mass.
[6]In the nuclear engineering literature, ξ is the mean gain in *lethargy* per generation. We will not here undertake the detailed calculations that make it desirable to employ the concept of lethargy.

For hydrogen, $\xi=1$ by Eq. 5.9 (noting that the second term approaches zero as A approaches 1) and $n=\ln(2\times10^6)=14.5$. The same analysis shows that $n=92$ for ^{12}C ($\xi=0.158$) and $n=1730$ for ^{238}U ($\xi=0.0084$). This highlights the key advantage of hydrogen as a moderator and shows how difficult it would be to thermalize the neutrons if reliance were placed on elastic scattering with uranium.

5.2.2 Moderating Ratio

In a moderator, there is competition between elastic scattering, which produces the desired reduction in energy, and capture, which means the loss of the neutron. The competition depends on the relative scattering and capture cross sections. It is usually expressed in terms of *macroscopic* cross sections Σ (in cm^{-1}), which are related to the microscopic cross sections σ (in cm^2), as follows:

$$\Sigma = N\sigma = \frac{\rho N_A}{M}\sigma, \tag{5.11}$$

where, again, N is the number of nuclei per cm^3, N_A is Avogadro's number, ρ is the density (g/cm^3), and M is the atomic mass (u). For example, for ^{12}C in the form of graphite, $\rho=1.6$ g/cm^3 and $\sigma_{el}=4.75$ b (thermal), where thermal cross sections for light elements are here taken from Ref. [6], part A. Therefore

$$\Sigma_{el} = \frac{1.6\times6.022\times10^{23}}{12}\times4.75\times10^{-24} = 0.38 \text{ cm}^{-1}. \tag{5.12}$$

This means that the average distance traversed, or *mean free path* between elastic scattering events, is $\lambda=1/\Sigma_{el}=2.6$ cm.

A figure of merit for a moderator is the *moderating ratio*, defined as

$$MR = \frac{\xi\Sigma_{el}}{\Sigma_c} = \frac{\xi\sigma_{el}}{\sigma_c}. \tag{5.13}$$

A good moderator is one in which ξ and σ_{el} are high and σ_c is low. Properties of various moderators are presented in Table 5.2.

An obvious choice for use as a moderator is hydrogen, in the form of ordinary ("light") water (^{1}H$_2$O). The oxygen in the water is relatively inert, having a very low capture cross

Table 5.2. *Properties of alternative moderators.*

Material[a]	A	ξ	n[b]	MR
H	1	1.000	14	
D	2	0.725	20	
H$_2$O		0.920	16	71
D$_2$O		0.509	29	5670
Be	9	0.209	69	143
C	12	0.158	91	192
^{238}U	238	0.008	1730	0.009

Source: Data are from Ref. [7], p. 324.
[a]Here, H denotes ^{1}H and D denotes ^{2}H.
[b]n=number of collisions to reduce the energy by a factor of 2×10^6 (see Eq. 5.10).

section and being much less effective than hydrogen as a moderator. However, ordinary hydrogen has the major drawback of having a large cross section (0.33 b at 0.0253 eV) for the capture reaction

$$n + {}^1H \rightarrow {}^2H + \gamma.$$

The moderating ratio for light water, as defined in Eq. 5.13, is 71 (see Table 5.2).

The moderating ratio for heavy water, D_2O (^2H_2O), is almost 100 times greater, despite the larger mass of deuterium.[7] This is because the neutron-capture cross section is much less in deuterium than in hydrogen (0.0005 b at 0.0253 eV). It was recognized early in the nuclear weapons programs of World War II that heavy water would be a very valuable substance in reactor development, and considerable effort was devoted to attacking heavy water production facilities in Norway, which were then under German control. The choice between light and heavy water remains unclear, with the former having the advantage of being very easily obtainable and the latter the advantage of being a better moderator.

Carbon, in the form of pure graphite, is a more effective moderator than light water, with a moderating ratio of 192, despite the large A and low ξ for ^{12}C. The reason for this, again, is a low neutron-capture cross section (0.0035 b at 0.0253 eV). However, in contrast to light water and heavy water, graphite cannot be used as a coolant as well as a moderator.

5.3 REACTOR KINETICS

5.3.1 Reactivity

If the multiplication exceeds unity by more than a small amount, the reactor power will build up at a rapid rate. For constant power, the effective multiplication factor must be kept at unity. Thus, a key quantity is the difference $k-1$. This quantity is usually expressed in terms of the *reactivity* ρ, where by definition

$$\rho = \frac{k-1}{k}, \tag{5.14}$$

and

$$k = \frac{1}{1-\rho}. \tag{5.15}$$

A runaway chain reaction is one in which k rises appreciably above unity or, equivalently, the reactivity ρ is appreciably greater than zero. A major aspect of reactor safety is the avoidance of such an excursion.

5.3.2 Buildup of Reaction Rate

The rate at which a reactor delivers power is proportional to the number of fission events per unit time, which in turn is proportional to the neutron flux, here designated as ϕ. To describe changes in the reaction rate, it is convenient to introduce the *mean neutron lifetime l*, defined as the average time between the emission of a fission neutron and its absorption. Fission

---

[7]The actual moderation ratio for a heavy-water reactor depends on the magnitude of the light-water impurity in the heavy water.

follows absorption with no time delay, and l is the average time between successive fission generations.

The change in the neutron flux per generation is $(k-1)\phi$, and the time rate of change in neutron flux is therefore

$$\frac{d\phi}{dt}=\frac{(k-1)\phi(t)}{l}=\frac{\phi(t)}{T},$$ (5.16)

where T is the *reactor period*, defined as

$$T=\frac{l}{k-1}.$$ (5.17)

The solution of Eq. 5.16 is an exponential buildup:

$$\phi(t)=\phi(0)e^{t/T}.$$ (5.18)

Thus, the flux in the reactor grows exponentially. It is important for the stability of the reactor that the growth rate is not too rapid—that is, that the period T be large. As is obvious qualitatively and follows from Eq. 5.17, this is achieved by keeping the multiplication constant k close to unity.

The reactor period also depends upon the mean neutron lifetime l. This time is the sum of the time required for moderation to thermal energies and the diffusion time, at thermal energies, before the neutron is absorbed. The magnitude of this time depends upon the type of reactor. For prompt fission neutrons in a water-moderated, uranium-fueled reactor, l is typically about 0.0001 sec (see, e.g., Ref. [7], p. 236). If k were to rise, for example, from 1.00 to 1.01, then $T=0.01$ sec, and in 0.1 second, by Eq. 5.18, ϕ would rise by a factor of 2×10^4.

5.3.3 Role of Delayed Neutrons

The instability suggested by the very short buildup times seen in the previous section brings us to the crucial role of delayed neutrons. For ^{235}U, as discussed in Section 4.3.2, about 0.65% of fission neutrons are delayed, i.e., $\beta=0.0065$. With delayed neutrons, a mean lifetime, averaged over prompt and delayed neutrons, can be taken to be (see, e.g., [7], p. 237)

$$l_{av}=(1-\beta)l+\beta\tau_e,$$ (5.19)

where τ_e is the effective mean lifetime for the delayed neutrons, averaged over all precursors of the delayed neutrons (ignoring the slowing down of the delayed neutrons themselves). As defined here,

$$\tau_e=\frac{1}{\beta}\Sigma\beta_i\tau_i,$$ (5.20)

where τ_i and β_i are the mean lifetime and fraction of emitted neutrons, respectively, for the ith delayed neutron group.

Evaluating Eq. 5.20 by summing over the neutron groups for thermal fission in ^{235}U, it is found that τ_e is about 10 sec (see, e.g., Ref. [9], p. 76). This gives an average mean lifetime for all neutrons of $l_{av}\approx0.1$ sec. This mean lifetime is almost 1000 times greater than the mean lifetime l for prompt neutrons and greatly slows the rate of buildup of the neutron flux.

A more complete analysis of the buildup rates with delayed neutrons shows that it is only valid to use the formalism of Eqs. 5.17 and 5.18, with l_{av} substituted for l, in the special case where the reactivity ρ is small compared to β [7, p. 245]. When ρ is large compared to β, then, as might be expected, the fact that some of the neutrons are delayed makes little difference, and the earlier result, without delayed neutrons, still applies. To keep the reactor stable against quick excursions, it is important to avoid this condition in reactor operation.

An important distinction can be made between the case where the reactor is supercritical without delayed neutrons and the case where delayed neutrons are required for criticality. We can distinguish these cases as follows:[8]

$$\rho \geq \beta: \text{ prompt critical}$$

$$0 < \rho < \beta: \text{ delayed critical.}$$

For stability, reactors are normally operated under delayed critical conditions, i.e., with $\rho < \beta$.

5.4 PRODUCTION OF PLUTONIUM: CONVERSION RATIO

With uranium fuel, ^{239}Pu is produced by the capture of neutrons in ^{238}U (see Section 3.1.1). This provides a means of obtaining ^{239}Pu for possible use in weapons or in other reactors. It also contributes fissile material which is consumed in the reactor before the fuel is removed, supplementing the original ^{235}U in the fresh fuel. The rate of ^{239}Pu production is described in terms of the *conversion ratio*. In this section, we calculate the conversion ratio for thermal reactors.

If we specialize to uranium fuel, the conversion ratio is defined as

$$C = \frac{N(238)}{N(235)}, \tag{5.21}$$

where $N(238)$=number of capture events in ^{238}U, producing ^{239}Pu and $N(235)$=number of absorption events in ^{235}U, destroying ^{235}U. If $C=1$, as much ^{239}Pu is produced as ^{235}U is destroyed, and the threshold for "breeding" is reached.

To evaluate Eq. 5.21, it is convenient to normalize to a single absorption event in ^{235}U, setting $N(235)$ at unity. Then Eq. 5.21 reduces to

$$C = N(238) = N_{fast} + N_{th}, \tag{5.22}$$

where N_{fast} and N_{th} are the number of capture events in ^{238}U for fast (nonthermal) neutrons and thermal neutrons, respectively.

To calculate C, we follow the consequences of an absorption event in ^{235}U. Each absorption leads to the production of η fission neutrons, where η is the value for pure ^{235}U. For simplicity, we will ignore the fast-fission enhancement factor ϵ, assuming it to be close to unity (see the four-factor formula, Eq. 5.5). As a further simplification, we will assume that all absorption in ^{235}U occurs at thermal energies. The multiplication factor of Eq. 5.5 then reduces to

---

[8]It can be seen that $\rho = \beta$ is a dividing point by separating the multiplication factor k into a prompt and delayed part, $k = k_p + k_d$, where $k_d = \beta k$. Then $k_p = k(1 - \beta)$ and, by Eq. 5.15, the condition $\rho = \beta$ is equivalent to the condition $k_p = 1$.

$$k=\eta_F f p. \tag{5.23}$$

It is a good approximation to assume that all neutrons that do not reach thermal energies are captured in ^{238}U in the resonance region. The number of fast-neutron ^{238}U capture events can then be expressed in terms of the resonance escape probability p:

$$N_{fast}=(1-p)\eta. \tag{5.24}$$

The remaining neutrons reach thermal energies, where the fraction f is captured in uranium. The number of thermal–neutron capture events in ^{238}U is the product of the number of neutrons captured in uranium $\eta f p$ and the fraction of these that are captured in ^{238}U:

$$N_{th}=\eta f p \frac{(1-x)\sigma_a(238)}{(1-x)\sigma_a(238)+x\sigma_a(235)}, \tag{5.25}$$

where σ_a is the thermal–neutron absorption cross section, and x is the fractional abundance of ^{235}U. For ^{238}U, absorption involves only capture; for ^{235}U, it involves both capture and fission.

The fraction on the right-hand side of Eq. 5.25 gives the ^{238}U share of the thermal absorption in the fuel. With the aid of Eq. 5.3, this fraction can be expressed as

$$\frac{(1-x)\sigma_a(238)}{(1-x)\sigma_a(238)+x\sigma_a(235)}=\frac{\eta_F}{\eta}\left(\frac{\eta}{\eta_F}-1\right). \tag{5.26}$$

For criticality ($k=1$), Eq. 5.23 requires that $\eta_F f p=1$. Combining this requirement with Eqs. 5.25 and 5.26, it follows that

$$N_{th}=\frac{\eta}{\eta_F}-1. \tag{5.27}$$

The total conversion ratio C is given by adding the components in Eqs. 5.24 and 5.27:

$$C=(1-p)\eta+\left(\frac{\eta}{\eta_F}-1\right). \tag{5.28}$$

As a specific example, we adopt the parameters cited earlier (Section 5.1.3) for a conventional light water reactor ($p=0.87$, $f=0.71$, $\eta=2.075$, and $\eta_F=1.65$). It then follows that

$$C=0.27+0.26=0.53.$$

Thus, for this case, roughly one ^{239}Pu nucleus is produced for every two ^{235}U nuclei consumed.

The actual conversion ratio at a given time in the history of the reactor will depend upon the magnitude of the ^{235}U fraction x. As this quantity decreases from $x=0.03$, absorption in ^{238}U becomes relatively more important for thermal neutrons, η_F decreases, and C rises. The example cited above is for an intermediate value of x. At a later time the conversion ratio will be still higher. A commonly cited overall figure for a light water reactor is about 0.6 (see, e.g., Ref. [7], p. 86). Reactors using moderators with a higher moderating ratio (heavy water or graphite) in general will have a high thermal utilization factor f and therefore can have a lower resonance escape probability p. This means a greater production of ^{239}Pu—i.e., a higher conversion ratio C.

The limiting condition for breeding is $C=1$, where for a thermal reactor, C is given by Eq. 5.28. A necessary, but not always sufficient, condition for breeding (specialized to the case of $^{235}U+^{238}U$ fuel) is that $\eta \geqslant 2$. The criterion reduces to $\eta=2$ for an "ideal" reactor in which no neutron absorption occurs except in the uranium fuel itself—i.e., one in which the thermal utilization factor f is unity. In this idealized case, again assuming ϵ and P_L to be unity, the criticality condition $k=1$ becomes $p \eta_F=1$. Then, from Eq. 5.28, $C=1$ for $\eta=2$.

5.5 CONTROL MATERIALS AND POISONS

5.5.1 Reactor Poisons

Some of the fission products in a reactor have very high capture cross sections at thermal energies. For example, the thermal–neutron capture cross section is 2.65×10^6 b for xenon 135 (^{135}Xe) and 4.1×10^4 b for samarium 149 (^{149}Sm) (see, e.g., Ref. [9], p. 643).[9] Such nuclei are known as *poisons* because their presence in the reactor fuel creates an additional channel for the loss of neutrons and reduces the reactivity of the core. In terms of Eq. 5.5, they lower the thermal utilization factor f and the multiplication factor k.

The buildup of poisons is limited by both their radioactive decay (if they are not stable) and their destruction in the capture process. However, the longer the fuel remains in the reactor, the larger will be the total poison content. This is one of the factors that limits the length of time that fuel can be used in a reactor.

5.5.2 Controls

Control materials are materials with large thermal neutron-absorption cross sections, used as controllable poisons to adjust the level of reactivity. They serve a variety of purposes:

- To achieve intentional changes in reactor operating conditions, including turning the reactor on and off
- To compensate for changes in reactor operating conditions, including changes in the fissile and poison content of the fuel
- To provide a means for turning the reactor off rapidly, in case of emergency

Two commonly used control materials are cadmium and boron. The isotope ^{113}Cd, which is 12.2% abundant in natural cadmium, has a thermal–neutron capture cross section of 20,000 b. The isotope ^{10}B, which is 19.9% abundant in natural boron, has a thermal absorption cross section of 3800 b. (In this case, absorption of the neutron leads not to capture but to the alpha-particle emitting reaction, $n+^{10}B\rightarrow\alpha+^7Li$, but the effect on the reactivity is the same as for capture.) Thus, cadmium and boron are effective materials for reducing the reactivity of the reactor.

Cadmium is commonly used in the form of control rods, which can be inserted or withdrawn as needed. Boron can be used in a control rod, made, for example, of boron steel, or it can be introduced in soluble form in the water of water-cooled reactors. It is also possible to use other materials.

A sophisticated control variant is the so-called *burnable poison*, incorporated into the fuel itself. During the three years or so that the fuel is in the reactor, the ^{235}U or other fissile

---

[9]The capture cross sections for these nuclei are not well fit by the $1/v$ law. The cross sections specified here are the measured values for a neutron velocity of 2200 m/sec (see Section 3.4.3).

material in the fuel is partially consumed, and fission product poisons are produced. The burnable poison helps to even the performance of the fuel over this time period. It is a material with a large absorption cross section, which is consumed by neutron capture as the fuel is used. The decrease in burnable poison content with time can be tailored to compensate for the decrease in ^{235}U and the buildup of fission fragment poisons. Boron and gadolinium can be used as burnable poisons.

5.5.3 Xenon Poisoning

Reference has already been made to the buildup of poisons as the fuel is used. These are fission products that have large neutron-absorption cross sections. A particularly important example is xenon 135 (^{135}Xe), the cause of the "xenon poisoning" that for a brief time threatened operation of the first large U.S. reactors in World War II, which were designed to produce ^{239}Pu.[10]

The first large reactor for this purpose went into operation in September 1944. After a few hours of smooth operation, the power level of the reactor began to decrease, eventually falling to zero. It recovered by itself within about 12 h and then began another cycle of operation and decline. The possibility of fission-product poisoning had previously occurred to John Wheeler, who had participated in the project. From the time history of the poisoning cycle, and the known high absorption cross section of xenon, Wheeler and Fermi "within a couple of days . . . discovered the culprit."[11]

The phenomenon was due to the production of $A=135$ precursors, including ^{135}I, which decay into the strongly absorbing ^{135}Xe, which in turn decays into a stable (or long-lived) product with a small absorption cross section. In anticipation of possible difficulties, the reactor had been built in a way such that more uranium could be added to the reactor. This enabled it to sustain criticality even in the face of the xenon poisoning. Thus, the wartime plutonium production was not greatly delayed. Nonetheless, the event dramatized a phenomenon that remains an important consideration in reactor design.

Using subsequently refined determinations of half-lives and cross sections, the xenon poisoning can be described in more detail in terms of the chain of $A=135$ isotopes:[12]

$$\text{Te } (T=19.2 \text{ sec}) \rightarrow \text{I } (T=6.61 \text{ hr}) \rightarrow \text{Xe } (9.09 \text{ hr}) \rightarrow \text{Cs } (3\times10^6 \text{ yr}).$$

The main fission product in this chain is tellurium 135 (^{135}Te), which almost immediately decays to iodine 135 (^{135}I). This is the chief source of ^{135}Xe, although a small amount (about 1/20th the amount from ^{135}I) is produced directly. Thus, initially there is a negligible concentration of ^{135}Xe. On a time scale of about 7 h, the concentration begins to build. If this is sufficient to shut down the reactor, as was the case, then the concentration eventually declines with a 9-h half-life. Reactor operation can resume, and the cycle can repeat.

The effectiveness of ^{135}Xe as a poison, despite the very small amounts present, stems from its extremely high cross section: 2.65×10^6 b at 0.0253 eV. In a normally operating reactor, the amount of ^{135}Xe is held in check by both its own radioactive decay and its destruction by neutron capture. If the reactor is shut down suddenly, the ^{135}Xe level will increase for a period, as the ^{135}I decay continues but the destruction by neutron capture stops. The early restart of the reactor is then more difficult. This effect, and imprudent measures taken to

---

[10]A vivid description of these events is given in Ref. [10], pp. 557–560.
[11]See Ref. [11], pp. 29–30.
[12]See, for example, Ref. [12], p. 195 for relative fission yields.

overcome it, were important contributors to the Chernobyl nuclear reactor accident (see Section 12.3.2).

REFERENCES

1. Bertrand Goldschmidt, *Atomic Rivals*, translated by George M. Temmer (New Brunswick, N.J.: Rutgers University Press, 1990).
2. Rudolf Peierls, "Reflections on the Discovery of Fission," *Nature* 342 (1989): 853–854.
3. Victoria McLane, Charles L. Dunford, and Philip F. Rose, *Neutron Cross Section Curves*, vol. 2 of *Neutron Cross Sections* (New York: Academic Press, 1988).
4. Anthony V. Nero, Jr., *A Guidebook to Nuclear Reactors* (Berkeley: University of California Press, 1979).
5. Alvin M. Weinberg and Eugene P. Wigner, *The Physical Theory of Neutron Chain Reactors* (Chicago: University of Chicago Press, 1958).
6. S. F. Mughabghab, M. Divadeenam, and N. E. Holden, *Neutron Resonance Parameters and Thermal Cross Sections*, Part A, $Z=1-60$ and Part B: $Z=61-100$, vol. I of *Neutron Cross Sections* (New York: Academic Press, 1981 and 1984).
7. James J. Duderstadt and Louis J. Hamilton, *Nuclear Reactor Analysis* (New York: John Wiley, 1976).
8. Emilio Segrè, *Nuclei and Particles*, 2nd ed. (Reading, Mass.: W. A. Benjamin, 1977).
9. John R. Lamarsh, *Introduction to Nuclear Engineering*, 2nd ed. (Reading, Mass.: Addison-Wesley, 1983).
10. Richard Rhodes, *The Making of the Atomic Bomb* (New York: Simon and Schuster, 1986).
11. Alvin M. Weinberg, *The First Nuclear Era: The Life and Times of a Technological Fixer* (New York: AIP Press, 1994).
12. D. J. Bennet and J. R. Thomson, *The Elements of Nuclear Power*, 3rd ed. (Essex: Longman, 1989).

Chapter 6

Types of Nuclear Reactors

6.1 SURVEY OF REACTOR TYPES

6.1.1 Uses of Reactors

Nuclear reactors were first used for the production of ^{239}Pu. Subsequently, there have been many other applications, the most prominent being the generation of electricity. Further uses have been to propel ships (mostly naval vessels), to produce radioisotopes, and to a limited extent, to supply heat. Numerous additional reactors have been built for teaching or research, much of the latter involving the study of the properties of materials under neutron bombardment and the intrinsic properties of neutrons and other subatomic particles.

In some cases, applications have been combined. For example, the N reactor at Hanford and the Chernobyl-type reactors were used for both ^{239}Pu production and electricity generation. There has also been limited use of waste heat from reactors to produce hot water or steam for industrial applications, for heavy water production, and for desalination. These reactors have been primarily in the USSR and Canada [1]. The total thermal capacity of these facilities is only 5 gigawatts, which is about 0.5% of the total worldwide thermal capacity of electricity-generating nuclear reactors.

For the most part, reactors have been single-purpose, designed and used solely for weapons production, for electricity generation, for vessel propulsion, or as a neutron source. We are here directly concerned only with the use of reactors for electricity generation.

Even so, the separation from the issue of ^{239}Pu production is not complete. Inevitably, if the reactor fuel contains ^{238}U, then ^{239}Pu will be produced following neutron capture in ^{238}U, as discussed in Section 3.1.1. In breeder reactors, the production of ^{239}Pu is one of the chief goals, with the ^{239}Pu intended for use as fuel for further reactors. In nonbreeders, which means almost all of the world's operating reactors, ^{239}Pu is a by-product. We will return in Chapter 14 to the link between power reactors and the possible use of their ^{239}Pu for bombs.

6.1.2 Classifications of Reactors

Thermal Reactors and Fast Reactors

In previous chapters, reference has been made to the thermalization of neutrons in reactors— i.e., to the reduction of the neutron energies to thermal energy. Reactors designed to operate in this fashion are termed *thermal* reactors. However, it is also possible to operate a reactor with "fast" neutrons, at energies in the neighborhood of 1 MeV or higher. These reactors are called *fast-neutron* reactors or just *fast* reactors. The only prominent example of a fast reactor is the liquid-metal breeder reactor (see Section 6.3.3).

Homogeneous and Heterogeneous Reactors

We have been tacitly assuming that the reactors under consideration are what are sometimes known as *heterogenous*, in which the fuel, coolant, and moderator (if any) are distinct physical entities. All reactors used today for power generation are of this form. However, in the early days of nuclear power there was considerable exploration of an alternative configuration, the *homogeneous* reactor, defined as "a reactor whose small-scale composition is uniform and isotropic" [2, p. 378]. Homogeneity can be achieved if the fuel is in liquid form, where the liquid is circulated for heat transfer to a steam generator.

One variant of this was known as the *aqueous homogenous reactor* because the fuel was mixed with water (H_2O or D_2O). In the so-called homogeneous reactor experiment, two small reactors, called HRE-1 and HRE-2, were built at Oak Ridge National Laboratory (ORNL) in the 1950s.[1] For HRE-2, the fluid was uranyl sulfate (UO_2SO_4) in heavy water (D_2O), with the uranium highly enriched in ^{235}U. This program had the potential of developing a thermal breeder reactor, but although HRE-2 operated uninterruptedly for over 100 days at 5 MWe, some difficulties developed, and the program was dropped in favor of alternative liquid-fuel projects.

One alternative was the *molten-salt reactor*. The fluid was a mixture of fluoride compounds, including the fissile component $^{235}UF_4$ and the fertile component ThF_4. After initial operation, $^{233}UF_4$ was successfully tried as an alternative to $^{235}UF_4$. Like HRE-2, this reactor was designed to be a thermal breeder reactor. Again, there was initial success in the reactor operation, but a decision was made in the 1960s to abandon development of thermal breeders in favor of fast breeder reactors.[2] A further homogenous reactor approach, a *liquid-metal thermal breeder* using uranium compounds in molten bismuth, was also investigated but was abandoned without construction of even a test reactor.

At present there is no prospect in sight of a major program to develop homogenous reactors, and we will not consider them further. Instead we will restrict consideration to heterogenous reactors, which are so dominant that it is unusual to include the specification "heterogenous." Nonetheless, some interest remains in homogenous reactors, particularly in molten-salt reactors (see, e.g., Ref. [5]). It would be advisable, as one focuses on reactors of the sort in actual use or in immediate prospect, to remember that on a longer time scale a

---

[1]For a description of the history of this program and the ORNL program on molten-salt breeders, see Chapter 6 of Ref. [3]. Technical aspects of fluid-fuel reactors are also discussed in Ref. [4], pp. 403–413.

[2]The fast breeder reactor program was subsequently sharply reduced, with the centerpiece of the U.S. fast breeder reactor program, the Clinch River Breeder Reactor, abandoned in 1975.

wide array of variants are possible, some already explored and perhaps still others that have not as yet been given extensive consideration.[3]

6.1.3 Components of Conventional Reactors

Overall

Any generating plant consists of an array of structural components and a system of mechanical and electrical controls. In a nuclear plant, there are special demands upon structural integrity and reliability. In addition, a nuclear reactor is characterized by the use of specialized materials, some aspects of which were already discussed in Chapter 5. In standard reactors, these are the fuel itself, the coolant, the moderator, and neutron-absorbing materials used to control the power level. A main distinction between different types of reactors lies in the differences in the choices of fuel, coolant, and moderator.

Fuels

There are few nuclides that can be used as reactor fuels. The paucity of possible candidates can be seen by examining the properties of the naturally occurring heavy elements:

Uranium (atomic number Z=92). This is the main fuel in actual use, especially ^{235}U which is fissile. In addition, ^{238}U is important in reactors, primarily as a fertile fuel for ^{239}Pu production, and ^{233}U could be used as a fissile fuel, formed by neutron capture in ^{232}Th.

Protactinium (Z=91). The longest-lived isotope of Pa (^{231}Pa) has a half-life of 3.3×10^4 yr, and therefore there is essentially no Pa in nature. Further, there is no stable $A=230$ nuclide that could be used to produce ^{231}Pa in a reactor.

Thorium (Z=90). Thorium is found entirely as ^{232}Th, which is not fissile (for thermal neutrons). It can be used as a fertile fuel for production of fissile ^{233}U.

   Between thorium (Z=90) and bismuth (Z=83), the isotope with the longest half-life is ^{226}Ra ($T=1600$ years), and therefore there are no fuel candidates, quite apart from the issue of fissionability. By the time the atomic number is as low as 83, the threshold for fission is much too high for a chain reaction to be conceivable (see, e.g., Ref. [6] p. 574). Thus, uranium and thorium are the only natural elements available for use as reactor fuels. In addition, ^{233}U and ^{239}Pu can be produced from capture on ^{232}Th and ^{238}U in reactors. This means that the nuclides listed in Table 4.1 exhaust the practical possibilities for reactor fuels. Of these, only ^{235}U is both fissile and found in nature in useful amounts.

   Restricting consideration to uranium fuel, there remains a number of options as to the form of the fuel. Reactors can operate over a considerable range of enrichments in ^{235}U. Enrichment to a concentration of 3% is somewhat typical in reactors used for electricity generation, but there is a trend towards higher enrichments and greater burnup of the fuel. In heterogenous reactors of the sort being considered, the fuel is solid. For the most part it is in an oxide form, UO_2, but metallic fuel is a possibility and has been used in some reactors. The fuel usually is in cylindrical pellets with typical dimensions on a centimeter scale, but some designs for future reactors are based on fuel in submillimeter microspheres embedded in

---

[3]There has been speculation about a quite different sort of molten-salt reactor, driven by a proton accelerator. If pursued, this would represent a radical departure from the sorts of reactors that have been built to date (see Section 13.5.3).

graphite. The goal here is enhanced ruggedness at high temperatures.[4]

Moderators

As discussed in Section 5.2, a moderator is required if the reactor is to operate at thermal neutron energies. This means that most operating reactors use moderators, with the fast breeder reactor the exception. The options for moderating materials are limited:

Hydrogen (Z=1). The isotopes ^{1}H and ^{2}H are widely used as moderators, in the form of light (ordinary) water and heavy water, respectively.

Helium (Z=2). The isotopes ^{3}He and ^{4}He are not used, because helium is a gas, and excessive pressures would be required to obtain adequate helium densities for a practical moderator.

Lithium (Z=3). The isotope ^{6}Li (7.5% abundant) has a large neutron-absorption cross section, making lithium impractical as a moderator.

Beryllium (Z=4). ^{9}Be has been used to a limited extent as a moderator, especially in some early reactors. It can be used in the form of beryllium oxide, BeO. However, beryllium is expensive and toxic.

Boron (Z=5). The very large neutron-absorption cross section in ^{10}B (20% abundant) makes boron impossible as a moderator.

Carbon (Z=6). Carbon in the form of graphite is widely used as a moderator. It is important that the graphite be pure, free of elements that have high absorption cross sections for neutrons.

There are no advantages in considering elements heavier than carbon. The effectiveness for moderation decreases with increasing mass, and there are no counterbalancing advantages in other properties. Again, therefore, there is a limited list of candidates: light water, heavy water, graphite, and beryllium. Any of these can be used with uranium enriched in ^{235}U. With natural uranium, it is not possible to use light water and it is common practice to use heavy water or graphite, which have particularly high moderating ratios (see Table 5.2).

Coolants

The main function of the coolant in any generating plant is to transfer energy from the hot fuel to the electrical turbine, either directly or through intermediate steps. During normal reactor operation, cooling is an intrinsic aspect of energy transfer. However, in a nuclear reactor, cooling has a special importance, because radioactive decay causes continued heat production even after the reactor is shut down and electricity generation has stopped. It is still essential to maintain cooling to avoid melting the reactor core, and in some types of reactor accidents (e.g., the accident at Three Mile Island) cooling is the critical issue.[5]

The coolant can be either a liquid or a gas. For thermal reactors, the most common coolants are light water, heavy water, helium, and carbon dioxide. The type of coolant is

---

[4]In particular, this fuel is for high-temperature gas-cooled reactors.

[5]If the fuel is designed to operate at high enough temperatures, cooling can be provided by radiation from the fuel, and maintaining the flow of coolant would not be essential under accident conditions. However, at present no operating reactors are designed for such high temperatures.

commonly used to designate the type of reactor. Hence, the characterization of reactors as light water reactors (LWRs), heavy water reactors (HWRs), and gas-cooled reactors (GCRs).

The coolant may also serve as a moderator, as is the case for LWRs and HWRs. In gas-cooled reactors, the density of the coolant is too low to permit it to serve as a moderator, and graphite is used. Fast reactors, in which fission is to occur without moderation, require a coolant that has a relatively high atomic mass number (A). Generically, these reactors are termed liquid-metal reactors, because the coolant is a liquid metal, most commonly sodium ($A=23$).

Control Materials

As discussed in Section 5.5.2, control materials are needed to regulate reactor operation and provide a means for rapid shutdown. Boron and cadmium are particularly good control materials because of their high cross sections for the absorption of thermal neutrons. These control materials are usually used in the form of rods. Control rods for pressurized water reactors (PWRs) commonly use boron in the form of boron carbide (B_4C) or cadmium in a silver–indium alloy containing 5% cadmium. Control rods for boiling water reactors (BWRs) commonly use boron carbide [7, p. 715]. In addition, boron may be introduced into the circulating cooling water to regulate reactor operation.

6.1.4 World Inventory of Reactor Types

Reactor Sizes

The capacity of a reactor can be described in terms of its electrical output, typically in megawatts of electricity (MWe), or its total energy production rate, typically in megawatts of thermal power (MWt). For present reactors, the efficiency of conversion of heat to electricity is about 32%. At this efficiency, a typical large reactor could be characterized as either a 1000-MWe reactor or a 3125-MWt reactor. The designation in terms of electrical output is by far the more common; when the notation is not explicit, it is usually safe to assume for a power reactor that "1000 MW" means 1000 MWe.

The first reactors had capacities well below 100 MWe, but there was a rapid transition to 1000-MWe reactors and larger. The move to larger size was motivated by the desire to capture economies of scale. Some analysts suggest that this escalation proceeded too rapidly, especially in the United States, and was responsible for some of the difficulties encountered in achieving short construction times and reliable operation.

Using the data in Table 6.1, the mean capacity of all reactors in operation worldwide, as of the end of 1994, was about 800 MWe. At the extremes, a class of older British gas-cooled reactors has capacities of 50 MWe, while the standard PWRs recently completed in France have capacities of about 1300 MWe, and the latest reactors under construction in France have a design capacity close to 1450 MWe.

While recent commitments in France, Japan, South Korea, and (for the one reactor being built) the United Kingdom are for large reactors, considerable attention is being given to smaller reactors, especially in the United States. While going to smaller reactors means sacrificing economies of scale, some advantages can be regained if a number of identical units are placed at the same site. Smaller individual units involve a lesser commitment of resources at one time, allow for faster construction schedules, and make it possible to have

Table 6.1. *World totals for nuclear reactors in operation and under construction, as of December 31, 1994.*

Type	Number Oper.	Number Cons[a]	Capacity (GWe) Oper.	Capacity (GWe) Cons[a]	Usual Fuel[b]	Mod-erator	Cool-ant	First devel.
PWR	245	39	215.7	36.8	UO_2 enr	H_2O	H_2O	USA
BWR	92	6	75.9	6.0	UO_2 enr	H_2O	H_2O	USA
PHWR[c]	34	16	18.6	7.9	UO_2 nat	D_2O	D_2O	Canada
LGR	15	1	14.8	0.9	UO_2 enr	C	H_2O	USA/USSR
GCR	35	0	11.7	0	U, UO_2[d]	C	CO_2	UK
LMFBR	3	4	0.9	2.4[e]	UO_2+PuO_2	None	Liq Na	Various
Total	424	66	338	54				

Sources: Capacity data are from Ref. [9]. Fuel and country data are from Ref. [10], p. 67.
[a]Reactors nominally under construction or on order; some may never be completed.
[b]Fuel designations: enr=enriched in ^{235}U, nat=natural.
[c]Includes one 148-MWe HWLWR in Japan.
[d]Natural U used for GCR; enriched UO_2 used for AGCR.
[e]The Superphenix reactor is included in this total (see Section 6.3.3).

more of the plant built at the factory rather than at the final site. Further, the impact of an accident is less for a smaller reactor; more debatably, the probability of an accident may also be less.

Types of Reactor

A variety of different reactors is in use in the world today, although there was greater diversity in the early days of reactor design. Table 6.1 lists the types of nuclear power plants in operation at the end of 1994 as well as those reported to be under construction or on order, as tabulated in *Nuclear News* [9]. The dominant reactor is the light water reactor (LWR), which uses ordinary water as both the coolant and moderator and enriched uranium in UO_2 fuel. There are two types of light water reactor: the pressurized water reactor (PWR) and the boiling water reactor (BWR). Together, they account for 86% of the world's present generating capacity and 79% of the capacity nominally being built or on order. The main types of reactors in past or present use for electricity generation are

PWR. The pressurized water reactor accounts for almost two-thirds of all capacity and is the only LWR used in some countries, for example, France, the former Soviet Union, and South Korea.

BWR. The boiling water reactor is a major alternative to the PWR, and both are used in, for example, the United States and Japan.

PHWR. The pressurized heavy water reactor uses heavy water for both the coolant and moderator and operates with natural uranium fuel. It has been developed in Canada and is commonly referred to as the CANDU.[6] CANDU units are also in operation in India and are being built in Romania and South Korea.

LGR. The light-water-cooled, graphite-moderated reactor uses water as a coolant and graphite for moderation. The world's only currently operating LGRs are the RBMK

[6]This acronym stands for Canadian deuterium uranium and has an obvious double meaning.

reactors in the former Soviet Union. There were four such units at the Chernobyl plant at the time of the accident there, two of which are still in operation.

GCR. The gas-cooled, graphite-moderated reactor uses a CO_2 coolant and a graphite moderator. Its use is largely limited to the United Kingdom; it is sometimes known as the Magnox reactor. A larger second-generation version is the advanced gas-cooled, graphite-moderated reactor (AGCR).

HTGR. The high-temperature gas-cooled reactor uses helium coolant and a graphite moderator. The only operating HTGR in the United States (Fort St. Vrain) has been shut down, and there are no operating HTGRs elsewhere.

LMFBR. The liquid-metal fast breeder reactor uses fast neutrons and needs no moderator. A liquid metal is used as coolant, now invariably liquid sodium. There are four liquid-metal reactors in operation (two in France and one each in Kazakhstan and Russia) and several others under construction, including one in Japan.

HWLWR. The heavy-water-moderated, light-water-cooled reactor is an unconventional variant of the heavy water reactor, and only one has been in recent operation, a 148-MWe plant in Japan.

It is seen in Table 6.1 that the number of reactors nominally under construction or on order at the end of 1994 (66) was small compared to the number in operation (424). Many of these reactors do not represent firm commitments with assured completion, although some programs, especially those in France, Japan, and South Korea, appear to be secure. If all were completed, these reactors would represent an addition of about 16% to world generating capacity. Individual reactor sizes differ widely. For example, the reactors under construction include five 220-MWe PHWRs in India and four PWRs in France having listed capacities of 1450 MWe or 1455 MWe.

The dominance of light water reactors, for both plants in operation and those under construction, is seen in Table 6.1. These reactors were first developed in the United States, in both the PWR and BWR configurations, and have become dominant in almost all other major nuclear countries. The main exceptions are Canada, the United Kingdom, the former Soviet Union (FSU), and India. Even in the United Kingdom and, with one exception, in the FSU, the reactors under construction are LWRs.

History of Reactor Development

Immediately after World War II the three pioneering countries in nuclear reactor development were the United States, Canada, and the United Kingdom. Each went in a different direction.

The first United States power reactors, beyond plutonium-producing or experimental reactors, were built for submarines, not for civilian electricity generation. The earliest were a PWR for the submarine *Nautilus*, commissioned in 1955, and a sodium-cooled reactor for the submarine *Seawolf*. The *Seawolf* reactor had difficulties, and sodium-cooled reactors were abandoned by the navy in 1956 in favor of light water reactors [8, p. 423]. The navy PWR program provided the foundation for the development of PWRs for electricity generation, starting with the 60-MWe reactor at Shippingport, Pennsylvania, in 1957.

As was noted in Section 1.2, during the 1950s a varied group of reactors was ordered in the United States. These even included a small fast breeder reactor (Fermi I) in Michigan, which went into operation for a few years starting in 1966. However, since the mid-1960s the only power reactors put into operation in the U.S. have been PWRs and BWRs, with the sole exception of the trouble-plagued Fort St. Vrain HTGR in Colorado, which has now been shut down. The commercial BWRs were an outgrowth of a program of experimental BWR development carried out in the mid-1950s at Argonne National Laboratory.

The United Kingdom and Canada followed routes that did not require enriched uranium. The United States had a monopoly on uranium enrichment at the time, and although it presumably would have provided enriched uranium to such close allies, there may have been a reluctance on their part to become dependent. The United Kingdom program began very early, with two 50-MWe reactors at Calder Hall in 1956. These were GCRs, with graphite moderation and CO_2 cooling. They differed from most later reactors in the world in that they used uranium metal for the fuel, not uranium dioxide (UO_2). They gained the name Magnox, because the fuel pin cladding material was a magnesium alloy called Magnox [10, p. 165]. The GCRs that were built later had increasing size, up to 420 MWe. From the mid-1970s until the recent construction of the Sizewell B reactor, the few new plants brought on line in the United Kingdom have been AGCRs in the 600–700 MWe range. Like the Magnox reactors, they use graphite moderation and CO_2 cooling, but their fuel is enriched UO_2.

On the whole, after a fast start the British reactor program has moved fitfully, with indecision abetted by the abundance of North Sea oil and a strong commitment to domestic coal. In recent years, only one reactor was under construction in the United Kingdom, the 1188-MWe Sizewell B reactor [9], going into operation in 1995. Interestingly, it is a PWR, selected after prolonged study, adding further to the dominance of LWRs in the world nuclear picture.

The Canadian nuclear program offers the main successful alternative to the LWR among reactors now in operation or under construction. This program has involved only one type of reactor, the pressurized heavy water reactor (PHWR), known as the CANDU (Canadian deuterium uranium). Use of a deuterium moderator enabled Canada to use natural, rather than enriched, uranium. This was an attractive option for a country that had sophisticated engineering capabilities but no enrichment facilities and that wanted to be independent of outside suppliers of enriched uranium [11]. CANDU reactors regularly place among the world leaders in capacity factor. A moderate further expansion of the number of PHWRs is in progress in several countries, including India, Romania, and, for some of its reactors, South Korea. In Canada, these reactors have varied relatively little in size, starting at 525 MWe and most recently built at 881 MWe. India has emulated the Canadian example of reliance on PHWRs, although at the smaller size of 220 MWe. The first of these were constructed under Canadian supervision, but subsequently India has assumed independent responsibility. Romania also has opted entirely for Canadian PHWRs, but none of these is in operation as yet.

The other major dissenter from LWRs had been the Soviet Union, but this is now changing. The Soviet program started with light-water-cooled, graphite-moderated reactors (LGR), the best known of which are the Chernobyl RBMK-1000 reactors. These were used for both plutonium production and electricity production, as was the now-closed Hanford N reactor in Washington State. Plants listed as being under construction in the FSU are almost all PWRs, with the exception of one additional LGR at Kursk in Russia [9]. Thus, with the exception of the Canadian PHWR, what worldwide growth there is in nuclear power is now almost all in the form of LWRs. It is not clear whether this is because of intrinsic technical and economic advantages or because of historical and commercial forces. Some sophisticated opinion favors the CANDU, but the LWR won out in most of the world marketplace, including countries such as the United Kingdom and USSR, which were not particularly subject to United States pressure.

6.2 LIGHT WATER REACTORS

6.2.1 PWRs and BWRs

The two types of LWR in use in the world are the pressurized water reactor (PWR) and the boiling water reactor (BWR). The difference between them, as embodied in the names, is in the condition of the water used as coolant and moderator. In the PWR, the water in the reactor vessel is maintained in liquid form by high pressure. Steam to drive the turbine is developed in a separate steam generator. In the BWR, steam is provided directly from the reactor. These differences are brought out in the schematic representation of the two types of reactors in Fig. 6.1.

Under typical conditions in a PWR, temperatures of the cooling water into and out of the reactor vessel are about 292 °C and 325 °C, respectively, and the pressure is about 155 bar [7, p. 713].[7] For the BWR, typical inlet and outlet temperatures are 278 °C and 288 °C, respectively, and the pressure is only about 72 bar. The high pressure in the PWR keeps the water in a condensed phase; the lower pressure in the BWR allows boiling and generation of steam within the reactor vessel.

Neither the PWR or BWR has an overwhelming technical advantage over the other, as indicated by the continued widespread use of both. Among the major LWR users, the United States, Japan, and Germany use both types, while France and Russia use only PWRs in the LWR part of their programs. Overall, the number of PWRs in operation is significantly greater than the number of BWRs, and PWRs also have a large lead in reactors listed as under construction. The future is not clear-cut, however, and in Japan 3 BWRs were under construction at the end of 1994, compared to only 1 PWR [9]. In the discussion below, we will emphasize the PWR in giving specific illustrations, but to some extent we will consider both.

6.2.2 Components of a Light Water Reactor

The containment structure and enclosed components for a typical PWR and a typical BWR are shown schematically in Figs. 6.2 and 6.3.[8] The most conspicuous difference between the PWR and the BWR is the absence of a steam generator in the BWR. Another difference is the presence in the bottom of the BWR containment of a pool of water, the suppression pool, for condensing escaped steam. In essence, however, the two reactors are similar.

At the heart of the reactors, literally and figuratively, is the reactor core, contained within the reactor pressure vessel. The pressure vessel encloses three vital components:

The fuel itself, contained in many small fuel rods comprising the reactor core

The surrounding water, acting as coolant, moderator, and heat-transfer agent

---

[7]1 bar$=10^5$ newton/m$^2=0.987$ atm.
[8]These diagrams are copied from a draft version of Ref. [12].

Boiling water reactor (BWR)

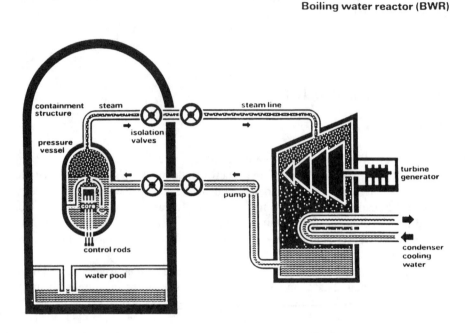

Pressurized water reactor (PWR)

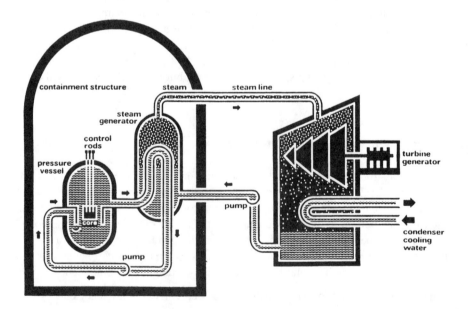

Figure 6.1. Schematic representation of BWR and PWR systems, emphasizing the difference in the means for providing steam to the steam turbine (adapted from figures provided by the U.S. Council on Energy Awareness).

Control rods, used to maintain the reactivity at the desired level

The reactor pressure vessel is a massive cylindrical steel tank. Typically for a PWR, it is about 12 m (40 ft) in height and 4.5 m (15 ft) in diameter [13, p. 304]. It has thick walls, about 20 cm (8 in), and is designed to withstand pressures of up to 170 atmospheres.

A second major component, or set of components, is the system for converting the reactor's heat into useful work. In the BWR, steam is used directly from the pressure vessel to

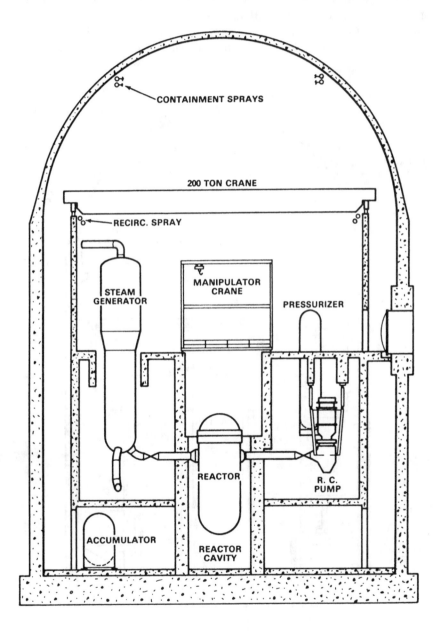

Figure 6.2. Schematic diagram of containment building with enclosed reactor vessel and steam generator for an illustrative PWR: the 781-MWe Surry plant. The output of the steam generator drives the turbines, external to the containment building. (From Ref. [12], p. 4-4.)

drive a turbine. This is the step at which electricity is produced. In the PWR, primary water from the core is pumped at high pressure through pipes passing through a heat exchanger in the steam generators. Water fed into the secondary side of the steam generator is converted into steam, and this steam is used to drive a turbine. The secondary loop is also closed. The exhaust steam and water from the turbine enter a condenser and are cooled in a second heat exchanger before returning to the steam generator.

The cold side of the condenser heat exchanger represents the tertiary loop for the PWR. In principle, this loop need not be closed, and the condenser cooling water could circulate to and from a river or the ocean. More commonly, the condenser output is circulated through a cooling tower, where it is cooled (in still another heat exchanger). Water in the cool side of

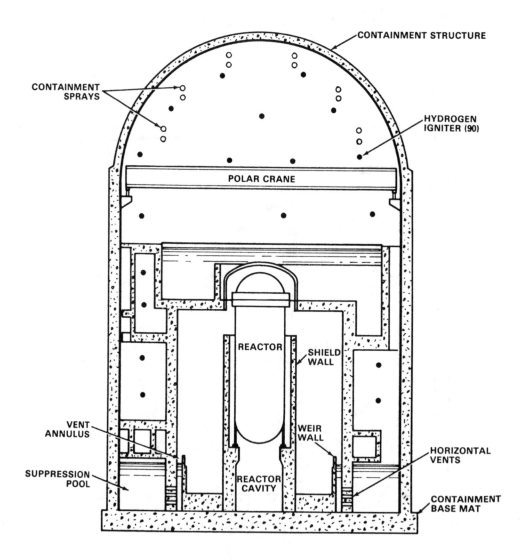

Figure 6.3. Schematic diagram of containment building with enclosed reactor vessel for an illustrative BWR: the 1142-MWe Grand Gulf plant. Steam from the reactor vessel drives the turbines, external to the containment building. (From Ref. [12], p. 4-40.)

the cooling tower heat exchanger is partially turned into steam, and the ultimate heat sink becomes the air. There is some exchange of water and loss of heat as the water lost in steam is made up by water drawn from, say, a river.

Cooling towers became an ominous symbol during the Three Mile Island accident, but they themselves are quite benign. The steam that may be seen rising from a cooling tower is isolated from water passing through the reactor core by three heat exchangers (two in a BWR). Such cooling towers are not unique to nuclear power plants and are used in other places where it is necessary to dissipate large amounts of heat, including coal-fired plants.

The pressure vessel and the steam generators are contained within a massive structure, the *containment building*, commonly made of strongly reenforced concrete. In some designs, the concrete containment is lined with steel; in others, there is a separate inner steel containment vessel. The containment is intended to retain activity released during accidents and is also believed capable of protecting a reactor against external events such as an airplane crash. In the Three Mile Island accident, the containment very successfully retained the released radioactivity, although it may be noted that the physical structure was not put fully to the test, since there was no explosion or buildup of high pressures. At Chernobyl there was no containment, with disastrous results. In principle, if a reactor is sufficiently protected against accidents, a containment is unnecessary. Nonetheless, it is widely thought to be an essential safety feature, providing an important redundancy.

6.2.3 PWR Reactor Cores

We here consider the specific characteristics of a reactor core based on a Westinghouse Corporation PWR design, but the gross features are similar for all large LWRs.[9] The reactor fuel is in the form of cylindrical uranium oxide (UO_2) pellets, about 0.8 cm in diameter and 1.35 cm in length. The pellets are placed in tubes—called fuel rods or fuel pins—made of zircaloy, a zirconium alloy (98% Zr, 1.5% Sn, and small amounts of other metals [13, p. 234]) selected on the basis of structural strength and low neutron absorption. The zircaloy cladding is thin, 0.06 cm. The fuel rod is typically 3.7 m (12 ft) in length and 1.0 cm in diameter. There is some free space within the fuel rod to allow for the expansion of the fuel pellets and to accommodate gaseous fission products such as xenon and krypton. The expansion is due to both increased temperature and the replacement in the fuel of one uranium atom by two fission-product atoms. Noble gases produced as fission products may be trapped as bubbles in the fuel or may escape from the fuel into the gap between the fuel and cladding.

A 17×17 array of fuel rods forms a "fuel bundle" or *assembly*. Although this would allow 289 fuel rods in an assembly, positions are left open in some assemblies for the insertion of control rods or measuring instrument rods. The total core contains 193 assemblies and 50 952 fuel rods. Some 53 of the assemblies have spaces for clusters of 20 control rods, which can be moved in and out within the assembly. These control rods are made from a silver–indium–cadmium alloy.

As the ^{235}U is consumed in the reactor, the reactivity of the fuel decreases. This is compensated for in several ways. Control rods, initially used to limit the reactivity, are partly withdrawn, or soluble poison concentrations in the cooling water are reduced. Further, fuel assemblies are replaced on a rotating basis. Thus, although a particular fuel assembly may

---

[9]Most of the detailed numbers in this paragraph are based on Westinghouse Corporation information, as reported in Ref. [13], especially Table 9.1. They are specific to this particular Westinghouse design; others differ in detail.

remain in the reactor for three years, one-third of the core can be changed every year. (Recently, cores have been designed to have longer times between fuel changes.)

6.3 BURNERS, CONVERTERS, AND BREEDERS

6.3.1 Characterization of Reactors

As discussed in Section 5.1.1, the condition for a chain reaction is that for every neutron initiating fission in one generation, one or more neutrons initiates fission in the next generation. If, in addition, another fissile nucleus is produced for every ^{235}U atom consumed, then there is no decrease in the amount of nuclear fuel available. This is the principle of the breeder reactor.

The *conversion ratio C* (or *breeding ratio B*) is defined as the ratio of the rate of production of fissile nuclei to the rate of consumption of fissile nuclei (see Eq. 5.21).[10] For uranium fuel, this is the ratio of ^{239}Pu produced to ^{235}U consumed. If the conversion ratio is small, the reactor is sometimes called a *burner*; if the conversion ratio is between about 0.7 and 1.0, it is called a *converter*; and if it exceeds unity, the reactor is called a breeder (see, e.g., Ref. [13] p. 458). In other usage, the term converter denotes a reactor in which different nuclides are burned and produced (e.g., ^{235}U and ^{239}Pu), while a breeder reactor is one in which a nuclide is burned and a greater amount of the same nuclide is produced (e.g., ^{239}Pu).

6.3.2 Achievement of High Conversion Ratios in Thermal Reactors

Difficulty of Reaching a Conversion Ratio of Unity with ^{235}U

As discussed in Section 5.4, the limiting condition for a breeder reactor is that the conversion ratio C be at least 1. This means that η, the number of neutrons produced for each neutron absorbed in ^{235}U, must be 2 or more.[11] For thermal neutrons absorbed in ^{235}U, η=2.075 (see Table 5.1). Were there no losses, this would suffice for breeding: 1 neutron for continuing the chain reaction, 1 neutron for production of ^{239}Pu, and 0.08 neutrons free to be "wasted." Such efficient utilization cannot be achieved, however, because there is absorption in the moderator and other nonfuel materials, as well as escape of neutrons from the core. Therefore, a ^{235}U-fueled thermal breeder reactor is not practical. Nonetheless, the production of ^{239}Pu is significant, increasing the overall energy output from the reactor fuel beyond that gained from the ^{235}U alone.

Potential of ^{233}U for a Thermal Breeder Reactor

The number of neutrons produced is significantly higher for ^{233}U (η=2.296) than for ^{235}U, and there have been serious suggestions for developing ^{233}U thermal breeders. These date to as early as 1945, in work done by Eugene Wigner's group in Chicago [3, ch. 6]. A cycle is envisaged in which ^{233}U is produced initially in a reactor with ^{235}U as the fissile fuel and ^{232}Th as the fertile fuel. Subsequently, a ^{233}U–^{232}Th cycle could in principle be self-sus-

---

[10]Sometimes a distinction is made in terminology, with conversion ratio used when $C<1$ and breeding ratio used when $C>1$.

[11]The distinction between η and the commonly cited parameter ν is discussed in Section 5.1.2.

taining. Not only is η higher for ^{233}U than for ^{235}U, but the capture cross section is signifi-cantly higher for ^{232}Th than for ^{238}U at thermal energies, making the conversion ratio higher than in the ^{235}U–^{238}U cycle (see Table 5.1).

However, while thermal breeders based on ^{233}U are in principle possible, and preliminary exploratory work towards their development was done at Oak Ridge National Laboratory in the 1950s, thermal breeders were abandoned in favor of the fast breeder reactor. It is conceivable that interest in thermal breeders could revive, but at present the relatively few breeder reactor development programs in the world are all based on fast breeders.

High Conversion Ratios without Breeding

Before turning to fast breeder reactors, it may be noted that even if breeding is not achieved with thermal reactors, a high conversion ratio can still be desirable. One motivation could be plutonium production. Another motivation is the extension of fuel resources. As the conver-sion ratio increases, the energy output increases for a given original ^{235}U content. A high conversion ratio means a high ratio of capture in ^{238}U to absorption in ^{235}U. This must be accomplished without losing criticality. Greater losses of neutrons to ^{238}U can be compen-sated for by smaller losses in the moderator.

The use of carbon instead of light water as a moderator is favorable on two counts, if a high conversion ratio is desired (in addition to the advantage that with a carbon moderator it is possible to use natural uranium). Because carbon is a less effective moderator than water, more collisions are required to reach thermal energies, and therefore there is more possibility of neutron capture in ^{238}U at intermediate energies. Further, because of the low neutron-capture cross section in ^{12}C (see Section 5.2.2), the loss of thermal neutrons to absorption will be less for carbon than for light water.[12] Together, this means a higher conversion ratio. It may be noted that reactors designed with production of weapons-grade plutonium in mind, as either the main or an auxiliary function, have been mostly graphite-moderated. Examples include the Windscale plant in England, the plutonium production reactors at Hanford, and the RBMK reactors built in the USSR.

The conversion ratio is higher in a heavy water reactor than in a light water reactor for much the same reasons as for a graphite moderator. Again, heavy-water is a less effective moderator than light water, and it has a lower capture cross section for thermal neutrons. For both graphite-moderated and heavy water reactors, there have been suggestions that the ^{232}Th–^{233}U cycle be used, to further increase conversion and extend the life of the uranium fuel, even without breeding.

6.3.3 Fast Breeder Reactors

Plutonium as Fuel for Fast Breeders

The fast breeder reactor relies on a chain reaction in which the neutrons are not thermalized but instead produce fission at relatively high energies. If ^{239}Pu is the fissile fuel, this has the advantage of a relatively large fission cross section σ_f and a relatively large number of neutrons per fission event (high ν and η). For 1.0-MeV neutrons on ^{239}Pu, $\sigma_f = 1.7$ b, and the

---

[12]In terms of the formalism introduced in Chapter 5, this means that carbon leads to a lower resonance escape probability p and a higher thermal utilization factor f than does light water (see the four-factor formula, Eq. 5.5). Criticality can still be maintained ($k = 1$), and C will be greater (see Eq. 5.28).

ratio (α) of the capture cross section to the fission cross section is less than 0.03. The low value of this ratio means that almost all absorption in ^{239}Pu leads to fission ($\eta \doteq \nu \doteq 3.0$) [14]. In contrast, the fission cross section in ^{235}U is 1.2 b at 1 MeV, and η is only 2.3. With about three neutrons per fission, breeding with ^{239}Pu is obtained if 1 neutron is used for continuing the chain reaction, 1^{+} for breeding, and 1^{-} for losses. It has been shown that it is quite possible to achieve this.

To avoid thermalization of the neutrons, fast breeder reactors use a coolant with high mass number A to reduce moderation. Liquid metals have the best combination of high A and favorable heat-transfer properties, and the fast breeder reactors in actual use or under construction are *liquid-metal fast breeder reactors* (LMFBR). The standard choice for the coolant is liquid sodium (^{23}Na).

The fuel is made of pellets of mixed plutonium and uranium oxides, PuO_2 (about 20%) and UO_2 (about 80%). Uranium depleted in ^{235}U is commonly used, it being available as a residue from earlier enrichment. The fission cross section for ^{239}Pu is between 1.5 and 2.0 b over virtually the entire fast-neutron region (from 0.01 to 6 MeV) [14], and therefore fission in ^{239}Pu is much more important than fission in ^{238}U (see Fig. 4.1). The most probable fast-neutron reactions in ^{238}U are inelastic scattering, which produces lower energy neutrons, and capture, which produces ^{239}Pu.

Status of Fast Breeder Programs

The main incentive for the development of fast breeder reactors is the extension of uranium supplies. A fast breeder economy would extract much more energy per tonne of uranium than is obtained from other reactors, e.g., the LWRs. Further, with more energy per unit mass, it becomes economically practical to use more expensive uranium ores, increasing the ultimate uranium resource.

However, during the 1980s and early 1990s the rate of expansion in the use of nuclear power fell far short of earlier expectations, there was little pressure on uranium supplies, and interest in fast breeder reactors declined. Further, the initial fast breeder reactors proved to be more expensive than alternatives such as the LWR or HWR. Nonetheless, some development of breeders has continued, in part to develop the technology as an insurance against future needs.

In a later turn of the argument, it has been pointed out that a liquid-metal fast reactor (LMR) can be used to *destroy* unwanted plutonium and other heavy elements in weapons stockpiles or nuclear wastes. In this reversal of motivation, the LMR would be used to consume unwanted plutonium, rather than to produce plutonium as a fuel. There is flexibility in this, because as LMR technology and facilities are developed, they could be turned to either purpose. However, if the driving fear is concern about misuse of plutonium, it may be decided to dispose of plutonium from weapons stockpiles in ways that do not involve expanding a technology that is closely related to potential plutonium production.

France has led in the development and deployment of breeder reactors, with two completed reactors, the 233-MWe Phenix, put into operation in 1973, and the 1200-MWe Superphenix at Creys-Malville, which first generated electricity in 1986. The Superphenix was shut down for almost two years beginning in May 1987 because of leaks in the sodium-filled spent-fuel storage tank. Although it resumed some operation in 1989, troubles recurred, and from 1989 through the summer of 1995 operation of Superphenix has been intermittent and trouble-plagued, with long periods of shutdown [15]. The emphasis has been shifted from operation of Superphenix as a breeder reactor to its use for study of the safety of sodium-cooled fast reactors and the destruction of heavy elements in fast reactors. Russia and Kazakhstan each have one LMFBR in operation [9]. Japan has completed a 280-MWe prototype LMFBR, but its operation was set back in December 1995 by a leak of the sodium coolant and a fire-like chemical reaction of sodium with the air [16].

The United States breeder reactor program has been marked by indecision and opposition, with successive projects started and abandoned. The latest apparent casualty was the main U.S. breeder-related project of the past decade—the investigation at Argonne National Laboratory of fast LMRs as part of the *integral fast reactor* program (see Chapter 13). Advocates of this program stressed its potential to offer a high degree of safety against reactor accidents and to destroy nuclear wastes in an on-line process. The breeding potential was often secondary in these arguments, and the planned LMR need not have operated as a breeder, namely, with a conversion ratio greater than unity. Nonetheless, the basic configuration of the system was similar to that of a breeder reactor. Culminating several years of debate, most funding for this project was dropped from the federal budget for the 1995 fiscal year.

Two considerations particularly contributed to the difficulties of the breeder program: (1) the slowness in nuclear power growth pushed back the time at which uranium shortages might make breeders necessary; and (2) there was concern that the availability and large-scale use of ^{239}Pu might increase dangers from terrorism and, by example, of nuclear weapon proliferation (see Section 7.4.2 and Chapter 14). These factors made the program a vulnerable target, at a time when there was significant opposition to any federal assistance for projects to advance nuclear power.

6.4 THE NATURAL REACTOR AT OKLO

A remarkable discovery was made in 1972 by French scientists analyzing uranium extracted from the Oklo uranium mine in Gabon. The uranium was depleted in ^{235}U, sometimes by large amounts, although normally the isotopic ratio in uranium is nearly constant over the surface of the Earth. It was soon suspected and then demonstrated that this isotopic anomaly was due to a natural uranium chain reaction occurring more than a billion years ago. Conclusive evidence in support of this explanation was provided by the relatively high abundance of intermediate-mass nuclei, the rare earths, which are characteristic fission products but are normally not found in large abundance in nature (see, e.g., Ref. [17]).

The scenario, as it has been recreated, puts the event about 1.8 billion years ago. At that time, the isotopic abundance of ^{235}U exceeded its present value by the factor $\exp(\Delta\lambda t)$, where $\Delta\lambda$ is the difference in the decay constants of the two isotopes, and t is the time since the event. The decay constants of ^{235}U and ^{238}U are 0.985×10^{-9} yr^{-1} and 0.155×10^{-9} yr^{-1}, respectively (see Table 2.1). Therefore, at $t=1.8\times10^9$ yr, $\Delta\lambda t=1.49$, and the isotopic abundance has changed by a factor of 4.4, putting the past enrichment at slightly above 3%. (This is strikingly close to the enrichment used in modern PWRs.) The intrusion of water, acting as a moderator, apparently initiated a chain reaction, which appears to have simmered for at least several hundred thousand years. In this explanation, the reaction did not occur earlier because the concentrated uranium deposits had been only recently formed by the leaching of rocks and the precipitation of the dissolved uranium.

Aside from having posed an intriguing scientific puzzle, with a very interesting explanation, the Oklo event is considered by some to have significance as a test of the motion of fission products through the ground. For the most part, these products have moved very little over a period of more than one billion years. This could have implications for the rate of movement of fission products in buried nuclear wastes. The Oklo example cannot be used as an all-embracing guide because differences in the chemical form of the product and in the type of rock formation may vitiate an extrapolation from Oklo to the behavior of an individual modern waste-disposal site. However, Oklo illustrates that under some circumstances, nuclear wastes will not migrate appreciably from a waste emplacement site.

REFERENCES

1. H. Barnert, V. Krett, and J. Kupitz, "Nuclear Energy for Heat Applications," *IAEA Bulletin* 33 (1) (1991): 21–24.
2. Alvin M. Weinberg and Eugene P. Wigner, *The Physical Theory of Neutron Chain Reactors* (Chicago: University of Chicago Press, 1958).
3. Alvin M. Weinberg, *The First Nuclear Era: The Life and Times of a Technological Fixer* (New York: AIP Press, 1994).
4. J. Smith, "Novel Reactor Concepts," in *Nuclear Power Technology,* vol. 1 of *Reactor Technology,* edited by W. Marshall (Oxford: Clarendon Press, 1983), pp. 390–415.
5. Uri Gat and H. L. Dodds, "The Source Term and Waste Optimization of Molten Salt Reactors with Processing," in *Future Nuclear Systems: Emerging Fuel Cycles & Waste Disposal Options, proceedings of Global '93,* (La Grange Park: American Nuclear Society, 1993), 248–252.
6. Emilio Segrè, *Nuclei and Particles,* 2nd ed. (Reading, Mass.: W. A. Benjamin, 1977).
7. Ronald Allen Knief, *Nuclear Engineering: Theory and Technology of Commercial Nuclear Power,* 2nd ed. (Washington, D.C.: Hemisphere Publishing Company, 1992).
8. Richard G. Hewlett and Jack M. Holl, *Atoms for Peace and War, 1953–1961* (Berkeley: University of California Press, 1989).
9. "World List of Nuclear Power Plants," *Nuclear News* 38, 3 (March 1995): 27–42.
10. D. J. Bennet and J. R. Thomson, *The Elements of Nuclear Power,* 3rd ed. (Essex: Longman, 1989).
11. J. S. Foster and E. Critoph, "The Status of the Canadian Nuclear Power Program and Possible Future Strategies," *Annals of Nuclear Energy* 2, 11/12 (1975): 689–703.
12. U. S. Nuclear Regulatory Commission, *Reactor Risk Reference Document,* NUREG-1150, vol. 1, Draft for Comment (Washington, D.C.: NRC, 1987).
13. F. J. Rahn, A. G. Adamantiades, J. E. Kenton, and C. Braun, *A Guide to Nuclear Power Technology* (New York: John Wiley, 1984).
14. Victoria McLane, Charles L. Dunford, and Philip F. Rose, *Neutron Cross Section Curves,* vol. II of *Neutron Cross Sections,* (New York: Academic Press, 1988).
15. "France: Superphénix up again," *Nuclear News* 38, no. 14 (November 1995): 37–38.
16. "A Sodium Leak at Monju," *Nuclear News* 39, no. 1 (January 1996): 13.
17. Michel Maurette, "Fossil Nuclear Reactors," *Annual Review of Nuclear Science* 26 (1976): 319–350.

Chapter 7

Nuclear Fuel Cycle

7.1 STEPS IN THE NUCLEAR FUEL CYCLE

The *nuclear fuel cycle* is the progression of steps in the utilization of fissile materials, from the initial mining of the uranium (or thorium) through the final disposition of the material removed from the reactor. It is appropriately called a cycle because in the general case some of the material taken from a reactor may be used again, or "recycled."

The extreme case is the *breeding cycle*, in which there is more fissile material in the spent fuel removed from the reactor than in the fresh fuel put into the reactor. An intermediate case is the *reprocessing cycle*, in which plutonium and uranium in the spent fuel are retrieved for subsequent use but in amounts insufficient to be more than a supplement to the fuel derived directly from fresh ore. Finally, there is the *once-through* (or "throw away") cycle, in which the spent fuel is treated as waste and then discarded, without further use in reactors. Even in the once-through cycle, however, until the spent fuel is disposed of in some irretrievable fashion, there remains the potential for using it in reprocessing or breeding cycles.

A schematic picture of the fuel cycle is shown in Fig. 7.1, which indicates alternative paths, with and without reprocessing [1]. At present, all United States commercial reactors and the majority of reactors worldwide are operating with a once-through fuel cycle, although some countries, particularly France, have large-scale reprocessing programs with some reuse of plutonium from spent fuel.

The steps in the fuel cycle that precede the introduction of the fuel into the reactor are referred to as the *front end* of the fuel cycle. Those that follow the removal of the fuel from the reactor comprise the *back end* of the fuel cycle. At present there is essentially no back end to the fuel cycle in the United States, as virtually all commercial spent fuel is accumulating in cooling pools or storage casks at the reactor sites. Implementation of a spent fuel disposal plan, or of a reprocessing and waste disposal plan, would represent the "closing" of the back end of the fuel cycle. This closing is viewed by many to be an essential condition for increasing the use of nuclear power in the United States—and perhaps even for its continued use beyond the next several decades.

Key aspects of the fuel cycle will be surveyed in the remainder of this chapter. The fuel cycle will be discussed particularly in the context of light water reactors, in view of their dominance among world nuclear reactors. The main aspects of this discussion are relevant to other types of reactors as well. A more extensive treatment of the crucial step in closure, waste disposal, will be given in Chapters 8–10.

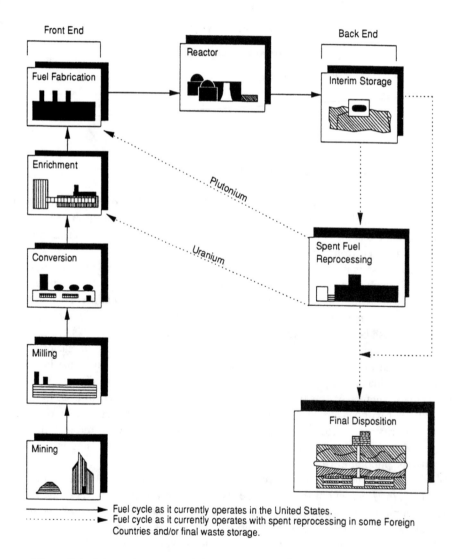

Figure 7.1. Schematic picture of the nuclear fuel cycle. (From Ref. [1], p. 53.)

7.2 FRONT END OF THE FUEL CYCLE

7.2.1 Uranium Mining and Milling

Uranium Deposits in the Earth's Crust

The concentration of uranium varies greatly among different geological formations. The average concentration in the Earth's crust is about 3 parts per million (ppm) by weight, but extremes extend from under 1 ppm to over 500 000 ppm. The highest concentrations are seen in some pitchblende deposits. The term *uranium ore* is sometimes reserved for rocks with a uranium content of 350 ppm or higher. At these concentrations, the content is expressed in terms of the uranium grade, in percent. Thus, 350 ppm corresponds to a grade of 0.035%. The higher the grade, the less ore that must be extracted, which in general leads to lower costs. Ores below a grade of 0.05% are considered low-grade ores and have not been widely needed. Of course, the ultimate criterion for mining is overall cost, not grade *per se*, and open-pit mining has utilized ores down to 0.04% [2, p. 411]. The average grade of uranium used in the United States in recent years (1987–1991) has been about 0.24% [3].

Uranium in rocks is mostly in the form of a uranium oxide, U_3O_8. The U_3O_8 is extracted by crushing the rock and leaching with acid, and is then recovered from the liquid and dried. The product is concentrated U_3O_8, called *yellowcake*.[1]

Radon Exposures from Uranium Mining and Mill Tailings

In the early days of uranium mining, little attention was paid to radiation safety. In the Middle Ages, long before uranium had been identified as an element, metal miners in southern Germany and Czechoslovakia contracted lung ailments, called Bergkrankheit ("mountain sickness"). Modern doctors have attributed the ailment to lung cancer caused by a chance high uranium content in the rock formations being mined. The ^{238}U decays to ^{226}Ra and then to radon gas (^{222}Rn) and its radioactive progeny. Inhalation of these "radon daughters" leads to lung cancer.

As one would expect, the problem of radon exposure is more extreme in uranium mines than in other sorts of mines. It became a particularly serious problem, for example, in the United States, Czechoslovakia, and Canada when large-scale uranium mining was begun in the 1940s to meet the demands of weapons and nuclear power programs. By the late 1950s steps were initiated in the United States to reduce radon exposures, mainly through better ventilation, and by the 1970s the average exposures of uranium miners had become quite low (lower than that from indoor radon in many homes). But a good deal of damage had already been done, and there is unambiguous evidence of increased lung cancer fatalities among uranium miners.

The residues of the milling operation, representing the remainder of the ore after extraction of the U_3O_8 are the *mill tailings*. All of the uranium progeny, starting with ^{230}Th, are present in the tailings.[2] The radionuclide ^{230}Th has a half-life of 75 400 years and thus sustains the remainder of the uranium series for a long period of time. This means continuous production of radon, with some escaping to the atmosphere. Of course, these steps do not increase the rate of radon production above what it would have been without mining, but the

---

[1] U_3O_8 is not yellow in its pure form. Yellowcake is about 85% U_3O_8 [4, p. 241], and the yellow color results from another uranium compound in the ore.

[2] The ^{234}U remains in the yellowcake, and the radionuclides between ^{238}U and ^{234}U are short-lived.

radon in the tailings can more readily reach the atmosphere than can radon in underground ore. At one time this was viewed by some as constituting an important environmental hazard, and it is still deemed necessary to take remedial measures to limit radon emissions from the tailings (using overlying layers of material to impede radon escape). However, interest in the issue has diminished, as it has become obvious that exposures from "normal" indoor radon pose a much more serious problem, in terms of both the number of people impacted and the magnitude of the radon concentrations to which they are exposed.[3]

7.2.2 Enrichment of Uranium

Degrees of Enrichment

Natural uranium has an abundance (by number of atoms) of 0.0055% ^{234}U, 0.720% ^{235}U, and 99.275% ^{238}U.[4] The fissile nuclide in thermal reactors is ^{235}U. For reactors that require uranium with a higher isotopic fraction of ^{235}U than is found in natural uranium, *enrichment* is necessary. This is of course the case for light water reactors.

Fuel used in LWRs in recent years has been enriched to ^{235}U concentrations ranging from under 1% to over 4%, with typical enrichments in the neighborhood of 3% [6, Table 11]. This material is known as slightly enriched uranium, in contrast to the highly enriched uranium used for nuclear weapons and submarine reactors. Within the core of a given reactor, enrichments vary with the location of the fuel assemblies. Further, there are variations among reactors, and as discussed below in the context of the burn-up of fuel, there is a general trend towards using fuel with higher initial enrichments.

The products of the enrichment process are the enriched material itself and the depleted uranium, sometimes called *enrichment tails*. Typically, enrichment tails have in the neighborhood of 0.2% to 0.35% ^{235}U remaining [7, p. 7]. (Here and in the remainder of the discussion of uranium isotopic enrichment, we will follow the common practice of describing the ^{235}U fraction in percent by weight rather than, as is common in many other scientific applications, specifying isotopic abundances in percent by number of atoms.[5]) This depleted material is even less radioactive than natural uranium and is sometimes used in special applications, such as armor-piercing shells, where the high density of uranium is useful ($\rho \approx 19$ g/cm^3).

Methods for Enrichment

There are a variety of approaches to enrichment, each taking advantage of the small mass difference between ^{235}U and ^{238}U. In the most used of these processes, it is necessary to have

---

[3]For a comparison of the hazards from mill tailings and indoor radon, see, e.g., Ref. [5].

[4]The ^{234}U arises as a member of the ^{238}U series, with an abundance relative to ^{238}U that is inversely proportional to the half-lives of the two isotopes (2.45×10^5 yr and 4.468×10^9 yr, respectively). The presence of the ^{234}U is usually ignored, because corrections on the order of 10^{-4} or less that would result are irrelevant.

[5]These descriptions of isotopic abundance are related by the expression $w = [(1-\delta)/(1-x\delta)]x$, where, specialized to the case of uranium, w is the ratio of ^{235}U mass to total uranium mass, x is the ratio of the number of ^{235}U atoms to the total number of uranium atoms, and δ is the ratio of the ^{238}U$-^{235}$U atomic mass difference to the ^{238}U atomic mass. For low enrichments (with $\delta = 0.0126$ for uranium), $w \approx 0.987x$, and there is little difference between the two formulations. For natural uranium, $x = 0.00720$ and $w = 0.00711$.

the uranium in gaseous form. For that purpose, the U_3O_8 is chemically converted to gaseous uranium hexafluoride, UF_6. The leading enrichment methods in terms of past or anticipated future use are[6]

Gaseous diffusion. The average kinetic energy of the molecules in a gas is independent of the molecular weight M of the gas and depends only upon the temperature. At the same temperature, the average velocities are therefore inversely proportional to $\sqrt{M}$. For uranium in the form of UF_6, the ratio of the velocities of the two isotopic species is 1.0043.[7] If a gas sample streams past a barrier with small apertures, a few more ^{235}U molecules than ^{238}U molecules pass through the barrier, slightly enriching the gas in the lighter isotope. The ratio of $^{235}U/^{238}U$ before and after passing the barrier is the enrichment ratio α. Its ideal or maximum value is given by the velocity ratio: $\alpha = 1.0043$. However, one cannot calculate the number of stages of diffusion needed to achieve a given enrichment merely in terms of powers of α, because the ideal value is not achieved in practice and because it is necessary to continually recycle the less enriched part of the stream. Typically, if one starts with natural uranium (0.71%) and with tails depleted to 0.3%, it is found that about 1200 enrichment stages are required to achieve an enrichment of 4% [10, p. 36].

Centrifuge separation. Any fluids—liquid or gaseous—are separated by mass in a high-speed centrifuge. It is possible, therefore, to enrich UF_6 using several stages of centrifuges. The power requirement per unit output is much less for centrifuge separation than for diffusion separation.

Aerodynamic processes. These processes exploit the effects of centrifugal forces, but without a rotating centrifuge. Gas—typically, UF_6 mixed with hydrogen—expands through an aperture, and the flow of the resulting gas stream is diverted by a barrier, causing it to move in a curved path. The more massive molecules on average have a higher radius of curvature than do the lighter molecules, and a component enriched in ^{235}U is preferentially selected by a physical partition. The process is repeated to obtain successively greater enrichments. The gas nozzle process was developed in Germany as the *Becker* or *jet nozzle* process. A variant with a different geometry for the motion of the gas stream, the so-called *Helikon* process, has been developed and used in South Africa.

Electromagnetic separation. When ions in the same charge state are accelerated through the same potential difference, the energy is the same, and the radius of curvature in a magnetic field is proportional to $\sqrt{M}$. Thus, it is possible to separate the different species magnetically. This separation can be done with ions of uranium, and so conversion to UF_6 is not in principle necessary. Overall, this approach gives a low yield at a high cost in energy, but it has the advantage of requiring a relatively straightforward technology.

Laser enrichment. The atomic energy levels of different isotopes differ slightly.[8] This effect can be exploited to separate ^{235}U from ^{238}U, starting with uranium in either atomic or molecular form. For example, in the atomic vapor laser isotope separation (AVLIS) method, the uranium is in the form of a hot vapor. The ionization energy to remove an

---

[6]Detailed discussions of these methods are given in, for example, Refs. [8] and [9].

[7]The atomic mass of fluorine (F) is 19.00 u.

[8]This "isotope effect" was responsible for the discovery of ^2H. It arises for two reasons: (a) the atomic energy levels depend upon the reduced mass of the electrons, which differs from the mass of a free electron by an amount proportional to m_e/M, where m_e is the electron mass and M the atomic mass; and (b) the energy levels of heavy atoms depend in small measure on the overlap between the wave functions of the innermost electrons and the nucleus, with differences between isotopes due to differences in their nuclear radii.

electron is 6.2 eV. Lasers precisely tuned to the appropriate wavelength are used to excite ^{235}U atoms, but not ^{238}U atoms, to energy levels that lie several eV above the ground state. An additional laser is used to ionize the excited ^{235}U atoms.[9] The ionized ^{235}U atoms can be separated from the un-ionized ^{238}U atoms by electric and magnetic fields. It is not possible to obtain complete enrichment in a single stage. The ionization of ^{235}U is not complete, and some ^{238}U is ionized by collisions and is collected with the ^{235}U. Nevertheless, the fractional abundance of ^{235}U can be increased by a factor of the order of ten in the first stage [8, p. 163]. Thus, a single pass suffices to provide uranium that is sufficiently enriched for nuclear reactors. The cost in energy of laser enrichment is lower than those of other enrichment methods. However, despite these advantages and the conceptual simplicity of the method, a highly sophisticated laser technology is required, and to date there are no facilities for laser separation of commercial fuel.

During World War II, not knowing which methods would be most effective, the United States embarked on both diffusion and electromagnetic separation, as well as still another method that was later discarded (namely, thermal diffusion, which exploits temperature gradients). The electromagnetic separation technique was abandoned in the U.S. after World War II and was widely considered to be obsolete. However, it was found in 1991, after the Gulf War, that Iraq had been secretly using this approach in an attempt to obtain enriched uranium for nuclear weapons.

In the United States since World War II, the enrichment program has relied on gaseous diffusion, as did early European programs. The DOE operates two large gaseous diffusion enrichment facilities, one in Paducah, Kentucky, and one in Portmouth, Ohio. Together, they represent over 40% of the world enrichment capacity [11, p. 149]. Outside the United States, there are major enrichment facilities in France and Russia and smaller ones in a number of other countries, including the Netherlands, Germany, Japan, and South Africa. Many of the facilities outside the U.S. use gas centrifuges rather than gaseous diffusion.

The capacity for enrichment has developed more rapidly than demand, and there is no urgent pressure, as of 1995, for additional facilities. However, should there be a major expansion in nuclear power, laser separation may play an important role, because, once mastered, the laser technique is expected to be relatively inexpensive. On the down side, there are some fears that laser separation may make it easy for small countries or well-organized terrorist groups to enrich uranium, or perhaps even plutonium, for use in nuclear weapons.

Separative Work

In a ^{235}U enrichment process, there are three streams of material: the input or *feed*, the output or *product*, and the residue or *tails*. The system will operate with a cascade of steps, with the enrichment of the product increasing successively in each step.[10] As the enrichment cascade progresses, the tails from an intermediate stage have a higher ^{235}U concentration than the original feed material, and these tails can profitably be returned to the cascade. There are different strategies for reusing the tails of successive steps to maximize the efficiency of the process, including an "ideal cascade" (see, e.g., Ref. [4]).

---

[9]The number of lasers and their wavelengths are determined by both the energy level structure of uranium atoms and the availability of suitable lasers.

[10]The logical structure of the system is similar to that of fractional distillation, and some of the formalism was developed in the 19th century by Lord Rayleigh [4, p. 649].

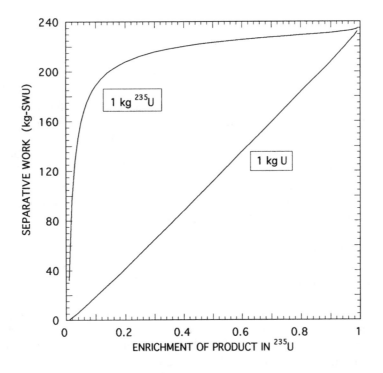

Figure 7.2. Separative work (in kg-SWU) as a function of ^{235}U enrichment for the production of 1 kg of U and 1 kg of ^{235}U. (The initial feed concentration is $w_F = 0.00711$ and the assumed tails concentration is $w_T = 0.0025$.)

The difficulty of carrying out uranium enrichment, as measured, for example, by the relative energy required in the diffusion process, is described by a quantity known as the *separative work*.[11] Separative work has the dimensions of mass and is specified in *separative work units* (SWU), as kg-SWU or tonne-SWU. Fig. 7.2 shows the separative work required to produce 1 kg of enriched uranium product and 1 kg of ^{235}U in the form of enriched uranium, for different degrees of final enrichment.

It is seen from Fig. 7.2 that it requires more separative work per kg ^{235}U to enrich uranium from 0.7% to 5% (158 SWU) than to carry it the rest of the way to 95% enrichment (an additional 76 SWU). Even 3% enriched uranium fuel is more than "halfway" to 95% enrichment. This could make the moderately enriched uranium produced for reactors a somewhat attractive initial material for the production of uranium for weapons (see Chapter 14).

---

[11]In a separation process, the mass are in the ratios

$$\frac{M_F}{M_P} = \frac{w_P - w_T}{w_F - w_T} \quad \text{and} \quad \frac{M_T}{M_P} = \frac{w_P - w_F}{w_F - w_T},$$

where M_F, M_P, and M_T are the masses and w_F, w_P, and w_T are the ^{235}U concentrations (by weight) of the feed, product, and tails, respectively. This result follows from the conservation of total mass and ^{235}U mass: $M_F = M_P + M_T$ and $w_F M_F = w_P M_P + w_T M_T$. The separative work in an isotopic enrichment process is

$$\Delta V = M_P V_P + M_T V_T - M_F V_F,$$

where V is the *value function*, defined for each component as $V = (1 - 2w)\ln[(1-w)/w]$. (For the derivation leading to this result, see, e.g., Ref. [4], Ch. 12.)

Although the formalism was developed in the context of gaseous diffusion enrichment methods, it is used for other processes as well. Separative work serves as a general measure of what is achieved in the enrichment. For an individual process, it serves also as a measure of relative energy consumption. Different processes vary greatly in their energy consumption. For example, gaseous diffusion uses 2.5 MWh per kg-SWU, while the gas centrifuge uses about 1/50th as much energy and laser isotope separation methods still less [7, p. 530].

It is of interest, in the spirit of what is known as "net energy analysis," to compare the energy required to enrich uranium with the energy obtained from it. The production by the diffusion process of 1 kg of ^{235}U in the form of uranium enriched to 3% in ^{235}U, with a ^{235}U concentration in the tails of 0.25%, requires 127 kg-SWU (see Fig. 7.2) or about 320 MWh. The reactor output from 1 kg of ^{235}U is about 1 MWyr or 8760 MWh (see Section 7.3.2). Therefore, enrichment requires roughly 4% of the energy output of a reactor. While this is the largest single energy input to the production of nuclear power, it does not significantly reduce the net positive energy balance from nuclear power.

7.2.3 Fuel Fabrication

Most nuclear fuel used in light water reactors is in the form of uranium oxide (UO_2). This is not a single compound but rather a mixture of oxides (UO_n), where n on average ranges from 1.9 to 2.1 [12, p. 226]. The enriched UF_6 is chemically converted into UO_2. The UO_2 is processed into a fine powder and is then compacted and sintered to form rugged pellets. During sintering, the oxygen content of the fuel can be adjusted. The pellets are corrected in size, to close tolerances, by grinding.[12] The pellets are loaded into zircaloy fuel pins that are arranged in a matrix to form the fuel assembly. As discussed in Section 6.2.3, the reactor core consists of a large number of such assemblies.

7.3 FUEL UTILIZATION

7.3.1 Uranium Consumption and Plutonium Production in LWRs

The central purpose of the fuel cycle is the consumption of fuel in the reactor and the extraction of energy. An overall measure of the energy obtained from the fuel is its *burn-up*, i.e., the energy output per unit mass of fuel. The burn-up is commonly specified in megawatt-days or gigawatt-days of thermal output per metric tonne of initial heavy metal (MWDT/MTIHM or GWDT/MTIHM).[13] For the reactors in use in the U.S. today, the "heavy metal" initially present in fresh fuel is uranium. (A distinction is made between the initial and final fuel because some of the uranium mass is converted into fission fragments, which are not heavy metals.)

In uranium-fueled reactors, most of the energy comes from the fission of ^{235}U. However, as discussed earlier, production of ^{239}Pu also plays a significant part in the energy economy of the reactor. In conjunction with ^{239}Pu production, heavier plutonium isotopes are also formed, and the total energy output is the sum of the energy from the fissile plutonium isotopes plus the energy from ^{235}U and, to a much lesser extent, ^{238}U.

Table 7.1 lists the main uranium and plutonium nuclides (other than ^{238}U) in PWR fuel at the time of discharge from the reactor, following a normal residence time of about three years. The abundances reflect the destruction and creation of nuclides by neutron reactions

---

[12]For a discussion of the details of these processes, see Ref. [12], Section 7.5.

[13]This is a cumbersome unit for repeated use, and we here represent the GWDT/MTIHM in a more compact form as GWd/t (gigawatt days per tonne), with the fuller meaning to be inferred as a matter of convention.

Table 7.1. *Composition of typical LWR discharge fuel.*

Nuclide	Fission Properties	% of Fuel by Weight
Initial input		
^{235}U	fissile	3.20
Output		
^{235}U	fissile	0.85
^{236}U	*Note a*	0.38
^{239}Pu	fissile	0.55
^{240}Pu	fertile	0.23
^{241}Pu	fissile	0.15
^{242}Pu	*Note a*	0.05
Fission products		3.8

Source: Percent of fuel data are from Ref. [12], p. 458, for a burn-up of 35.0 GWd/t.
a ^{236}U and ^{242}Pu are neither fissile nor fertile in the usual sense, but act mainly as weak "poisons" (see text).

during this time. The numbers are for the specific case of an initial ^{235}U enrichment of 3.2% and a burn-up of 35.0 GWd/t [12, p. 458].

The results in Table 7.1 can be found with some precision with the aid of a computer model, which traces the full history of production and consumption of the nuclides involved. It is difficult to calculate these numbers without a reasonably complex model, because one has to consider both the interplay of the various nuclides and the behavior of the cross sections over the full neutron energy spectrum. However, it is possible to make some qualitative comments about the main features of Table 7.1:

Fission products. It may seem surprising that the fission product mass exceeds that of the destroyed ^{235}U. However, in addition to fission in ^{235}U, there is fission of the two fissile plutonium isotopes, ^{239}Pu and ^{241}Pu, which are produced by neutron captures starting in ^{238}U.

^{236}**U**. Neutron absorption in ^{235}U leads to capture in 14% of the events (see Table 5.1). Therefore, the destruction of ^{235}U should be accompanied by the production of a significant amount of ^{236}U. The isotope ^{236}U is more a weak poison than a fertile fuel, because at thermal energies it has a modest capture cross section (σ_γ=5.1 b) and a negligible fission cross section. Its capture product, ^{237}U, decays with a rather short half-life (6.75 days) to ^{237}Np, which has a large capture cross section and, again, a negligible fission cross section [13].[14] The situation for ^{242}Pu is qualitatively similar.

^{239}**Pu**. Assuming a conversion ratio in the fuel of 0.6 (see Section 5.4), the reduction in the concentration of ^{235}U from 3.20% to 0.85% would give a ^{239}Pu concentration of 1.4% with no ^{239}Pu losses. However, some of the ^{239}Pu is lost due to fission and some lost due to neutron capture with the resultant production of ^{240}Pu. Neutrons from the fission of ^{239}Pu will also be captured in ^{238}U, creating additional ^{239}Pu beyond that resulting from ^{235}U fission alone.

^{240}**Pu**. The capture cross section is relatively high in ^{239}Pu, so that neutron absorption results in the production of ^{240}Pu as well as in fission. ^{240}Pu acts as a fertile fuel for the production of ^{241}Pu, but the thermal absorption cross section in ^{240}Pu is relatively low (290 b), and ^{240}Pu is not consumed as rapidly as is ^{239}Pu.

It can be seen from Table 7.1 that the chain of plutonium isotopes does not stop at ^{240}Pu

---

[14]Logically, the story should be pursued through further capture steps, but on doing so one finds no important addition to fission yields.

Table 7.2. *Energy per unit mass from fission of uranium, at different degrees of enrichment.*

Category	Enrichment, w (%)	Energy per Unit Mass [a] (J/kg)	GWd/t	Tonne per GWyr(e)[b]
Pure ^{235}U	100	8.21×10^{13}	950	1.2
Enriched U	3.2	2.63×10^{12}	30.4	38
Natural U	0.711	5.84×10^{11}	6.8	169

[a]Thermal energy, assuming fission of *all* ^{235}U and no fission of other nuclides.
[b]Assuming a thermal conversion efficiency of 32%.

but continues to ^{242}Pu.[15] The plutonium isotopes up to atomic mass number $A = 242$ have half-lives that are long compared to normal exposure periods (the shortest of these is 14.4 yr, for ^{241}Pu), so the material remains plutonium, unlike the case for capture in ^{238}U, which is quickly followed by β^- decay from ^{239}U to ^{239}Np and then to ^{239}Pu. Two of the plutonium isotopes in the spent fuel (^{239}Pu and ^{241}Pu) are fissile and therefore are useful fuels if reprocessed.

7.3.2 Energy From Consumption of Fuel

Energy per Unit Mass from Fission of ^{235}U

The fission of a ^{235}U nucleus corresponds on average to the release of 200 MeV (3.20×10^{-11} J), including the associated contributions from capture gamma rays but omitting any contributions from plutonium fission (see Section 4.4.2). The number of ^{235}U atoms per kg of uranium, n, is given by $n = 1000 \times w N_A / M$, where w is the fraction of ^{235}U in the uranium (by weight), N_A is Avogadro's number, and M is the atomic mass of ^{235}U. For example, one kilogram of natural uranium ($w = 0.00711$) has 1.822×10^{22} nuclei of ^{235}U. The available fission energy per unit mass of uranium, for different degrees of enrichment, is given in Table 7.2 for a few illustrative cases, assuming fission of all ^{235}U and ignoring losses due to capture in ^{235}U and gains from fission in ^{239}Pu and other nuclides.

The rate of energy production in a prototypical reactor is 1000 MWe, or, at 32% thermal efficiency, 3125 MWt. One gigawatt-year of electric power therefore corresponds to a thermal output of 1141 GWd(t) or 9.86×10^{16} J. Assuming complete fission of all the ^{235}U and ignoring fission in plutonium, this would correspond to a fuel requirement per GWyr(e) of 1.20 tonnes of ^{235}U or 37.5 tonnes of uranium enriched to 3.2% in ^{235}U. The corresponding thermal burn-up is 30.4 GWd/t, as indicated in Table 7.2.

Energy per Unit Mass from Fuel

The model used in the preceding paragraphs is unrealistic, because it omits many crucial factors that significantly modify the amount of ^{235}U required by a reactor. These include the following:

1. Not all of the ^{235}U is consumed in the reactor. For example, for the case described in Table 7.1, the ^{235}U enrichment is 3.2% for the fresh fuel and 0.85% for the spent fuel. This means a consumption of only 73% of the ^{235}U.

2. About 14% of the thermal neutron-absorption reactions in ^{235}U result in capture rather than fission.

---

[15]The plutonium buildup stops here, because for ^{243}Pu the half life is 5.0 h; its decay product is ^{243}Am (T=7370 yr).

3. Fission in ^{239}Pu (and, to a lesser extent, ^{241}Pu) provides a substantial additional energy source. This reduces the input requirement significantly, with the amount dependent on the length of the fuel's exposure to neutrons.

4. There is a small contribution from fast neutron fission in ^{238}U.

These effects all change the number of fission events, reducing the number of fissions in ^{235}U and adding fission in plutonium isotopes and even ^{238}U. The overall consequence can be estimated, for the case described in Table 7.1, by comparing the fission product yield to the original ^{235}U content. The mass of the fission products is 19% greater, implying a correction factor of 1.19 to the burn-up calculated above. This raises the previously calculated burn-up of 30.4 GWd/t to a "corrected" value of about 36 GWd/t.

This result is in reasonable agreement with the quoted burn-up of 35 GWd/t specified in Table 7.1 for this case. A value of 35 GWd/t is also a good representation of present PWR performance. In particular, the average burn-up in 1992 in the United States was approximately 28.8 GWd/t for BWRs and 36.4 GWd/t for PWRs [6, p. 20].[16]

The burn-up in GWd/t can be translated into the fuel requirement per year. As above, an electrical output of 1 GWyr at 32% efficiency corresponds to a thermal output of 1141 GWd. If we assume, for example, a burn-up of 35 GWd/t and a ^{235}U enrichment of 3.2%, the total requirement per GWyr is 32.6 tonnes of enriched uranium or about 1.04 tonnes of ^{235}U. Past practice has been to use the fuel on a three-year cycle, so in this case the core would have a total of about 100 tonnes of enriched uranium, one-third of which is discharged each year.

The precise uranium requirements for a given energy output depend upon details of the fuel cycle, as discussed above. Nonetheless, the following equivalence, as established in the discussion above, is useful for approximate estimates of the magnitudes:

$$1 \text{ tonne of } {}^{235}\text{U} \rightarrow 1 \text{ GWyr(e) } (approximate).$$

Trends in Burn-Up of LWR Fuel

In general, there is a trend towards higher burn-up, which means a lower fuel discharge per GWyr. In the early 1970s the average burn-up for PWR spent fuel was in the neighborhood of 20 GWd/t. Typical burn-ups rose to 25–30 GWd/t in the late 1970s and reached 36 GWd/t in 1992 [6, p. 20]. Looking to the future, it is expected that the median PWR fuel burn-up for standard assemblies will be about 43 GWd/t in the year 2000 and 51 GWd/t in 2010 [11, p. 144]. Even in 1991 almost one-quarter of the spent fuel had a burn-up in excess of 40 GWd/t [14, p. 36], and some did considerably better.

The burn-up depends upon the power density in the fuel and the length of time the fuel is kept in the reactor. Unless the conversion ratio is very high, a high burn-up requires a high initial enrichment in ^{235}U. Model calculations carried out at Oak Ridge National Laboratory described a fuel cycle that gives a burn-up of 60 GWd/t [15, p. 2.4-3]. The initial fuel has a 4.7% ^{235}U concentration, the reactor power level is 33.3 megawatts per tonne, and the fuel is used in the reactor for three 600-day cycles.

High enrichment alone is not sufficient for high burn-up. In addition, the fuel rods must be rugged enough to withstand the added neutron bombardment. This depends both on the fuel

---

[16]The average burn-up is systematically less in BWRs than in PWRs because burn-up in the former is less uniform along the length of the fuel rods. The water at the bottom of the BWR tank has a higher density and is a better moderator than the steam-water mixture at the top. This means that the maximum burn-up is achieved at the bottom of the rod, with a smaller burn-up higher up on the rod. The maximum acceptable burn-ups are about the same for the PWR and BWR, but the PWR has a more uniform profile along the length of the rod and therefore a greater average burn-up.

itself and on the composition of the zirconium alloy used for the cladding. Further, to compensate for the initial high reactivity with high enrichment, burnable poisons are used with the fuel.

With higher burn-up for the same total energy output, there is a smaller mass and volume of spent fuel. There is also less consumption of uranium, because a larger fraction of the ^{235}U is consumed and there is a greater production of ^{239}Pu. In addition, longer cycles before removing the fuel mean less reactor time lost for refueling.

7.3.3 Uranium Ore Requirement

The amount of uranium ore required to operate a reactor depends upon the burn-up achieved, the initial enrichment, and the amount of ^{235}U lost in the enrichment process. The mass M_F of natural uranium used as feed input to the enrichment facility is related to the mass M_P of enriched uranium produced in the fuel by the expression $M_F/M_P = (w_P - w_T)/(w_F - w_T)$, where w_P, w_T, and w_F are the enrichments of the fuel, the tailings, and natural uranium, respectively. For $w_P = 3.2\%$, $w_T = 0.2\%$, $w_F = 0.711\%$, and $M_P = 33$ tonnes, the natural uranium requirement is 194 tonnes/GWyr. Thus, in round numbers, a once-through PWR fuel cycle requires about 200 tonnes per GWyr.

This result is of course dependent on particular assumptions about the enrichment of the fuel and the operation of the reactor. Thus, it is to be taken as an illustrative value for the fuel requirements for PWRs, not an exact calculation of the performance of any particular reactor. We can expect that in the future, PWRs will have higher values of the burn-up and somewhat smaller fuel requirements.

7.3.4 Alternative Fuel Cycles

If the fuel cycle is different, fuel requirements will differ from those cited above. For a consistent comparison, we cite a set of estimates for a number of fuel cycles, in terms of the number of tonnes of natural uranium required for a 1000-MWe reactor operating for 30 yr at a 70% capacity factor [16, p. 180]:

LWR: 4620 tonnes without recycling of reprocessed spent fuel and 3000 tonnes with recycling

HTGR: 3320 tonnes (no recycling)

HTGR or PHWR with 90% conversion ratio: 1360 tonnes (with recycling)

Breeder reactor: 35 tonnes (recycling intrinsic)

Of course, in the cycles cited above, the key to reducing uranium requirements is to utilize ^{238}U by conversion to ^{239}Pu. In the HTGR without recycling, this results from the high conversion ratios characteristically obtained with a graphite moderator (see Section 6.3.2). With recycling of the spent fuel, there is a further gain. This is carried to the limit in a breeding cycle, where it is possible to return as much fuel as was used originally (or more).

7.4 BACK END OF FUEL CYCLE

7.4.1 Handling of Spent Fuel

Initial Handling of Reactor Fuel

Periodically, a portion of the fuel in the reactor is removed and replaced by fresh fuel. The removed fuel is the spent fuel. In typical past practice, an average sample of fuel remained in the reactor for three years, and approximately one-third of the fuel was removed each year, with a shutdown time for refueling and maintenance of up to about two months. Efforts are being made to extend the interval between refueling operations and to reduce the time for refueling.

When the spent fuel is first removed from the reactor, the level of radioactivity is very high. Each radioactive decay involves the release of energy, which immediately appears as heat, so the fuel is thermally hot as well as radioactively "hot." Independent of the reprocessing question, the first stage is the same, namely, allowing the fuel to cool both thermally and radioactively. The cooling of the fuel normally takes place in water-filled cooling pools at the reactor site.

Originally, it was planned to keep the spent fuel at the reactor for roughly 150 days and then to transfer it to handling facilities at other locations. The nature of the next step in principle depends upon whether the fuel is to be disposed of as waste or reprocessed. However, as yet there has been no "next step" in the U.S. because no away-from-reactor facilities have been developed. Instead, almost all of the fuel has remained in the cooling pools, in some cases for more than 20 yr.

In the absence of alternatives, some U.S. utilities are beginning to transfer older fuel rods from cooling pools to air-cooled (dry storage) casks at the reactor site. This may provide a workable temporary solution to the continuing impasse in implementing a national waste disposal program. However, it is only a stopgap because utilities cannot be counted on to be both willing and able to supervise the spent fuel for prolonged periods of time.

Disposal or Storage of Spent Fuel

For many years it had been assumed that all United States civilian nuclear waste would be reprocessed, but U.S. reprocessing plans have been abandoned. Instead, official plans now call for disposing of the spent fuel directly, retaining for several decades the option of retrieving it. Disposal of the fuel is conceptually simple. The fuel remains in solid form and is put in a protective container. It then is transported in secure casks to either a permanent or an interim site; in the latter case, the waste is moved to a permanent repository.

In *retrievable* storage systems, the permanent sealing of the repository is deferred for a period of time, allowing the spent fuel to be recovered should this be desired. Thus, the reprocessing option is not foreclosed, and the potentially retrievable spent fuel need not ultimately be a "waste." There are several motivations for maintaining retrievability: (a) it allows for remedial action in case surprises are encountered in the first decades of waste storage that require modifying the fuel package or the repository; (b) it keeps open the option of recovering plutonium from the fuel; and (c) it allows the recovery of other materials deemed useful—for example, fission products for use in medical diagnosis and therapy or in the irradiation of food or sewage sludge.

7.4.2 Reprocessing

Extraction of Plutonium and Uranium

The alternative to disposing of the spent fuel is to reprocess it and extract at least the uranium and plutonium. In reprocessing, the spent fuel is dissolved in acid and the uranium and plutonium chemically removed for use in new fuel.

Most U.S. plans for reprocessing assumed that 99.5% of the U and Pu would be removed. The remainder constitutes the high-level waste. In the traditional plans, this includes almost all of the nonvolatile fission products, 0.5% of the uranium and plutonium, and almost all of the remaining heavy elements, particularly neptunium ($Z=93$), americium ($Z=95$), and curium ($Z=96$). The uranium represents most of the mass of the spent fuel, but the fission products contain most of the radioactivity.

Extraction can be more complete than in the original U.S. thinking. The current French program exceeds the 99.5% goal, separating out more than 99.9% of the uranium and 99.8% of the plutonium [17, p. 28]. There is no essential reason to limit extraction to plutonium and uranium, although the former represents the valuable fuel and the latter the bulk of the mass. It is possible to extract other radioisotopes as well, either because they are deemed pernicious as components of the waste or because they are useful in other applications.

The so-called *minor actinides*—namely, neptunium, americium, and curium—have been of particular interest. These actinides include long-lived products whose removal would decrease the long-term activity in the waste. Therefore, there is ongoing consideration of the separation of the minor actinides for return to a reactor where they would be consumed in neutron reactions. Of course, if the chief goal is safety, it is necessary to balance the benefits from decreased activity in the wastes against the increased hazards of handling and processing the radioactive materials when they are still very hot.

The residue of reprocessing constitutes the wastes. They are to be put in solid form for eventual disposal. The standard method is to mix the high-level waste with borosilicate glass and to contain the solidified glass in metal canisters. Although other solid waste forms have been suggested, borosilicate glass has been used in the French nuclear program and had figured prominently in the original U.S. plans for reprocessing commercial wastes. It still may be used for the sequestering of already reprocessed military waste, such as that at Hanford.

The fuel manufactured from the output of the reprocessing phase is generally a mixture of plutonium oxides and uranium oxides. It is called a *mixed-oxide fuel*, or MOX. Such MOX fuel is being used in some French reactors. Due to differences in the nuclear properties of ^{239}Pu and ^{235}U, the use of MOX fuel as a replacement for UO_2 fuel in most standard LWRs is limited to about 30% of the total reactor fuel load [18]. Some LWRs, however, have been designed to accommodate a full load of MOX fuel.[17]

Status of Reprocessing Programs

Until the late 1970s reprocessing had been planned as part of the United States nuclear power program. A reprocessing facility at West Valley, New York, was in operation from 1966 to 1972, with a capacity of 300 tonnes of heavy metal per year. This is enough, roughly speaking, for the output of 10 large reactors. There were plans for further facilities at Morris (Illinois) and Barnwell (South Carolina), which would have substantially increased the reprocessing capacity. However, all these plans have been abandoned.[18]

In part, the abandonment was impelled by technical difficulties. There had been high worker exposures at West Valley, and the plant was shut down in 1972; plans to remodel and expand it were later aborted. When the Morris plant was first tested with nonradioactive materials, it did not perform reliably, and the General Electric Co., which was building the

---

[17]See Chapter 14 for a further discussion of MOX fuel, in the context of the burning of plutonium from dismantled nuclear weapons.

[18]For a discussion of this history, see Ref. [19].

plant, decided there were serious difficulties. The Barnwell plant moved ahead until the early 1980s, but it faced problems of meeting escalating standards on permissible radioactive releases.

These difficulties might have been surmounted had there been a belief that reprocessing was needed. But the fundamental motivation for reprocessing began to slip away. There was no prospective near-term shortage in uranium supply, and uranium prices were low enough to remove the economic incentive for reprocessing. Further, an important body of opinion had developed in the U.S. against reprocessing, on the grounds that it might make plutonium too readily available for diversion into destructive devices.

This view was expressed in the report *Nuclear Power Issues and Choices*, sponsored by the Ford Foundation and authored by an influential group of national science policy leaders. In its conclusions on reprocessing, the report stated:

> the most severe risks from reprocessing and recycle are the increased opportunities for the proliferation of national weapons capabilities and the terrorist danger associated with plutonium in the fuel cycle.

> In these circumstances, we believe that reprocessing should be deferred indefinitely by the United States and no effort should be made to subsidize the completion or operation of existing facilities. The United States should work to reduce the cost and improve the availability of alternatives to reprocessing worldwide and seek to restrain separation and use of plutonium [20, p. 333].

Consistent with this line of thinking, the Carter administration decided in 1977 to "defer

Table 7.3. *Major plants for reprocessing commercial nuclear fuel, in operation or under construction.*

Plant	Year of first operation	Capacity (t/yr)[a]	
		1994	2005
France			
Marcoule (UP1)	1958	400	0
La Hague (UP2)[b]	1966	800	800
La Hague (UP3)	1990	800	800
India			
Tarapur	1982	≈150	≈150
Kalpakkam			125[c]
Japan			
Tokai	1981	90	90
Rokkasho-mura	≈2002	0	800
Russia			
Chelyabinsk	1978	400	400
Krasnoyarsk[d]	2005?	0	1500
United Kingdom			
Sellafield (B205)	1964	1500	1500
Sellafield (THORP)	1995	0	≈700

Sources: Data are from Refs. [18] and [22] to [27].
[a]The capacity is given in tonnes of heavy metal per year.
[b]UP2 was upgraded and redesignated as UP2-800, with full capacity reached in 1994 [23].
[c]This would represent Phase I of the project at Kalpakkam; Phase II would bring the capacity to 1000 tonnes/yr [25].
[d]Construction was halted on the partially completed Krasnoyarsk plant, but Russia has announced plans to resume construction [27].

indefinitely the commercial reprocessing and recycling of the plutonium produced in U.S. nuclear power programs" [21, p. 54]. Work on U.S. reprocessing plants was phased out, culminating in the closing of Barnwell at the end of 1983 [19, p. 124].

Nonetheless, reprocessing has moved ahead in other countries. Major expansions of reprocessing capacity were completed, or nearing completion, in the United Kingdom and France in 1994; a large reprocessing facility (800 tonnes per yr) is being built in Japan; and a very large plant (1500 tonnes per yr) has been under construction, with interruptions, in Russia. Table 7.3 lists the reprocessing plants in operation or under construction. It does not include plants that have been shut down, such as those in Belgium, Germany, and the United States.

The largest reprocessing programs are in France and the United Kingdom, handling both domestic and foreign fuel. France has the most fully developed fuel cycle. Although much of the present reprocessing capacity in France is devoted to foreign orders, a program of reprocessing and plutonium recycling of domestic fuel is underway and expanding. As of 1993 four 900-MWe reactors used mixed-oxide (MOX) fuel elements, in a combination of PuO_2 and UO_2, for 30% of their fuel. It is planned that use of MOX fuel will increase, as fabrication facilities are increased and the performance of MOX fuel is further improved [17]. With the forthcoming completion of a large reprocessing plant, Japan is also planning to make increased use of MOX fuel [18].

7.4.3 Waste Disposal

All countries with announced plans for disposing of high-level radioactive wastes are planning on eventual disposal in deep geologic repositories, typically excavated caverns or holes in an environment where the intrusion of water is deemed unlikely. Many of the plans for these permanent disposal facilities include a period during which the waste could still be retrieved.

Deep geologic disposal has been the favored course in U.S. thinking since the first attempts to formulate plans. There have been continuing efforts to locate and design a suitable facility. A site at Yucca Mountain in Nevada was selected in 1987 as the candidate for a U.S. repository, but the site characterization work to establish the site's suitability has proceeded slowly. The current goal is to have a facility ready to receive waste by 2010, if the Yucca Mountain site is found to be suitable. (See Chapters 8 to 10 for further discussion of waste disposal.)

7.5 URANIUM RESOURCES

7.5.1 Specification of Resources

Uranium Prices

The magnitude of uranium resources can be specified in terms of the amount of uranium oxide (U_3O_8) or the amount of natural uranium. Usually these amounts are expressed in short tons of U_3O_8 or in tonnes of uranium, where 1 ton of U_3O_8=0.7693 tonnes of U.[19] To use the units in which uranium prices are usually couched, 1 kg of U is equivalent to 2.60 lb of U_3O_8, and therefore a price of $100 per lb of U_3O_8 is equivalent to $260 per kg of U.

---

[19]The molecular weights of U_3O_8 and U are 842.09 and 238.03, respectively, and 1 ton=0.9072 tonnes.

Uranium Prices and Electricity Costs

Uranium prices are now very low compared to, say, the prices projected 20 years ago. Instead of rising, they have dropped markedly in recent years, in large measure due to the lag in the expansion of nuclear power. U.S. prices peaked in 1978 at an average of $43 per lb of U_3O_8 [3]. By 1994 average market prices had fallen to roughly $9 per pound of U_3O_8 ($23 per kg of U) [28, p. 32]. Most of the uranium used by U.S. utilities in 1994 was of foreign origin, with Canada the largest supplier.

The relationship between uranium price and the contribution of uranium fuel costs to electricity costs depends upon the effectiveness of fuel utilization. As was discussed in Section 7.3.2, we can take 200 tonnes of uranium per GWyr as an approximate input for LWRs. (It will probably be less in the future, because of expected improvements in fuel burn-up.) Then, for example, fuel at $100/kg translates into a cost of $20 million per 8.76×10^9 kWh, and the cost of the uranium is 0.23 ¢/kWh.

A useful overall equivalence for LWR uranium costs is

$$\$100 \text{ per lb of } U_3O_8 = \$260 \text{ per kg of } U \approx 0.59 \text{ ¢/kWh}.$$

The 1992 price of U ($21 per kg) corresponds to a contribution to the cost of electricity of about 0.05 ¢/kWh. This is roughly 1% of the total cost of electricity from nuclear power plants. Thus, uranium would remain "affordable" even with a very large increase in uranium prices. For example, uranium costs still would amount to only 1 ¢/kWh with uranium at $440 per kg of U. This is more than 20 times recent uranium prices, but the resulting increase in electricity costs would still be small compared to the disparity in electricity prices in different parts of the U.S.

It should be noted that only the costs of the U_3O_8 are considered here. Fuel costs as seen by a utility are normally defined to include not only the cost of U_3O_8, but also the costs for conversion to UF_6, enrichment, fuel fabrication, and waste disposal.[20] However, these are not dependent on the grade of the original uranium ore. Rather, they make the ore cost itself relatively less significant. Typical fuel costs have been about 0.7 ¢/kWh in recent years (see Table 15.1).

Another perspective on the "affordability" of uranium at different prices can be obtained by comparing it to fluctuations in oil prices. Changes in the price of oil of $10 per barrel have not been uncommon since 1973. This corresponds to a change in electricity costs of about 1.6 ¢/kWh.[21] Put another way, the above-cited uranium at $440/kg is as "affordable" as an increase in oil costs of about $6 per barrel, if oil is used for electricity generation.

7.5.2 Magnitude of Uranium Resources

Classification of Resources

In resource tabulations, such as those of the OECD document *Uranium 1993: Resources, Production and Demand* [30], it is common to specify resources in terms of the degree of confidence one can place in the availability of the resource. So-called conventional resources

---

[20]A 1 mill/kWh charge for a waste disposal reserve fund is normally included as part of the fuel cost.

[21]In the United States in 1993 about 163 million barrels of oil were used to generate 100 billion kWh (net), corresponding to a rate of slightly over 600 kWh/barrel [29].

are divided into four groups: "reasonably assured," "estimated additional" (in two categories of differing uncertainty), and "speculative." This OECD terminology has also been adopted in U.S. DOE publications [31, p. 3]. In addition, there are unconventional resources, such as marine phosphates [30, p. 29] as well as uranium from seawater.

The several categories are further specified in terms of costs. These are often divided into resources available in different price ranges, with the usual maximum taken to be $260/kg of U ($100/lb of U_3O_8). As discussed above, there is no compelling reason to regard this price as an upper limit for economically interesting resources. However, such a price is so far above current prices that there is little motivation for a critical evaluation of uranium resources above $260 per kg.

Terrestrial Resources of Uranium

Table 7.4 lists estimated conventional uranium resources for selected countries, as determined from national reports of these resources. The listed countries are those believed to have the greatest uranium resources.

On the basis of the data in Table 7.4, there might be some temptation to say that world resources total about 18 million tonnes, including the speculative resources. However, this is likely to be misleading for a number of reasons, pointing in different directions:

- Not all countries are included in Table 7.4.

- The OECD tabulations on which much of Table 7.4 is based are for Category A resources only up to $130 per kg of U. A price maximum of $130 per kg does not represent a long-term ceiling, because considerably higher prices are affordable.

- Not all speculative resources are assured.

- Because of the slack demand for uranium, there has been little motivation for an exhaustive study of resources.

In view of both geological and economic uncertainties, it is not certain that a meaningful total can as yet be established for world uranium resources. It seems reasonable, however, to say that the useful resources, assuming practices not very different from present ones, lie between 10 and 30 million tonnes.

Table 7.4. *Geographical distribution of uranium resources, based on national estimates for selected countries (in million tonnes of uranium).*

Region[a]	Category A[a]	Category B[b]	Sum
Australia	0.9	3.9	4.8
Former Soviet Union (FSU)	1.8	2.0	3.8
United States	1.7	1.3	3.0
China		1.8	1.8
Mongolia	0.1	1.4	1.5
South Africa	0.3	1.1	1.4
Canada	0.6	0.7	1.3
Sum of listed countries	5.4	12.2	17.6

Sources: Data are from Ref. [30], except Category B resources for FSU are from Ref. [31].
[a]Category A corresponds to reasonably assured and estimated additional resources at costs up to $130/kg of U.
[b]Category B corresponds to speculative resources at unspecified costs; in the past it has been common to consider costs up to $260/kg of U.

With a requirement of 200 tonnes of U per GWyr, resources of, say, 20 million tonnes would meet the needs of LWRs for about 100 000 GWyr. Present world generation is roughly 250 GWyr/yr. Thus, a resource of this magnitude could sustain five times the present rate of nuclear generation for 80 yr. Such an increase does not appear to be imminent, and there is therefore little pressure on uranium resources at this time.

However, if one contemplates a large expansion extending over 50 to 100 yr, uranium supplies could become a constraint for generation with LWRs. Reducing the resource estimate to 10 million tonnes or raising it to 30 million tonnes would not drastically change the overall qualitative conclusion that with LWRs there is sufficient conventional uranium for the next several decades, but possibly insufficient uranium for the duration of a heavily nuclear 21st century.

In addition to the conventional resources usually considered, there are unconventional uranium sources available at higher costs. One such source is Chattanooga shale in the United States. The amount is estimated to be 5 millions tons of U_3O_8 (equivalent to 3.8 million tonnes of U), at an average concentration of about 60 ppm U [2, p. 411]. There are also large deposits in marine phosphates, including an estimated 6.5 million tonnes in Morocco [30, p. 30].

A cautionary note has been sounded in a comprehensive energy analysis carried out by the International Institute for Applied Systems Analysis (IIASA), in which it was pointed out that with very low-grade uranium resources, it will be necessary to mine large amounts of ore, losing one of the main advantages of nuclear power over coal. The product is pejoratively termed "yellow coal" [32, p. 120]. The IIASA report finds that uranium at 70 ppm requires slightly more material handling than coal.[22] At present in the United States the average uranium concentration in the ore being used is about 2400 ppm (see Section 7.2.1), so uranium still retains a large advantage over coal in terms of volume of material handled.

7.5.3 Uranium from Seawater

Uranium from seawater represents a very large potential additional resource, but probably at a price considerably higher than that of any resource considered above. The amounts available are very large. The volume of seawater in the oceans is about 1.4×10^{21} liters, with an average concentration of 3.2 parts per billion (see Section 2.3.3). This corresponds to 4 billion tonnes of ^{238}U. On the other hand, the energy content is very dilute, and vast amounts of water would have to be processed to extract uranium.

The cost of seawater extraction was quoted in a 1964 estimate at $306/lb of U or about $800/kg of U [2, p. 412]. Studies of seawater extraction have continued, particularly in Japan, and there has apparently been a substantial reduction in real (constant dollar) cost. A project under the auspices of the Metal Mining Agency of Japan with funding from the Ministry of International Trade and Industry has led to pilot-plant extraction of uranium. The status was described as follows, in a report on *Nuclear Power in Japan*:

The plant was operated continuously from 1986–1987 to produce 15.4 kg of yellow cake (U_3O_8). Basically, the process included absorption of the uranium on substances such as titanium oxide, followed by removal with an acid, and subsequent recovery of the uranium by ion exchange techniques. An economic assessment revealed that the cost of uranium recovered from a 1000-ton/yr commercial plant would be approximately $260/lb. Ultimately, with an improved sorbent, a goal of about $100/lb would not be impossible, but it is recognized that even this

---

[22] At this uranium concentration, the ^{235}U concentration is only 0.5 ppm.

would be an order of magnitude greater than the present uranium price. Impressive university research continues [33, p. 22].

A cost of $260/lb of U_3O_8 ($680 per kg of U) corresponds to an electricity cost of 1.5 ¢/kWh. As discussed in the preceding section, in terms of fuel for electricity generation, this cost for uranium is equivalent to an additional cost of about $10 per barrel of oil. For comparison, it may be noted that the refiner acquisition cost of crude oil rose from 1972 to 1981 by over $30 per barrel (in constant 1987 dollars) [34, p. 163].

7.5.4 Implications for Breeder Reactors

The issue of breeder reactors is closely connected to that of resources. The costs and requirements for uranium depend in part upon the fuel cycle used. We have here been assuming a relatively inefficient once-through cycle. At the other extreme, for a breeder reactor fuel cycle, the uranium requirements are much reduced, and virtually no extraction cost is prohibitive. Under these circumstances, seawater and other low-grade uranium sources offer a quasi-infinite supply of nuclear fuel.

There is no great pressure to make a decision on breeder reactors, because present uranium resources, with or without uranium from seawater, appear adequate to sustain an expanded nuclear economy for many decades. Within this time span, other technologies may prove to be practical and preferable, such as solar photovoltaic power or fusion power. However, there is no certainty connected with either fusion or renewable sources of electricity, nor is it certain that extraction of large amounts of uranium from seawater will prove economically practical. For such reasons, there are advocates for keeping a breeder reactor program alive.

Two policy questions arise here. The first has to do with the pace of breeder reactor development. Given that there is no immediate need for breeder reactors, what sort of development and pilot-plant testing program, if any, should be maintained? The second has to do with the disposition of the plutonium removed from dismantled nuclear weapons. Alternatives include saving the plutonium, to permit a jump start for a future breeder cycle, or rendering it inaccessible to avoid its possible diversion for weapons purposes. At present the trend in the United States is towards some version of the second alternative (see Ref. [21]).

REFERENCES

1. U.S. Department of Energy, *World Nuclear Capacity and Fuel Cycle Requirements 1992*, Energy Information Administration report DOE/EIA-0436(92) (Washington, D.C.: U.S. DOE, 1992).
2. De Verle P. Harris, "World Uranium Resources," *Annual Review of Energy* 4 (1979): 403–422.
3. John Geidl and William Szymanski, "United States Uranium Reserves and Production," paper presented at *Global '93*, Seattle, Wash., September 1993.
4. Manson Benedict, Thomas H. Pigford, and Hans Wolfgang Levi, *Nuclear Chemical Engineering*, 2nd ed. (New York: McGraw-Hill, 1981).
5. Ahmad E. Nevissi and David Bodansky, "Radon Sources and Levels in the Outside Environment," in *Indoor Radon and Its Hazards*, edited by D. Bodansky, M. A. Robkin, and D. R. Stadler (Seattle, University of Washington Press, 1987): 42–50.
6. U.S. Department of Energy, *Spent Nuclear Fuel Discharges from U.S. Reactors 1992*, EIA service report SR/CNEAF/94-01 (Washington, D.C.: U.S. DOE, 1994).
7. Ronald Allen Knief, *Nuclear Engineering: Theory and Technology of Commercial Nuclear Power* (Washington, D.C.: Hemisphere Publishing, 1992).
8. Allan S. Krass, Peter Boksma, Boelie Elzen, and Wim A. Smit, *Uranium Enrichment and Nuclear Weapon Proliferation* (London: Taylor & Francis, 1983).
9. J. H. Tait, "Uranium Enrichment," in *Nuclear Power Technology*, vol. 2, *Fuel Cycle*, edited by W. Marshall (Oxford: Clarendon Press, 1983), 104–158.

10. R. E. Leuze, "An Overview of the Light Water Reactor Fuel Cycle in the U.S.," in *Light Water Reactor Nuclear Fuel Cycle*, edited by R. G. Wyneer and B. L. Vondra (Boca Raton, Fla.: CRC Press, 1981).

11. U.S. Department of Energy, *World Nuclear Capacity and Fuel Cycle Requirements 1993*, Energy Information Administration report DOE/EIA-0436(93) (Washington, D.C.: U.S. DOE, 1993).

12. F. J. Rahn, A. G. Adamantiades, J. E. Kenton, and C. Braun, *A Guide to Nuclear Power Technology* (New York: John Wiley, 1984).

13. S. F. Mughabghab, M. Divadeenam, and N. E. Holden, *Neutron Resonance Parameters and Thermal Cross Sections*, Part B:Z=61–100, vol. I of *Neutron Cross Sections* (New York: Academic Press, 1984).

14. U.S. Department of Energy, *Integrated Data Base for 1992: U.S. Spent Fuel and Radioactive Waste Inventories, Projections and Characteristics*, report DOE/RW-0006, rev. 8 (Oak Ridge, Tenn.: Oak Ridge National Laboratory, 1992).

15. U.S. Department of Energy, *Characteristics of Potential Repository Wastes*, report DOE/RW-0184-R1, vol. 1 (Oak Ridge, Tenn.: Oak Ridge National Laboratory, 1992).

16. Hans A. Bethe, "Energy Supply," in *Physics Vade Mecum*, edited by Herbert L. Anderson (New York: American Institute of Physics, 1981).

17. J.-Y. Barre and J. Bouchard, "French R&D Strategy for the Back End of the Fuel Cycle," in *Future Nuclear Systems: Emerging Fuel Cycles & Waste Disposal Options, Proceedings of Global '93* (La Grange, Ill.: American Nuclear Society, 1993): 27–32.

18. Pierre M. Chantoin and James Finacune, "Plutonium as an Energy Source: Quantifying the Commercial Picture," *IAEA Bulletin* 35, no. 3 (1993): 38–43

19. Luther J. Carter, *Nuclear Imperatives and Public Trust: Dealing with Radioactive Waste* (Washington, D.C.: Resources for the Future, 1987).

20. *Nuclear Power Issues and Choices*, report of the Nuclear Energy Policy Study Group, chaired by Spurgeon M. Keeny, Jr. (Cambridge, Mass.: Ballinger, 1977).

21. National Academy of Sciences, *Management and Disposition of Excess Weapons Plutonium*, report of the Committee on International Security and Arms Control (Washington, D.C.: National Academy Press, 1994).

22. Frans Berkhout and Harold Feiveson, "Securing Nuclear Materials in a Changing World," *Annual Review of Energy and the Environment* 18 (1993): 631–665

23. Organization for Economic Cooperation and Development, Nuclear Energy Agency, *Nuclear Energy Data 1995* (Paris: OECD, 1995).

24. Simon Rippon, "Europe: The Nuclear Scene in the mid-1990s," *Nuclear News*, 37, no. 11 (September 1994): 52 ff.

25. Charles E. Foreman, "Nuclear Fuel Reprocessing," *NUEXCO Monthly Report*, no. 286 (Denver: NUEXCO, 1992): 29–38.

26. "Chemical separation starts at U.K.'s THORP plant," *Nuclear News* 38, no. 3 (March 1995): 45.

27. "Russian reprocessing plant completion approved," *Nuclear News* 38, no. 3 (March 1995): 46–47.

28. U.S. Department of Energy, *Uranium Institute Annual 1994*, Energy Information Administration report DOE/EIA-0478(94) (Washington, D.C.: U.S. DOE, 1995).

29. U.S. Department of Energy, *Monthly Energy Review, March 1994*, Energy Information Administration report DOE/EIA-0035(94/03) (Washington, D.C.: U.S. DOE, 1994).

30. Organization for Economic Cooperation and Development, *Uranium 1993: Resources, Production and Demand*, joint report of OECD Nuclear Energy Agency and IAEA (Paris: OECD, 1994).

31. U.S. Department of Energy, *Uranium Industry Annual 1991*, Energy Information Administration report DOE/EIA-0478(91) (Washington, D.C.: U.S. DOE, 1992).

32. International Institute for Applied Systems Analysis, *Energy in a Finite World: A Global System Analysis*, Wolf Häfele, program leader (Cambridge, Mass: Ballinger, 1981).

33. Ersel A. Evans, "Nuclear Fuel Cycle," in *JTEC Panel Report on Nuclear Power in Japan*, K. F. Hansen, ch. (Baltimore: Japanese Technology Evaluation Center, Loyola College in Maryland, 1990).

34. U.S. Department of Energy, *Annual Energy Review 1992*, Energy Information Administration report DOE/EIA-0384(92) (Washington, D.C.: U.S. DOE, 1993).

Chapter 8

Nuclear Waste Disposal:
Magnitudes of Waste

8.1 CATEGORIES OF NUCLEAR WASTE

8.1.1 Origins of Nuclear Waste

The operation of a nuclear reactor creates a large amount of radioactive material, produced through fission and neutron capture. The level of radioactivity in the reactor fuel remains very high even after the reactor is shut down. Some of the activities are long-lived, and it is necessary to cope for an extended period with the problems arising from the radioactivity in the spent fuel or in reprocessed material derived from the spent fuel.

These products are generically referred to as nuclear wastes, although some of the radio-nuclides have potentially useful applications and it has been argued that the spent fuel is a resource, not a waste product. Nonetheless, the attribute of spent fuel that is of greatest current significance is the waste aspect, not the resource potential, and its designation as "waste" conforms to the dominant official and informal usage.

The waste disposal problem is the problem of handling these radionuclides in a safe manner. The requirement for safekeeping extends over many centuries. The necessary degree of isolation from the environment, the time period over which isolation must be maintained, and the methods to be employed are all matters of controversy.

8.1.2 Military and Civilian Wastes

As discussed in Chapter 1, the first nuclear reactors were those built during World War II to produce plutonium for weapons. In order to extract plutonium, it is necessary to reprocess the spent fuel, first converting it to liquid form. This both increases the volume of the residue and puts it in a form that can more readily escape into the environment. This residue was originally stored as a liquid in large underground, single-walled tanks.

Given the pressures of wartime development, there was no well-engineered, long-term plan for the permanent disposal or storage of these wastes. Weapons production continued and increased after World War II, with large programs at the DOE's Hanford (Washington) and Savannah River (South Carolina) facilities, but disposal plans were still not developed in a timely fashion. As a result, there were mishaps, including large leaks from some tanks at Hanford during the 1970s.[1] Concern remains about possible future leaks and conceivable, if unlikely, chemical explosions.

In recent years intense consideration has been given to methods for isolating and disposing of these wastes, but the earlier casual treatment has complicated matters. To date the wastes have caused no known harm to human health, and it is not clear that there is any realistic prospect of major future harm. However, their ultimate disposal is a complex matter. Some estimates have placed the price of a complete cleanup of the sites in the range of hundreds of billions of dollars, but the actual cost will depend greatly upon the methods that are employed and the stringency of the cleanup requirements.

Spent fuel from nuclear reactors used on naval vessels constitutes a second, smaller category of military wastes. For the most part, this fuel has been reprocessed at the DOE's Idaho National Engineering Laboratory (INEL). (The product is included with other military wastes as "DOE wastes" in Table 8.1.) The activities and physical volumes involved are considerably less than those at Hanford and Savannah River from weapons production [2, p. 44].

Civilian or *commercial* wastes are those produced by reactors built for commercial electricity generation. The amount of radioactivity produced in this manner to date is much greater than that for the military wastes, because more reactors have been involved, operating over longer total periods. The volume is much less, however, because the wastes have remained as spent fuel. This is the key distinction between "military" and "civilian" wastes. Military wastes, with the possible exception of some of the spent fuel from submarines, have been reprocessed to extract plutonium, and they have not yet been turned into a stable, solid form. Civilian wastes have remained as solid fuel rods, and thus they are much easier to handle than the reprocessed military wastes.

The large volume and the way in which they have been handled make military wastes a problem comparable to or greater than that of civilian nuclear wastes, even though their total radioactivity is less. However, the focus here will be on commercial wastes, because their successful disposal is crucial to the future of nuclear power. The issues relating to military wastes are fundamentally irrelevant to nuclear power, however desirable it would be to resolve them expeditiously and economically.

Military wastes do have a practical relevance, however, in that much of the public does not differentiate between the two categories of waste. And even if the distinction is recognized, it is still easy to assume that errors and problems in one area portend similar errors and problems in the other. Of course, there is a great difference in the histories. Military wastes not only were reprocessed and put in liquid form, but they were reprocessed at a time when speed was thought to be urgent and the dangers small. Now the situation is quite the reverse: Speed is virtually irrelevant to current planning for the handling of civilian spent fuel, and there is great attention to possible dangers.

---

[1]The hazards from these leaks were analyzed in a 1976 study published by the National Academy of Sciences[1]. Despite the large magnitude of the leaks, the study panel concluded "that there has not been in the past, and is not at present, any significant radiation hazard to public health and safety from waste-management operations at Hanford" [1, p. 2].

8.1.3 High- and Low-Level Wastes

Nuclear wastes are sometimes divided into *high-level* and *low-level* wastes, along with a separate category of *transuranic wastes*:

High-level wastes (HLW) are the highly radioactive fission and capture products of the nuclear fuel cycle. They may be in the form of either spent fuel or liquid and solid products from the reprocessing of spent fuel.[2] (In some alternative definitions, spent fuel is not considered to be a "waste" and is in a category by itself.)

Transuranic wastes (TRU) are wastes that do not qualify as high-level wastes but contain more than 0.1 microcuries of long-lived ($T>20$ yr) transuranic alpha-particle emitters per gram of material. The term *transuranic* refers to elements of atomic number greater than 92. These are formed in neutron-capture chains, starting with uranium.

Low-level wastes (LLW) are the remaining radioactive materials that arise from the operation of nuclear facilities, as well as residues from medical and industrial use of radionuclides.[3]

We will restrict the discussion here to consideration of high-level wastes, because the amounts of activity are the greatest for these and the potential hazards consequently the greatest.

8.1.4 Inventories of U.S. Nuclear Wastes

Accumulated Inventories of Nuclear Wastes

Inventories of U.S. commercial and military wastes as of the end of 1991 are summarized in Table 8.1. Despite the precision with which the activities are specified, the data provide only a nominal snapshot in time. The inventories are changing due to the production of new spent fuel in reactors and the radioactive decay of existing inventories. With minor exceptions, military wastes derive from the production of plutonium and tritium for nuclear weapons at Hanford and Savannah River and the reprocessing of fuel from naval nuclear reactors at Idaho Falls.

It is seen that the spent fuel from commercial reactors contains 96% of the total activity of the combined wastes but only 0.2% of the volume. This disparity between commercial and military wastes stems from the fact that the former is in solid form in compact fuel assemblies, while the latter is admixed with a good deal of other material in liquid or solid form. The problems in the disposal of spent fuel arise from the high total activity, but they are mitigated by the small volume of material to be handled.

The total mass of spent fuel stored as of the end of 1991 corresponded to about 23 700 metric tonnes of initial heavy metal (MTIHM) [2, p. 25].[4] All but about 3% of the spent fuel is presently stored at the reactor sites, with most of the remainder stored at a site in Morris,

---

[2]This definition, with the inclusion of spent fuel, corresponds to the official definition expressed in 10 CFR 60.2 of the Code of Federal Regulations [3, Part 60].

[3]Wastes with relatively low levels of activity are categorized on the basis of their activity per unit volume, with different NRC disposal criteria for each class (see Part 61 of Ref. [3]). For "emergency" disposition, low-level wastes are defined as radioactive materials that are not high-level wastes and that are classified as low-level wastes by the NRC (see Part 62 of Ref. [3]).

[4]This is equivalent to about 27 000 tonnes of UO_2. In discussions of nuclear waste disposal the "mass" considered is almost always the mass of the initial heavy metal.

Table 8.1. *Inventories of commercial and military nuclear wastes in the United States as of December 31, 1991.*

Category of waste	Volume (1000 m³)	Activity (MCi)	Thermal power (MW)
Commercial wastes			
Spent fuel[a]	10	23 245	87.6
High-level waste	2	26	0.08
Low-level waste	1423	6	0.03
Total	1435	23 277	87.7
Military wastes[b]			
High-level waste	395	971	2.8
Low-level waste	2816	13	0.02
Transuranic waste	255	3	0.04
Total	3466	987	2.9
Sum	4900	24 264	91

Source: Data are from Ref. [2], p. 14.
[a]The quoted volume is the volume of the fuel assemblies, not of the fuel alone.
[b]These are called "DOE wastes" in Ref. [2]; they consist almost entirely of wastes from military programs. These data omit small amounts of spent fuel stored at DOE sites, primarily at INEL. The mass of this fuel (in MTIHM) is roughly 1% of the mass of the commercial spent fuel listed above [2, Appendix A].

Illinois, where a reprocessing plant had been planned but was not completed. Roughly two-thirds of the fuel is from PWRs and one-third from BWRs.

Future Inventories of Nuclear Wastes

At the present level of nuclear generating capacity, about 2000 tonnes of spent fuel are discharged each year from commercial reactors. Future rates of spent fuel discharge will depend upon the growth or contraction of nuclear capacity. Waste repository needs are sometimes discussed with reference to a projection for the year 2030 in which it is assumed that there will be no new orders for reactors and that reactor shutdowns will start to be numerous in about 15 or 20 yr. In a 1992 version of this scenario, the cumulative spent fuel inventory in 2030 was projected to be 87 700 MTIHM [4, p. 14], with an activity of 32 100 MCi and a heat output of 116 MW [2, p. 28]. In a 1993 version of the same general scenario, but assuming a more rapid decline in nuclear capacity, the projection for 2030 was 84 500 MTIHM [5, p. 15].[5]

These projections for the year 2030 could be too low, if there is a resumption of nuclear reactor construction, or too high, if nuclear power is phased out precipitously. Nonetheless, they set a scale for storage needs over the next few decades. A total of 85 000 tonnes is comparable to, but somewhat exceeds, the capacity now allocated to spent fuel in the one repository under active consideration, at Yucca Mountain. This repository, if it is constructed, is to be limited to a total of 70 000 tonnes of waste (measured in MTIHM), of which 63 000 tonnes are allocated to spent fuel (see Section 9.5.3).

The remaining 7000 tonnes of capacity is allocated to high-level military waste. This waste is still in many different forms, including liquid, sludge, and salt cake. It is anticipated that a large fraction of this waste will be incorporated into borosilicate glass in a form that could be put into permanent storage. This material would correspond to roughly 8100

---

[5]In this scenario, net nuclear generation rises slightly from 71 GWyr in 1992 to 72 GWyr in 2005 and then starts dropping rapidly, falling to 4 GWyr by 2030 [5, p. 9].

Table 8.2. *Assumed reactor characteristics and annual spent fuel discharges, in model for large light water reactors.*

Parameter		Data Source[a]	Magnitudes	
			BWR	PWR
A	Reactor capacity, thermal (MWt)	*	3800	3800
B	Reactor capacity, electric (MWe)	*	1250	1250
C	Thermal efficiency (%)	B/A	32.9	32.9
D	Average thermal power (MWt/MTIHM)	*	25.9	37.5
E	Fuel in reactor (MTIHM)	A/D	146.7	101.3
F	Average fuel burn-up (GWd/t)	*	27.5	33.0
G	Days of full-power irradiation	(F/D)	1062	880
H	Refueling cycle (full-power days)[b]	(G/n)	265.5	293.3
I	Assumed capacity factor (CF) (%)	*	80	80
J	Refueling cycle for 80% CF (days)	(H/I)	332	367
K	Duration of refueling cycle (years)	J/365.25	0.908	1.004
Input fuel per GWyr				
L	Uranium (tonnes)	(E/nK)	40.369	33.647
M	^{235}U (tonnes)	*	1.110	1.077
N	^{235}U enrichment (% weight)	M/L	2.75	3.20
Discharged spent fuel per GWyr (tonnes)				
P	Total heavy metal (tonnes)	*	39.233	32.376
Q	Total uranium (tonnes)	*	38.872	32.048
R	Total plutonium (tonnes)	*	0.340	0.307
S	Total U+Pu (tonnes)	(Q+R)	39.212	32.354
T	^{235}U (tonnes)	*	0.304	0.266
U	U and Pu fissioned (tonnes)	L−P	1.136	1.271

[a]Parameters taken from Table 1 of Ref. [7] are designated with an asterisk (*). Other quantities are calculated from these parameters (e.g. C=B/A, where A, B, and C correspond to the designations in the left-hand column). When values are given in Ref. [7], but also can be calculated from earlier numbers, the arithmetic prescription is in parentheses.
[b]The assumed irradiation periods are 4 yr for the BWR (n=4) and 3 yr for the PWR (n=3).

MTIHM.[6] Thus, as with civilian wastes, the planned Yucca Mountain capacity will not quite suffice for the expected amount of military wastes.

8.2 WASTES FROM COMMERCIAL REACTORS

8.2.1 Mass and Volume per GWyr

Spent Fuel

The amount of spent fuel discharged per year varies among reactors, depending upon the reactor size, the design of its fuel elements, and its operating history during the year. Table 8.2 presents a summary description drawn from an extensive 1980 Oak Ridge National Lab study [7, p. 3]. The data are for 1250-MWe reactors operating at an 80% capacity factor and

---

[6]This estimate is from Ref. [6], p. 5-5. It is sensitive to assumptions as to the fraction of the defense wastes that will be vitrified.

Table 8.3. *Physical characteristics of typical LWR fuel assemblies.*

Parameter	BWR	PWR
Fuel element array	8×8	17×17
Fuel elements per assembly	63	264
Other elements per assembly[a]	1	25
Uranium mass (kg)	183.3	461.4
UO$_2$ mass (kg)	208.0	523.4
Zircaloy and hardware mass (kg)	66.7[b]	134.5
Total assembly mass (kg)	275	657.9
Assembly length (m)	4.470	4.059
Assembly cross-sectional area (m^2)	0.0193	0.0458
Nominal assembly volume (m^3)	0.0864	0.186
Average burn-up (GWd/t)	27.5	33.0
Uranium mass per GWyr (MTIHM)	40.369	33.647
Waste output per GWyr[c]		
Number of assemblies	220	73
Mass of assemblies (tonnes)	60	48
Volume of assemblies (m^3)	19	14

Sources: Unless otherwise indicated, the parameters are taken from Table 3 of Ref. [7]. These parameters have been reaffirmed as reference characteristics for LWR fuel assemblies in Table 1.8 of Ref. [2]. Average burn-up data and uranium mass per GWyr data from Table 1 of Ref. [7], for a reactor operating at a thermal efficiency of 32.9%.
[a]These are for control and instrumentation; in the BWRs, control rods are between assemblies rather than incorporated in them.
[b]Some obvious inconsistencies in Table 3 of Ref. [7] have been addressed by using the version of the data presented in Table 4-2 of Ref. [8].
[c]Calculated from preceding data.

from which fuel is removed on an annual cycle. These reactors are somewhat larger than the U.S. average, the assumed capacity factor is better than typical, and the assumed burn-up level is now commonly exceeded. Nonetheless, these results are reasonably representative of current performance.

As seen in Table 8.2 (line L), something in the neighborhood of 35 tonnes of uranium is used per reactor year, with the actual amount being slightly more for BWRs and slightly less for PWRs, because of differences in burn-up. In each case the heavy-metal content decreases by a little more than 1 tonne (line U), due to fission of uranium and plutonium.[7]

The reactor fuel is packaged in fuel assemblies, which come in many configurations. Parameters for typical BWR and PWR fuel assemblies are given in Table 3 of Ref. [7] and summarized in Table 8.3. As reflected in Table 8.3, the BWR assemblies are smaller but more numerous than the PWR assemblies. About 220 fuel assemblies are removed per GWyr for BWRs and 70 for PWRs. Spent fuel assemblies are now for the most part stored in cooling pools at the reactor sites, but the plan is to transfer them to dry canisters for further storage or underground disposal (see Chapter 9).

For both PWRs and BWRs, the reactor fuel contains between 75% and 80% of the total assembly mass, and the zircaloy cladding and support hardware the remainder. The UO$_2$ fuel, with a density of about 10 g/cm^3, only accounts for a relatively small fraction of the volume.

---

[7]For pure ^{235}U fission, assuming the deposition of 200 MeV per fission, it requires 1.21 tonnes to provide 1 GWyr of electric power at the 32.9% thermal efficiency of Table 8.2. This is in approximate agreement with the numbers in line U of Table 8.2 and in good agreement with their average.

The actual assembly volume, taken from its outer dimensions, is 3 to 4 times greater than that of the fuel alone. The total volume of the assemblies discharged per GWyr is about 19 m^3 for BWRs and about 14 m^3 for PWRs.

Reprocessed Waste

There are no present plans to reprocess commercial spent fuel in the United States. However, reprocessing was long assumed as part of the U.S. waste program; it is being actively carried out in France and elsewhere, and it may at some future time play a role in the U.S. waste disposal program.

Reprocessing of commercial spent fuel subtracts uranium and adds glass. One 1983 projection called for incorporating the reprocessed waste into a borosilicate glass form, with the wastes from 2.1 MTIHM formed into a canister 0.3 m in diameter and 3 m high [8, p. 35]. For the 33.6 MTIHM discharge from the model PWR (see Table 8.3), this would require 16 canisters per GWyr. The total volume of the canisters, neglecting the need for spacing between them, would be roughly 3.4 m^3 per GWyr. The density of borosilicate glass is about 3 g/cm^3, and the total mass is about 10 tonnes. Thus, in this model the mass and volume of the wastes in glass form are significantly less than those of the spent fuel. Of course, most of the mass of the original spent fuel is in the uranium and plutonium which was removed.

8.2.2 Radioactivity in Waste Products

Activity of Spent Fuel

The cooling of spent fuel following removal from the reactor is shown in Fig. 8.1. The activity and heat output are plotted as a function of the time since discharge for the fuel from

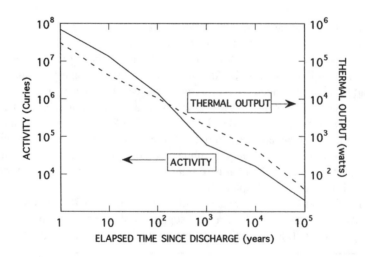

Figure 8.1. Decay of spent fuel from 1 GWyr of reactor operation, for burn-up of 40 GWd/t (29 MTIHM): activity and thermal output as a function of time since discharge.

Table 8.4. *Activity of selected radionuclides in spent fuel at various times after discharge from reactor, for a burn-up of 40 GWd/t and a thermal efficiency of 31.8%.*

Nuclide	Half-life (yr)	E^a (MeV)	Activity (Ci per GWyr)					
			1 yr	10 yr	100 yr	1000 yr	10^4 yr	10^5 yr
^{90}Sr	28.8	0.196	2.41×10^6	1.94×10^6	2.28×10^5			
^{90}Y	≪1	0.933	2.41×10^6	1.94×10^6	2.28×10^5			
^{99}Tc	211 k	0.085					0.43×10^3	321
^{137}Cs	30.1	0.171	3.47×10^6	2.82×10^6	3.53×10^5			
^{137}Ba	≪1	0.664	3.27×10^6	2.67×10^6	3.33×10^5			
^{237}Np	2.14 M	4.857						44
^{239}Pu	24.1 k	5.101				1.11×10^4	8.70×10^3	660
^{240}Pu	6.56 k	5.155				1.52×10^4	5.88×10^3	
^{241}Pu	14.35	0.005	4.25×10^6	2.75×10^6	0.36×10^5			
^{241}Am	432	5.539			1.34×10^5	3.18×10^4		
Sum[b]			15.8×10^6	12.1×10^6	13.1×10^5	5.82×10^4	15.0×10^3	1025
Total[b]			69.9×10^6	13.5×10^6	14.2×10^5	6.10×10^4	16.3×10^3	2000

Sources: Half-lives are from Ref. [11]. The symbols *k* and *M* denote 10^3 and 10^6, respectively. Activities are derived from Ref. [9], Tables 2.4.6 and 2.4.13, for 3.72% ^{235}U enrichment.
[a]*E* is the average energy release per disintegration, excluding neutrino energy, as given in Ref. [2], Table B.1. [Note: 1 MeV/disintegration corresponds to 5.93 kW/MCi.]
[b]The "sum" is the sum for radionuclides listed in the table; the "total" is the sum for all radionuclides, including those not listed.

1 GWyr of reactor operation.[8] The activity and thermal output fall off in roughly the same fashion, but the fractional drop is more rapid for the activity because, on average, the beta-particle emitters have shorter half-lives and less energy deposition than do the alpha-particle emitters (see the next section).

When a reactor is just shut down, there are extremely high levels of activity. The actual activity at the moment of shutdown depends on the recent history of the reactor operation, and is dominated by the decay of short-lived radionuclides. These radionuclides, especially those with half-lives of the order of hours or days, are important for the immediate heat budget of the reactor after shutdown, and their role is crucial in reactor accidents in which there is difficulty in maintaining the flow of cooling water (or other coolant). However, they are not part of the waste disposal problem, because their activity falls to negligible levels while the spent fuel is still in the cooling pools at the reactor sites.

After one year the total radioactivity of the spent fuel is about 1% of the activity at discharge [7], and after a total of 10 yr it has decreased by more than another factor of five. Ten years can be considered to be roughly the time at which the waste disposal problem begins, because the earlier high levels of radioactivity and heat production make it unfavorable to attempt to dispose permanently of the wastes much sooner than 10 years after removal from the reactor.[9]

---

[8]The data are taken from Ref. [9], Tables 2.4.6 and 2.4.7. The specific case shown is for a burn-up of 40 GWd/t for a PWR with an initial ^{235}U enrichment of 3.72% and a fuel mass of 29 MTIHM. A normalization from burn-up in megawatt-days thermal (MWd) to a net electrical output of 1 GWyr is made using a thermal efficiency of 31.8% (the value for U.S. reactors in 1991 [10]).
[9]If reprocessing takes place, it may occur well before 10 yr, perhaps after only 150 days. It is more difficult to handle the wastes if reprocessing is undertaken so soon, but plutonium isotopes, which would decay into undesired long-lived products, are thereby extracted.

Table 8.4 shows the radionuclides most responsible for the activity in the wastes for the period from 1 year to 100 000 yr after removal from the reactor [7]. During the first few centuries, the most important contributors to the total activity are strontium 90 (^{90}Sr) and cesium 137 (^{137}Cs). Each has a half-life in the neighborhood of 30 years, and each decays into a short-lived product, essentially doubling its contribution to the total activity of the waste. More particularly, ^{90}Sr decays to yttrium 90 (^{90}Y) ($T=64$ hr), and ^{137}Cs decays 95% of the time to an excited state of barium 137 (^{137}Ba), which emits a 0.66-MeV delayed gamma ray.[10]

A striking feature of the data in Table 8.4 is the dominance of a very few radionuclides during the most critical period for waste handling, from 10 yr to 10 000 yr after discharge. At each of the times, at least 90% of the total activity is accounted for by only three parent nuclides. During the first few hundred years, the dominant parent nuclei are ^{90}Sr, ^{137}Cs, and ^{241}Pu. At successive later times, the important nuclei are ^{241}Am (a decay product of ^{241}Pu), ^{240}Pu, ^{239}Pu, ^{99}Tc, and ^{237}Np.[11]

The Impact of Reprocessing

The nature of the long-term waste storage problem would be substantially changed if reprocessing were undertaken to remove most of the plutonium from the waste. Plutonium removal would also reduce the contributions from americium 241 (^{241}Am) and neptunium 237 (^{237}Np); most of their abundance in the spent fuel is due to the β^- decay of ^{241}Pu to ^{241}Am, followed by alpha-particle decay to ^{237}Np. To use reprocessing to reduce the abundance of ^{241}Am and ^{237}Np, it is necessary to reprocess the fuel promptly, before an appreciable fraction of the ^{241}Pu ($T=14$ yr) has decayed.

It is also possible to remove and package separately the strontium and cesium, although this is not an essential goal of reprocessing. The strontium and cesium can be encapsulated, and the integrity of their containers is of importance for only a few hundred years, up to perhaps 600 yr, because their activity drops by roughly a factor of 1000 in each 300 yr.

Although the separation of these isotopes from the wastes reduces the activity of the main body of the wastes, it involves far more handling of radioactive material than does direct disposal of the spent fuel. It also creates new pools of radioactive material. The plutonium can be returned to reactors to provide additional energy and be consumed in the process. However, the increased availability of plutonium for diversion to use in weapons has caused strong objections to the reprocessing of spent fuel and the extraction of plutonium (see Chapter 14).

8.2.3 Heat Production

Although the radioactive waste hazards arise mainly from possible biological damage caused by the emitted radiations, there is also a significant issue of heat generation. On occasion, it has been suggested that the heat from the wastes could be used constructively as a heat

---

[10]The gamma-ray emitting excited state of ^{137}Ba is a so-called isomeric state, with a decay half-life of 2.55 min. Usually, gamma rays are emitted within less than 10^{-9} sec following beta or alpha decay and are not considered an additive contribution to the activity, i.e., to the number of decays. However, for delayed emission from isomeric states there is customarily a separate counting. The ground state of ^{137}Ba is stable. ^{137}Cs decays 5% of the time to this ground state.

[11]The data of Table 8.4 are for the specific case of a burn-up of 40 GWd/t. The activity and thermal output of the fuel vary with burn-up, although by specifying the activity in terms of the energy output (MCi per GWyr), the variation is somewhat less than when the activity is specified in terms of the mass of fuel (MCi per MTIHM).

source; for example, for direct warming of arctic installations. However, until there is more confidence in our ability to retain the wastes safely within their containers, it is unlikely that such applications will be considered prudent. The main interest in heat generation at present is in the demand that it creates for cooling the waste containers, with either water or air, and in the problems that the total thermal output create for a waste repository.

The amount of heat generation per unit activity depends upon the energy deposition per decay. For orientation purposes, the following conversion factor is useful:

$$\text{1 megacurie} \rightarrow 5.93 \text{ kW per MeV per disintegration.}$$

An order-of-magnitude estimate of the heat generation per curie is obtained by assuming an average energy deposition of 1 MeV per disintegration. With this rough average, 1 MCi corresponds to a heat generation on the order of 6 kW.

This is an overestimate for most beta decays and a substantial underestimate for all alpha decays. In a typical beta decay, the maximum beta-particle energy is about 1 MeV, but there is a distribution of energies, and on average more than half of the energy is carried off by neutrinos.[12] Thus, typical energy depositions for beta decay are well under 1 MeV. For all alpha-particle activities of interest, the alpha-particle energy exceeds 4 MeV, and this energy is all captured in the fuel within 0.1 mm of the site of the decaying nuclide.

For the first several decades after the fuel is removed from the reactor, the most important activities are beta decays. At later times alpha-particle activities become more important, unless the alpha-particle emitters are removed in reprocessing. The actual heat output for current inventories of wastes is about 4 kW/MCi for spent fuel and 3 kW/MCi for high-level wastes from reprocessing (see Table 8.1). Both include samples broadly distributed in time since discharge from the reactor, within the limitation that the greatest possible time is less than about 50 yr.

The heat output for spent fuel with a burn-up of 40 GWd/t is displayed in Figure 8.1 as a function of time since discharge. It is 4.3 kW/MCi after one year and 31 kW/MCi after 1000 yr. At 1000 yr over 95% of the activity is due to three alpha-particle emitters: ^{239}Pu, ^{240}Pu, and ^{241}Am. However, by then the total heat output is only about 2 kW for the waste from 1 GWyr, and heat production is no longer an important matter. The subsequent drop at still later times of the thermal output per unit of activity is due to the increased relative importance of some beta emitters, such as ^{99}Tc.

After 10 yr of cooling, the fuel from one GWyr of reactor operation has a heat output of 42 kW, or 1.47 kW/tonne. The heat output falls substantially with time, dropping to 0.37 kW/tonne after 100 yr and to 0.07 kW/tonne after 1000 yr. From some perspectives these are not very great heat outputs, and after 10 yr or so spent fuel can be cooled by convective air flow. However, the combined output of a large mass of fuel creates a significant heat output, which becomes a major consideration in the design of an underground repository with tens of thousands of tonnes of fuel. One way to reduce the peak temperature reached in a repository is to postpone the placement of wastes in it. In the well developed Swedish plans for a waste repository, a long preliminary cooling time is planned, using interim storage.

---

[12]A precise determination of the energy deposition in beta decay requires: (a) the establishment of the actual average beta energy; (b) a correction for the energy input from gamma rays produced in beta decay to an excited state; and (c) a (very small) correction for the energy lost by escape of beta particles and gamma rays from the waste into the surroundings.

8.3 HAZARD MEASURES FOR NUCLEAR WASTES

8.3.1 Exposures from Direct Contact

The spent fuel assemblies are placed, with remote handling, into cooling pools. They eventually are to be transferred from the cooling pools into protective inner canisters and outer casks. The canisters and casks provide substantial shielding, essentially as much as one wants if a price is paid in weight. Therefore, radiation exposure from the spent fuel is not a critical issue in ordinary handling of the fuel, once it is initially secured in a canister after removal from the cooling pool.[13]

However, the high radiation levels from unshielded spent fuel provide important protection against theft by people who do not have elaborate protective equipment and facilities. For that reason, data on exposures are of interest in the context of weapons proliferation. We here cite values from an article on the disposition of plutonium from dismantled nuclear weapons [13, pp. 204–5].

The largest radiation exposures from spent fuel that is between, say, one year and 100 yr old, are from 0.66-MeV gamma rays emitted in the beta decay of ^{137}Cs. The dose from these gamma rays at the surface of an unshielded spent fuel assembly is about 260 Gy per hour for fresh fuel and 130 Gy per hour after 30 yr of decay. At 1 m from the center of assembly (on a line perpendicular to its axis), the dose after the fuel has cooled for 30 yr would be about 8 Gy per hour. Thus, a one-hour exposure would give a dose in excess of the lethal dose of 4–5 Gy, making an unshielded assembly of spent fuel exceedingly dangerous for many decades. (See Section 14.2.5 for further discussion.)

Of course, in considering the radiation hazards from nuclear wastes, the chief concern is not with direct exposures of terrorists or others. It is with the possible exposure of the general public many centuries hence, through the escape of radionuclides from waste repositories into the biosphere.

8.3.2 Water Dilution Volume

Motivation for Introducing the Water Dilution Volume

Although the amount of activity, i.e., the number of decays per unit time, gives some indication of the importance of a radionuclide from the standpoint of risk to human health, activity per se is not the crucial quantity. In addition, the radiation dose for radionuclides deposited in the body depends on the type and energy of the emitted particles, whether the nuclides are quickly excreted from the body upon ingestion (or inhalation) or retained, and, if retained, for how long and in what organs.

Just these considerations go into determining the *maximum permissible concentration* of radionuclides in water.[14] The maximum permissible concentration has been established as the maximum level acceptable for drinking water. Such levels have been tabulated by both the International Commission on Radiation Protection and the U.S. Nuclear Regulatory Commission.[15]

The potential hazard represented by a given radionuclide is frequently indicated in terms

---

[13]See Section 9.1.3 for a brief discussion of exposures from fuel packaged for transportation.

[14]The maximum permissible concentration is closely related to the annual limit of intake (ALI). (See Section 2.6.3.)

[15]The NRC standards, effective 1994, are specified in Appendix B to §20.1001–20.2402 of Ref. [12].

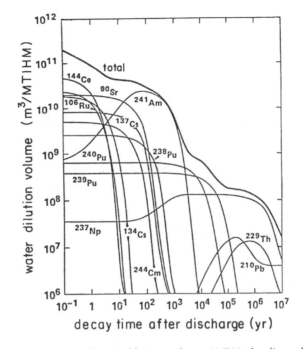

Figure 8.2. Illustration of the use of water dilution volume: WDV of radionuclides in PWR spent fuel as a function of time, based on ICRP values for maximum permissible concentrations. (From Ref. [8], p. 32.)

of the *water dilution volume* (WDV). The water dilution volume (in m³) is the amount of water required to dilute the radionuclide to the maximum permissible concentration. It equals the amount of the radionuclide being considered (in Bq or Ci) divided by the maximum permissible concentration (in Bq/m³ or Ci/m³). There is no implication here that the radionuclide *will* be so dissolved. However, the WDV is used in an attempt to provide an overall measure of the relative hazard associated with different radioactive samples.

Water Dilution Volume for Spent Fuel

To compare the hazards from the various radionuclides in the waste, it has been common to plot the WDV values for individual radionuclides as a function of time since discharge of the fuel. An example of such a plot is given in Fig. 8.2, which is from a 1983 National Academy of Sciences study [8]. The results are presented in terms of the amount of water needed (in m³) to dilute the wastes in 1 MTIHM of spent fuel.[16]

There are a number of interesting features in Fig. 8.2.

- Although the activity falls by a factor of more than 1000 in the first thousand years (see Fig. 8.1 and Table 8.4), the WDV falls more slowly. This is in part due to the fact that the longer-lived nuclides are alpha-particle emitters. Typically, alpha particles are more energetic than beta particles, by a factor of 5 to 10, and have a 20 times greater relative biological effectiveness.

---

[16]This fuel is assumed to have the characteristics of the PWR fuel described in Table 8.2, in particular a burn-up of 33 GWd/t.

- Over long time periods, ^{237}Np ($T=2.14\times10^6$ yr) becomes the most important of the waste nuclides, if ICRP values for the maximum permissible concentrations are used. Its hazard index (WDV) initially increases with time, because it is produced in the alpha-particle decay of ^{241}Am ($T=433$ yr), which in turn is largely the product of the beta-decay of ^{241}Pu ($T=14.35$ yr).

- There still is substantial uncertainty with respect to some of these parameters. In particular, a report of the National Council on Radiation Protection and Measurements has re-analyzed the available information on neptunium [14]. An earlier ICRP analysis (1980) had tightened limits on ^{237}Np by a factor of 340, making it a key radionuclide. The later NCRP analysis suggests a relaxation from the ICRP values by a factor of about 8. If this is accepted, the ^{237}Np curve in Fig. 8.2 should drop by a factor of 8.

To illustrate the way in which the WDV can be calculated, consider ^{137}Cs. Ten years after discharge, the ^{137}Cs activity is 8.2×10^4 Ci/MTIHM, for the reactor parameters used in the NAS study [8, pp. 28, 31]. For an annual limit of intake of 10^{-5} Ci of ^{137}Cs and an annual water intake of 0.73 m^3, the maximum ^{137}Cs concentration would be 1.37×10^{-5} Ci/m^3, corresponding to a WDV of 6×10^9 m^3/MTIHM.[17] This is in approximate agreement with the value plotted in Fig. 8.2.

Limitations of the Usefulness of the WDV

Despite its suggestiveness, the water dilution volume is not a true measure of relative hazard. The large changes made in recent years in some of the values for the maximum permissible concentration and the differences between the values attributed to the ICRP and the NRC [8] suggest that the numerical values are uncertain.

Even more fundamentally, use of the WDV is implicitly based on a faulty premise, namely, that the chance of ingestion is the same for all radionuclides. However, if the waste is exposed to water, different radionuclides will have very different rates of entering the water, speeds of transport with the water into the biosphere, and degrees to which they are concentrated in the food chain.

For example, ^{99}Tc has a small WDV in view of its slow rate of decay (long half-life) and its small energy release per disintegration, which leads to a relatively large maximum permissible concentration. However, the transport of technicium through the ground surrounding a potential repository site is unusually rapid. In comparing, for example, the relative hazards from ^{99}Tc and ^{237}Np, it is necessary to consider not only the maximum permissible concentration and WDV for each, but also the way in which they move through the ground and enter the biosphere. Aspects of these issues are discussed briefly in Section 9.2.3.

8.3.3 Comparisons between Wastes and Natural Uranium

Comparisons of WDV of Waste and Initial Uranium

In attempts to obtain a perspective on waste disposal hazards, it has been common to compare the WDV (or an equivalent measure of relative toxicity) of the spent fuel and of the

---

[17]The present occupational ALI is 10^{-4} Ci, corresponding to a dose limit of 50 mSv/year (see Table 2.7). At the time of the NAS report, the dose limit for the general public was 5 mSv/yr, corresponding to an ALI of 10^{-5} Ci. The assumed water intake is 2 liters per day.

uranium ore initially mined to fuel the reactor. However, this comparison has had more significance as a debating tool than as a pertinent measure.

In the 1970s it was frequently pointed out that after several hundred years the radiotoxic hazard measure (equivalent to the WDV) of the wastes falls below that of the original uranium ore. This was cited as demonstrating that the hazard posed by the wastes is small. However, the WDV of the wastes was substantially increased when the focus shifted in the 1980s from reprocessed wastes, with the plutonium removed, to spent fuel. In addition, an ICRP re-evaluation of maximum permissible concentrations for various radionuclides had the effect of decreasing the WDV for the parent uranium ore and increasing the WDV at long times for the wastes. Specific changes included an increase by a factor of 8 in the maximum permissible concentration for ^{226}Ra, a major contributor in the uranium ore, and decreases in the maximum permissible concentrations for several important actinides in the wastes, including a factor of 340 reduction for ^{237}Np [8, p. 31].

As a consequence of these changes, the calculated time interval for the relative toxicity of the spent fuel to fall to that of the parent ore increased greatly. Depending on which set of parameters was adopted, this time ranged from 10 000 yr to over one million years.[18] In contrast, for reprocessed wastes and using the NRC parameters of that time, the crossover was about 500 yr [8, p. 37].

There are a number of reasons to question the appropriateness of this comparison to the parent ore:

- It is tacitly assumed that radionuclides from the wastes and from the parent ore have equal probabilities of reaching the biosphere. There is no basis for such an assumption, given the differences in their location and physical form.

- The most hazardous components of the original ore, particularly the radium and the radon emitted from it, are not destroyed in the reactor but instead are transferred as mill tailings to disposal sites near the surface of the Earth.[19]

- A comparison to the parent ore represents an arbitrary constraint, presumably based on the premise that use of uranium for nuclear power should not increase the total hazard from uranium. But this is not a meaningful standard for comparison, because the radiation dose attributed to ingestion of natural uranium is exceedingly small (see Section 2.4.2).

Comparisons to Activity in Earth's Crust

A different perspective is obtained by comparing the wastes to the total uranium in the upper layer of the Earth's crust, not just to the small portion that is mined. However, some of the earlier cautions still apply. The probabilities for reaching the biosphere can differ greatly, and relative amounts of radioactivity do not alone determine relative hazards.

With this caveat, we compare the activity of the wastes placed in a geologic repository to the activity in the Earth's crust down to a depth of 1 kilometer. A 1-km layer provides a reasonable basis for comparison, because a repository is expected to be at least several hundred meters below the surface of the Earth. Over the area of the United States, this

---

[18]The smaller time corresponded to the values of the maximum permissible concentration then being used by the NRC; the longer time corresponded to values used by the ICRP.

[19]As pointed out in Chapter 7, this does not make mill tailings a significant hazard, if one compares radon exposures from the tailings to those received by the general population from air in typical homes.

activity totals about 10^{12} Ci (see Section 2.4.1). The activity in spent fuel after 1000 yr of decay is 6×10^4 Ci per GWyr (see Table 8.4). Thus, for example, a U.S. nuclear program of five times the present size (enough to provide all of present U.S. electricity consumption) continuing for 100 yr would mean an output of 3.5×10^4 GWyr and an activity of about 2×10^9 Ci (again, after 1000 yr of decay). This is about 0.2% of the natural activity in the top kilometer of the U.S. land area. Thus, on a gross overall basis, the wastes are not being placed in an otherwise pristine Earth.

There are, however, very large increases in the abundances of individual radionuclides, for example plutonium isotopes, and very large increases in total activity at the location of the disposal sites. To assess the magnitude of the hazard this represents, it is necessary to have a detailed picture of the methods to be used for radioactive waste disposal, including the form of the waste (spent fuel or reprocessed), the stability of the waste form and its containers, and the pathways for transfer of individual radionuclides from the disposal site into the biosphere. Aspects of these matters are discussed in Chapters 9 and 10.

REFERENCES

1. National Research Council, *Radioactive Wastes at the Hanford Reservation: A Technical Review*, report of the Panel on Hanford Wastes (Washington, D.C.: National Academy of Sciences, 1978).
2. U.S. Department of Energy, *Integrated Data Base for 1992: U.S. Spent Fuel and Radioactive Waste Inventories, Projections and Characteristics*, Report DOE/RW-0006, rev. 8 (Oak Ridge, Tenn.: Oak Ridge National Laboratory, 1992).
3. *Energy, U.S. Code of Federal Regulations*, title 10, parts 51 to 199 (1993).
4. U.S. Department of Energy, *World Nuclear Capacity and Fuel Cycle Requirements 1992*, Energy Information Administration report DOE/EIA-0436(92) (Washington, D.C.: U.S. DOE, 1992).
5. U.S. Department of Energy, *World Nuclear Capacity and Fuel Cycle Requirements 1993*, Energy Information Administration report DOE/EIA-0436(93) (Washington, D.C.: U.S. DOE, 1993).
6. M. L. Wilson et al., *Total-System Performance Assessment for Yucca Mountain—SNL Second Iteration (TSPA-93)*, report SAND93-2675 (Albuquerque, NM: Sandia National Laboratories, 1994).
7. A. G. Croff and C. W. Alexander, *Decay Characteristics of Once-Through LWR and LMFBR Spent Fuels, High-Level Wastes, and Fuel-Assembly Structural Material Wastes*, report ORNL/TM-7431 (Oak Ridge, Tenn.: Oak Ridge National Laboratory, 1980).
8. National Research Council, *A Study of the Isolation System for Geologic Disposal of Radioactive Wastes*, report of the Waste Isolation Systems Panel (Washington, D.C.: National Academy Press, 1983).
9. U.S. Department of Energy, *Characteristics of Potential Repository Wastes*, report DOE/RW-0184-R1, vol. 1 (Oak Ridge, Tenn.: Oak Ridge National Laboratory, 1992).
10. U.S. Department of Energy, *Monthly Energy Review, March 1993*, Energy Information Administration report DOE/EIA-0035(93/03) (Washington, D.C.: U.S. DOE, 1993).
11. Jagdish K. Tuli, *Nuclear Wallet Cards* (Upton, N.Y.: Brookhaven National Laboratory, 1995).
12. *Energy, U.S. Code of Federal Regulations*, title 10, parts 0 to 50 (1993).
13. F. Berkhout, A. Diakov, H. Feiveson, H. Hunt, E. Lyman, M. Miller, and F. von Hippel, "Disposition of Separated Plutonium," *Science and Global Security 3* (1993): 161–213
14. National Council on Radiation Protection and Measurements, *Neptunium: Radiation Protection Guidelines*, NCRP report no. 90 (Washington, D.C.: NCRP, 1988).

Chapter 9

Storage and Disposal of Nuclear Wastes

9.1 STAGES IN WASTE HANDLING

9.1.1 Overview of Possible Stages

There are a number of possible stages in waste handling, already alluded to in Chapter 7.

- Storage of spent fuel in cooling pools at the reactors.

- Dry storage of spent fuel at reactor sites.

- Reprocessing or transmutation of spent fuel.

- Interim storage of reprocessed waste or spent fuel at centralized facilities.

- Permanent disposal of spent fuel, reprocessed waste, or residues of transmutation, by placement in repositories or by other means.

- Transportation of spent fuel or reprocessed waste as it moves through the stages above.

Not all of these stages are essential to a coherent waste management program. Facilities for on-site dry storage and off-site interim storage are needed only if there are delays in implementing the later stages of disposal. In fact, these intermediate stages did not enter into the early thinking about waste disposal, when it was expected that prompt reprocessing would be the norm.

However, the present situation is quite different from that contemplated in the planning of the 1960s and 1970s. Reprocessing has been abandoned for U.S. commercial wastes, there

137

are no U.S. repositories for long-term storage, and there are no firm plans for interim storage. In practice, the spent fuel is kept in cooling pools next to the reactors and, on a small but growing scale, in dry storage casks at the reactor sites. Innumerable plans have been developed for waste management programs over the past 30 yr, but there has been little progress in the construction of centralized facilities.

A few other countries are somewhat more advanced in their programs than is the U.S. For example, France routinely reprocesses spent fuel, and Sweden has started operation of an interim storage facility as part of what appears to be a coherent long-term plan for disposal of spent fuel. However, no country has yet begun construction of a repository for long-term storage of high-level commercial wastes. While site investigations have been carried out in a number of countries, the earliest target dates for a working repository are the years 2008 for Germany and 2010 for the United States [1]. There is no assurance that either target will be met.

For many years the United States "nuclear establishment" felt no urgency about the waste disposal problem because it appeared to be easily soluble and not technically interesting. Now it is obvious that the problem is not easy to solve, but there is deep disagreement as to whether the main difficulties are technical or political.

9.1.2 Short-Term Approaches to Waste Handling

Storage of Spent Fuel in Cooling Pools

Originally it was expected that the spent fuel from nuclear reactors would be held in cooling pools, at the reactor sites for a brief time and that reprocessing would be carried out after about 150 days. Instead, in the United States the fuel has remained in these cooling pools and some pools have been holding fuel for over 20 yr. The capacity of the pools is limited, but to date no reactor has had its operation stopped by a shortage of cooling pool space. For many reactors, there is still considerable storage capacity remaining. At the end of 1993 the total capacity at reactor sites was sufficient for roughly 210 000 assemblies, and 95 000 assemblies were being stored [2, p. 40].

Nonetheless, individual reactors have faced a severe squeeze. The capacity for storage of spent fuel assemblies at the reactor site can be increased to a modest extent by modifying the geometric arrangement of the assemblies in the cooling pool. A much larger expansion can be achieved by using dry storage, and this has been the solution at a number of reactor sites.

Dry Storage of Spent Fuel at Reactor Sites

In dry storage systems, the spent fuel rods are transferred to special casks when the total activity and the heat output are reduced enough for air cooling to suffice. This solves the problem of limited cooling pool capacity and is an option that can be implemented without having to rely on agreements on transportation of nuclear wastes[1] or on the creation of other facilities. Some utilities have already begun using dry storage, with special casks to hold fuel that has previously cooled in pools for at least 5 or 10 yr. These casks are cooled by natural

---

[1]Transportation has been a stumbling block because of fears in some communities about the dangers of moving radioactive wastes. Casks have been developed that would appear to ensure the safe transport of spent fuel. Calculated risks are primarily from the ordinary damage of truck collisions, not from releases of radioactivity.

convective air flow, without pumps. They must be licensed by the Nuclear Regulatory Commission, which by late 1993 had approved six different cask designs [3].

The first such casks were installed at the Surry power station in Virginia in 1986 [4]. At the Surry facility, as of 1993, 18 storage casks were standing on a concrete pad, each holding 21 to 28 assemblies. The overall capacity of the facility is 84 casks.[2]

Several other utilities have followed suit, and there is a growing movement to adopt this short-term solution. For example, the Palisades reactor in Michigan (a 768-MWe PWR) was near the end of its remaining capacity by the end of 1991, and at the time it was concluded that operation would not be able to continue beyond 1993 without alternative storage [2, p. 15]. Dry storage was proposed and approved, using casks that are designed to hold 24 PWR assemblies.[3] Two casks were loaded with assemblies in May 1993, together holding 48 assemblies [6]. The site is authorized to have up to 25 casks, enough to hold the fuel that will be discharged through the year 2007.

A diagram of the cask used at the Palisades facility is shown in Fig. 9.1. The fuel assemblies are transferred underwater in the reactor cooling pool to a steel canister (called a basket). The canister is then brought to a decontamination area, where it is pumped dry, filled with helium, and sealed with redundant welded lids. The sealed canister is placed in the storage cask, which consists of an outer cylinder with 29-inch-thick concrete walls and a steel inner liner that is at least 1.5 inches thick. The canister is cooled by natural air convection in a gap between the canister and the cask liner.

The future of on-site storage may depend on resolution of legal and political issues. The Surry facility was installed in 1986 without any significant legal challenge. Legal actions to stop on-site dry storage at the Palisades reactor in 1993 failed in the Michigan courts. However, in Minnesota, where the laws are different, the courts decided that proposed dry cask storage at the Prairie Island nuclear plant needed the approval of the state legislature. After several months of debate, the Minnesota legislature in May 1994 approved a plan that gave the plant a reprieve by allowing a gradual installation of storage casks, but with extensive conditions that appear to point towards a long-term phasing out of nuclear power in Minnesota [7, 8]. One may anticipate a series of similar fights as other reactor operators attempt to adopt dry storage.

Interim Storage of Waste or Spent Fuel at Centralized Facilities

On-site spent fuel storage, even with dry casks, is only a temporary expedient. Pending a truly permanent solution, a supplementary step is to store the waste at centralized interim facilities. These have been variously referred to as *monitored retrievable storage* (MRS) and *away-from-reactor* (AFR) sites. There are many MRS designs. In typical schemes, the spent fuel assemblies would be placed in steel canisters (before shipping) and the canisters stored at the MRS site either above ground in casks or in holes near the surface of the ground. Cooling would be provided by natural air convection.

As MRS was originally conceived, several such facilities would be situated in the United States. No MRS facilities have been approved as yet, due to local opposition in candidate states (e.g., Tennessee in an early phase and New Mexico more recently). In qualified support

---

[2]The facility is limited to 811 tonnes of uranium, which may prevent using the full 84 casks. Assuming 0.46 tonnes of uranium per assembly (see Table 8.3), 811 tonnes corresponds to a little under 1800 assemblies.

[3]The reactor core at the Palisades plant has 204 assemblies [2, p. 15], and one-third of the core is discharged every 15 months [private communication from Mark Savage of Consumers Power]. Thus, each cask holds roughly the output for one-half year of operation. Details of the cask construction come primarily from Ref. [5].

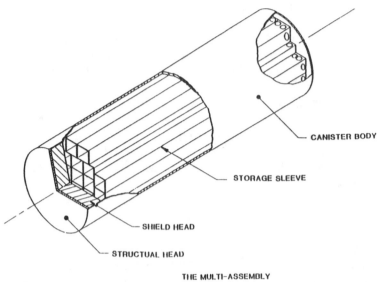

CANISTER BODY

STORAGE SLEEVE

SHIELD HEAD

STRUCTUAL HEAD

THE MULTI-ASSEMBLY
SEALED BASKET

Sectional View of a
Dry Fuel Storage Container

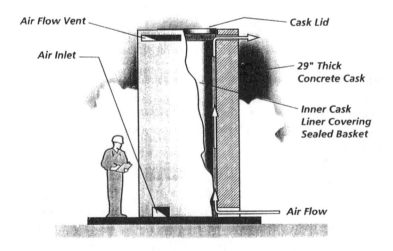

Air Flow Vent

Air Inlet

Cask Lid

29" Thick
Concrete Cask

Inner Cask
Liner Covering
Sealed Basket

Air Flow

Figure 9.1. Dry storage cask system used at the Palisades nuclear power plant. Top: canister (basket) for holding fuel assemblies. Bottom: storage cask for containing sealed canister. (Adapted from diagrams provided by Consumers Power.)

of proceeding with an MRS program, Congress in 1987 approved the investigation of sites for an MRS facility, subject to the following restrictions: there is to be only one such site, construction of the MRS facility will not begin until the NRC has issued a license for the construction of a repository for permanent disposal of high-level wastes, and the MRS facility can accept no more than 10 000 MTIHM until the permanent repository begins accepting wastes.

Such a MRS facility, if established, would have only a modest impact on the overall waste disposal problem. The capacity limit of 10 000 tonnes allows for the discharges from about five years of operation of U.S. reactors. This limit was in part a reflection of the U.S. view that interim storage is an *ad hoc* response, in lieu of establishing a permanent repository. In some other countries, it is considered to be an integral and desirable stage in the waste disposal process, because interim storage allows the fuel to cool before permanent disposal. Sweden in particular has a centralized interim facility that uses a pool rather than dry cask storage, and the plan is to let the fuel cool for 40 yr before placing it in a geologic repository [1, p. 49].

The most vigorous initial effort to move toward the establishment of an MRS facility came from the Mescalero Apache tribe in New Mexico, which applied for and received in October 1991 a $100 000 grant to undertake a preliminary study of the desirability of "hosting" an MRS facility on its reservation [9]. The reaction of the tribal leaders to the initial study was favorable, and in a next step, in August 1993, they applied for a $2.8 million DOE grant for a definitive study of the matter [10]. But the grant was opposed in Congress by a senator from New Mexico, and the Mescalero Indians and the operators of the Prairie Island nuclear plant began investigating a joint private MRS project [11]. In the meantime New Mexico authorities and congressional representatives have strongly opposed all MRS plans for the state and the Mescaleros tribe has oscillated in its attitude. Nonetheless, the project has drawn broad utility support and as of mid-1995 design efforts were underway preparatory to seeking NRC approval of licenses for construction and operation of the facility [12].

While it is not clear where the final power to make and carry out decisions in this case will lie, it is not hard to understand why there is a difference in viewpoint. The facility will generate little general commercial activity, but reportedly "the Mescaleros estimate that the project could generate as much as $2.3 billion in revenue over its 40-year lifetime" [12]. Such an amount may be a great attraction for the Mescaleros, who number 3400, but it is not a decisive inducement if the financial benefits are spread over the entire state of New Mexico.

9.1.3 Multi-Purpose Canisters

Role of a Multi-Purpose Canister

Secure containers, or casks, are needed in the various stages of waste handling: dry storage at the reactor site, transportation from one facility to another, storage at an MRS facility, and disposal at the ultimate site. If the same basic container can be used at each stage, handling procedures are simplified. To this end, there has been a program to design multi-purpose canisters (MPCs) suitable for use at all stages.

The MPC is not the entire container. At each stage it is placed in an outer cask, or overpack. Thus, for transportation it would be placed in a crash-resistant outer cask, and for final geologic disposal it would be placed in a corrosion-resistant container. But the spent fuel would remain sealed and isolated in the same MPC as the MPC is transferred to more specialized overpack casks.

The DOE decided in February 1994 to proceed with the final design and the eventual construction of MPCs. [13, p. 1-3]. Moving from the present "conceptual designs" of MPCs to completion of the first unit is scheduled to take until early 1998 [13, p. 6-4]. This time includes the issuance of design contracts, the obtaining of NRC approval, the decision on construction contracts, and the actual fabrication. In the schedule for the first MPC, only six months are allocated for fabrication, with the remaining time for the preparatory steps. Presumably, the first units would be used at reactor sites or at a monitored retrievable storage facility. Placement of MPCs in a waste repository would probably not take place until more than a decade later.

Details of Preliminary Multi-Purpose Canister Design

The first stages of use and the general configuration of an MPC are similar to containers for dry cask storage, but the canister itself is designed for a longer lifetime. The fuel assemblies are transferred underwater in the spent fuel pool into the MPC (see, e.g., Ref. [13]). The assemblies are held in place in the MPC with a honeycomb structure of aluminum alloy panels, which give physical support and conduct heat to the walls of the canister. The planned panels contain boron 10 (^{10}B) as a neutron absorber, to prevent the fuel mass from becoming critical, and they are clad with stainless steel for physical strength.

After removal from the pool, two (redundant) lids are welded to the top of the MPC. To provide shielding, especially during the welding operation, a shielding plug of depleted uranium or steel may be placed at the top of the canister before welding. Access ports through the lids allow the canister to be dried by vacuum pumping. It is then filled with an inert gas (perhaps argon or helium), and the access ports are sealed. For storage at the reactor site or in an MRS facility, the MPC is placed in a thick concrete cask and cooled by air convection.

In the preliminary planning, two MPC sizes have been considered: a large size for 21 PWR or 40 BWR assemblies and a smaller size with about one-half the capacity [13]. The larger unit would be a steel cylinder, 5 feet in diameter and about 16 feet long, with a 1-inch-thick stainless steel wall. The mass of this canister, plus the contained fuel and a surrounding transportation cask, is nominally 125 tons. The mass of fuel for the PWR assemblies would be about 10 MTIHM, which is on the order of one-third of the reactor output per GWyr.

For transportation, the MPC would be placed in a special cylindrical overpack designed to withstand severe accidents. An example is shown in Fig. 9.2. In one plan for the 125-ton MPC, the transportation cask walls are 15-cm (6-inch) thick, with successive layers of stainless steel, depleted uranium, lead, and stainless steel [13, p. 3-14]. The uranium and lead provide gamma-ray shielding. For further radiation protection, an outer 15-cm (6-inch) neutron shield is added with a thin stainless steel outer shell. The total cask is 1.55 m in inner diameter, 2.17 m in outer diameter, and 5.3 m in length. "Impact limiters" are placed at the ends for protection in case of an accident. The calculated dose rate at 2 m from the cask is 0.075 mSv/hr, and at peak spots at the surface it is 1.12 mSv/hr [13, p. 3-20].[4]

The MPCs would be brought to a permanent repository in transportation casks and then

---

[4]The dose rates with the 75-ton cask are somewhat higher, but they are still within the regulatory limits for transportation casks of 0.1 mSv/hr at 2 m and 2 mSv/hour at the surface [15, §71.47].

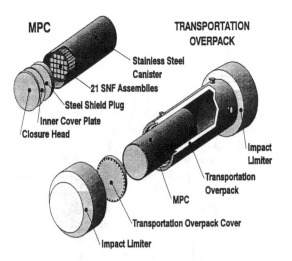

Figure 9.2. Large multi-purpose canister (MPC) and transportation overpack cask. (From Ref. [14], Fig. 2.6.)

transferred to disposal casks, where protection against physical impact is less important but resistance to corrosion is more important. These casks are discussed further in Section 9.3.2.

The estimated cost of a single large MPC, in a configuration for 21 PWR assemblies, is $354 000 [13, p. 4–7]. The electricity generated by the fuel in these assemblies is roughly 2.5×10^9 kWh.[5] Thus, the canister adds a cost of about 0.014 ¢/kWh. A target goal for nuclear power is roughly 4 ¢/kWh, and the canister represents well under 0.5% of the cost of nuclear electricity. This point is mentioned here because there has been a history in waste disposal planning of considerable, and perhaps excessive, attention to the cost of the waste packages.

9.2 DEEP GEOLOGIC DISPOSAL

9.2.1 Multiple Barriers in Geologic Disposal

Placement of radioactive wastes in deep geologic repositories has for many decades been the favored mode for disposal, both worldwide and in the United States (as mentioned in Section 7.4.3). This type of disposal relies on what is sometimes termed the "multiple barrier concept" to protect against the escape of radionuclides into the biosphere. In this description, the barriers are often divided into the *engineered system* and the *natural system*.

The repository and its contents, in particular the waste package, constitute the engineered system. The waste package consists of the solid waste (the spent fuel assemblies or the resolidified products of reprocessing) and the surrounding protective envelopes. These are usually concentric cylinders made of materials that will resist corrosion and prevent water from reaching the waste material. In some designs, additional protection against water

---

[5]This is based on 461 kg of uranium per assembly, a burn-up of 33 GWd/tonne, and a 33% thermal efficiency, giving 1.2×10^8 kWh per assembly (see Table 8.3).

intrusion is provided by backfill surrounding the waste package.[6] The waste package, backfill, and configuration of the repository together constitute engineered protective barriers.

The natural system is the surrounding rock through which water would move to the biosphere. It includes the rock out of which the repository is excavated and even the rock walls remaining in the repository. A good repository site is one for which the location and type of rock (a) prevent or limit the entry of water into the repository; (b) in the event of entry of water, slow the outward migration of water to the biosphere; (c) provide geochemical conditions favorable to low solubility of radionuclides and a low rate of corrosion of the waste package; (d) retard the motion of major radionuclides so that they move more slowly than the water; and (e) are at low risk of future disruption by earthquake, erosion, or other natural phenomena. Together, these attributes provide a series of natural barriers.

The engineered system cannot be designed independently of the natural environment, because factors such as the water flow rate, the water chemistry, and the heat conductivity of the medium strongly influence the choice of waste package design and the repository configuration. The close coupling between the waste package and its environment is sometimes addressed by dividing the study of the movement of radionuclides into "near field" and "far field" studies, where the near field includes the engineered system and the part of the natural system formed by the closely surrounding rock (see, e.g., Ref. [16]).

Any consideration of barriers to the escape of radionuclides must take note of the natural radioactive decay of the wastes, which progresses at very different rates for different species. Thus, in some design thinking the engineered barriers will provide protection against the escape of relatively short-lived radionuclides, notably ^{90}Sr ($T=28.8$ yr) and ^{137}Cs ($T=30.1$ yr), which decay to negligible proportions in under 1000 years. The natural barrier is then relied on for protection against the longer-lived radionuclides, such as ^{239}Pu ($T=24110$ yr).

Different countries place different degrees of emphasis on the engineered and natural barriers. In recent years, although this may be changing, primary emphasis in U.S. planning has been on the properties of the site, while in Sweden, which has a very ambitious plan for the waste package, reliance is placed on the engineered safeguards.

9.2.2 Alternative Host Rocks for a Geologic Repository

A large number of different types of rocks have been considered for waste repositories. There is no single overall "best" choice, as evidenced by the different choices made by different countries (see Section 9.5.2). Among the physical factors that would favor a particular type of rock are the extent to which water entry would be inhibited, the retardation of the flow of any escaped radionuclides, and heat conduction properties. We list below some of the candidate rocks that have been considered for repositories (see, e.g., Ref. [17]).

Bedded salt. Bedded rock salt was the initial candidate of choice. The existence of salt beds was taken as evidence that there had been no water intrusion for many thousands of years. Further, salt has high thermal conductivity, which would limit the temperature rise of the wastes. Salt melts at relatively low temperatures, and the waste would eventually be surrounded by a tight resolidified mass of salt. On the negative side, salt brine is highly corrosive and may attack the canister.

There had been a suggestion to use a salt bed in Lyons, Kansas, in the early 1970s, but this

---

[6]The backfill is sometimes taken to be part of the waste package.

plan was cancelled following strong local opposition and the realization that prior oil drilling had made water intrusion possible.[7]

Salt domes. Under some circumstances, the pressure on a thick bed of salt will cause some of the salt (which in general has a lower density than the surrounding rock) to break through the overlying material and rise upward to form a salt dome. One advantage of salt domes over bedded salt is a generally lower water content (see, e.g., Ref. [18], p. 201). The Gorleben site, a waste disposal site under consideration in Germany, is a salt dome.

Granite. Granite and similar rocks (granitoids) are very abundant. They are stable and generally homogenous, with low permeability to water movement. However, they are susceptible to fractures, which could provide paths for relatively rapid water flow. Granite is the choice in Sweden, the country with the most advanced planning for waste disposal, as well as in Canada.

Basalt. Basalt is an alternative rock formation, but an authoritative review has suggested that "a major reason for considering basalt for repositories is its abundance in federal land near Hanford, Washington, and the Idaho National Engineering Laboratory (INEL) and not its overall favorable characteristics" [17, p. 155].

Tuff. Tuff is the residue of material blown out of exploding volcanoes. At high temperatures, some of the material fuses to form "welded tuff," a material of low permeability. Tuff, both welded and unwelded, is present at Yucca Mountain, in the vicinity of the Nevada nuclear weapons test site. Tuff can be highly fractured, and a study of the fracture structure is an important component of the Yucca Mountain site characterization.

The suitability of a particular site depends not only upon the type of rock, but on location-specific aspects, including the history of past disturbance of the region (as at Lyons, Kansas), the thickness of the available rock layers, and the absence of valuable mineral resources. In practice, an important further consideration is the political acceptability of the site. As discussed in Chapter 10, an essentially political decision has been made in the United States to concentrate efforts on the Yucca Mountain tuff site.

Repositories may be in rock in a *saturated zone*, lying below the groundwater table, or in an *unsaturated zone*, lying above the water table. Except in arid climates, the water table usually lies too close to the ground surface to permit having a geologic repository in the unsaturated zone. Yucca Mountain, in the Nevada desert, is an exception, and the proposed repository would be in the unsaturated zone, well above the water table.

In the saturated zone, the gaps and pores in the rock are filled with water, although this need not prevent use of the site if the flow of water through the rock mass is at a slow rate. In the unsaturated zone, seepage of rainwater keeps the rock from being free of water. In particular, at the Yucca Mountain site it is reported that most of the pores in the rock in the unsaturated zone are more than half filled with water, although fractures in the rock are usually drained by capillary action [19].[8]

---

[7]The Waste Isolation Pilot Plant (WIPP), in a bedded salt site in New Mexico, was completed in 1991. It is to be used for military transuranic wastes only. However, local opposition and regulatory requirements have imposed long delays in the actual transfer of waste to the repository. It is not expected to open before 1998.

[8]For describing geologic media, some basic definitions have been succinctly expressed in Ref. [20], p. 4-9:

A fracture is a surface along which the cohesion of the rock has been lost. The deformation process is fracturing. The term is neutral and is neither related to the size of the structure or any displacement along the surface. If the displacement along the surface is insignificant (close to zero) it is a joint . . . . A fracture which has a component of displacement normal to the fracture surface is a fissure. A fracture with a significant component of displacement along the surface is a fault.

9.2.3 Motion of Water and Radionuclides through Surrounding Medium

Travel Times for Water and Waste

The main mode of transfer of radionuclides to the biosphere is through movement in the groundwater. The repository is likely to be at a depth of at least several hundred meters. Travel to the surface, where the radionuclides can be taken up by plants or enter the water supply, is not normally in a direct vertical path. More commonly, water flow paths are in a predominantly horizontal direction, and in an unsaturated zone they may be mainly downward. Typical travel distances from the repository to the biosphere are estimated to be in the neighborhood of tens of kilometers (see, e.g., Ref. [17], p. 253).

The rate of water travel depends upon the medium. If the medium is uniform, motion is through small spaces between grains. Fractures in the medium offer more rapid paths for water flow; the overall travel speed depends upon the size of the fractures and the degree to which they form a connected network. Taking a gross viewpoint, the detailed fracture structure is sometimes ignored and its character subsumed by considering the hydrological properties of the medium to be homogenous over a large scale. However, it is necessary to map the fracture structure of a potential repository site to avoid zones with a high density of fractures and to understand the extent to which the fractures that cannot be avoided will speed the escape of radionuclides.

Water travel time estimates are presented in Table 9.1 for hypothetical "reference" repositories in a number of geologic media. The values listed are mean values from Ref. [17], p. 253; maximum and minimum estimates are also given in Ref. [17], which in many cases differ widely from these means. Thus, the values are more useful for general orientation than

Table 9.1. *Estimated water travel time to biosphere for reference repositories in several geologic media and approximate retardation factors for radionuclides in these media. [Note: there are large uncertainties in these numbers (see text).]*

Element	Atomic number	Granite	Basalt	Tuff	Clay[a]	Salt
Water travel time (yr)						
Mean path length (km)		10	14[b]	6[c]		100
Linear velocity (m/yr)		0.01[d]	0.9	5.7		2
Travel time (1000 yr)		1000	15	1.2		40
Retardation factors, R^e						
Strontium (Sr)	38	200	200	200	200	10
Technetium (Tc)	43	5	5	5	5	5
Iodine (I)	53	1	1	1	1	1
Cesium (Cs)	55	1000	1000	500	1000	10
Lead (Pb)	82	50	50	50	50	20
Radium (Ra)	88	500	500	500	500	50
Uranium (U)	92	50	50	40	200	20
Neptunium (Np)	93	100	100	100	100	50
Plutonium (Pu)	94	200	500	200	1000	200
Americium (Am)	95	3000	500	1000	800	1000

Source: Ref. [17], pp. 147 and 253.
[a]Also refers to soil and shale.
[b]Corresponds to the smaller of two estimates of distance.
[c]The path and travel time are for travel in saturated tuff to a nearby well.
[d]Velocity not given in Ref. [17]; inferred from distance and time.
[e]These are values termed "suitably conservative" in Ref. [17], p. 147.

as realistic estimates for actual sites. For a given site, the detailed hydrological properties must be determined by very extensive "site characterization" studies.

Retardation Factors

The travel of ions through a given distance from the repository may take much longer than the travel of water. The ions move through the ground more slowly than does water due to processes designated, in their totality, as *sorption*.[9] These processes include the adsorption of ions on the surface of the medium and ion exchange between ions in the water and ions in the host rock. These processes retard the ions without permanently sequestering them. The retardation is parameterized in terms of a retardation factor R, which is the ratio of the velocity of ground water (v_{water}) to the average velocity of the ions (v_{ion}). R can be found from the ratio of the total number of ions per unit bulk volume N, to the number n being carried by the water, i.e., the number momentarily *not* trapped in the rock:

$$R = \frac{v_{water}}{v_{ion}} = \frac{N}{n}. \tag{9.1}$$

This expression can be rewritten as:

$$R = 1 + \frac{N-n}{n} = 1 + K_d \frac{\rho(1-\epsilon)}{\epsilon}, \tag{9.2}$$

where ϵ is the porosity of the medium, ρ is the density of solid rock, and K_d is the *sorption coefficient*, or *distribution coefficient*, defined as the ratio of the number of ions per unit rock mass attached to the rock, $(N-n)/\rho(1-\epsilon)$, to the number per unit water volume in the water, n/ϵ.[10] In Equation 9.2, K_d depends upon the chemical properties of the ion–water–rock system and ρ/ϵ on the physical properties of the rock.

Equation 9.2 sometimes is approximated by the expression [17, p. 195]

$$R = 1 + 10K_d, \tag{9.3}$$

where K_d is in units of cm^3/g. This is a conservative approximation, because commonly $\rho > 2$ g/cm^3 and $\epsilon < 0.2$. Thus, Equation 9.3 often underestimates the retardation factor.[11]

Determination of K_d, and hence R, depends on experimental measurements. In the words of Konrad Krauskopf,[12] the experimental results "show a discouraging lack of agreement from one laboratory to another" [21, p. 64]. Pending better experimental consistency, or understanding of the reasons for lack of consistency, the results of these measurements are mainly useful in providing "qualitative generalizations." Table 9.1 gives estimates by Krauskopf as "suitably conservative" values for the 1983 NAS study [17, p. 147]. They fall between the extremes in a range of values presented in later work of Krauskopf [21, p. 65].

Extensive site-specific research is required to establish with any high degree of confidence the actual travel times of wastes from the repository to the biosphere. In quantitative consid-

---

[9]We here follow in part the discussion in Ref. [21], p. 62ff.
[10]The porosity of the medium is the fraction of volume that is "empty" (or filled with water), i.e., the pore volume per unit bulk volume.
[11]For $\rho = 2$ g/cm^3 and $\epsilon = 0.2$, the coefficient of K_d in Equation 9.2 is 8, not 10. However, in more typical cases the coefficient will exceed 10.
[12]Professor of geology at Stanford University and then chairman of the Board on Radioactive Waste Management of the National Research Council.

eration of actual repositories, the numbers given in Table 9.1 should be replaced by results of detailed local studies. Such studies are being carried out for the Yucca Mountain site for the different types of tuff at the site. Recent results for the retardation factors, based on estimated sorption coefficients K_d, are larger than those of Table 9.1 for plutonium and smaller for technetium and neptunium, but the uncertainties in the parameters remain large [22, pp. 9–13].

However, even approximate data illustrate the relevance of the retardation factors in understanding the role of individual radionuclides:

Plutonium. It is often believed that ^{239}Pu (T=24 110 yr) is a major long-term threat. But with a retardation factor of 200 in tuff, its travel time for a 6-km distance is 240 000 yr, and in 10 half-lives it will have decayed to a negligible level.

Americium. The high retardation factor for americium makes ^{241}Am (T=432 yr) of negligible concern, although it looms large for the first several thousand years in terms of activity and water dilution volume.

Technetium. The low retardation for technetium makes ^{99}Tc (T=211 000 yr) a relatively important player, although its importance is limited by its modest total activity and its small release of energy per disintegration (see Table 8.4).

Neptunium. If the time horizon over which waste hazards are calculated extends up to say, one million years, then even with a substantial retardation factor a very long-lived isotope can reach the biosphere. In consequence, in some calculations of hazards, ^{237}Np (T=2.14 million yr) becomes a particularly pernicious radionuclide [22,23].

These "conclusions" from Table 9.1 are not well-established points. The uncertainties in the quoted retardation factors and in the overall models in which they are used are too great. Among the complications is the possibility that at high concentrations of dissolved material the adsorption capacity of the rock may become saturated, reducing the retardation factor. In view of the large effect of retardation as a modifier of the motion of radionuclides, it is important to determine the water travel time and the retardation factors in the environment of the specific waste repository being considered, with attention to issues of water chemistry, solute concentrations and the possible pathways through the bulk rock and through fractures in the rock.

9.2.4 Thermal Loading of the Repository

One of the key decisions in the planning of a geologic repository is the choice of the desired temperature profile for the repository region, as a function of time and location. The temperature profile is controlled by the density with which the heat-dissipating wastes are placed in the repository and the time delay before their placement. The overall plan is referred to as the "thermal loading" strategy. For a fixed waste inventory, this strategy determines the size of the repository.

The heat output of spent fuel is on the order of 1 kW/tonne, with the actual magnitude depending upon the burn-up of the fuel and on the time of cooling since the fuel was removed from the reactor. The thermal load can be expressed in terms of an (initial) *areal power density*, in kilowatts per unit area, or an *areal mass loading*, in tonnes per unit area. For example, at 1 kW/tonne, an areal mass loading of 1 tonne per acre corresponds to an areal power density of 1 kW/acre.

Delaying the burial of the fuel decreases the initial power density, due to the decay of short-lived radionuclides, and limits the peak temperatures reached by the fuel and its surroundings. However, any initial cooling of the fuel has relatively little effect on the long-term temperature profile of the repository viewed on a time scale of hundreds or thousands of years. The long-term, large-scale temperature history of the repository is dominated by the long-lived radionuclides and is controlled by the spacing between canisters, i.e., by the areal mass loading [24].

Depending upon the thermal load, one can achieve either a *below-boiling* regime or an *above-boiling* regime.[13] In the former case, the temperature in the rock surrounding the repository always remains below the boiling point of water. In the latter case, the temperature rises above the boiling point, perhaps for hundreds or thousands of years. An advantage of the higher-temperature option is that the heat will vaporize water and drive it away from the wastes, thereby protecting the waste containers. It also means that less area is required for the repository. An advantage of the lower-temperature regime is that it keeps the rock in a better known and more predictable condition and one with less thermal stress.[14]

A decision as to the optimal thermal loading must consider the temperature dependence of the properties of the wastes and container materials, including their corrosion rates, solubility, and sorption coefficients. It also requires an evaluation of the heat and fluid flow through the repository under different thermal conditions. The problem is complex, because the temperature profile in the repository influences the movement of water and steam and may also impact the physical structure of the rock. The effects are coupled, with water and steam providing mechanisms for heat transfer.

There have been substantial differences in the approaches taken on thermal loading by different countries. The original "baseline" United States plan for Yucca Mountain was to design the repository to be in the above-boiling mode for roughly 300–1000 yr [25, p. 28]. However, studies are being made of other thermal loading approaches for the Yucca Mountain repository, ranging from those in which there is no boiling to ones in which repository temperatures remain above the boiling point for more than 10 000 yr. Studies of thermal loading at Yucca Mountain are discussed further in Section 9.5.3.

In Sweden, it is planned to operate the repository in the below-boiling mode, keeping the maximum temperature no higher than 80 °C, even for the hottest canister [20, p. 3-5]. The temperature is to be held down by having a low areal mass density for the fuel and by allowing the fuel to cool for at least 40 yr before placing it in the repository.

9.3 THE WASTE PACKAGE

9.3.1 General Consideration of Waste Package Design

Relation of the Waste Package to Its Environment

The design of the waste package is tailored to the choice of site. The waste package—and for reprocessed wastes, the waste form—are selected to minimize corrosion of the canister and leaching of the waste. The rate of corrosion or leaching of any particular material depends upon the chemical composition of the water attacking it, which depends on the type of rock through which the water has traveled. Therefore, the choice of potential waste packaging materials is specific to the chosen site.

---

[13]These options are discussed, for example, in Refs. [19] and [25].
[14]It also makes possible the use of a heat-sensitive backfill, such as bentonite (see Section 9.3.1).

Waste Form

If the wastes are disposed of in the form of spent fuel, there is no option as to the waste form. The fuel is taken as it comes, usually in the form of pellets of UO_2 contained in long, thin-walled fuel pins. These are held in the reactor in assemblies with up to several hundred fuel pins (see Section 6.2.3). The assemblies remain the basic physical unit in the handling of the spent fuel. For protection, assemblies are placed in cylindrical metallic canisters. Typical designs have called for canisters holding from only a few assemblies, in some of the early planning [26, p. 183], up to 21 PWR assemblies, in one version of the currently planned multi-purpose canisters (see Section 9.1.3). The main protection of the fuel comes from the surrounding canister and its overpack.

For reprocessed waste, a good deal of attention has been given to finding an optimal waste form. The basic criterion is resistance against leaching under conditions of varying temperatures and of possible radiation damage from alpha particles emitted by waste products. The preferred waste form has long been borosilicate glass (see Section 7.4.2). France is currently converting reprocessed liquid wastes into borosilicate glass and is expanding its reprocessing facilities. Similarly, the United Kingdom is building a large vitrification plant for the same purpose. Borosilicate glass is preferred because there is experience with its properties and it is not difficult to produce. It remains the choice in the planning of most countries that have reprocessed wastes [27].

However, there may be better alternatives, and many have been suggested. Among these are varieties of ceramics, including one that mimics natural rocks, called Synroc. But ceramics may be more difficult to make on a large scale than borosilicate glass. Small amounts of Synroc have been made and tested in a number of countries, with encouraging results.

Components of the Waste Package

Typical components of a waste package include [28, p. 13]:

Spent fuel assemblies. The spent fuel remains packaged in assemblies, which are held in place by a support structure within the canister. (Alternatively, the contents might be reprocessed fuel in borosilicate glass cylinders.)

Stabilizer. This is material within the canister that provides mechanical support and protection to the assemblies, prevents criticality, and provides for heat transfer. It might be in the form of a powder or sand [16, p. 63]; the aluminum panels in the planned multi-purpose canisters perform the same functions (see Section 9.1.3).

Canister. The canister is the container holding the waste plus stabilizer; it is typically cylindrical.

Overpack. Further physical and corrosion protection can be provided by an overpack, in the form of a concentric cylinder surrounding the canister.

There are many candidate materials for the waste container. Metals are favored. Among these, particular consideration has been given to copper, steel, and a variety of alloys based on titanium, nickel, or iron. The most crucial property is the resistance of the material to corrosion under the expected repository conditions. Corrosion may be either generalized, with uniform corrosion over the entire surface, or localized at pits or cracks, which develop in the corrosion process. In many cases, this localized corrosion is the more troublesome phenomenon, and there have been intensive, but as yet not fully conclusive, efforts to establish rates for both general and localized corrosion for the various combinations of metal and the surrounding environment.

Placement of the Waste Packages

A typical underground repository design is a large cavern, honeycombed with tunnels, called "drifts." In the early thinking about Yucca Mountain, for example, each tunnel had a series of boreholes for emplacing the waste packages; a cylinder containing spent fuel assemblies or solid reprocessed waste would be placed in each borehole and the top portion of the borehole refilled and sealed. Assuming that multi-purpose canisters are used, this arrangement would have to be modified to accommodate their greater diameter and heat output. In fact, a quite different geometry may be used for the MPCs, in which they are placed horizontally on the floor of the tunnel, sitting on rail tracks. This would simplify their handling and provide flexibility in moving them. These two configurations are depicted schematically in Fig. 9.3.

In some designs there is no extensive further protection against contact with water, other than the natural dryness of the environment plus the effects of repository heating. Alternatively, the cavity around the package can be filled with a *backfill* to impede water movement. A common choice is bentonite, a material made largely of clays. Bentonite is a material that swells when water enters it, impeding the flow of water towards or away from the waste containers. Further, it adsorbs many radionuclides, retarding their migration even more than that of the water itself. However, bentonite may not be effective if subjected to high temperatures, and thus it may be more suitable for the relatively cool environment of the planned Swedish repository than for Yucca Mountain, where a hotter environment is tentatively envisaged. Materials known as *buffers* can also be added to the backfill to condition the chemical composition of any water moving through the backfill.

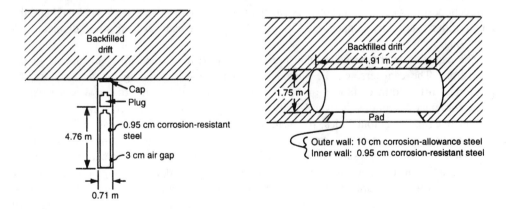

Figure 9.3. Alternative containers and emplacement geometries for deep geologic disposal of spent fuel or reprocessed wastes. Left: thin-walled container in vertical borehole; right: thick-walled container in horizontal drift. The backfill in this illustration is coarse rock, used to prevent the drift from collapsing. (From Ref. [22], Figure ES-3.)

9.3.2 Waste Package Designs in U.S. Planning

A typical canister in the orientation originally contemplated for Yucca Mountain is shown in the left half of Figure 9.3. This canister has a 0.95-cm-thick wall made of a corrosion-resistant steel alloy. One possibility is an alloy with a very high nickel content, known as Alloy 825 [22, p. 4-2]. The canister has an outer diameter of 0.71 cm and could hold about 2 tonnes of spent fuel—4 PWR assemblies or 10 BWR assemblies. The durability of these canisters depends upon the conditions of moisture, water chemistry, and temperature in their neighborhood. In some calculations it has been concluded that 10% of the canisters may fail within the first 500 yr in one repository temperature regime and within about 2500 yr in another [22, p. 14-47]. However, radionuclides that escape will move very slowly away from the canisters to the accessible environment, providing overall protection for much longer times.

A larger, thicker-walled container is illustrated in the right half of Figure 9.3. Although a multi-purpose canister system is not a prerequisite for such "robust" containers, an MPC fits naturally into planning for a thick-walled system. The MPC would be transferred at the repository site from the transportation container into a disposal container. In one such design, the disposal container would have two barriers. The inner barrier would be made from a corrosion-resistant alloy (for example, Alloy 825), 0.95 to 2.54 cm thick. The outer barrier would be made of steel, with possible thicknesses of 10, 20, or 45 cm [29, pp. 89–93]. In this design, the container would be approximately 2 m in diameter. A large MPC could hold 21 PWR assemblies.

9.3.3 Lifetime of the Waste Package

Durability and Thickness of the Waste Package

If wastes are placed in a sufficiently impenetrable container, then the waste problem has been solved. The most ambitious plan in this regard is the one adopted by Sweden. The design calls for putting the spent fuel in thick-walled canisters, using a copper outer wall to take advantage of the low corrosion rate of copper in the water of the host rock, which is granite. A recent design specifies concentric cylinders, the inner with a 5-cm-thick steel wall and the outer with a 5-cm-thick copper wall [30, p. 16]. The entire canister is 88 cm in diameter and 5 m long.

The canisters in the Swedish plan are to be placed in vertical holes in tunnels in the repository. The canisters will be surrounded on all sides, including top and bottom, by bentonite in the form of blocks or powder. The calculated lifetime of the waste package, under ordinary conditions, is over 1 000 000 yr [20, p. 5-50]. There are scenarios that give earlier releases of radionuclides; for example, the disruption of individual canisters due to rock motion produced by a growth of glaciers. Nonetheless, the engineering system is expected to provide very substantial protection for periods of time far in excess of 1000 yr.

In contrast, most United States planning has been predicated on a package lifetime of only 1000 yr, with the natural geologic barrier providing protection thereafter. The decision between thin and thick overpacks is crucial to the overall design. In U.S. planning until recently, emphasis was placed on thin-walled containers, in part because the containers were not expected to be subjected to high pressures, because thin walls suffice for corrosion resistance in a dry environment, and because the cost is less when there is less material in the overpack [31, p. 4].

Even for MPC containers, some early discussions have been in the context of a 1000-yr requirement, with the slight modification of pointing out that a 0.95-cm inner barrier and a 10-cm outer barrier would each independently provide containment for 1000 yr, "a defense in depth philosophy" [29, p. 89]. However, the emphasis may change towards taking advantage of the potential for longer-lasting containers.

Thick container walls offer important advantages. They provide resistance against corrosion for long times, as well as self-shielding against the radiation from the enclosed wastes. The reduced radiation level means less difficulty in keeping exposures of workers low. It also reduces dissociation by radiolysis of any surrounding water into oxygen and hydrogen, both of which can contribute to corrosion.

Role of NRC Requirements

The historic preference for thin-walled canisters among U.S. designers arose at least in part because of NRC regulations that were interpreted as setting a limit of 1000 yr on the time for which the designer can assume the waste package to remain intact. Thereafter, it has been commonly believed, reliance would have to be placed solely on the natural barriers. NRC requirements for the engineered barrier system, as stated in Title 10 of the Code of Federal Regulations (10CFR), include the provision:

> Containment of HLW within the waste packages will be substantially complete for a period to be determined by the Commission taking into account the factors specified in §60.113(b) provided, that such period shall be not less than 300 yr nor more than 1000 yr after permanent closure of the geologic repository [15, §60.113(a)].

This statement was promulgated in final form in 1983 and remained unchanged in the 1993 edition of 10CFR [15]. Not unnaturally, this stipulation has often been interpreted to mean that in designing a repository system, it cannot be assumed that the waste package will provide containment for more than 1000 yr; no matter how long the waste package might actually last, the rest of the system must be designed on the premise that the package loses its integrity after 1000 yr. This interpretation appears to have influenced much repository planning, and it has led to the adoption of a thin overpack in many designs.

This interpretation of NRC regulations may have been incorrect. The Nuclear Regulatory Commission issued a "clarification" on July 27, 1990, in the form of a Staff Position paper, which stated that the 1000-year stipulation is not a limit [32]. Instead, according to this document, the 300- to 1000-year stipulation

> is a *minimum* performance requirement which is not intended, and should not be interpreted, as a cap on the waste package lifetime or a limitation on the credit that can be taken (in engineered barrier system and overall repository performance assessments) if the waste package is designed to provide containment in excess of 1000 yr.

Further, it states that "the waste package may be designed for a longer lifetime and such longer lifetime may be considered in evaluations."

The ambiguous wording in the 10CFR60.113 provisions seems to have evolved from the need to establish a somewhat flexible *minimum* lifetime for the waste package. Unfortunately, the wording was such that the minimum became a *de facto* maximum, with cost-conscious design aimed at achieving the target level and nothing much better. The clarification provided by the Staff Position paper may have been stimulated by the Nuclear Waste Technical Review Board (NWTRB), which since 1990 has repeatedly stressed the desirability of having a waste package that would provide protection for more than the stipulated 1000 years [33, p. 40].[15]

Quite apart from the question of the interpretation to be given to the provisions of 10CFR60.113, by the NRC or the courts, there is the question of how the DOE will ultimately evaluate the merits of the case. Recent studies of the Yucca Mountain repository have extended the time horizon for analysis from 10 000 yr to 1 million yr [22,23]. If it turns out that the federal agencies give a high priority to protection for 1 million years, the durability of the container becomes less important, because no plausible container for Yucca Mountain can be counted on to provide protection for so long a time. If the focus is on 10 000 yr, when the levels of activity are much greater, then the container thickness can be very important.

9.4 ALTERNATIVES TO DEEP GEOLOGIC DISPOSAL

9.4.1 Variants of Geologic Disposal

In addition to excavated caverns, which everywhere remain the adopted approach, several other sorts of geologic sites have been considered [34, p. 1.16], although none of these are serious contenders at the moment:

Deep hole disposal. The wastes would be placed in holes at depths on the order of 10 km. The wastes would then be below the level where water circulation is expected. However, it would be difficult to handle large volumes in this manner.

Rock melt. High-level wastes in liquid form would be put into an underground cavity where, at high concentrations and confined volumes, the surrounding rock would melt. Subsequent cooling and solidification, perhaps after about 1000 yr, would trap the wastes in a well-sealed environment. Of course, the wastes would not be as well trapped at early times, while the material is liquid. Further, it would be difficult if not impossible to retrieve the wastes, even at times shortly after disposal.

Remote islands. Given local opposition to waste disposal sites near populated areas, it has been suggested that geologic repositories be placed on remote islands. It is questionable whether indeed this would overcome environmental objections, and additional transportation problems would be created both during construction of the repository and when the waste is deposited.

9.4.2 Sub-Seabed Disposal

The leading alternative to geologic disposal has been *deep seabed* or *sub-seabed* disposal (SSD) (see, e.g., Refs. [35] and [36]). The deep seabed, at places where the ocean is several thousand meters deep, has been formed from the deposition of sediments over millions of years. The seabed is in the form of a water-saturated clay layer, on the order of 50 m thick. Its physical properties as a site for high-level waste burial were described in enthusiastic terms in a 1994 National Academy of Sciences study on the disposition of plutonium from dismantled nuclear weapons:

The deep ocean floor in vast mid-ocean areas is remarkably geologically stable; smooth, homog-

---

[15]See Section 10.1.2 for a discussion of the roles of the NWTRB and other administrative and advisory bodies that deal with radioactive waste disposal.

enous mud has been slowly building up there for millions of years. The concept envisioned for HLW [high-level wastes] was to embed it in containers perhaps 30 m deep in this abyssal mud, several kilometers beneath the ocean surface . . . the mud itself would be the primary barrier to release of the material into the ocean, because the time required for diffusion of radionuclides through this mud would be very long [37, p. 200].

In sub-seabed disposal, the wastes would be put in canisters of relatively small diameter (30 cm), and the canisters would be placed in the ocean floor by either letting appropriately shaped canisters fall freely and bury themselves in the sediment or by putting the canisters in pre-drilled holes. Estimates of the lifetime of a canister in the environment of the enclosing mud range from "several centuries" [35, p. 28] to "a few thousand years" [37, p. 200]. The lifetime obviously would depend on the material used and its thickness.[16]

If SSD ever becomes a serious possibility from a political standpoint, it would be important to have further data on the durability of potential containers and the rate of motion of radionuclides through the clay. This would help determine the nature of the containers needed to avoid migration of radionuclides to the seabed surface, which otherwise might add significantly to the existing natural radiation exposures received by organisms living near the ocean floor.[17]

Whatever the merits of sub-seabed disposal from a technical standpoint, the obstacles at the moment appear to be formidable. For many decades geologic disposal appeared to be sufficiently practical and straightforward that there seemed to be little need for intensive study of alternatives. In the meantime, a worldwide visceral opposition has developed against ocean dumping of any material deemed potentially harmful, and this opposition has been codified into laws and treaties.

On the international level, the London Dumping Convention of 1972 banned the ocean dumping of high-level radioactive wastes, and in 1993 many participants, including the United States, voted to extend the ban to low-level wastes [37, p. 202]. The prevailing view is that the ban on ocean dumping applies to sub-seabed disposal as well, and this interpretation may be made explicit. In any event, no country appears to be ready to risk the worldwide criticism that would arise should it decide to pursue sub-seabed disposal.

Nonetheless, the objections to sub-seabed disposal appear to be more institutional than technical, and one may anticipate calls for further examination of this option should serious difficulties develop with geologic disposal. It may also turn out that countries with less available land area than the United States and Russia may one day press for the acceptance of sub-seabed disposal.

9.4.3 Partitioning and Transmutation of Radionuclides

Much of the calculated long-term waste hazard comes from a limited set of heavy radionuclides, with half-lives ranging from hundreds of years to a million or so years. Exposure of selected radionuclides to high neutron fluxes, produced in a nuclear reactor or by an accelerator, could transmute them into nuclides with much shorter or much longer half-lives. Either outcome could reduce the hazards from radiation exposure.

For example, a possible candidate for transmutation is ^{237}Np ($T=2.14\times10^6$ yr), which

---

[16]Presumably, multilayer canisters are possible if needed.

[17]In one study for reprocessed wastes the exposure due to the wastes was found to be very small [38, pp. 380–386], but it would be desirable to have additional quantitative studies, including ones for spent fuel. (We here tacitly accept the importance of avoiding, say, doubling the dose to these organisms, although that premise also warrants examination.)

becomes a relatively important component of the wastes after long periods of time (see Section 8.3.2). Successive neutron captures in ^{237}Np leads to the production of ^{239}Np, which decays with a 2.4-day half-life to ^{239}Pu. The ^{239}Pu can be consumed by further neutron bombardment. Assuming that the resulting fission products present less of an overall hazard than does ^{237}Np, this is a net gain. Troublesome fission products can also be converted into less troublesome products by neutron capture. In such cases, chemical separations are necessary to allow the partitioning of selected groups of radionuclides into different waste streams.

The issue of transmutation arises in several different contexts:

Destruction of plutonium from dismantled nuclear weapons. A limited specific goal for transmutation is to aid in the disposal of plutonium extracted from dismantled nuclear weapons. (This issue is discussed further in Section 14.5.3.) Many plans that would use reactors or accelerators have been suggested. One alternative is to combine plutonium with uranium to form a mixed-oxide fuel (MOX), which can be substituted for some or all of the UO_2 fuel used in most reactors (see Section 7.4.2). Here the purpose is not to reduce nuclear waste hazards *per se*, but to reduce the weapons proliferation hazard represented by inventories of ^{239}Pu. Putting the plutonium in a reactor consumes some ^{239}Pu and mixes the rest with ^{240}Pu and highly radioactive wastes. These changes reduce the potential usefulness of the plutonium for weapons production, especially for clandestine weapons production carried out without elaborate facilities.

Reduction of waste disposal hazards in existing spent fuel. In principle, existing spent fuel can be reprocessed and separated chemically into different waste streams for transmutation by neutron bombardment. This could be accomplished with a wide variety of reactors.

Reactor fuel cycles with on-line or integrated destruction of waste products. It is possible to design reactor systems to include waste destruction as an integral part of the fuel cycle, rather than as an afterthought. In this spirit, one proposed system, the Integral Fast Reactor, recycles all actinides and therefore substantially reduces the eventual residue of plutonium and other transuranic elements (see Section 13.4.1). In a still more ambitious suggestion, involving an accelerator-driven fission system, all the waste products are continuously recycled and selected nuclides destroyed (see Section 13.5.3).

The possible gains from reducing the inventories of undesirable radionuclides by transmutation must be balanced against the possible disadvantages, in cost and risk, of the required additional treatment and handling. There is a trade-off between the possible leakage of intensely radioactive materials due to human or equipment failures today and the possible leakage of attenuated wastes in the distance future. But present technology and ingenuity would probably allow for the virtual elimination of any targeted radionuclide, were the goal deemed sufficiently important.

As yet there are no well-defined plans to proceed in the transmutation direction, in either the United States or elsewhere. However, many different sorts of reactors, some coupled with accelerators, have been suggested as suitable for transmutation. The issue is particularly pressing in the case of plutonium from dismantled nuclear weapons, and we will return briefly to that issue in Chapter 14.

9.4.4 Extraterrestrial Disposal with Rockets

In a scheme that most observers consider frivolous, nuclear species that are particularly pernicious, in terms of half-life or biological impact, would be extracted from the waste and propelled into space, perhaps into the Sun or out of the solar system. There are no serious plans to adopt such a strategy. In view of the possibility of rocket failure, it would be difficult to establish that disposal via rocket creates less danger than does deep underground storage. Nonetheless, if convincing methods could be advanced for ensuring that failed rockets would drop their payloads without damage, for future disposal, this would provide a means for truly "permanent" disposal.

9.4.5 Polar Disposal

It was at one time suggested that a waste canister placed on polar ice shelves would burrow down into the ice by the action of its own heat, and disappear harmlessly. This suggestion was not taken very seriously even before issues of protection of the Antarctic came to the fore, and it is not a plausible option today.

9.5 STATUS OF LONG-TERM DISPOSAL PLANS

9.5.1 Options

The nuclear policy in the United States, and other countries as well, has long been based on the assumption that ultimate disposal of high-level wastes would be in deep geologic repositories. It is difficult to make a definitive technical assessment of the relative advantages and disadvantages of sub-seabed and geologic disposal. The wastes are less accessible in sub-seabed disposal than in geologic disposal, so retrieval in the former case is nearly impossible. Whether this is seen as an advantage or disadvantage depends on whether the focus is on possible future beneficial uses of the radionuclides or on the dangers of theft or accidental intrusion.

Advocates of one or another approach often couch their case in terms of adequate technical acceptability and superior political acceptability. Geologic disposal is politically difficult because of opposition from people living in the vicinity of any contemplated site. Obtaining international agreement on ocean disposal is probably even more difficult, because of widespread opposition to anything that might be termed "ocean dumping." Thus, deep geologic disposal remains the leading prospect. If the obstacles to both approaches appear too formidable, and if interim storage is deemed unsatisfactory, there may be a turn towards transmutation. Such a turn, if it occurs, could be regarded as an act of prudence, desperation, or both.

9.5.2 Geologic Disposal Plans in Different Countries

Plans for geologic disposal being pursued in a number of countries, including the United States, are summarized in Table 9.2, taken from a 1992 survey prepared by the U.S. Nuclear Waste Technical Review Board (NWTRB) [1, Table 3.1]. Even for this limited group of countries, there are striking differences in approach. Three different types of rocks are to be used, but this in part reflects differences in geologic conditions. On aspects where there is more freedom of choice, there is a division between wastes as spent fuel and as reprocessed vitrified wastes. The projected lifetimes of the planned waste packages vary greatly, reflecting differences in their construction and in the underlying strategy guiding the planning of the waste package.

Table 9.2. *Status of official high-level waste disposal plans and projections in selected countries, 1992.*

	Canada	Germany	Sweden	Switz.	U.S.
Spent fuel by year 2000 (tonnes)	27 000[a]	8950[b]	4400	2000	40 000
Spent fuel or reprocessed waste?	SF	both[c]	SF	RW[c]	SF
Type of host rock	granite	salt	granite	note d	tuff
Host rock saturated?	yes	yes	yes	yes	no
Specific site being studied?	no	yes	no	no	yes
Thermal loading for boiling?	no	yes	no	no	yes[e]
Minimum fuel cooling time (yr)	decades	none	40	40	10
Projected package lifetime (yr)	⩾500	note f	1 million	100 000[g]	300–1000[e]
Underground research facility (date)	1986	1967	1990	1983	1996–7
Target date for repository	>2025	2008[h]	2020	>2020	2010

Source: Table 3.1 of Ref. [1].
[a]The CANDU reactors use natural uranium and have a low burn-up; hence the high fuel mass.
[b]Includes spent fuel from West Germany only.
[c]Spent fuel from Germany and Switzerland is being reprocessed in France and Great Britain.
[d]Switzerland hopes to share in a multinational site.
[e]Below-boiling alternatives and longer-lived packages have been considered in more recent U.S. studies (see, e.g., Ref. [23]).
[f]German plans indicated for spent fuel only: 500–1000 years
[g]Swiss time estimate takes credit for waste package and bentonite buffer.
[h]German target date could be met only if Gorleben site is approved.

Some indication of the degree of progress in the overall program is the existence of an underground test facility for studies of the properties of the surrounding rock. The United States lags behind many of the other countries in this regard. In addition, the NWTRB points out that "the United States is the only country [of the group surveyed in Table 9.2] that does not have an integrated and widely accepted plan for long-term interim storage of all spent fuel and high-level waste" [1, p. 40].

None of the sites in the countries listed in Table 9.2, or in any other countries, is scheduled to be ready much before the year 2010, and neither of the earliest listed target dates, for Germany and the United States, is certain to be met. With so long a time interval before the actual completion of a repository, it is possible that many of these plans will change. In the United States, for example, there have been pressures to design waste packages that will last longer than the indicated 1000 yr (see Section 9.3.3).

Table 9.2 omits some of the major nuclear countries. Among these, France is pursuing a program in which the high-level wastes are the vitrified residues of reprocessing, currently in the form of borosilicate glass. They are to be put in a deep geologic site after several decades of cooling. The expected date for this placement is about 2020. Representatives of the French Commission for Atomic Energy speak with great confidence of the efficacy of their program:

> vitrification offers a stable and safe solution, long-since mastered, to package high-activity wastes. The behavior of this embedding ... lead to predict a minimum lifetime of 10 000 yr without significant change. The mean package integrity lifetime will most likely reach a million years .... The know-how necessary to design, construct, and implement deep sites is classical and available [39, pp. 31–32].

Nonetheless, despite this apparent confidence, the French authorities plan prior tests in underground laboratories before proceeding with the final repository.

The NWTRB also briefly reviewed waste disposal planning in Japan. According to their report:

Because of the complex nature of geologic conditions in Japan, the Japanese are investigating a multibarrier concept with a massive, robust engineered barrier system. The Japanese have a vigorous program of engineered barrier and generic research in hydrology and geochemistry, yet they have not taken the first step toward repository site selection [1, p. 9].

From other discussions in NWTRB reports, it would appear that the board tacitly applauds the emphasis on a robust engineered barrier.

9.5.3 Yucca Mountain Nuclear Waste Repository

Physical Features of Repository

The only site in the United States being actively considered for the permanent disposal of nuclear wastes is at Yucca Mountain in Nevada. The designation of Yucca Mountain as the sole site for intensive study came through Congressional action in 1987 (see Section 10.1.3).

Yucca Mountain is located in the Mohave Desert in southern Nevada, about 120 km northwest of Las Vegas and about 50 km east of Death Valley in California. The Yucca Mountain site is adjacent to the Nevada Test Site, used in the past for nuclear weapons testing, and the Nellis Air Force Bombing Range. Part of it would extend into land assigned to these facilities. The climate is very dry: Rainfall at Yucca Mountain averages about 17 cm per year, of which more than 97% is estimated to be evaporated back into the atmosphere before it can penetrate deeply into the ground (see Refs. [40] and [41]).

Fig. 9.4 displays both the location of the site and the orientation of the repository at the site. The proposed placement of the repository is in a rock layer of welded tuff, known as the Topapah Spring Member. This formation is roughly 320–350 m thick [41, p. 203]. The repository would lie in an unsaturated zone, with early plans placing it "approximately 225 m above the water table and approximately 350 m below the ground surface" [23, p. 3-1]. The main pathways by which radionuclides might reach the outside environment are by downward flow from the repository to the saturated zone below the water table, and thence by flow in the saturated zone away from the site.

The repository has been described as occupying about 1520 acres (6.2 km^2) and honey-

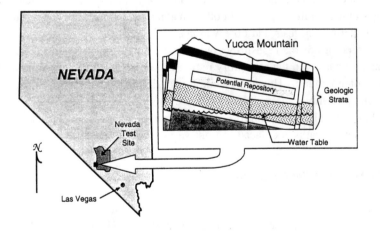

Figure 9.4. Sketch of the Yucca Mountain site. Left: location of site in Nevada. Right: orientation of the repository at the site. The different shadings represent different layers of tuff, with different hydrological properties. (From Figure ES-1 of Ref. [22].)

combed with about 100 km of tunnels [25, p. 27]. The actual area to be used and the configuration of tunnels cannot be established without further investigations of the site, as well as decisions as to the thermal loading of the wastes and the type of canister to be used.

The Waste Inventory

The capacity of the repository is to be limited to 70 000 tonnes of fuel, or more specifically, to 70 000 tonnes of initial heavy metal (MTIHM). This is enough to accommodate most but not all of the spent fuel that would be accumulated by the year 2030 if there are no new reactor orders and no intensification of weapons production (see Section 8.1.4). This limit was established in the Nuclear Waste Policy Act of 1982 [42, Sec. 114]. The total is expected to be composed of 63 000 tonnes of commercial spent fuel and 7000 tonnes (MTIHM equivalent) of reprocessed wastes from the defense program [25, p. 27].

The total number of waste canisters depends upon the type and size used. For disposal using MPCs, two 1993 scenarios give projections that assume on average about 7000 canisters, each holding about 9 tonnes of fuel. The average projected burn-up is about 38 GWd/t (see Refs. [22], p. 5-1, and [23], p. 2-49).[18] At this burn-up, the 63 000 tonnes of spent fuel would have provided about 2000 GWyr of electrical output, equivalent to a 30-yr output at the 1993 annual rate of 70 GWyr.

Site Characterization

The purpose of site characterization is to evaluate the suitability of the site as a repository, through research and testing both in the laboratory and at the site location.[19] The on-site studies require extraction of rock cores from the region of the repository and excavation of an extensive tunnel system through the region.

The goal is to understand the potential travel of fluids and radionuclides through the repository, from the waste package to the biosphere. Any such travel depends on physical and chemical properties of the tuff, including the fracture structure in the repository site. It is also necessary to understand the mechanisms for water entry and exit in the unsaturated zone over wide ranges in temperature. These and other matters are analyzed with theoretical modeling as well as experimental tests and observations.

Normal properties of the repository could be altered by violent events. The likelihood of significant earthquakes in the repository region can be assessed by investigating the site for existing or potential faults. However, while it is desirable to avoid large faults, some ground displacement probably can be tolerated if the engineered features of the repository are suitable. For example, multi-purpose canisters on the floor of a drilled tunnel are less vulnerable than are canisters placed in tight vertical boreholes.

Thermal Loading of the Repository

A major issue to be resolved in the site characterization of the Yucca Mountain repository is the choice of an optimal thermal loading strategy (see Section 9.2.4). A wide range of

---

[18]These results consider the combined contributions of BWR and PWR fuel, with the PWR fuel representing about two-thirds of the total.

[19]Site characterization is mandated and broadly defined in the Nuclear Waste Policy Act of 1982 [42].

loadings has been considered, with concentration on areal thermal loadings of 28.5, 57, and 114 kW/acre [22,23]. The corresponding areal mass densities depend upon the age of the fuel and the burn-up of the fuel.[20]

The baseline United States strategy for Yucca Mountain has for many years been to design the repository to operate in the above-boiling mode for roughly 300–1000 yr [25, p. 28]. With the fuel expected at Yucca Mountain, this could be achieved at a loading of 57 kW/acre, using an area of about 1140 acres [22, p. 4-15].

Studies have been made for a wide range of alternative thermal loading approaches, ranging from those in which there is no boiling to ones in which the repository temperature remains above the boiling point for more than 10 000 yr. A below-boiling regime is achieved for the entire repository if the areal power loading is halved, to 28.5 kW/acre [23, p. A-6]. However, this would double the required area to over 2000 acres, which exceeds the planned repository size.

Time Schedule for Yucca Mountain Repository

A schedule for the Yucca Mountain project was announced in 1991 that called for the issuance of a draft environmental impact statement in 1999 and presentation of a final recommendation to the President in the year 2001 [43]. This would complete site characterization. If the site is approved, construction would begin in 2004, and waste would start to be placed in the repository in 2010. The 2010 target date was reaffirmed in August 1994 by the new director of DOE's civilian waste disposal program [44]. However, as discussed further in Section 10.1, there has been a long history of delays in the U.S. program for handling radioactive wastes, and there is no general confidence that a site at Yucca Mountain actually will be completed by 2010. The immediate implications for nuclear power of a failure to meet this schedule would depend upon the extent to which it is possible to implement the dry storage of spent fuel at reactor sites or create centralized interim storage facilities.

9.5.4 Total Performance Assessments for Yucca Mountain

History of Performance Assessments

One outcome of site characterization is the development of data that will make it possible to carry out a *total-system performance assessment* (TSPA).[21] At present the data are not available to permit a final TSPA for Yucca Mountain, but there has been a series of preliminary assessments.

An analysis was published by the EPA in 1985, before the Yucca Mountain site was designated, as a prelude to setting standards for future high-level waste repositories [45]. The EPA performance criteria were described as follows, for a hypothetic repository for 100 000 MTIHM:

> The Agency has chosen to base the containment requirements on a population risk level of no more than 1000 premature cancer deaths over 10 000 yr [45, p. 7-13].

---

[20]For orientation, we note that for a burn-up of 33 GWd/t, the thermal output of the fuel is 1.16 kW/tonne after 10 years of cooling, 0.74 kW/tonne after 30 yr, and 0.46 kW/tonne after 60 yr [24, p. 429].

[21]The TSPA, if carried out with a probabilistic approach, is analogous to the probabilistic risk assessments that have been used for evaluating reactor safety during the past 20 yr (see Section 11.4.2).

The EPA study considered potential repositories with a view to estimating whether this standard could be met. Generic geologic properties were assumed.

For a repository in tuff, the expected number of cancer fatalities over a period of 10 000 yr was calculated to be three. These arose from human intrusion through drilling and from fault motions in the ground. It was further estimated that there was only a 0.02% probability of exceeding 1000 fatal cancers over the 10,000-year period [45, p. 8-27]. This satisfied the EPA's goal at a high level of probability and apparently established that the EPA's standards on radionuclide releases were achievable, despite being stringent.

After the designation of the Yucca Mountain site, more detailed performance assessments have been undertaken—including an "initial" study, TSPA 1991, and a "second iteration," TSPA 1993. There are several versions of these studies. Sandia National Laboratories (SNL) has prepared versions of both TSPA 1991 [46] and TSPA 1993 [22], and INTERA Inc. has prepared a version of TSPA 1993 [23].[22] These studies were carried out in the spirit of a probability risk assessment.

Mechanisms for Radionuclide Release

Four general mechanisms for the release of radionuclides have been considered in the TSPA studies:

Groundwater flow and radionuclide transport (aqueous release). The major mode for release of radionuclides is by transport in groundwater, in particular downward transport to the saturated zone below the water table and subsequent lateral transport to the biosphere.

Gas flow and radionuclide transport. Some radionuclides are released in gaseous form and then move upward through the repository. The most important of these, in the context of the EPA requirements, is ^{14}C in the form of carbon dioxide.[23] Although the amounts of ^{14}C are relatively small, gases escape from the repository much more quickly than do materials carried by water.

Human intrusion. The contemplated intrusion is by drilling, and it is assumed to occur randomly in space and time. It is therefore necessary to make assumptions about the frequency of drilling in the area and the chance that a waste canister will be hit by a drill bit.

Volcanic disruptions. Material from volcanic activity could damage the canisters, allowing radionuclides to be carried to the environment by the flowing molten rock.

The relative importance of these mechanisms has been viewed differently in successive studies. The 1985 EPA study did not consider gaseous release mechanisms, and it concentrated on a 10 000 yr time period, during which aqueous releases are small. Thus, human intrusion and fault motions in the ground were found to be the largest sources of risk—although still very small in magnitude.

In the later TSPA studies, the gaseous flow of ^{14}C was found to pose the greatest problem over the first 10 000 yr, or at least the greatest violation of EPA radionuclide release

---

[22]INTERA Inc. is a commercial firm that serves as a subcontractor on the Yucca Mountain project, engaged in performance assessments.

[23]The ^{14}C is produced primarily by neutron bombardment of ^{14}N, which is an impurity in the fuel and cladding.

standards. Further, the time horizon was extended far beyond 10 000 yr, and over these longer periods the effects of waterborne radionuclides become greater. In consequence, human intrusions and volcanic eruptions became relatively less important; in the SNL studies, neither was found to be a major contributor to the total releases,[24] and neither was even included in the INTERA study [23]. These disruptive events will not be considered further here.

Inputs to the TSPA Calculations

In implementing a system assessment, certain properties of the repository are known and can be taken as given conditions. In particular, for a repository that will hold 63 000 tonnes of commercial fuel and 7000 tonnes of reprocessed defense wastes, there can be only small variations in the inventory of radionuclides, because the composition of the initial fuel and the burn-up of the fuel are defined by actual practice to within rather tight ranges.

Other parameters can vary widely but are under the control of the system designers. These include the thermal loading of the repository and the physical construction of the canisters. The particular set of design parameters chosen for an individual scenario can be taken to be the defining properties for that scenario. The scenario emphasized in both the SNL and INTERA TSPA 1993 calculations was for a thermal loading of 57 kW/acre and a disposal cask with a 10-cm outer wall and a 0.95-cm inner wall, but other cases were also considered.

Still other parameters are physical quantities that depend upon the geological and chemical properties of the site and the waste, and over which one has no control. They describe, for example, the future rainfall in the region, the structure of the rock and the rate of motion of water and gas through it, and the solubility and sorption coefficients for the various nuclides. The values of the individual parameters can be estimated, but it is most realistic to assume that there is a range of possible magnitudes. In a Monte Carlo calculation, a distribution function assigned to each parameter specifies the relative probability of it taking on different numerical values. An individual calculational run adopts a set of these parameters and then might calculate, for example, the number of curies of ^{237}Np that reach the accessible environment as a function of time. For each such run, the parameters are "randomly" chosen, on the basis of a weighting from the distribution function. A probability distribution for ^{237}Np release is obtained by combining the results of a large number of these calculational runs.

Crucial processes, such as the corrosion of the canisters and the movement of water through the rock matrix and fractures, are only incompletely understood. They can be described by alternative, approximate models. The final calculational results are therefore probability distributions for the magnitude of the radionuclide releases and radiation doses, for a particular choice of design parameters and calculational models.

Comparison to EPA Limits on Radionuclide Releases

The results of these calculations are compared to EPA limits on releases of radionuclides to the accessible environment.[25] For each radionuclide, a limiting release is specified, corre-

---

[24]For example, it was estimated that there was only about a 0.02% chance of a significant volcanic event over the next 10 000 yr in the vicinity of Yucca Mountain, and even in the case of such an event relatively little radioactive material would be carried to the surface by the flowing magma [46, Ch. 7].

[25]These release limits are set forth in Part 191 of Title 40 of the Code of Federal Regulations, particularly §191.13 and the associated Appendix [47]. Although EPA requirements are not at present binding upon the Yucca Mountain site, they may be suggestive of future requirements (see Section 10.1.4).

sponding to approximately 1000 cancer fatalities over 10 000 yr from the release of that radionuclide alone [45, p. 7-13]. A normalized "EPA sum" is obtained by calculating for each radionuclide the ratio of its predicted actual release to its EPA release limit, and summing these ratios for all radionuclides.

If the normalized EPA sum is unity or less, then the EPA calculation limit of 1000 cancer fatalities in the 10 000-yr period is satisfied. The EPA repository requirement stipulated that the chance that the EPA sum exceeds unity must be less than 10% and the chance that it exceed ten must be less than 0.1%.

Performance During the First 10 000 Years

Standards promulgated to date have only specified repository performance during the first 10 000 yr. For the cases examined, the only scenarios that violate the EPA limits on the releases of radionuclides during this time involve the escape of ^{14}C through gas flow to the surface. Here, calculations in some models indicate that the chance of having an EPA sum greater than unity exceeds the stipulated limit of 10% (see Refs. [22], p. 18-10, and [23], p. 3-30). However, the total amount of ^{14}C that can be released is small compared to the amount of ^{14}C that exists naturally in the atmosphere, and the reasonableness of the EPA criterion for ^{14}C has been questioned (see Section 10.1.4).

The ^{14}C release is reduced if the container has a 20-cm wall instead of the reference 10-cm wall, and it is eliminated with a 45-cm wall [23, p. 3-30]. It is also reduced at thermal loadings above and below the reference loading of 57 kW/acre. At 28.5 kW/acre the corrosion of the container is relatively slow, because the water temperature is low; at 114 kW/acre the corrosion of the container is reduced, because the high temperatures keep water away for an extended time period.

Aqueous transport of radionuclides to the environment outside the repository is negligible over the first 10 000 yr, due to the slow movement of water through and from the repository, and the retardation of most radionuclides. The INTERA analysis for TSPA 1993 found that "virtually all (greater than 99.99%) of the release to the accessible environment is the result of ^{14}C" [23, 4-16]. According to the INTERA calculation there is a 90% probability that the EPA sum will be less than 10^{-6}. According to the SNL calculation, in their least favorable model, there is a 90% probability that this sum will be less than 10^{-4} [22, 14-43].

With these small releases, the doses will be negligible during the first 10 000 yr. Even for ^{14}C, the only nominal offender, the maximum individual dose has been estimated to be less than 0.0005 mSv per year (see Section 10.1.4). The expected dose rates from aqueous releases will be even lower.

Performance Over Longer Time Periods

Radionuclide releases were also calculated in the INTERA study for a 100 000-yr period, and again compared to the EPA limits. On this time scale nuclides other than ^{14}C can escape from the repository region, in particular ^{99}Tc, which has a low retardation factor. Over this long period the EPA sum exceeds unity for most of the cases considered, except for a container with a 45-cm wall.

The EPA limits discussed above refer to the magnitudes of radionuclide releases, which in turn were derived from estimates of the cumulative population dose. It is also possible to evaluate repository performance in terms of the maximum individual doses that might be incurred. The time horizon for such calculations has been extended to 1 000 000 yr in both versions of TSPA 1993—far beyond the original EPA horizon of 10 000 yr.

Doses were calculated for persons who obtain all their water from sources in the accessible environment near the site, defined as places more than 5 km from the site. It is assumed that individual water consumption is at the rate of 2 liters per day. The main contributor to the exposure, after very long times, is ^{237}Np. Although neptunium is substantially retarded as it travels through the ground, over a period of hundreds of thousands of years some models indicate that it could be transported down to the saturated zone and through the saturated zone to the outside environment [22, p. 18-11]. High concentrations of ^{237}Np could result if the ^{237}Np is carried by a small volume of water.

It was found that this could lead to high radiation doses at times well in excess of 100 000 yr. Estimates differ as to the magnitude of the potential dose, extending from roughly 1 mrem (0.01 mSv) per year in one SNL model to 10 rem (100 mSv) in another [46, p. 18-11]. In the INTERA calculation, the maximum dose is close to 40 rem (400 mSv) per year [23, p. 3-60]. These doses can be contrasted with the EPA's present drinking-water standard of 4 mrem (0.04 mSv) per year.

The calculations tacitly assume that the radionuclides are neither detected nor removed.[26] No estimates were given of the *likelihood* of people drawing upon this water source or of collective population doses. In qualitative terms, it is pointed out:

> The calculated peak dose rates for Yucca Mountain are high because of relatively little dilution, but at the same time there is relatively low probability of someone being exposed to those high doses because of the aridity. The whole issue of individual vs. population doses enters at this point. Peak individual doses may be high while population doses are low. (Keep in mind, though, that the dose calculations are very preliminary, and the high individual doses calculated in this study could decrease upon further examination) [22, p. 19-8].

There is a serious question as to whether it is reasonable to consider exposures as far into the future as 100 000 or 1 000 000 yr. However, if they are deemed to be of sufficient interest to warrant further investigation, then it would be valuable to continue studies of groundwater flow in order to obtain better estimates of the range of possible doses. It would also be valuable to explore the effects of modifying the repository configuration at Yucca Mountain—for example, the effects and predictability of alternative backfills. Further, to understand the implications of high individual doses, it is crucial to have estimates of the number of people potentially impacted.[27]

Utility of Performance Assessments

These performance assessments are to be viewed as preliminary, and the projections of TSPA 93 may be modified as future assessments are made with improved data and methodology. The significance of successive iterations lies not only in their results, but also in the guidance they give for repository design and for the planning of further assessments. It is to be hoped that in the end these assessments will provide a believable semi-quantitative description of the spectrum of risks that would be created by the repository.

---

[26]Detection and removal would be quite possible if technological capabilities do not fall below present levels.

[27]If, for example, 10 000 people receive annual doses of 200 mSv per year, the resulting population dose is 2000 person-Sv. If one assumes there is no change in cancer treatments and one accepts the current risk estimate of 0.05 fatalities per Sv (see Section 2.5.4), this would imply 100 cancer fatalities per year.

REFERENCES

1. Nuclear Waste Technical Review Board, *Sixth Report to the U.S. Congress and the U.S. Secretary of Energy* (Arlington, Va.: NWTRB, December 1992).
2. U.S. Department of Energy, *Spent Nuclear Fuel Discharges from U.S. Reactors 1993*, service report SR/CNEAF/95-01 (Washington, D.C.: U.S. DOE, 1995).
3. "Late News in Brief," *Nuclear News* 36, no. 14 (November 1993): 86.
4. Betsy Tomkins, "Onsite Dry Spent-Fuel Storage: Becoming More of a Reality," *Nuclear News* 36, no. 15 (December 1993): 35–41.
5. "Dry Cask Storage of Spent Nuclear Fuel" and related information documents, Consumers Power, Palisades Nuclear Plant, Michigan (1994).
6. "Late News in Brief," *Nuclear News* 36, no. 8 (June 1993): 22.
7. "Legislature: Prairie Island ISFSI Can Proceed," *Nuclear News* 37, no. 8 (June 1994): 26.
8. "PUC Asks NSP to Plan for Power Alternatives," *Nuclear News* 37, no. 9 (July 1994): 25.
9. "Late News in Brief," *Nuclear News* 34, no. 14 (November 1991): 25.
10. Matthew L. Wald, "Nuclear Storage Divides Apaches and Neighbors," *New York Times*, November 15, 1993, p. A8.
11. Elaine Hirao, "NSP Teams Up With Mescaleros to Try Siting Commercial MRS," *Nucleonics Week* 35, no. 6 (February 10, 1994): 14–15.
12. "Utilities sign to support Mescaleros storage project," *Nuclear News* 38, no. 9 (August 1995): 84.
13. U.S. Department of Energy, Office of Civilian Radioactive Waste Management, *Multi-Purpose Canister System Evaluation*, report DOE/RW-0445 (Washington, D.C.: U.S. DOE, 1994).
14. TRW Environmental Safety Systems, *A Preliminary Evaluation of Using Multi-Purpose Canisters Within the Civilian Radioactive Waste Management System*, rev. 0 (Vienna, Va.: TRW, March 1993).
15. *Energy, U.S. Code of Federal Regulations*, title 10, parts 51 to 199 (1993).
16. Organization for Economic Co-operation and Development, *The Status of Near-Field Modelling*, proceedings of a technical workshop Cadarache, France, May 1993 (Paris: OECD, 1993).
17. National Research Council, *A Study of the Isolation System for Geologic Disposal of Radioactive Wastes*, report of the Waste Isolation Systems Panel (Washington, D.C.: National Academy Press, 1983).
18. A. G. Milnes, *Geology and Radwaste* (London: Academic Press, 1985).
19. Thomas A. Buscheck and John J. Nitao, *The Impact of Repository Heat on Hydrological Behavior at Yucca Mountain*, preprint UCRL-JC-115798 (Livermore, Calif.: Lawrence Livermore National Laboratory, January 1994).
20. Swedish Nuclear Power Inspectorate (SKI), *SKI Project-90*, SKI technical report 91:23 (Stockholm: SKI, 1991).
21. Konrad B. Krauskopf, *Radioactive Waste Disposal and Geology* (London: Chapman and Hall, 1988).
22. M. L. Wilson et al., *Total-System Performance Assessment for Yucca Mountain—SNL Second Iteration (TSPA-93)*, report SAND93-2675 (Albuquerque, N.M.: Sandia National Laboratories, 1994).
23. Robert W. Andrews, Timothy F. Dale, and Jerry A. McNeish, *Total-System Performance Assessment-1993: An Evaluation of the Potential Yucca Mountain Repository* (Las Vegas: INTERA, 1994).
24. Thomas A. Buscheck and John J. Nitao, "Repository-Heat-Driven Hydrothermal Flow at Yucca Mountain, Part 1: Modeling and Analysis," *Nuclear Technology* 104 (December 1993): 418–448.
25. Nuclear Waste Technical Review Board, *Fifth Report to the U.S. Congress and the U.S. Secretary of Energy* (Arlington, Va.: NWTRB, June 1992).
26. Office of Nuclear Waste Isolation, *Conceptual Waste Package Designs for Disposal of Nuclear Waste in Tuff*, technical report ONWI-439 (Pittsburgh: Westinghouse Electric, 1983).
27. J. L. Zhu and C. Y. Chan, "Radioactive Waste Management: World Overview," *IAEA Bulletin* 31, no. 4 (1989): 5–13.
28. Office of Nuclear Waste Isolation, *Engineered Waste System Design Specification*, technical report ONWI-423 (Pittsburgh: Westinghouse Electric, 1983).
29. TRW Environmental Safety Systems, *Mined Geologic Disposal System Multi-Purpose Canister Design Considerations Report* (Las Vegas: TRW, September 1993).
30. Swedish Nuclear Fuel and Waste Management Company (SKB), *Activities 1993* (Stockholm: SKB, 1993).
31. R. D. McCright, *An Annotated History of Container Candidate Material Selection*, report UCID-21472 (Livermore Calif.: Lawrence Livermore National Laboratory, 1988).
32. Robert M. Bernero, "Clarification of the 300–1000 Years Period for Substantially Complete Containment of High-Level Wastes with the Waste Package under 10 CFR 60.113(a)(1)(ii)(A)," Staff Position 60-001 (Washington, D.C.: Nuclear Regulatory Commission, July 27, 1990).
33. Nuclear Waste Technical Review Board, *First Report to the U.S. Congress and the U.S. Secretary of Energy* (Washington, D.C.: NWTRB, 1990).
34. U.S. Department of Energy, *Management of Commercially Generated Radioactive Waste, Final Environmental Impact Statement*, report DOE/EIS-0046F (Washington, D.C.: U.S. DOE, 1980).
35. Edward L. Miles, Kai N. Lee, and Elaine M. Carlin, "Nuclear Waste Disposal Under the Seabed," *Policy*

Papers in International Affairs, no. 22 (Berkeley: Institute of International Studies, University of California, 1985).

36. R. Kilho Park, Dana R. Kester, Iver W. Duedall, *et al.* eds., *Wastes in the Ocean*, vol. 3: *Radioactive Wastes and the Ocean* (New York: John Wiley, 1983).

37. National Academy of Sciences, *Management and Disposition of Excess Weapons Plutonium*, report of the Committee on International Security and Arms Control (Washington, D.C.: National Academy Press, 1994).

38. David F. McVey, Kenneth L. Erickson, and William E. Seyfried Jr., "Thermal, Chemical, and Mass Transport Processes Induced in Abyssal Sediments by the Emplacement of Nuclear Wastes: Experimental and Modelling Results," in *Wastes in the Ocean*, vol. 3: *Radioactive Wastes and the Ocean*, edited by R. Kilho Park *et al.* (New York: John Wiley, 1983).

39. J.-Y. Barre and J. Bouchard, "French R&D Strategy for the Back End of the Fuel Cycle," in *Future Nuclear Systems: Emerging Fuel Cycles & Waste Disposal Options, Proceedings of Global '93* (La Grange, Ill.: American Nuclear Society, 1993), 27–32.

40. U.S. Department of Energy, *Environmental Assessment Overview, Yucca Mountain Site, Nevada Research and Development Area, Nevada*, report DOE/RW-0079 (Washington, D.C.: U.S. DOE, 1986).

41. Anthony Buono and Larry R. Hayes, "Overview of the Geology and Hydrology of the Yucca Mountain Area and a Summary of the U.S. Geological Survey On-Going Studies for the Yucca Mountain Site Characterization Project," in *Waste Management '91*, vol. II, edited by Roy G. Post (Tucson: University of Arizona, 1991), 201–213.

42. 97th Congress, *Nuclear Waste Policy Act of 1982*, Public Law 97-425 (January 7, 1983).

43. U.S. Department of Energy, Office of Civilian Radioactive Waste Management, "Secretary of Energy Approves Project Decision Schedule," *OCRWM Bulletin*, report DOE/RW-0317P (July–August 1991): 4–9.

44. U.S. Department of Energy, Office of Civilian Radioactive Waste Management, "OCRWM Director Gives Congress a Program Update," *OCRWM Bulletin*, report DOE/RW-0448 (Summer/Fall 1994).

45. U.S. Environmental Protection Agency, *High-Level and Transuranic Radioactive Wastes, Background Information Document for Final Rule*, EPA report 520/1-85-023 (Washington, D.C.: U.S. EPA, 1985).

46. R. W. Barnard *et al.*, *TSPA 1991: An Initial Total-System Performance Assessment for Yucca Mountain*, Sandia report SAND91-2795 (Albuquerque, N. M.: Sandia National Laboratories, 1992).

47. *Protection of the Environment, U.S. Code of Federal Regulations*, title 40, parts 190 to 299 (1993).

Chapter 10

Administrative and Policy Issues in Nuclear Waste Disposal

10.1 FORMULATION OF U.S. WASTE DISPOSAL POLICIES

10.1.1 Early History

Planning for the disposal of nuclear wastes in the United States has undergone many changes—in both administrative leadership and overall strategy—in the 50 yr since the wastes began to be generated. The power to make decisions has moved from the tight control of a single federal body into an arena in which many governmental and even nongovernmental groups have significant, and sometimes conflicting, roles. On the technical side, the basic assumption that the wastes would be reprocessed has been abandoned, and the originally favored geological formation for disposal, bedded salt, has been bypassed. It has been extremely difficult to find disposal sites, because public trust in the federal government and its scientists is insufficient to overcome local fears. Some of this history is sketched in the succeeding sections, along with a summary of the current institutional status of nuclear waste disposal.

Responsibility for the handling of nuclear wastes in the United States initially resided with the Atomic Energy Commission (AEC), which was established in 1946 to take charge of all aspects of the U.S. nuclear energy program, including nuclear weapons and nuclear power. Although nuclear waste management came under the AEC's purview, it was not addressed as a matter of high priority. The problems did not appear to be very great, and there were more pressing and technically interesting matters that demanded attention. On the military side, there were the difficulties of simultaneously building U.S. weapons capabilities, coping with the dangers created by the nuclear arms race between the U.S. and the USSR, and restraining the spread of nuclear weapons to other countries. On the nonmilitary side, there were the challenges of fostering peaceful nuclear research throughout the world and establishing a nuclear electric power industry.

It was recognized that it would be necessary to handle wastes emerging from both the civilian and military programs, but there was implicit confidence that this could be done successfully when the need became more pressing. However, the low priority originally given to defense wastes has led to serious difficulties (see Section 8.1.2). A number of reactors were operated at the Hanford reservation in Washington and at the Savannah River facility in South Carolina, primarily to produce plutonium for nuclear weapons. The residues from reprocessing of the fuel were not well handled, at least by present standards, and successors to the AEC face large and uncertain cleanup operations.

There have been no comparable problems with civilian wastes from nuclear reactors. There were no pressures to do anything in haste, and the spent fuel has been left in cooling ponds at the reactor sites.[1]

There was at first a consensus that the spent fuel would be reprocessed as soon as reprocessing plants were built. The plutonium and uranium would be extracted for use in other reactors, and additional isotopes possibly extracted for specialized applications in medicine and industry. The remaining wastes would be solidified and eventually transferred to a federal repository for permanent disposal. A standard plan for solidification was to add the liquid product of reprocessing to a melt of borosilicate glass. When the glass cooled, the wastes would be entrapped within a stable glass matrix.

It was widely expected that the eventual repository would be in bedded salt. The use of a bedded salt repository had been suggested by a National Academy of Sciences (NAS) Panel as early as 1957, and this recommendation was reaffirmed in a 1970 NAS study, which spelled out the advantages of salt, although it also suggested some further studies and indicated the importance of taking care in the selection of specific salt sites [1].

The first repository site selected by the AEC was in salt beds located near Lyons, Kansas. It was designated in 1970 in response to pressures from Idaho's governor and senators to find a location to receive wastes that had been recently transferred to the National Reactor Testing Site near Idaho Falls.[2] The NAS report gave a qualified endorsement of the Lyons site but emphasized the need for further studies—for example, the identification of possible holes from earlier oil or gas well drilling. As studies proceeded, in accordance with the NAS recommendation and spurred by local objections to the repository, it was found that prior commercial activity at the site had created cavities and boreholes in the salt that compromised the site's safety. Fear grew that it was vulnerable to both physical collapse and the intrusion of water. A combination of possible technical difficulties and growing opposition within Kansas doomed the site, and in 1971 the AEC abandoned plans for its development.

Since the early 1970s the U.S. nuclear waste program has been in disarray. There have been innumerable studies and plans, but every specific proposal has encountered intense opposition, especially from people in the vicinity of the proposed facility and their political representatives. The Atomic Energy Commission was itself abolished in 1974. Its roles of developing nuclear energy and of regulating it were separated and given to two newly formed organizations. As their names suggest, the Energy Research and Development Administration (ERDA) assumed the development role, and the Nuclear Regulatory Commission (NRC) assumed the regulatory one. In turn, ERDA was replaced by a new cabinet-level entity in 1977, the U.S. Department of Energy. The Environmental Protection Agency (EPA) in 1976 took on the task of developing radiation protection standards for nuclear wastes [3, p. 1-3].

---

[1]This has been a largely uneventful solution to date, but nuclear plants are gradually running out of capacity in these ponds, and it remains to be seen the extent to which attempts to expand the capacity through on-site dry storage are accepted or thwarted (see Section 9.1.2).

[2]For a summary account of this action and subsequent developments, see Ref. [2]. This reference is an excellent source of information on historical and institutional aspects of waste management.

The organizational changes have done little to advance the formulation and implementation of waste disposal planning. The separation of the NRC from ERDA, and later the DOE, was made in part to forestall imprudent actions. That goal may have been achieved, but the goal of moving forward with a consistent and coherent policy has not been met. Overall, it has been very difficult for the federal government to implement a nuclear waste policy. Under the weight of outside pressures, there was for many years something close to administrative paralysis, punctuated by occasional preemptive Congressional actions.

10.1.2 Organizations Involved in Waste Management Policy

Federal Agencies

Three federal agencies have particularly important roles in nuclear waste management, as defined and limited by Congress. These are:

Department of Energy (DOE). The Department of Energy, through its Office of Civilian Radioactive Waste Management (OCRWM), has primary responsibility for building and operating waste management facilities, and for research that provides a technical basis for evaluating waste management options and problems.

Environmental Protection Agency (EPA). The EPA has overall responsibility for establishing standards for radionuclide releases and radiation exposures arising from the operation of waste management facilities.[3]

Nuclear Regulatory Commission (NRC). The NRC has the responsibility for licensing waste management facilities and for setting forth standards for construction and operation of these facilities. Its standards must be consistent with EPA standards.[4]

Beyond this trio, important roles are played by the U.S. Geological Service and the Department of Transportation.

Federal Advisory Bodies

In addition to the main federal agencies with administrative responsibilities, there are a large number of groups and organizations with official advisory status. Some of these report to the individual agencies, some to Congress, and some to the public at large. They include bodies with broad responsibilities, of which nuclear waste studies are a relatively small component, as well as a number of boards and committees established specifically to advise on nuclear waste issues. These include:

National Academy of Sciences (NAS). The NAS was established by congressional charter in 1863 as a self-governing body to provide scientific advice to the federal government. The National Academy of Engineering and the Institute of Medicine were established more recently under the NAS charter. The National Research Council is a joint arm

---

[3]These regulations are set forth in Part 191 of Title 40 of the *U.S. Code of Federal Regulations* (40CFR191), entitled "Environmental Radiation Protection Standards for Management and Disposal of Spent Nuclear Fuel, High-Level and Transuranic Radioactive Wastes" [4].

[4]The NRC regulations are contained in Title 10 of the *U.S. Code of Federal Regulations* [5], including parts dealing with geologic disposal (Part 60), transportation (Part 71), and temporary storage at the reactor or in a monitored retrievable storage facility (Part 72).

of these three organizations, and it serves as the organizing entity for academy studies, which are published by the National Academy Press. An example of a major study under the council's aegis is *A Study of the Isolation System for Geologic Disposal of Radioactive Wastes* [6]. The Energy Policy Act of 1992 gave the NAS a central role in making findings and recommendations for the formulation of regulations to govern the Yucca Mountain nuclear waste repository [7, Sec. 801].

Nuclear Waste Technical Review Board (NWTRB). The NWTRB was established under the terms of the 1987 Amendment to the Nuclear Waste Policy Act (NWPA) as an independent entity within the Executive branch. It is appointed by the President from a group of persons nominated by the National Academy of Sciences as "eminent in a field of science or engineering, including environmental sciences" [8, p. A-2]. Its mandate is to "evaluate the technical and scientific validity of activities undertaken by the Secretary [of Energy]," with a focus on the activities necessary to establish a waste repository. Results of its work are contained in a series of Reports to Congress and the Secretary of Energy, and are also made available to the public.[5]

Office of Technology Assessment (OTA). The OTA served as the analytic arm of Congress, studying technological matters. An example of an OTA study on nuclear wastes is the report *Managing the Nation's Commercial High Level Radioactive Waste* [9].[6]

General Accounting Office (GAO). The GAO is an arm of Congress that reports to it on fiscal and managerial aspects of federal programs. This mandate gives the GAO the scope to study the overall progress and effectiveness of these programs. An example of a GAO study on nuclear wastes is the report *Operation of Monitored Retrievable Storage Facility Is Unlikely in 1998* [10].

Advisory Committee on Nuclear Wastes (ACNW). The ACNW was established by the Nuclear Regulatory Commission in 1988 to provide independent advice to the NRC on matters pertaining to the NRC's responsibilities for licensing and regulating future nuclear waste facilities.

States and Indian Tribes

It is a widely held fear that in attempting to address the national nuclear waste disposal problem, the federal government is willing to ignore local interests. To address this concern, Congress included in 1982 legislation extensive requirements for consultation with affected states and Indian tribes. Although in the end Congress retained a prerogative for federal overriding of state objections, the states were given both time and funds to formulate possible objections.

In addition to the powers explicitly granted by Congress, the states and Indian tribes have access to the courts, and on frequent occasions they have taken legal action to forestall federal actions. Thus, for example, Minnesota, Texas, and Vermont joined in a suit objecting to EPA waste repository standards. Since the designation of the Yucca Mountain site, Nevada has used both the courts and other means in an effort to stop the project. At one stage, local denial of water permits forced the Yucca Mountain administration to truck water from California to carry out some preliminary investigations of the site.

---

[5]Critiques contained in these reports are extensively cited in Chapter 9.
[6]The OTA was disbanded by Congress in 1995, in a highly controversial action.

Private Individuals and Organizations

Private individuals and groups also have taken an active interest in waste disposal policy and have input through public hearings, normal political processes, and legal actions. Public opposition has, for example, prevented the investigation of potential granite repository sites in Maine and Michigan.

To date, most intervention by individuals and private organizations, aside from industry organizations, has been directed towards preventing the development of specific waste disposal facilities. To counter this trend, the Office of Civilian Radioactive Waste Management (OCRWM) has undertaken extensive programs to make information available to the public and to schools, in the belief that the information provided would allay concerns over waste disposal studies and plans.

In recognition of the important role of local opinion, some commentators have suggested that public participation and acceptance be elevated to the point where it becomes the key to determining the siting of waste facilities:

> Congress should mandate that no community be forced to accept a repository against its will and then establish a broad-based participatory process as a means of developing greater trust. The public and representatives of all affected parties should participate in all stages of the siting and development process [11, p. 47].

Further text highlights the intended far-reaching implications of the words *all stages*. Obviously, this would be a time-consuming course, but the authors see no need for haste, pointing out that with dry cask storage "wastes from commercial nuclear reactors can be stored safely for a lengthy period at current sites," i.e., at the reactors.

This may be a realistic course, and in the end it may be more effective than the alternative of attempting to force the Yucca Mountain site upon Nevada. However, the proposed expedient of on-site storage highlights the complexities of the waste disposal problem, because utilities may face considerable difficulties in obtaining approval for on-site storage plans, especially if no permanent repository is in imminent prospect (see Section 9.1.2). More broadly, a requirement of local approval may give disproportionate weight to short-term local priorities over long-term national priorities.

10.1.3 Congressional Preemption of the Site Selection Process

Nuclear Waste Policy Act of 1982

In response to the failure of the federal waste program to reach any clear and final decisions, Congress enacted legislation in 1982 intended to put the program on a new schedule. The congressional mandate was embodied in the Nuclear Waste Policy Act (NWPA) of 1982 [12]. Key provisions were:

Geologic sites. Geologic repository sites were confirmed as the leading choice for nuclear waste disposal.

Designation of candidate sites. The Secretary of Energy was directed to nominate five sites for initial study. Following study and consultation, the Secretary was to recommend to the President by January 1, 1985, three of the five sites for intensive study as possible sites of the first waste repository. This study process is termed site characterization.

Recommendation of selected site. On the basis of information developed in the site characterizations and upon recommendation of the DOE, the President was called upon to recommend to Congress one site as qualified for development of a repository. This Presidential recommendation was to be made by March 31, 1987, with an option for a one-year extension.

Nuclear Regulatory Commission. The NRC was called upon to approve or disapprove an authorization for construction of a repository at the selected site by January 1, 1989.

Nuclear Waste Fund. A Nuclear Waste Fund was established by assessing a fee of 0.1¢/kWh on electricity generated at nuclear reactors. This fund is to be used to cover the expenses of developing waste repository facilities.[7]

States and Indian tribes. Throughout the process there are extensive mandated consultations with the affected states and Indian tribes, and opportunities to indicate disapproval. However, in the end the authority to make decisions was retained by the President and Congress.

The schedule outlined above was for a first repository. The goal of this schedule was to have an operating repository by 1998. A parallel schedule, with a small lag, was also specified for a second repository.

The first step, the DOE's selection of the three choices for site characterization, was completed in May 1986, after lengthy hearings and well behind schedule. The sites chosen were a basalt site at Hanford, a bedded salt site in Texas, and a tuff site at Yucca Mountain in Nevada. The Hanford and Yucca Mountain sites were on federal land, the former part of the Hanford Reservation in Washington and the latter overlapping the Nevada Test Site, which was used for nuclear weapons testing. These recommendations engendered a great deal of controversy, including lawsuits initiated by the selected states.

With the difficulties encountered at each stage, it became apparent that the schedule outlined in the NWPA of 1982 would not be met. Not only was there no prospect that decisions would be reached on a schedule that would permit the President to recommend a single site in 1987, but it was also not clear that a decision could be made that would hold up against future lawsuits challenging the decisions and the process by which they were reached.

Nuclear Waste Policy Amendments Act of 1987 and Designation of Yucca Mountain

The matter was resolved, in a legislative sense at least, when Congress adopted the Nuclear Waste Policy Amendments Act of 1987, which designated Yucca Mountain as the *sole* site for characterization for a possible high-level waste repository [17]. Thus, instead of having

---

[7]Net nuclear generation from 1991 through 1993 averaged 6.1×10^{11} kWh per year, resulting in an annual input to the fund of about $600 million. By mid-1993 cumulative receipts of the fund, including investment income, were $7.3 billion [13, p. 47]. A balance of over $4.0 billion remained, expenditures having been limited by Congress in light of the federal budget deficit. Funding limitations led to the curtailment of important parts of the program, including the study of waste packages [14, p. 6] and some site characterization tests [15, p. 8]. The funding situation improved, at least briefly, with the appropriation for high-level waste disposal of $522 million for the 1995 fiscal year [16]. This was up sharply from $381 million for FY 94 and $142 million for FY 90. However, it appears likely that the funding level will drop sharply for FY 96.

three site characterization studies proceeding in parallel (at least two of which faced severe local opposition), there would be a single study in a place where opposition was thought to be the least.

This Congressional decision did not resolve all disputes, and the subsequent history has been one of continued conflict between federal authorities and representatives of Nevada. It did, however, reduce what had been a dispute with several states to a dispute with a single state.

A further major delay was introduced in November 1989, when the Secretary of Energy issued a "Reassessment of the Civilian Waste Management Program" in a Report to Congress [18]. This called for a new start on the Yucca Mountain studies, because, as interpreted by *The New York Times*, "the DOE lacked confidence in its studies so far" [19]. A new schedule was announced that called for the repository to open by "approximately 2010" if Yucca Mountain was found to be suitable.

10.1.4 EPA Regulations Governing Waste Disposal

Responsibilities of the EPA and NRC Under the NWPA of 1982

The responsibility of the EPA and NRC for regulating disposal facilities for civilian radioactive wastes was confirmed in the NWPA of 1982 [12, Sec. 121]. The EPA's mandate under the NWPA is to promulgate standards for protection of the off-site environment from releases of radioactive materials from repository sites. The NRC's mandate is to establish technical criteria for approving the construction, operation, and closure of repository sites. The EPA and NRC were given roughly equivalent deadlines of one year for the promulgation of standards, i.e., until January 1984, but there has been a long series of delays in developing EPA standards that could hold up under court challenges.

1985 Version of 40CFR191

Following passage of the NWPA of 1982, the EPA developed radiation protection standards for waste disposal sites, embodied in the regulations of 40CFR191 (see Section 10.1.2).[8] Initially, these standards were to become effective as of November 18, 1985. Key provisions of 40CFR191 included the following:

- During the handling and storage of spent fuel and wastes, prior to permanent disposal at a repository, there shall be "reasonable assurance" that no member of the general public will receive an annual whole-body dose equivalent in excess of 25 mrem (0.25 mSv). (This limit was included in Subpart A of 40CFR191.)

- Over a period of 10 000 yr releases of radionuclides from the repositories to the outside environment must conform to specified limits. These limits were set so that the rate of fatal cancers in the surrounding population is expected to be under 1000 over the 10 000 yr period, i.e., an average of less than one cancer fatality per decade [20, p. 38069]. (These limits were specified in Subpart B of 40CFR191.)

- Over a period of 1000 yr no member of the general public is to receive a dose in excess

---

[8]Some of the history of the formulation of these regulations and the methodology behind them are discussed in an EPA publication on "Background Information" [3].

of 25 mrem (0.25 mSv) per year due to releases from the repository. It is to be assumed that the person at risk drinks 2 liters of water per day from sources in the neighborhood of the repository site. (This limit was specified in Subpart B of 40CFR191.)

Judicial Overturn of 40CFR191

The stipulation of these time limits, of 1000 or 10 000 yr, led to considerable controversy. Some critics believed that setting a 10 000-yr limit was neither realistic nor necessary. Others attacked the seeming inconsistency in Subpart B of setting a 10 000-yr limit for general protection and a 1000-yr limit for individual protection. In particular, the National Resources Defense Council (NRDC), along with several states and several other environmental organizations, brought suit against the EPA on grounds that included the claimed inconsistency between these two time criteria and the fact that the EPA had not considered the applicability of the Safe Drinking Water Act to their standards.

In an opinion handed down by the First Circuit Court of Appeals in 1987, the court ruled in favor of the NRDC and its associates. Subpart B of 40CFR191 was "vacated and remanded" [21]. Grounds for this action included the court's opinion that the "EPA did not consider the interrelationship of the high-level waste rules and. . . the Safe Water Drinking Act" and that the EPA had "not provided an adequate explanation for selecting the 1000-yr design criterion" for individual protection.

Congressional Action in 1992

The EPA returned to the task of framing regulations that would meet the objections of the court. In the meantime, indicating what would appear to be a lack of confidence in the EPA and NRC, Congress in 1992 diluted their power to formulate the regulations that would govern Yucca Mountain. In the Energy Policy Act of 1992 [7, Sec. 801], Congress imposed the requirement that the EPA and NRC consult with the National Academy of Sciences before formulating the regulations for Yucca Mountain. The NAS was to advise on "reasonable standards of protection." The act called for NAS recommendations by December 31, 1993, but after passage of the act the NAS demurred, saying that the time allowed was too short, and the final NAS report was not issued until mid-1995 (see Section 10.1.5).

A second congressional action, the Waste Isolation Pilot Plant Land Withdrawal Act of 1992 [22, Sec. 8], called on the EPA to promulgate its revised regulations by April 30, 1993, for sites other than the excluded Yucca Mountain site. In particular, the regulations would apply to the Waste Isolation Pilot Plant in New Mexico, which was constructed for the deposit of transuranic wastes from defense programs.[9]

1993 Version of 40CFR191

The new version of 40CFR191, promulgated by the EPA on December 20, 1993, includes the following changes from the earlier 40CFR191 [23]:

• Extends the individual protection criterion from 1000 yr to 10 000 yr, removing the difference between the two time limits of the original version.

---

[9]The disposal of high-level wastes, including spent fuel, is specifically prohibited at WIPP [22, Sec. 12].

- Lowers the annual dose limit for an exposed individual from 25 mrem (0.25 mSv) to 15 mrem (0.15 mSv).

- Adds a Subpart C that requires that "underground sources of drinking water in the accessible environment" satisfy the requirements of the Safe Drinking Water Act, as codified in limits on radionuclide concentrations specified in 40CFR, Part 141. These are designed to limit the dose to an individual from drinking water to 4 mrem (0.04 mSv) per year.

While the stipulations of 40CFR191 were to have no official applicability to Yucca Mountain, pending the recommendations of the NAS study, they are cited in the Sandia National Laboratories TSPA 93 study as "relevant to the form a new individual-dose standard might take" [24, p. 2-13]. They served as a motivation for the study of potential exposures from drinking water.

The ^{14}C Problem

Preliminary assessments of the Yucca Mountain site have suggested that the most likely violation of the proposed 40CFR191 release limits would come from the escape of ^{14}C (see Section 9.5.3). The recommendations of the NAS study (see Section 10.1.5) may make the issue moot, but this potential "violation" has been a matter of major concern (see, e.g., Ref. [25]) and provides a striking example of conflict between what some believed were imperatives of a strict regulatory regime and others believed to be the imperatives of common sense.

About 28 000 Ci of ^{14}C ($T=5730$ yr) are produced annually in the atmosphere by the interaction of cosmic-ray neutrons and ^{14}N, leading to a global inventory of ^{14}C of 2.3×10^8 Ci, mostly in the oceans. The atmospheric inventory, in the absence of nuclear weapons tests, is about 3.8×10^6 Ci [26].[10] The resulting dose to individuals is about 1×10^{-5} Sv per year (1 mrem/yr), corresponding to a global collective dose of roughly 1×10^5 person-Sv. According to the linearity hypothesis for radiation effects, this dose implies about 5000 cancer fatalities per year, or roughly 50 million fatalities from *natural* ^{14}C over the 10 000 year period being considered for waste repositories (see Section 2.5.4).[11]

Some ^{14}C is produced in nuclear reactors, primarily through neutron reactions with ^{17}O in the cooling water and with ^{14}N impurities in the fuel and cladding. This leads to release to the atmosphere of about 10 Ci per year for a 1000-MWe reactor and the accumulation of roughly 1 Ci/MTIHM in spent fuel assemblies [26]. The ^{14}C releases from reactors are generally regarded as trivial in the context of the much larger amount of natural ^{14}C in the atmosphere.

However, the ^{14}C in the wastes created a regulatory stumbling block. The EPA, in a 1985 study of limits for radionuclide emissions, used a worldwide risk factor of 0.058 fatalities in 10 000 yr per Ci of ^{14}C released, and imposed a release limit of 100 Ci per 1000 MTIHM, corresponding to 580 fatalities for a hypothetical repository with 100 000 MTIHM [3, p. 7-16]. At the time, the EPA was focusing on repositories in saturated media. It concluded that there would be no excessive releases to the environment for water transport of ^{14}C, but it ignored the possible escape of ^{14}C as a gas in the form of CO_2 [27]. Those analyses have been superceded by later studies that considered the possible motions of gases through the

---

[10]Weapons testing increased the atmospheric concentration substantially, but with the cessation of U.S. and Soviet above-ground testing the concentration has been returning towards the equilibrium value through mixing with the oceans.

[11]We here assume a world population of 10 billion, which may be less than the actual future population.

rock at Yucca Mountain and concluded that there was an appreciable possibility that with gaseous transport the 40CFR191 criterion for ^{14}C would be violated (see Section 9.5.3).

The full inventory of 70 000 Ci of ^{14}C (for 70 000 MTIHM) is only about 2% of the atmospheric inventory and about 0.03% of the total inventory (including oceans and biosphere) with which the atmosphere mixes. It is estimated that a release of the entire inventory, if spread over a few hundred years or more, would increase the atmospheric concentration by less than 0.1%, corresponding to an additional 1×10^{-8} Sv/yr (0.001 mrem/yr) dose from ^{14}C for the average individual [28]. The dose for the maximally exposed individual has been put at 0.05 mrem/yr (0.0005 mSv/yr) in one estimate [28] and 0.01 mrem/yr (0.0001 mSv/yr) in another [29].

At either level, the maximum dose from ^{14}C is far below the EPA limit of 15 mrem/yr (0.15 mSv/yr). The additional 1×10^{-8} Sv in the average annual dose would add only 0.0003% to the dose from natural radiation. But the total releases and the calculated world-wide cancer fatalities could well exceed the EPA limits.

Overall, the ^{14}C issue exemplifies the conflict that sometimes arises between a sense of reasonable proportions and the strictures of a regulatory system.

10.1.5 Recommendations of the 1995 NAS Study

The Nature of the Recommended Standard

According to the Congressional mandate of the Energy Policy Act of 1992 [7], EPA standards for a repository at the Yucca Mountain site were to be held in abeyance pending recommendations from the National Academy of Sciences. Two of the key questions that the NAS was called upon to address were (a) "whether a health based standard based on doses to individual members of the public....will provide a reasonable standard of protection;" and (b) "whether it is possible to make scientifically supportable predictions of the probability....of human intrusion over a period of 10 000 years." The results of the NAS study were published in the summer of 1995 as the National Research Council report *Technical Bases for Yucca Mountain Standards* [30]. In brief, the answer to question (a) was *yes* and the answer to (b) was *no*. However, the significance of the report extends well beyond the answers to these specific questions.

The Recommended Nature of the Health Standard

Protective standards for a nuclear waste repository can be couched in alternative ways. Possibilities include limits on the release of radionuclides, on the maximum individual dose, on the average dose for a group of individuals, or on the maximum number of fatalities. Such limits are coupled. For example, the limits that the EPA had earlier proposed on radionuclide releases were intended to satisfy the limit of 1000 fatalities over 10 000 yr.

The NAS panel recommended "the use of a standard that sets a limit on the risk to individuals" from radiation exposure [30, p. 4]. For example, a plausible choice, as reflected in the NAS discussion of precedents, might be an incremental risk in the neighborhood of 10^{-5} per year. Such a risk limit would then be translated into a radiation dose limit, based upon an assumed dose-response relationship. For example, for a cancer fatality risk factor of 0.05 per Sv (see Section 2.5.4), the corresponding dose limit would be 0.2 mSv (20 mrem) per year. If the scientific consensus on the risk factor changes, the dose limit would be recalculated, but the fundamental risk level (10^{-5} per year in this example) need not be reevaluated.

The NAS panel recommended that the limit apply to a "critical group," rather than to a hypothetical maximally exposed individual or to a very broad population. The critical group for which the dose would be calculated should be "representative of those individuals in the population who, based on cautious, but reasonable, assumptions have the highest risk....resulting from repository releases" [30, p. 53]. This group is intended to be "relatively small," without great variations in dose within the group. Alternatives for defining the critical group were discussed, but the actual definition is left to the EPA.

Duration of Protection

In previous EPA standards, the period of protection had been set at 10 000 yr on the premise that the risk decreases with time as radionuclides decay and that projections beyond 10 000 yr would be highly uncertain. However, recent studies of the Yucca Mountain site suggest that the maximum radiation doses might not occur until after several hundred thousand years, because the repository would provide adequate confinement of radionuclides for earlier times (see Section 9.5.4). With such calculations in mind, the NAS report suggests that the time horizon be extended to the neighborhood of one million years. Specifically it is recommended that "the assessment be conducted for the time when the greatest risk occurs, within the limits imposed by long-term stability of the geologic environment" [30, p. 7], with the suggestion that the appropriate time frame for consideration is "on the order of 10^6 yr" [30, p. 9]. Assessment of compliance—the extent to which the repository would actually satisfy the established standards—was deemed "feasible for most geologic aspects of repository performance" over this time scale.

For want of better, it is suggested that calculations be based on patterns of working and farming similar to those that obtain today. The possibility that future generations will have made progress in preventing or curing cancer was not discussed.

The Problem of Human Intrusion

The NAS report concluded that it was not possible to predict either the likelihood that systems to prevent human intrusion into the repository will be successful over extended time periods or the probability that intrusions will actually occur. However, the report recommended that the consequences of such intrusions be studied. For specificity, it was suggested that the focus be on the integrity of the repository following the drilling of a single borehole through a waste canister and into an underlying aquifer. The calculated risk to the critical group from such intrusion should be no greater than from "normal" long-term releases from the repository.

Remaining Tasks

The charge of the NAS panel was to recommend the *nature* of the limit. It is then left to the EPA to set the *magnitude* of the limit as well as to define the critical group to which the limit applies. The determination of an appropriate risk standard (e.g., 10^{-5} per year or some alternative) is to be made after a rulemaking process which allows for public input [30, p.

49]. The EPA will then have the further task of deciding upon an appropriate factor (e.g., 0.05 per Sv) to convert the risk standard into a dose standard.

If, after the Department of Energy and its contractors complete the on-going assessment of the Yucca Mountain site, the DOE is satisfied that the repository will meet the EPA standards, the DOE will seek a license from the Nuclear Regulatory Commission. Whatever decisions are reached by these several agencies may be expected to be challenged in appeals to the administration, to Congress, or to the courts.

Many observers have doubted that the Yucca Mountain site would be ready to accept wastes by the year 2010, if ever. Some of these suspicions may be allayed by the confidence the NAS report appears to place on the potential validity of long-term repository assessments. It is also possible, however, that the NAS has created a formidable barrier in suggesting a time horizon as long as one million years and a risk standard as low as 10^{-5} per year for members of a small critical group.

10.2 UNDERLYING POLICY ISSUES

10.2.1 Responsibilities to the Future

Waste disposal planning is undertaken in the context of universal agreement that we have a responsibility to future generations. At its root the responsibility is a moral one, and the evolving recommendations and regulations seek to give concrete substance to the moral demands. But defining this responsibility, which is part of the question of "intergenerational equity," is very difficult, and agencies that grapple with these issues do so without the benefit of well-established guiding principles. As described in one review:

> Within the United States, and across the world, there has been and is today only limited, piecemeal consideration of how society should deal with those of its activities which have the potential to pose very long term risks far into the future. The absence of broad philosophic guidelines to deal with such issues makes decision making that much more difficult [31, p. 1].

We cite here some of the questions that arise in considering efforts to give appropriate weight to society's responsibilities to the future:

- For how long and at what level do our responsibilities extend? Are we to be concerned about people 100 yr hence, 1000 yr hence, 1 000 000 yr hence? Is a death in the distant future of as much concern as a death tomorrow?

- What is our picture of our descendants? If technological capabilities continue to advance over the next centuries, and technological memory is not lost, our descendants will find it relatively easy to detect and avoid any dangerous releases of radioactivity. If medical science advances to the point of a cancer cure, the dangers will be greatly reduced, even if radiation exposures are not avoided. On the other hand, if technological memory is lost, what social cataclysms might cause this loss? Are the projected effects of nuclear waste disposal of any consequence in such a context?

- In the thinking about intergenerational responsibility, is it reasonable to consider nuclear waste disposal in isolation? What coupling should there be to the consequences of the use of fossil fuels? Use of fossil fuels at present rates will deplete the world's readily available oil and natural gas within about a century, and coal within several centuries. At the same time, according to many scientific assessments, combustion of these fuels may produce significant climate changes. Are future generations more endangered by our production of nuclear wastes or by our consumption of fossil fuels? Are there realistic alternatives that might allow us to escape this choice?

These questions do not figure prominently in analyses of nuclear waste disposal. They have no crisp answers, and there is no accepted methodology for converging upon common conclusions. However, such questions should not be avoided, if there is to be responsible consideration of our intergenerational responsibilities.

10.2.2 Waste Disposal as a Surrogate Issue

Despite the emphasis we have placed on them, the technical and even the philosophical issues involved in waste disposal may not be central to ending the disputes over waste management. It has been suggested that for some opponents of nuclear power, the waste disposal issue is more a tool than a driving substantive matter. The then-chairman of the Sierra Club, Michael McCloskey, is quoted by Luther Carter in the 1987 book *Nuclear Imperatives and Public Trust* as expressing the view:

> I suspect many environmentalists want to drive a final stake in the heart of the nuclear power industry before they will feel comfortable in cooperating fully in a common effort at solving the waste problem. . . . Their concern would arise from the possibility that a workable solution for nuclear waste disposal would make continued operation of existing plants more feasible, and even provide some encouragement for new plants [2, p. 431].

This suggests that some opponents of nuclear power will oppose any waste disposal plan, independent of its intrinsic merits, because they see other compelling objections to nuclear power. This obviously complicates the matter of implementing any specific waste disposal plan, especially given the existing administrative and legal mechanisms for creating delays.

10.2.3 Technological Optimism and Its Possible Traps

On the pro-nuclear side of the debate, it is possible that technological optimism has led nuclear advocates into difficult positions. In one argument, nuclear advocates in the 1970s often pointed out that after a few centuries of radioactive decay the hazards of nuclear wastes (essentially the water dilution volume) become less than the hazards from the original ore used to generate the waste. In some estimates, this would be achieved after only 300 yr. As discussed above (Section 8.3.3), with the abandonment of reprocessing and changes in hazard estimates for individual radionuclides, the time period for this crossover to occur is now taken to be much longer, and the rhetorical situation has therefore changed. It was never a very appropriate criterion, but the concept has been implanted in some discussions of hazards.

In a different manifestation of technological optimism, there was a willingness to impose and accept very rigorous standards of protection, lasting for 10 000 yr, because they were deemed by regulatory agencies and many nuclear advocates to be achievable. Given the defense in depth of the waste package and the geological environment, there were optimistic expectations that fulfillment of the demanding requirements could be assured.

However, it may be difficult to establish a clear scientific consensus that this assurance is absolute. Although the TSPA 1993 analyses concluded that exposures from a Yucca Mountain repository would be very low for the first 10 000 yr, some scientists believe that the behavior of metals and of geological sites cannot be conclusively demonstrated over so long a time period. Thus, the establishment of standards for 10 000 yr could be a means of producing failure, in that a court may plausibly conclude that the requisite level of proof has not been met. The difficulties may be compounded if the time horizon is extended to 1 000 000 years in accord with the advice of the 1995 NAS report that this is a time over which "compliance assessment is feasible for most physical and geological aspects of repository performance" [30, p. 6].

10.2.4 Congressional Primacy

In the end, the practical determining factor in United States waste policy has been Congress. Congress has shown an interest, even an eagerness, to be involved in the details of nuclear waste policy formulation. In carrying out this role, Congress has not adhered to a single policy, but has adjusted the policy under the pressure of existing circumstances. It has at times acted cavalierly, as it did in 1987 when it threw out the elaborate procedures of the NWPA of 1982 and designated Yucca Mountain as the one location for site characterization.

The decision by Congress to call upon the National Academy of Sciences to frame recommendations for future waste disposal standards reflected an attempt to achieve a deliberate, measured approach. But it is not yet possible to foresee the impact of the NAS recommendations. These recommendations leave the most difficult decisions to the future, and the recommendations and their implementation are both subject to challenge.

In the end, Congress's future decisions may be determined more by its overall attitudes towards nuclear power than by issues intrinsic to nuclear waste disposal itself. Assuming deep geologic disposal remains the favored course, it probably will be possible for Congress to find scientific and moral backing for either the conclusion that it is prudent to move ahead with Yucca Mountain, or the conclusion that it is irresponsible. The choice of the advice to be accepted will probably be strongly influenced by the perceived need for nuclear power.

10.2.5 The Neglected Focus: A Defined Danger

The ongoing controversies about nuclear waste disposal have had something of the aspect of shadowboxing, because there has been no clear picture of the danger to be avoided. This can be contrasted with other facets of the nuclear debate. For nuclear reactor accidents, as analyzed for many years before Chernobyl and illustrated by Chernobyl, there are plausible, specific scenarios for very serious consequences. The issue then becomes estimating the likelihood of one or another of these scenarios. With nuclear bombs, there are the clear hazards of explosive blast and radioactive contamination, and there are compelling reasons to avoid conditions under which fissile material can become available to terrorist groups or hostile countries. In these cases, the problems can be discussed with reference to identifiable dangers, with unambiguously serious consequences.

With nuclear wastes, on the other hand, the discussion often is disconnected from meaningful concerns. Much of the attention is directed towards compliance with regulatory requirements. But it is not always possible to find a convincing justification for those requirements. The ^{14}C case is a particularly egregious example of what can happen when the standard for an individual radionuclide bears no relationship to a meaningful danger. In discussing the rationale for the 40CFR191 regulations, one knowledgeable observer[12] has stated [32]:

---

[12]Terry Lash, formerly director of the Illinois Department of Nuclear Safety and appointed in 1994 as the Director of the DOE's Office of Nuclear Energy.

> It is my belief that the EPA document never has been and is not today a health-based document. I believe it was and is a technology-forcing document, like somebody at EPA might do to encourage the best available technology.

Setting demands for the "best available technology," without regard to the actual dangers avoided, may not only be misleading but also could be destructive. Instead of forcing a technology to be better, excessively ambitious demands could have the perhaps unintended effect of forcing it to be abandoned,

The establishment of standards for future radiation doses is obviously central to the issue of repository acceptability. However, at some point a requirement for very low levels of exposure over very long periods of time can become decoupled from society's actual concerns. As an aesthetic matter, there may be an attractive simplicity in requiring that for 10 000 or 1 000 000 yr no individuals will receive doses greater than 0.2 mSv per year. But aside from aesthetics, it would be a perplexing standard. Society is essentially indifferent to much greater exposures today. The average dose of a person in the United States from natural sources is 15 times this level and we shrug off substantial variations above and below this average. The limit on occupational exposures is still higher, at present 50 mSv per year. Here, we are talking about ourselves, our families, and our friends, today.

Given our casual acceptance of natural doses of about 3 mSv per year for *everyone*, it is hard to believe that we care if a few of the *most exposed* of our descendants receive additional doses of 0.2 mSv per year. Similarly, the EPA's earlier standard of 1000 possible deaths in 10 000 yr from low-level radiation is difficult to reconcile with our acceptance of thousands of deaths per year in industrial accidents, an estimated 10 000 or more possible deaths per year from indoor radon, and close to 50 000 deaths per year from automotive accidents.[13] It is also hard to understand why the estimated risk to the maximally exposed small group of our remote descendants should be held to, say, 10^{-5} per year from a repository when the average person in the United States today faces much higher risks from many sources, both anthropomorphic and natural.

A deeper concern than risks at the levels of the standards being discussed by the EPA and the NAS, is the possibility that the nuclear wastes may cause a much greater disaster—whether it be termed a medical, environmental, or ecological disaster. We want strong evidence that the waste repository cannot cause severe harm. A reformulation of the issue with attention to the criteria for defining "severe harm" would add a useful focus to the discussion.

Human activities do create sobering possibilities of real disasters, starting with the still-present danger of a nuclear war. In the nuclear realm, far behind nuclear war in magnitude but still very serious, would be a terrorist's explosion of a nuclear weapon or a major nuclear reactor accident. Potential disasters loom in other energy-related areas as well. At projected consumption rates, the world's readily available oil and natural gas will be depleted within a century, and the coal within several centuries. If future generations fail to find adequate replacements in timely fashion, there could be serious economic, and possibly military, turmoil as societies scramble to secure what resources are available. There is also the possibility that combustion of fossil fuels will lead to climate changes with severe environmental consequences.

Could nuclear waste disposal, as envisaged in current planning, add to the roster of disaster scenarios? Recent studies do not suggest that this is possible, and instead the focus

---

[13]The deaths from radiation are termed "possible" because of uncertainties as to the effects of radiation at the low dose rates being considered (see Section 2.5.3).

has been placed on consequences that are very distant in time, involve small doses, and impact only a relatively few people.

A middle ground, between the exploration of disasters and the exploration of minor effects, might be to ensure that we do not create radiation hazards for future generations that are greater than those that were present before the beginning of the nuclear era. As emphasized in Chapter 2, the Earth has never been a radioactively pristine environment. Natural radionuclides have led to high exposures, for example, to miners in uranium-rich areas, to radium watch dial painters, and to people living in houses with high radon concentrations. It would be desirable to reduce the likelihood of comparable exposures in the future, but despite the individual tragedies involved, society has not treated these exposures as disasters.

Ensuring that our descendants do not receive greater exposures from nuclear wastes than present and past generations have received from natural radioactivity is made easier by continuing radioactive decay. For the first thousand or so years, when the activity levels are highest and the wastes have the greatest potential for harm, there is considerable confidence in the integrity of the engineered and natural barriers and in their ability to prevent the escape of radionuclides into the environment. Further into the future, when the possibilities for escape are greater, the radionuclide inventories are smaller.

In the end, society may decide not to attempt to extend a protective umbrella far into the future, in view of the great uncertainties about the behavior of repositories over very long time periods and about the nature and capabilities of future societies. However, if the distant future is considered, the crucial question is whether *a significantly large* number of people can receive *significantly large* doses. Inevitably, subjective judgments will enter into the definition of "significantly large," but couching the issue in these terms would frame the problem in a fashion related to the fundamental concerns.

As repository performance assessments are carried out with improved data and models and, it is to be hoped, with criteria attuned to realistic concerns, it will become possible to evaluate the acceptability of the risks posed by the repository. A "favorable" evaluation would not guarantee closure on the waste disposal issue. There would remain people who will contest the criteria for acceptability, who doubt expert assessments, who use the waste disposal issue as a tactical tool in a larger debate, or who prefer alternatives to geologic disposal. Nonetheless, a positive evaluation might create a sufficient consensus to proceed with the Yucca Mountain repository. A negative evaluation would force an intensified examination of alternatives—other geologic sites, sub-seabed disposal, transmutation, or interim storage for an extended time period.

The last of these alternatives may prove to be the most likely. The development of interim storage facilities, either at the Mescaleros site in New Mexico or elsewhere, would meet the demands of nuclear utilities that are under pressure to reduce their accumulating inventories of spent fuel. In addition, for better or worse, it would avoid achieving an ultimate resolution of the "waste disposal problem."

REFERENCES

1. National Research Council, *Disposal of Solid Radioactive Wastes in Bedded Salt Deposits*, report of the Committee on Radioactive Waste Management (Washington, D.C.: National Academy of Sciences, 1970).
2. Luther J. Carter, *Nuclear Imperatives and Public Trust: Dealing with Radioactive Waste* (Washington, D.C.: Resources for the Future, 1987).
3. U.S. Environmental Protection Agency, *High-Level and Transuranic Radioactive Wastes: Background Information Document for Final Rule*, report EPA-520/1-85-023 (Washington, D.C.: U.S. EPA, 1985).
4. *Protection of the Environment, U.S. Code of Federal Regulations*, title 40, parts 190 to 259 (1993).
5. *Energy, U.S. Code of Federal Regulations*, title 10, parts 51 to 199 (1993).

6. National Research Council, *A Study of the Isolation System for Geologic Disposal of Radioactive Wastes*, report of the Waste Isolation Systems Panel (Washington, D.C.: National Academy Press, 1983).

7. 102d Congress, *Energy Policy Act of 1992*, Public Law 102-486, title VIII-High Level Radioactive Waste (October 24, 1992).

8. Nuclear Waste Technical Review Board, *First Report to the U.S. Congress and the U.S. Secretary of Energy* (Washington, D.C.: NWTRB, 1990).

9. Office of Technology Assessment, *Managing the Nation's Commercial High-Level Radioactive Waste*, report OTA-O-171 (Washington, D.C.: U.S. Congress, 1985).

10. U.S. General Accounting Office, *Operation of Monitored Retrievable Storage Facility Is Unlikely by 1998*, report GAO/RCED-91-194 (Washington, D.C.: U.S. GAO, 1991).

11. James Flynn, Roger Kasperson, Howard Kunreuther, and Paul Slovik, "Time to Rethink Nuclear Waste Storage," *Issues in Science and Technology* VIII, no. 4, (1992): 42–48.

12. 97th Congress, *Nuclear Waste Policy Act of 1982*, Public Law 97-425 [H.R. 3809] (January 7, 1983).

13. U.S. Department of Energy, Office of Civilian Radioactive Waste Management, "Secretary Proposes a Nuclear Waste Revolving Fund," *OCRWM Bulletin*, report DOE-RW-0421 (Summer 1993): 47.

14. Nuclear Waste Technical Review Board, *NWTRB Special Report to Congress and the Secretary of Energy* (Washington, D.C.: NWTRB, 1993).

15. Nuclear Waste Technical Review Board, *Underground Exploration and Testing at Yucca Mountain: A Report to Congress and the Secretary of Energy* (Washington, D.C.: NWTRB, 1993).

16. Nuclear Energy Institute, "Special Report: Nuclear Waste," *Nuclear Energy* (2d quarter, 1994): 20–31.

17. U.S. Department of Energy, Office of Civilian Radioactive Waste Management, "Congress Amends Nuclear Waste Policy Act of 1982," *OCRWM Bulletin*, report DOE-RW-0153 (December 1987/January 1988): 1–3.

18. U. S. Department of Energy, Office of Civilian Radioactive Waste Management, "DOE Submits Report to Congress on Reassessment of the Civilian Radioactive Waste Management Program," *OCRWM Bulletin*, report DOE-RW-0227 (November/December 1989): 1.

19. Matthew L. Wald, "U.S. Will Start Over on Planning for Nevada Nuclear Waste Dump," *New York Times*, November 29, 1989: p. 1.

20. "Environmental Standards for the Management and Disposal of Spent Nuclear Fuel, High-level and Transuranic Radioactive Wastes," *Federal Register* 50 (October 7, 1985): 38066–38089.

21. *National Resources Defense Council, Inc. v. U.S. Environmental Protection Agency*, 824 F.2d 1258 (U.S. Court of Appeals, 1st Cir. 1987).

22. 102d Congress, *Waste Isolation Pilot Plant Withdrawal Act*, Public Law 102-579 [H.R. 776] (October 30, 1992).

23. "Environmental Radiation Protection Standards for the Management and Disposal of Spent Nuclear Fuel, High-Level, and Transuranic Radioactive Wastes," *Federal Register* 58 (December 20, 1993): 66398.

24. M. L. Wilson *et al.*, *Total-System Performance Assessment for Yucca Mountain—SNL Second Iteration (TSPA-93)*, report SAND93-2675 (Albuquerque, N.M.: Sandia National Laboratories, 1994).

25. Electric Power Research Institute, *EPRI Workshop 2-Technical Basis for EPA HLW Disposal Criteria*, proceedings of conference held in February 1992, EPRI report TR-101257 (Palo Alto, Calif.: EPRI, 1993).

26. Benjamin Ross, "The Technical Basis for Regulation of Gas-Phase Releases of Carbon-14," in *EPRI Workshop 1-Technical Basis for EPA HLW Disposal Criteria*, proceedings of conference held in September 1991, EPRI report TR-100347 (Palo Alto, Calif.: EPRI, 1993), 159–172.

27. U. Sun Park and Chris G. Pflum, "Requirements for Controlling a Repository's Release of Carbon-14 Dioxide; The High Costs and Negligible Benefits," in *High Level Radioactive Waste Management*, vol. 2, *Proceedings of the International Topical Meeting*, Las Vegas, Nevada, 1990 (La Grange, Ill.: American Nuclear Society, 1990), 1158–1164.

28. R. A. Van Konyenburg, "Gaseous Release of Carbon-14: Why the High Level Waste Regulations Should be Changed," in *High Level Radioactive Waste Management*, vol. 1, *Proceedings of the 2d Annual International Conference*, Las Vegas, Nevada, 1991 (La Grange, Ill.: American Nuclear Society, 1991), 313–319.

29. W. B. Light, E. D. Zwahlen, T. H. Pigford, P. L. Chambré, and W. W.-L. Lee, "Release and Transport of Gaseous C-14 from a Nuclear Waste Repository in an Unsaturated Medium," in *Scientific Basis for Nuclear Waste Management XIV*, edited by Teofilo A. Abrajano, Jr. and Lawrence H. Johnson (Pittsburgh: Materials Research Society, 1991), 863–870.

30. National Research Council, *Technical Bases for Yucca Mountain Standards*, report of Committee on Technical Bases for Yucca Mountain Standards (Washington, D.C.: National Academy Press, 1995).

31. David Okrent, "On Intergenerational Equity and Policies to Guide the Regulation of Disposal of Wastes Posing Very Long Term Risks," report UCLA-ENG-22-94 (Los Angeles: UCLA School of Engineering and Applied Science, 1994).

32. Terry Lash, *EPRI Workshop 1-Technical Basis for EPA HLW Disposal Criteria*, proceedings of conference held in September 1991, EPRI report TR-100347 (Palo Alto, Calif.: EPRI, 1993), 394.

Chapter 11

Nuclear Reactor Safety

11.1 GENERAL CONSIDERATIONS IN REACTOR SAFETY

11.1.1 Assessments of Reactor Safety

It can be variously argued that nuclear reactors are extraordinarily safe or extraordinarily dangerous. The former conclusion can be reached if one looks at the past record of plants outside the former Soviet Bloc. The latter conclusion can be reached if one looks at the Chernobyl accident and takes it as a harbinger of future accidents elsewhere.

For commercial reactors in the "western" world, which accounts for by far the larger part of this experience, the safety record is excellent. As of the end of 1994 these reactors had a cumulative operating experience of over 6000 reactor-years, of which over 1900 reactor-years were logged by U.S. reactors.[1] There has been no accident in a commercial nuclear power plant, including the 1979 Three Mile Island (TMI) accident in the United States, that has caused a substantial radiation exposure to any member of the general public or a *known* radiation death of a nuclear plant worker. Nor has there been any such accident worldwide for reactors moderated by water, including LWRs and HWRs.

However, if one goes beyond western commercial reactors or beyond water-moderated reactors, there are two exceptions to this record. The first is often ignored: the 1957 Windscale accident in a British plutonium-producing graphite-moderated reactor that led to some significant exposures. The second was much greater in impact and has received far more attention: the Chernobyl accident in the Soviet Union in 1986. It can be argued, correctly, that the graphite-moderated Chernobyl reactor was of unusual design, and the circumstances

---

[1]The number of reactor-years is extrapolated from December 31, 1992, data and includes the contribution from commercial reactors that are no longer in operation [1, Table 7].

of that accident could not be repeated in the standard LWRs and HWRs used outside the USSR. However, it can also be argued, correctly, that no reactor has a truly zero chance of an accident and that Chernobyl vividly demonstrated that a major reactor accident can potentially impact tens of thousands of people, including those living hundreds and even thousands of miles from the accident site.

Assessments of reactor safety involve estimates of both the probability of accidents and their severity. In the remainder of this chapter we will explore some of the general issues involved in achieving and evaluating nuclear safety. In the following chapter we will look at some of the failures, cases where accidents did in fact occur. We will be interested in both the consequences of the accidents and the light they may shed on nuclear safety issues.

11.1.2 The Nature of Reactor Risks

Categories of Reactor Accidents

There are two main categories of nuclear reactor accidents, each illustrated by one of the two major accidents to date, the Chernobyl and Three Mile Island accidents:[2]

Criticality accidents. These are accidents in which there is a runaway chain reaction within at least part of the fuel. In a light water reactor of normal design, such accidents are highly improbable, due to negative feedbacks and shutdown mechanisms. They are less unlikely in other types of reactors, given sufficient design flaws. The Chernobyl accident started as a criticality accident, although much of the energy release was from a steam explosion following the disruption of the core.

Loss-of-coolant accidents. If the coolant flow is disrupted, then even if the chain reaction is promptly stopped, there remains a large heat output due to the radioactive decay of the reactor fuel. Unless this heat is removed by one or another of the cooling systems, the temperature of the fuel will rise and the core eventually will melt. Under conditions of overheating and possible excess steam pressure, radioactive materials can escape from the reactor pressure vessel and perhaps from the outer reactor containment. The TMI accident was a loss-of-coolant accident. There was substantial core melting but no large escape of radioactive material from the containment.

Energy Sources in Nuclear Accidents

One reason it is necessary to emphasize safety in the design and operation of nuclear reactors is the high energy density of the fuel. The energy per unit mass of nuclear fuel far exceeds that of any chemical fuel. This accounts for the effectiveness and destructiveness of nuclear weapons and, without adequate safety measures, the potential destructiveness of nuclear reactors. Of course, weapons are designed to achieve destructiveness, while reactors are designed to avoid it.

The difference between nuclear and chemical fuels stems from the difference in the energy release per reaction. In fission, the release is about 200 MeV per event, or nearly 1 MeV per nucleon. In chemical reactions, the energy release is on the order of 1 eV per atom. Fossil fuels, such as gasoline, all have an energy content per unit mass of roughly 5×10^7 joules per

---

[2]The Windscale accident does not fit into either of these categories (see Chapter 12).

kilogram (J/kg).[3] The most potent chemical fuel, hydrogen, has an energy content of about 1.4×10^8 J/kg. In contrast, ^{235}U has an energy content of 8×10^{13} J/kg, roughly one million times that of the chemical fuels. Even "dilution" of ^{235}U to 3% in typical LWR fuel leaves a very high energy content per unit mass, and the possibility of a very large energy release.

In an accident situation, a substantial energy output continues even after the chain reaction is stopped, due to the radioactivity of the fuel. In addition, reactor accidents can be made more severe by violent auxiliary events, in particular, a hydrogen or steam explosion.

In normal LWR operation, the coolant temperature is about 330–350 °C, and the zircaloy cladding of the fuel rods is at only a slightly higher temperature.[4] If the coolant flow is interrupted and the temperature of the cooling water rises, there may be oxidation of the cladding in a chemical reaction with the water, producing ZrO_2 and H_2 as well as further heat. This reaction begins to proceed rapidly as temperatures approach 900 °C. The production of hydrogen creates the potential for an explosion if oxygen is available, via the reaction $2H_2 + O_2 \rightarrow 2H_2O$. During the TMI accident, which will be discussed in Chapter 12, there was a detonation of hydrogen within the concrete containment building, but it was not sufficiently violent to cause damage. There was also fear at the time of a hydrogen explosion inside the reactor pressure vessel, but this was not a real possibility because of insufficient oxygen (see Section 12.2.2).

Rapid contact between very hot or molten fuel and the cooling water can result in the generation of large amounts of steam and a steam explosion. At one time, it had been feared that should fuel from a damaged core fall into the cooling water, the generation of steam might be violent enough to breach the pressure vessel. Further study has concluded that so violent a steam explosion is probably "physically impossible" in an accident in which the fuel gradually melts due to a loss of coolant, as at TMI [2, p. 259]. However, there can be greater damage if the fuel is finely fragmented, as can happen in a rapidly evolving criticality accident. This was the case at Chernobyl, where the start of a runaway chain reaction was followed by the disruption of some of the reactor fuel and a large steam explosion [*ibid*].

Aftermath of an Accident

Nuclear accidents pose particular problems because of the persistent effects of radioactivity. At Three Mile Island, the continued production of heat led to the fear that the accident might progress further, with the release or ejection of radioactive material from the reactor containment. At Chernobyl, there was a very large release of radioactive material, and the dispersed debris has created hazards that will last for many years.

This may be contrasted with the situation in other sorts of accident. Once a dam breaks or a natural gas facility explodes, the damage is done, and society feels moderately secure in coping with the aftermath. There may be many more immediate fatalities than in a nuclear accident,[5] but when the accident is over, it is usually deemed to be over. There is seldom investigation of possible lingering consequences. With nuclear accidents, serious consequences may persist for a long period of time, from both cancers appearing long after the initial exposure and continued exposures produced by the contaminated ground. This aftermath is not ignored, by either governmental units involved or the general public.

---

[3]The term "energy content" is shorthand here for energy release in combustion or fission.

[4]See, e.g., Ref. [2], pp. 51 and 261.

[5]For example, explosions in liquid-natural-gas tanks and the associated fires killed 130 people in Cleveland, Ohio, in 1944 and 40 workers on Staten Island in New York in 1973 [3, p. 162]; both cases exceeded the prompt fatalities at Chernobyl (see Section 12.3.4).

These factors, plus speculations as to still worse things that might have happened, put nuclear accidents in a special category of societal concern and make it particularly urgent that they be avoided. There can be debates as to the effect of an accident on the health of the public. There is no doubt, however, that nuclear accidents are disastrous to the nuclear industry.

11.1.3 Institutional Responses

From the onset of commercial nuclear power, there was concern about nuclear safety both in official circles and among manufacturers and operators of nuclear reactors. However, the TMI and Chernobyl accidents have led to an intensified effort on the part of public and private institutions to avoid the occurrence of an accident. The intensity of this effort has increased because of fear that any accident will reflect discredit on all of nuclear power. This concern is encapsulated in the phrase: "An accident anywhere is an accident everywhere."

In the United States, the Nuclear Regulatory Commission has the legal responsibility for enforcing standards. After TMI the nuclear industry established the Institute of Nuclear Power Operations (INPO) to coordinate and monitor efforts to make reactor operation more reliable and safe. Internationally, there are long-standing reactor safety programs operated by the International Atomic Energy Agency—for example, the IAEA's International Nuclear Safety Advisory Group (INSAG)—and by the Nuclear Energy Agency of the OECD. The World Association of Nuclear Operators (WANO) was established after Chernobyl as an international counterpart of INPO [4, p. 27].

Overall, while countries such as the United States and France have their own strong national programs, reactor safety is seen as an international problem, both technically and politically. In that spirit, particular efforts have been devoted in recent years to providing technical advice to the countries of the former Eastern Bloc, in an effort to help improve their reactors, including both PWRs and the Chernobyl-type RBMK reactors.

11.1.4 Means of Achieving Reactor Safety

General Requirements

Underlying any approach to safety, for any sort of equipment, are high standards in design, construction, and reliability of components. In nuclear reactors, the prospect of accidents has led to intense efforts to achieve high standards. Individual components of the reactor and associated equipment must be of a codified high quality. As described in an OECD report:

> In the early years of water reactor development in the USA, a tremendous effort was put into development of very detailed codes and standards for nuclear plants, and these were widely adopted by other countries where nuclear plants were initially built under US licences [4, p. 62].

U.S. efforts have since been supplemented by parallel efforts in other countries and by the IAEA. In addition, nuclear reactor safety philosophy has developed to include a number of characteristic special features, some of which are listed in the succeeding paragraphs.[6]

---

[6]The discussion here loosely follows the organization used in Ref. [5], p. 9 *ff*.

Inherent Safety Features

Inherent safety features are those that are incorporated in the reactor design in such a way that they follow from basic properties of matter and do not require the proper operation of any particular piece of equipment—merely that they obey the laws of physics. For example, if the fuel temperature rises there is a negative feedback due to the Doppler broadening of the absorption resonances in ^{238}U (see Section 11.2.2). Similarly, some reactor designs use the thermal expansion of the core at elevated temperatures to provide a negative feedback. Such feedbacks inherent in the design give an overall negative feedback that contributes importantly to safety. However, this does not mean they would necessarily suffice to override any possible increase in reactivity—for example, the large increases that would result from too great a withdrawal of the control rods.

Active and Passive Safety Systems

An *active* safety system is one that depends upon the proper operation of reactor equipment, such as pumps or valves. For example, the pumps and valves that control the water supply for emergency core cooling are parts of an active safety system.

Passive safety features are aspects of the engineering design that are arranged to come into play automatically, without the action either of the operators or of mechanical devices that might fail. The gravity-driven fall of a control rod would be a passive feature; even here, however, the release of the rod may be initiated by an active system.

Redundancy

Redundancy in safety systems can be achieved in a number of ways:

Identical units of the same type. Often more than one pump or motor is provided to perform a given safety task, although it is only necessary that one of these operates properly. It is particularly important in such cases to avoid common-mode failures, in which one accident could simultaneously disable all the units. To achieve this, among other demands, there must be adequate physical separation between the units and between the control systems for them.[7]

Diverse types of systems. An example of diversity in reactor safety design is the provision of different types of emergency core cooling systems, which act independently.

Defense-in-Depth

A special kind of redundancy is sometimes singled out as being "at the heart of nuclear safety." This is the reliance on *multiple barriers*, or *defense-in-depth*, which is described as "a hierarchally ordered set of different independent levels of protection" [6, p. 109]. The principle of defense-in-depth is seen in considering the barriers that prevent or minimize exposures due to the release of radioactivity from a reactor:

---

[7]After the Browns Ferry fire in 1975, it was recognized that multiple wiring systems, intended for redundancy, were carried in the same cable trays and therefore were all disabled at the same time. A simple solution is to use different paths for redundant cabling.

- The UO_2 fuel pellets retain most radionuclides, although some gaseous fission products (the noble gases and, at elevated temperatures, iodine and cesium) may escape.
- The zircaloy cladding of the fuel pins traps most or all of the gases that escape from the fuel pellets.
- The pressure vessel and closed primary cooling loop are designed to retain nuclides that escape from the fuel pins due to either defects in individual pins or, in case of accident, overheating of the cladding.
- The heavy outer reactor containment, with its associated safety systems, is designed to retain radionuclides that escape from the cooling system, or in the case of a very severe accident, from the pressure vessel.
- If these systems all fail and there is a significant release of activity to the outside environment, the population can be partially protected through evacuation. However, if radiation escapes the containment, then the system has been defeated, even if evacuation reduces the damage.

Steps taken to avoid the overheating of the fuel—in particular, the standard and emergency cooling systems—as well as systems to suppress overpressurization of the containment, can also be considered to be part of the defense-in-depth.

These barriers against radiation exposures have been put to a severe test in only two instances. In the TMI accident, the reactor containment was highly successful. In the Chernobyl accident, there was no containment as the term is understood in Western design practice, and there was a massive release of radioactive material to the outside surroundings; subsequent emergency evacuations reduced the exposure of people in the evacuation zone, but there was substantial exposure of the public nonetheless (see Section 12.3).

Defense-in-depth and the more explicit forms of safety redundancy represent a sophisticated version of the view that while it is likely that *something* will go wrong, it is highly unlikely that *everything* will go wrong. If the causes of the problems are uncorrelated, three barriers that each have a 1% chance of failure can form a system in which there is only one chance in one million of overall failure.[8]

11.1.5 Measures of Harm in Reactor Accidents

The most fundamental harm in reactor accidents is that done to individuals. This can be measured in terms of the individual radiation exposures, the total population exposure, the number of prompt fatalities caused by intense exposures, and the number of latent cancers and genetic defects caused by lower levels of exposure. Of these, prompt fatalities are the most dramatic and least ambiguous, and latent cancers are the most closely linked to overall harm. Thus, studies of reactor safety pay particular attention to these.

There is also the question of physical damage to the reactor plant and contamination of the surrounding environment, which may force evacuation of large regions. Plant damage was clearly the most important direct consequence of the TMI accident, and ground contamination was a major, perhaps in the end *the* major, consequence of Chernobyl.

Two other accident consequences are also commonly cited and are preconditions for those

---

[8]Another way of looking at reactor safety, also sometimes termed "defense in depth," is to divide it into phases of accident avoidance, accident correction or protection, and accident mitigation (e.g., Ref. [7], p. 339). Avoidance is achieved by proper design, maintenance, and operation. Accident correction is achieved by reliable safety systems that, for example, shut the reactor down promptly and alert the operators. Accident mitigation is achieved by, for example, restoration of lost cooling, an effective containment system, and, as a last ditch measure, evacuation of the immediately surrounding population.

already mentioned. One is the probability of core damage through the melting of part of the reactor core. The other is the probability of a large radiation release, involving the failure of barriers provided by the reactor pressure vessel and the reactor containment.

It is difficult to single out any one of these as *the* indicator of reactor safety. In practice, the most significant question may be that of core damage. First, any instance of core damage at least raises the possibility of significant radiation release. Second, such an accident deeply concerns and alarms the public. Finally, the cleanup expense after the core is damaged, even with no release of activity outside the reactor containment, is punitively expensive for the utility. Thus, much of the efforts in assessing and reducing reactor risks focuses on the possibility of core damage.

11.2 CRITICALITY ACCIDENTS AND THEIR AVOIDANCE

11.2.1 Control of Reactivity

In normal operation of a thermal reactor, prompt criticality is avoided. The reactivity of the system is kept low enough to make delayed neutrons crucial for criticality. Thus, even if the reactivity rises, the rates of increase of the neutron flux and of the power output are relatively slow. This gives time for the insertion of control rods, which have high neutron-absorption cross sections and will terminate the chain reaction. In case of a power excursion, the insertion of the control rods can be accomplished by the operator or automatically. The control rods sometimes are arranged to fall under gravitational force, so the reliance on complex control systems can be minimized. For faster insertion, there can be drive mechanisms.

Nonetheless, control rods are not perfect in their operation. Their insertion is not instantaneous; in fact, it may take something on the order of a second or more. There also can be a failure of the release mechanism or a blockage of the path through which the control rod drops. Redundancy can reduce the risks from the malfunction of individual control rods.

11.2.2 Reactivity Feedback Mechanisms

General

Inherent feedback mechanisms provide a rapid and certain response. For example, if the reactivity drops when the temperature of the fuel rises, this gives a negative feedback that enhances stability. With adequate negative feedback, as in standard LWRs, a rapidly developing criticality accident is highly unlikely and could occur only after a very improbable sequence of hardware failures and human errors.

We consider below the major feedback mechanisms that can act to enhance or in some cases reduce reactor safety.[9] Unless otherwise indicated, it will be assumed that the reactor considered is a standard LWR.

Fuel Temperature Feedback: Doppler Broadening

Although we have been tacitly treating the nuclei of the fuel as motionless targets undergoing bombardment by neutrons, this is not a precise description. The uranium nuclei are in

---

[9]This is not intended as a full listing of feedback mechanisms. Additional ones exist, both positive and negative (see, e.g., Ref. [7], p. 145 *ff.*), and must be taken into account in reactor design.

thermal motion, with an average speed that increases as the temperature increases. The result is to increase the effective cross section for neutron absorption in ^{238}U if the temperature of the fuel rises, through the Doppler broadening of the absorption resonances (see Section 3.2.3). The number of neutrons available for fission is reduced, and the reactivity and the reactor power output decrease.[10] This negative feedback comes into play quickly, reversing the rise in power output within less than 0.1 seconds [5, p. 8].

However, the fuel temperature feedback is not automatically negative in all types of reactors. If a fuel has relatively little ^{238}U and is primarily made of fissile material, then the main effect of Doppler broadening is to increase the rate of fission at nonthermal energies, giving a positive feedback. Thus, to keep the fuel temperature feedback negative, the fraction of fissile fuel in liquid-metal fast breeder reactors is kept below 30% [7, p. 146].

Void Coefficients

In an LWR, water is essential for moderating the reaction. If the water is removed—for example, if there is a pipe break and insufficient replacement water is provided—the moderation will be inadequate and the reactivity will drop, because with slower thermalization there will be more loss of neutrons, particularly through capture in ^{238}U. Loss of water in the reactor vessel is the limiting case of a "void."

The term *void coefficient* is usually applied to the replacement of liquid coolant by bubbles, as can occur in a BWR, a water-cooled graphite reactor, or a sodium-cooled reactor. The void coefficient is defined as the rate of change of the reactivity with change in the void fraction. A negative void coefficient means that the reactivity decreases as the volume of steam bubbles increases (i.e., the void fraction increases). The loss of water leads to two effects which contribute to a negative void coefficient: with less effective moderation (i.e., relatively less elastic scattering of neutrons by hydrogen), there will be increases in the resonance absorption of neutrons in ^{238}U and in leakage of neutrons from the reactor.[11]

However, water also acts as an absorber of slow neutrons, and too much water leads to too much absorption (a low thermal utilization factor f). Were this the dominant effect, then the void coefficient would be positive. Thus, there is a competition between the moderating and absorbing roles of water, with opposite feedback signs.

For BWRs, in which steam and water are both present, an increase in the steam content corresponds to less water. The moderating role is more important than the absorbing role, and an increase in steam content decreases the reactivity. Thus, the void coefficient is always negative for BWRs. In PWRs there is no direct void coefficient, but the thermal expansion of water has the same general effect of reducing moderation and providing a negative feedback.

The situation is more complicated for water-cooled, graphite-moderated reactors, and the sign of the feedback can go either way depending on the relative amount of water and graphite. The role of the water as moderator is less important, and the main effect of the water (aside from the intended function of cooling) can be to absorb neutrons. Loss of this water, by conversion to steam or otherwise, can increase the reactivity, i.e., the void coefficient is positive. This was the situation at Chernobyl. However, this is not intrinsic to all water-cooled graphite reactors. In particular, the N reactor formerly operating at Hanford had a negative void coefficient.

---

[10]In terms of the four-factor formula (Eq. 5.5), the resonance escape probability p is reduced.

[11]Referring to the five-factor formula (Eq. 5.4), these feedbacks correspond to a lower resonance escape probability p and a lower non-leakage probability P_L, respectively.

In sodium-cooled fast breeder reactors, the sodium plays only a small role as a moderator, but the small moderating role it plays decreases the reactivity, because the fission cross sections increase with energy for neutron energies in the neighborhood of 1 MeV. Thus, the thermal expansion of the sodium or the development of bubbles reduces the moderation, increases the average energy in the neutron spectrum, and increases the reactivity. At the same time, with less sodium in the path of a potentially escaping neutron, more neutrons can escape from the reactor reducing the reactivity. Overall, these competing effects may leave a sodium-cooled fast reactor with a positive void coefficient, and it is important that there be counterbalancing negative feedbacks (see Section 13.4.2).

11.3 HEAT REMOVAL AND LOSS-OF-COOLANT ACCIDENTS

11.3.1 Decay Heat from Radioactivity

The most serious danger in LWR accidents arises from the need to remove the heat produced by radioactivity during the period after reactor shutdown. The heat production is very substantial.

The magnitude of the initial rate of heat generation can be understood in terms of the total energy release in fission, as discussed in Section 4.4.2. On average, for each fission event, about 7.8 MeV is released in beta decay and 6.8 MeV in accompanying gamma decay, for a total of 14.6 MeV out of about 200 MeV, i.e., approximately 7% of the total energy release. Strictly speaking, this result is applicable only when equilibrium has been reached between the production of radionuclides and their radioactive decay. However, the initial activity is dominated by short-lived radionuclides with half-lives of several days or less. Thus, if a reactor has been operating at full power for, say, a month, the total activity reaches a value close to its equilibrium level. The activity just after shutdown is the same as the activity just before shutdown (treating shutdown as essentially instantaneous), and the initial thermal output from radioactive decay is 7% of the thermal output of the reactor during normal operation.

A more precise calculation must look at the individual radionuclides in detail and also consider the activity from capture products of ^{238}U. Such a treatment gives essentially the same result for the decay heat, about 7% of the thermal output of the reactor or about 20% of the electric output. Thus, at shutdown of a 1000-MWe reactor, the heat output is initially about 200 MW. It drops to about 16 MW after a day and about 9 MW after five days [8, p. S23]. Without cooling, these heat production rates are sufficient to melt the fuel.

11.3.2 Core Cooling Systems

During normal operation, reactor cooling is maintained by the flow of a large volume of water through the pressure vessel. This flow can be disrupted by a break in a pipe, failure of valves or pumps, or, in PWRs, a failure of heat removal in the steam generators. Such accidental disruptions of the normal cooling system are generically termed loss-of-coolant accidents (LOCAs). To guard against the overheating of the fuel in a LOCA, light water reactors have elaborate emergency core-cooling systems intended to maintain water flow to the reactor core.

A distinction is sometimes made between large and small LOCAs. The prototypical large LOCA is a break in the pipes carrying the primary cooling water to the reactor. In a large break, the pressure in the reactor vessel will be lost and a large amount of water will escape. The emergency core-cooling system (ECCS) then comes into play. Initially, replacement water is delivered from "accumulators" driven by nitrogen gas under pressure. Later, low-pressure pumps can provide additional water from external supplies. A large LOCA is a dramatic event, and much of the early concern about reactor safety focused on preventing such an accident and, if prevention failed, assuring an effective and independent ECCS.

A small LOCA may occur from a leak in the primary cooling loop or, as was the case for the initiating event in the TMI accident, from a problem in the secondary cooling loop. Loss of secondary flow means that heat cannot be removed in the heat exchanger from the primary loop. In such an event, the pressure in the reactor vessel may not be relieved, and it may be difficult to establish the flow of replacement water in the complex hydraulic environment created by the mixture of steam and water at high pressures. To cope with such circumstances, the ECCS has a high-pressure injection system to provide replacement water to the reactor vessel.

The effectiveness of the ECCS for both large and small LOCAs has been the subject of many studies, starting before and intensifying after the TMI accident. In addition to calculations and theoretical analyses, there have been extensive tests, particularly the loss-of-fluid test (LOFT) program at the Idaho National Engineering Laboratory. This program was carried out from 1978 to 1985 and involved simulated accidents on a specially built 50-MWt test reactor. This was an NRC facility, but tests were also carried out there for the Nuclear Energy Agency of the OECD. Analyses of the results of these tests of system performance under simulated accident conditions have led to improvements in equipment and procedures (see, e.g., Ref. [4], pp. 39–42.).

11.3.3 Release of Radionuclides from Hot Fuel

If either the normal or emergency core-cooling system operates properly, there will be no damage to the reactor core in case of a reactor malfunction, and no concern about release of radionuclides. However, if the cooling system fails to keep the cladding temperatures low enough to avoid melting, radionuclides will escape into the pressure vessel and into the primary cooling system.

The radionuclides include both fission products and actinides. They can be grouped according to differences in their volatility. The most volatile are the noble gases. These can diffuse out of the fuel into the fuel pins even at normal fuel temperatures. As the fuel temperature rises, damage to the fuel and the cladding causes release of additional elements. The most volatile of these are iodine and cesium. Some other radionuclides, in contrast, are quite refractory and are not released in substantial amounts even under extreme circumstances.

Thus, in one hypothetical accident presented as an example in an NRC study, the median fission product release from the fuel rods was close to 100% for the noble gases, 6% for iodine, 1% for cesium, and 0.05% for strontium [9, p. A-34]. While these particular values cannot be taken as precise measures of what would happen in a specific actual accident, they illustrate the main trends.

Although we have emphasized transport through the cooling system as the main avenue for radionuclide release, as was the case at TMI, there are other possibilities. In one extreme case, molten reactor fuel might settle in the bottom of the reactor vessel, melt through the vessel wall, and penetrate into the concrete base below. This is the essence of the "China syndrome." The main problem is not as extreme as the name might suggest. It comes from the generation of gases (such as CO_2 and others) in the interaction between the molten fuel and the concrete. This could produce an aerosol that carries nonvolatile radionuclides out of the fuel and into the atmosphere of the containment.

If radionuclides escape from the cooling system or from the reactor vessel, the next barrier is the containment structure. The integrity of the containment can be compromised by overpressure, most likely from the buildup of steam. To avoid this, there are containment cooling systems, either passive or active, intended to condense the steam. For example, PWRs commonly have spray systems for condensation, and BWRs have pools of water for pressure suppression. Some units also have refrigeration units. It is also possible, in case of an excessive buildup of pressure, to release gas from the containment through valves, with filters to remove radionuclides.

11.4 ESTIMATING ACCIDENT RISKS

11.4.1 Deterministic Safety Assessment

One approach to establishing and evaluating reactor safety is to establish strict criteria for reactor design and construction and to analyze the behavior of the resulting system for a variety of postulated failures. The more demanding of these scenarios are termed *design basis accidents*. The reactor performance is studied through experiments and computational models to investigate whether the safety systems are adequate to cope with a design basis accident. For example, one can postulate a break in a cooling system pipe and then examine whether the emergency core-cooling systems will provide alternative cooling.

This straightforward approach is called *deterministic safety assessment*, and it is useful in establishing and verifying design criteria for the reactor. A limitation of the approach is that it does not address the question of likelihoods. In particular, it does not consider the probability that the design basis accident will occur nor the probability that the safety system will work as intended. Obviously, a particular sequence of events is more serious if the initiating problems are relatively probable and the safety systems have a relatively high probability of failing.

11.4.2 Probabilistic Safety Assessment

PSA Implementation: Reactor Safety Study, WASH-1400

Answering questions about risk probabilities is not an easy matter in the case of nuclear reactors. For automotive safety, by contrast, it is relatively easy to answer questions about the chances of a fatal accident. One merely has to look at the annual fatality rate, subdivided if one wishes by type of car, road conditions, driver, and so forth. There are ample data on auto fatalities, and these lend themselves to extensive analysis. Thus, there is reasonable quantitative knowledge of the safety of automobiles and roads.

With no fatal accidents and, prior to Browns Ferry and Three Mile Island, with no major accidents of any sort in light water reactors, reactor safety was a matter of conjecture. There had been a 1957 Atomic Energy Commission study (known as WASH-740) on the possible consequences of an accident, but in the early 1970s there was still no good estimate of the *probability* of an accident. A major expansion of nuclear power was then expected in the United States and throughout the world, but although there were many intuitions as to the level of risk, there was no defensible quantitative analysis.

To address this issue, the Atomic Energy Commission sponsored an extensive study under the direction of Norman Rasmussen of the Massachusetts Institute of Technology. This study was issued in draft form in 1974 and in final form in 1975 under the institutional sponsorship of the Nuclear Regulatory Commission, which by then had assumed the regulatory functions of the disbanded AEC. The study is variously referred to as the Rasmussen report, the Reactor Safety Study (RSS), and WASH-1400 [10]. It was the first major study to combine in one analysis the probability and consequences of accidents, in order to assess the *risk* associated with reactor accidents. It was a limited study in that only one PWR and one BWR were analyzed in detail, although the results were often taken to be representative of the situation for other PWRs and BWRs.

The RSS was controversial from the moment the first draft appeared, and the controversies were never fully resolved. However, it is generally agreed that the study made a very important contribution in pioneering the application of methods of *probabilistic risk assessment* (PRA) to the analysis of nuclear reactor safety. In later terminology, especially in international usage, this approach has been called *probabilistic safety assessment* (PSA). The terms are often used interchangeably.[12]

In principle, this approach permits an objective estimate of the *absolute* risk of accidents. However, even if the data and analyses fail to give correct absolute risks, they can be useful in suggesting *relative* risks of different configurations and in pinpointing weaknesses. There are some who argue that the chief value at present of probabilistic safety assessments is in identifying places where safety improvements are needed. In this view, PSAs are more useful for improving reactor safety than for estimating it.

While improvements in reactor equipment and advances in analysis techniques have made the detailed numerical results of the RSS obsolete, they remain of historical significance, and the report itself remains a historic milestone. Subsequent to the TMI accident, numerous steps have been taken to improve reactor safety as well as to refine the analyses, with separate analyses carried out for individual reactors. The general methodology employed in the RSS has been retained and has served in part as a guide to places where reactor design improvements are needed.

Event Trees and Fault Trees

The PSA (or PRA) tools used in the RSS were event tree analyses and fault tree analyses. In an event tree analysis, one imagines the occurrence of some initiating event and traces the possible consequences. We illustrate in Fig. 11.1 the event tree for studying the consequences of a major pipe break, following which the emergency core-cooling system (ECCS) must operate successfully for damage to be avoided [10, Main Report, p. 55]. The worst case would be the electric power failing to operate, the ECCS not functioning, the fission product removal systems within the containment not operating, and the containment integrity being breached.

The probability that everything goes wrong in this sequence is shown in the bottom leg of the "basic tree" in Fig. 11.1. It is the product of five individual failure probabilities. In the "reduced tree," shown in the bottom part of Fig. 11.1, cognizance is taken of the fact that the probabilities are not independent. In particular, the bottom leg of the reduced tree, which

---

[12]For example, the NRC describes the analysis in its study NUREG-1150 as a PRA [9], while the (American) chairman of INSAG terms this study a PSA [11, p. 50].

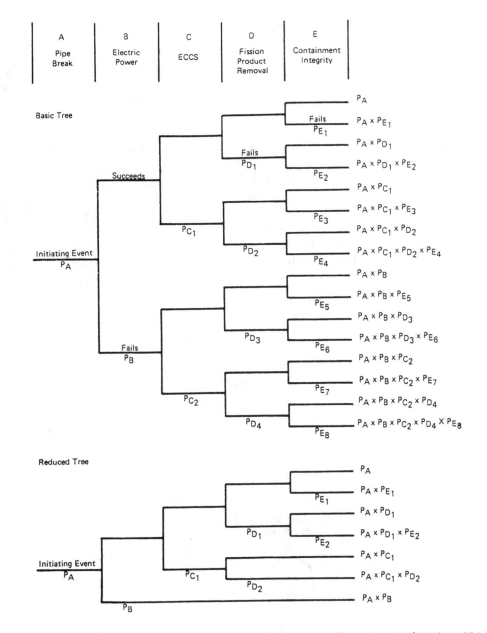

Figure 11.1. Simplified event trees for a large loss-of-coolant accident. (From Ref. [10], p. 55.)

bypasses three steps, is based on the assumption that without electrical power the other systems will also fail and the accident will proceed to the breaching of the containment.[13]

What is the probability that the electric power will fail? That question is answered in principle by a fault tree analysis, diagrammed in Fig. 11.2. For the electric power to fail,

---

[13]Figure 11.1 is a simplified version of the event trees that are actually used and is shown for illustrative purposes.

there must be a loss of *both* the off-site AC power (the standard source) and the on-site AC power (one or more emergency generators). The loss of AC power *or* the loss of DC power (required in this case to control the AC system) would mean that the safety systems would not operate.

In many cases, the individual ingredients for the event tree and fault tree analyses come from an extensive data base—for example, the rate of failure of a given type of valve or motor that may be widely used outside of the nuclear power industry. In other cases—for example, estimating the probability of human error—the input numbers are likely to be only rough surmises.

There is also the matter of *common-mode* failures, the fact that individual failure probabilities are not all independent. It is necessary, but not necessarily easy, to identify sequences in which the failure of one system enhances the likelihood of the failure of others. One such instance is illustrated in the diagram sketched in the bottom part of Fig. 11.1. Once failure of electric power is assumed, then all other systems are assumed to fail, greatly reducing the number of branches to be studied.

Combining the outcomes of the event tree analyses and the fault tree analyses gives the probability for an accident scenario. Some scenarios will represent accidents with large releases of radioactivity to the environment; others will represent small releases. The overall

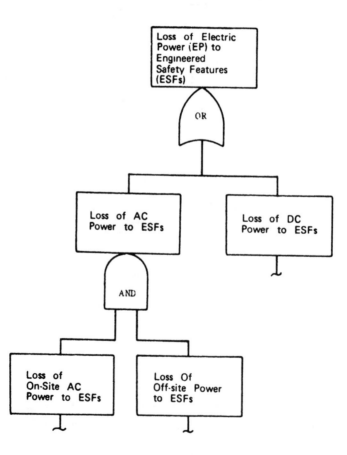

Figure 11.2. Fault tree for loss of electric power. (From Ref. [10], p. 56.)

results of the study can be embodied in graphs or tables in which the probability of an accident of a given or greater severity is plotted as a function of the severity of the accident.

Analysis Procedure in NUREG-1150

Another major step in the development of reactor safety analyses in the U.S. was the study leading to the publication by the NRC of the 1990 report *Severe Accident Risks: An Assessment for Five U.S. Nuclear Power Plants*, also known as NUREG-1150 [9]. Five LWRs were analyzed in detail for this study. These are in some sense typical of LWRs in the United States, but the reported results are specific to the individual reactors.

In this analysis an explicit distinction was made between *internal* and *external* events. Internal events are those due to malfunctioning of components of the reactor, including its control systems. External events are those initiated by things that happen outside the reactor, for example, earthquakes.[14] External events were analyzed for only two of the five reactors.

The analysis in NUREG-1150 was divided into several separate stages:

Accident frequency. The goal here is to estimate the probability that the reactor core is damaged. The starting point is to identify possible initiating events and assess their frequency. For events due to internal system failures, accident probabilities were determined through a combination of event tree and fault tree analyses. Human error and "dependent failures" (also known as common-mode failures) were included. For events due to external hazards—including earthquakes, fire, and aircraft impacts—somewhat analogous procedures were used, but the data base for the initiating events is not as good, and there is a greater chance of a simultaneous failure of several components.

Accident progression. Given damage to the reactor core, it is important to know what further damage occurs. Thus, probabilities were estimated for the breaching of the reactor vessel and for either breaching of the concrete containment or leaks through it.

Transport of radioactive material. Given damage to the fuel and to the reactor vessel or cooling system, there will be a transfer of radionuclides to the reactor building. Release to the environment then depends upon whether or not the containment fails. The noble gases are the most likely to be released, with virtually all escaping given sufficient fuel damage. For other radionuclides, the release rates depend upon the volatility of the element, ranging from high for iodine and cesium and low for ruthenium and strontium. The aggregate of total releases to the environment is called the *source term* (see Section 11.5).

Off-site consequences. The radiation doses received by people outside the reactor depend upon the magnitude of the source term, the transport by the plume of radioactive material, and the details of the pathways by which radiation exposures occur. Doses and health consequences were calculated for a variety of assumptions as to the evacuation of the surrounding population.

Integrated risk analysis. An overall integrated risk is found by summing over the array of probabilities for each of the various stages.

The ultimate result of this analysis is a probability distribution for the risk of occurrence vs. the magnitude of the consequence, for each adverse consequence of interest. Thus, the

---

[14]The distinction is not clean, and customarily loss of power from off-site sources is included as an internal event, while floods and fires within the plant are termed external events [9, p. 2-4].

result might be the probability distribution for exceeding various levels of population dose or of latent cancer fatalities. Although such a probability distribution cannot be fully represented by a single number, both medians and mean values are given in NUREG-1150 to provide some easily encapsulated overall perspective.[15]

The Role of Probabilistic Safety Assessment

In the original Reactor Safety Study, considerable emphasis was put on the absolute magnitude of the reactor accident risks. Uncertainties in the analysis were explicitly indicated, but there were criticisms that these had been underestimated. In NUREG-1150, the issue of uncertainties was featured more prominently.

Despite the difficulty of making precise estimates of reactor risk with PSA techniques, and the uncertainties that surround their results, they appear to offer the best available approach to risk estimation. As improvements are made in analysis methods, reactor accident scenarios, and input data on failure rates, there can be increasing confidence in the applicability of the results. However, ambivalence remains, as reflected in comments made in a report prepared by the Nuclear Energy Agency of the OECD:

> *Probabilistic safety assessment* (PSA) is a powerful technique for providing a numerical assessment of safety. It is being increasingly used as a guide for comparing levels of safety. As such it complements the deterministic approach to safety assessment, but it is not considered as an absolute measure of safety for regulatory purposes. . . .
>
> But the importance of PSA is not so much in the final answer that it gives for the chance of accidents. Its main value lies in the insights that are obtained in the process of the analysis. It will highlight those elements in a chain of events which contribute significantly to the probability of serious accidents—the weak links—and which if strengthened will therefore give a significant improvement in overall safety [4, p. 63].

There appears to be little dissent from the view that PSA studies give useful information on relative risks and on the identification of "weak links." However, their role in providing estimates of the absolute magnitude of the risks or specific criteria to be used by agencies such as the Nuclear Regulatory Commission in making regulatory decisions appears to be in flux.

11.4.3 Results of the Reactor Safety Study

Summary of Results

The results of the RSS included estimates of the probability distributions for a variety of forms of harm: early fatalities, early illness, latent cancer fatalities, thyroid nodules, genetic effects, property damage, and magnitude of the area in which relocation and decontamination would be required. The most serious consequences are those of latent cancer, the cancer deaths expected to occur eventually from cumulative radiation exposures. The exposures might be received mostly in the first few days or year following the accident, but the cancer fatalities would appear for three decades, generally starting after a latent period of 10 years.

The results of the analysis for various consequences were displayed in plots of their probability of occurrence per reactor-year (RY). The probability of an event that would

---

[15]The mean is in general higher than the median, because the probability distribution for a given consequence generally has a tail extending to high magnitudes.

cause more than 1 latent cancer per year (30 during 30 years) was about 3×10^{-5}/RY. The probability dropped to 2×10^{-6}/RY for more than 100 latent cancers per year and to under 10^{-8}/RY for more than 1000 latent cancers per year. Large uncertainties were indicated for both the probabilities and the number of cancer fatalities.

To provide perspective, the RSS also compared the risks from reactor accidents to those from other sorts of accidents or natural mishaps. For these other accidents, there are few data on latent effects. Perhaps the trauma of a nonfatal airplane accident increases one's chance of dying 30 years later, but this is not customarily included as a fatal consequence of airplane accidents. Thus, a direct comparison between nuclear power and other hazards is made simpler, although incomplete, if consideration is restricted to early fatalities. For a nuclear reactor accident, these would be primarily caused by very high early radiation exposures. In Fig. 11.3 the annual risks from 100 reactors are compared with the annual risks from other causes, such as airplane accidents and dam failures. For example, Figure 11.3 indicates that there is one airplane accident causing 100 or more fatalities every three years, while a nuclear reactor accident with this early toll is expected only once every 80 000 years.[16]

The probability of a core melt was estimated to be 5×10^{-5} per reactor-year [10, p. 135]. The most probable cause of a core melt was found to be not a break in the large pipes providing the main cooling water but rather an accumulation of smaller failures. This was surprising in view of prior prevailing beliefs. An upper bound on the core melt probability was put at 3×10^{-4}, or about 1 per 3000 reactor-years of operation.

Responses to the Reactor Safety Study

The RSS was received very differently by different groups. Nuclear power advocates greeted it enthusiastically as a vindication of their belief in nuclear safety. It was possible to draw all sorts of dramatic comparisons from it, and these were gleefully put forth: for example, that there was less chance of being killed by an accident in a nearby nuclear power plant than by an errant automobile, even if you were neither in a car nor crossing a street yourself. Nuclear opponents greeted the RSS with strong criticism and even scorn. It was not surprising, in their view, that a study sponsored and carried out by the "nuclear establishment" would conclude that nuclear power was safe.

An influential critique of the RSS was done by a special review group, commissioned by the Nuclear Regulatory Commission and chaired by Harold Lewis of the University of California at Santa Barbara [12]. The main conclusions of the Lewis report were: (1) the methodology used in the RSS was basically sound; (2) significant mistakes had been made, for example, in some of the statistical methods; (3) it was difficult to balance the instances of conservatism and nonconservatism; (4) the uncertainties were much greater than those quoted in the RSS; (5) the executive summary was misleading; and (6) the panel could not conclude whether the probabilities of a reactor core melt were higher or lower than those quoted in the RSS.

The Lewis report was regarded by some as a "repudiation" of the RSS, and the NRC backed away from using it as a guide for regulatory decisions. However, Lewis himself took a consistently "pro-nuclear" position in congressional testimony, stating that he felt "the plants are actually safer than stated in the Rasmussen report."[17] Lewis made this last statement, which reiterated earlier statements by him in this vein, in May 1979, shortly *after* the

---

[16]It should be noted that with accidents of this magnitude, the consequences other than early fatalities are likely to be much more severe for the reactor accident than for the airplane accident.

[17]References and further quotations are given in Ref. [13].

TMI accident. But TMI made such studies at least temporarily irrelevant. Quite apart from the merits and demerits of studies by academic scientists and engineers such as Rasmussen and Lewis, a significant fraction of the public concluded after TMI, and all the more after Chernobyl, that nuclear reactors were not safe enough for nearby siting. That conclusion has had a profound influence on the subsequent pace of nuclear power development.

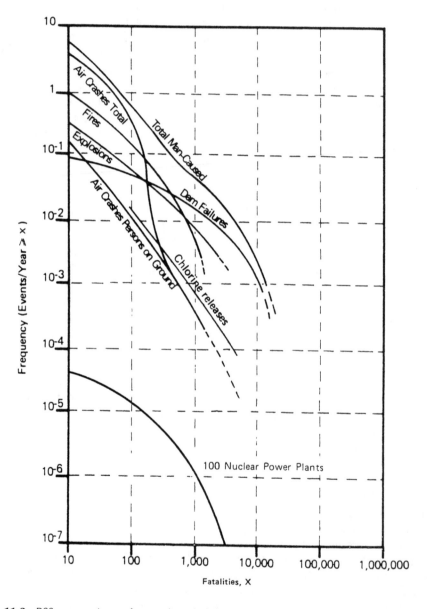

Figure 11.3. RSS comparison of annual probabilities of accidents causing X or more (early) fatalities: 100 nuclear reactors compared to other "man-caused" events. (From Ref. [10], p. 119.)

Hindsight on RSS Predictions

It is possible to look back with the benefit of hindsight on the RSS predictions.[18] As of the end of 1994 there had been roughly 1900 reactor-years of LWR operation in the United States. During this time the expected probability of a core melt was, from the discussion above, $1900 \times 5 \times 10^{-5} = 0.1$. Allowing for the stated uncertainty, the predicted upper bound was about 0.5. The actual number of core melts has been one (TMI).

The comparison given in the previous paragraph is not quite appropriate, however. The RSS was specific to reactors as they existed in the 1970s. Therefore, its results should be compared to reactor performance in that period. Subsequent studies, such as that of NUREG-1150, should be used for comparisons to more recent and presumably improved performance. As of the end of 1979 there had been less than 600 reactor-years of LWR operation in the U.S., and the "anticipated" accident rate above should be reduced by about a factor of three. With either base of comparison, one could infer that the average estimates of core melt probabilities given in the RSS were too low.

If one goes beyond average estimates, another interesting viewpoint emerges. As indicated above, the RSS studied certain reactors in detail, and these were taken as representative of all LWRs. In particular, the PWR analysis was based on the Surry 1 power plant, manufactured by Westinghouse. According to the RSS, using this as prototypical for all PWRs would "tend to overestimate, rather than underestimate the risk," because this was a relatively old plant and newer ones would on average be safer. However, as discussed in a subsequent study of the implications of the TMI accident, conducted under the auspices of the American Physical Society:

> The first reaction of many observers to the accident was that the Reactor Safety Study methodology was completely wrong because it had not predicted that type of accident would soon occur. The particular sequence. . . was calculated for the Surry facility. . . to have a frequency of once in 10^5 years. Yet. . . if the RSS procedures had been applied to a Babcock and Wilcox reactor like TMI-2, the methodology would have predicted a frequency of occurrence of one in 300 years. Babcock and Wilcox reactors had an operating history of about 30 reactor years. The differences stemmed from: (a) the pressure relief valve settings that caused the valve to be released before reactor scram and (b) the fact that the steam generators had a small heat capacity and dried out in ten minutes, compared with a time of about an hour calculated for the Westinghouse reactors such as Surry. . . if the methodology had been applied to the reactor at Three Mile Island, the plant-specific scenario differences might have been noted, modifications might have been made, and the accident perhaps avoided [14, p. S11].

In short, the trouble was not with the RSS analyses *per se*, but the failure to apply the analyses to all reactors, individually.

11.4.4 Results of NRC Analysis: NUREG-1150

Core Damage Probabilities

The NUREG-1150 study is more pertinent to the present situation, because it uses more advanced analysis techniques and considers reactors as they were after a period of considerable upgrading. Some results of the NUREG-1150 analysis are summarized in Table 11.1, for the five reactors studied.

---

[18]The RSS was for LWRs in the United States, and it is therefore appropriate to restrict comparison to their record.

Table 11.1. *Estimated probabilities of reactor accidents, for reactors studied in NUREG-1150. Probabilities are given as mean probabilities in units of 10^{-6} per reactor-year of operation.*

	Reactor Studied				
	Surry 1	Zion 1	Sequoyah 1	Peach Bottom 2	Grand Gulf 1
Reactor type	PWR	PWR	PWR	BWR	BWR
State located	VA	IL	TN	PA	MI
Capacity (MWe)	781	1040	1148	1100	1142
Start of commercial operation	1972	1973	1981	1974	1985
Core damage probability (10^{-6}/RY):					
Internal events	40	60^a	57	4.5	4
Seismic events^b	54			15	
Fires	11			20	
Large release probability^c (10^{-6}/RY):					
Internal events	4	6	7	2	1

Sources: Capacity data and commercial operation dates are from Ref. [15]. Internal events data are from Ref. [9], pp. 3-4, 7-4, 5-4, 4-4, and 6-5. Seismic events and fire data are from Ref. [9], pp. 3-4, 4-4. Large release probability data are from Ref. [9], p. 9-6.

^{a}This number reflects plant modifications after study was initiated [9, p. 7-4].

^{b}Seismic core damage probabilities are geometric means of results based on Livermore and EPRI earthquake hazard estimates. Individually, these give mean core damage probabilities (in units of 10^{-6}/RY) of 116 (Livermore) and 25 (EPRI) for Surry 1, and 77 (Livermore) and 3 (EPRI) for Peach Bottom [9, Table C.11.2].

^{c}Only given for internal events. A large release corresponds to early failure of the containment, or bypass of the containment with escape through the cooling system.

The mean calculated probability of core damage from internal events varied from 4×10^{-6} RY in the best case to 60×10^{-6} RY in the worst case, with a rough (arithmetic) average of about 3×10^{-5} RY. The probability of core damage was considerably lower for the two BWRs than for the PWRs, although the study cautioned that it would be "inappropriate" to conclude that this was true in all cases [9, p. 8-11]. Nonetheless, some advantages of BWRs were pointed out, particularly more redundancy in the emergency core-cooling systems.

The main causes of core damage differed for the different reactors [9, p. 8-3].[19] For two (Zion and Sequoyah), loss-of-coolant accidents were the most important factor. For two (Surry and Grand Gulf), loss of power (station blackout) was the main factor. For one (Peach Bottom), roughly equal responsibility was placed on loss of power and failure of control rod insertion during transient disturbances.

Core damage due to external causes was considered for only two of the reactors, Surry 1 and Peach Bottom. In both cases, seismic events and fires were the only significant external sources of risk. The data of Table 11.1 might suggest that the external risks are greater than the risks due to internal failure. However, such a conclusion may be premature. For one, the risk for seismic events is highly uncertain, with two analyses considered in NUREG-1150 differing substantially.[20] Further, the seismic risk distributions are very broad and are skewed so that the median risks are considerably lower than the mean risks [9, p. 8-6].

---

[19]A compact summary is given in Ref. [16], p. 3-7.

[20]In Table 11.1 the geometric mean is given of results based on earthquake probability estimates by the Electric Power Research Institute (EPRI) and by Lawrence Livermore National Laboratory. The mean core damage probability for the Livermore estimates exceeded the value for the EPRI estimates by a factor of 5 for Surry and a factor of 25 for Peach Bottom.

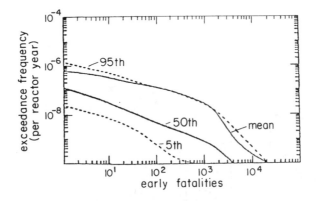

Figure 11.4. Probability of exceeding a given number of prompt fatalities in nuclear reactor accidents per reactor-year, labeled by percentile level. (Adapted from Ref. [9], Fig. 2.8.)

Consequences of Accidents for Human Health

Health effects were also calculated for these reactors. Some overall results are displayed in Figs. 11.4 and 11.5, in the form of distribution functions that give the probability of the specified consequence being exceeded. It is seen from Fig. 11.4 that very few prompt fatalities are expected. The mean expected frequency of an accident that causes one or more prompt fatalities is about 6×10^{-7}/RY, and the median expectation (50th percentile) is about 10^{-7}/RY. Even at the 95th percentile level, it is calculated that there will be fewer than two accidents in 10^6/RY that will cause one or more prompt fatalities.

Latent cancers from low-level radiation have a greater calculated potential for causing fatalities, as seen in Fig. 11.5. The doses for the most part are small and the calculated effects depend upon the model adopted for the effects of low-level radiation (see Section 2.5). A

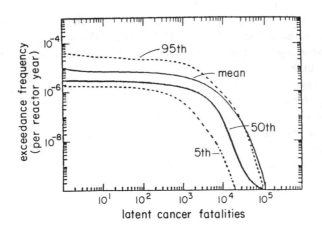

Figure 11.5. Probability of exceeding a given number of latent cancers in nuclear reactor accidents per reactor-year, labeled by percentile level. (Adapted from Ref. [9], Fig. 2.8.)

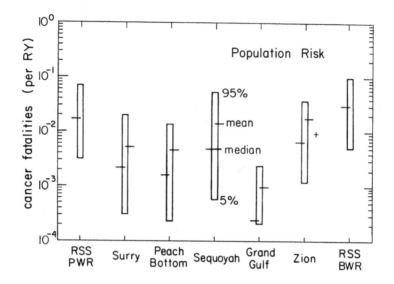

Figure 11.6. For surrounding population, cumulative risk of latent cancer death from internally initiated reactor accidents, per reactor-year. For Zion, the cross denotes a revised mean value, reflecting recent plant modifications; for the RSS only the median value is indicated. (Adapted from Ref. [9], Fig. 12.1.)

linear dose-response model, based in part on the BEIR-III study, was used [9, p. A-38]. In the resulting calculation, the mean expectation for 100 or more eventual cancer fatalities is roughly 10^{-5} per reactor-year. This would translate for 100 reactors to a 0.1% chance per year of an accident of this magnitude.

Figures 11.4 and 11.5 give composite representative results. There are considerable differences among reactors, as might be anticipated from the disparities seen in Table 11.1. For each of the reactors considered, there is less than one chance in 10^{6}/RY of a single early fatality. For Peach Bottom and Grand Gulf, there is less than one chance in 10^{7}/RY of such an accident [9, p. 11-3].

On the other hand, there are appreciably higher risks of latent cancers, as already seen in the comparison between Figs. 11.4 and 11.5. These risks are displayed in a somewhat different fashion in Figs. 11.6 and 11.7, which give the expectations for latent cancer fatalities for each of the reactors.

The mean risks to the population of latent cancer vary from about 10^{-3}/RY to 2×10^{-2}/RY, an improvement over the levels estimated in the RSS (also shown in Fig. 11.6). For 100 reactors, one would expect 0.1–2 latent cancers per year in the combined surrounding populations. Of course, the risk to a single individual is much less, as seen in Fig. 11.7. Even in the worst case, it is only 1 in 10^{8}/RY, a negligible risk if the estimates of NUREG-1150 are taken seriously, and if externally initiated events (particularly earthquakes) do not significantly add to the risk. This calculated level is far below the NRC "safety goal" of 2×10^{-6}/RY for individual risk.

Of course, it must be remembered that the calculated cancer fatality rates, as in Figs. 11.5 to 11.7, may somewhat underestimate the true effects if the assumed cancer risk coefficients are too small, or may greatly overestimate the true effects if there is little or no cancer induction at low doses.

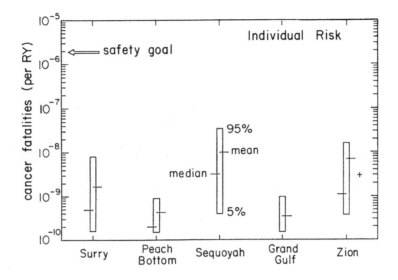

Figure 11.7. For an individual living in the surrounding region, risk of a latent cancer death from an internally initiated reactor accident, per reactor year. For Zion, the cross denotes a revised mean value, reflecting recent plant modifications. (Adapted from Ref. [9], Fig. 12.2.)

Seismic Risk

A striking aspect of Table 11.1 is the relative importance of seismic risks in the tabulated core damage frequencies. For the two reactors for which the seismic core damage probability was estimated, the indicated probabilities exceed those of internal events. However, there are very large uncertainties in these results.

Determination of the seismic core damage probabilities involves both the probability of earthquakes of various magnitudes at the reactor site, the so-called seismic hazard, and the ability of the reactor to withstand the resulting ground accelerations.[21] The NUREG-1150 calculation used seismic hazard assessments from both Livermore and EPRI. There is no conclusive method for predicting earthquake probabilities, and both the Livermore and EPRI studies relied upon an array of expert evaluations.[22]

Differences in these evaluations in the two studies led to substantially different mean results for the core damage probability and to broad probability distributions. For example, for the Livermore seismic hazards, the 5th and 95th percentiles in the core damage frequency distribution differ by more than a factor of 1000 for the Surry plant, and the median probability is only about one-eighth of the mean probability [9, p. 8-6]. The large width of the distribution makes any "average" result a very poorly defined quantity. Overall, the seismic risks estimated in the Livermore study exceeded those for the EPRI study (see footnote in Table 11.1).

---

[21]The methods used in these analyses are described in detail in Ref. [17] and Ref. [9], Section C. 11.

[22]The EPRI and Livermore studies were both part of a major program undertaken in the late 1980s to assess seismic hazards, in the region of the U.S. to the east of the Rocky Mountains, where the large majority of the reactors are situated. These hazards are ultimately couched in terms of site-specific probability distributions for ground acceleration.

Subsequent to the publication of NUREG-1150, a new Livermore study of seismic hazards was carried out, which in an overall sense moved the Livermore results in the direction of the EPRI results [18]. The disparities between different seismic hazard evaluations stimulated a comprehensive new study on the methodology for seismic hazard analysis, supported by the DOE, NRC, and EPRI [19]. The study is expected to lead to a new round of hazard analyses employing the suggested methodologies. However, while some of the disparities may be reduced, it is anticipated that significant uncertainties will still remain in the prediction of earthquake probabilities and in the resulting reactor hazard estimates.

Despite the emphasis here on probability risk assessments for evaluating reactor earthquake risks, this is not the chief approach adopted by the NRC. Instead, *seismic margin* methodology is being used for some present and all future reactors. This method is somewhat less demanding in terms of the analysis required and may be a reasonable approach given the large uncertainties in estimating earthquake probabilities. The starting point is the specification of the so-called *safe shutdown earthquake* (SSE).[23] This is an earthquake whose magnitude is based on the "maximum earthquake potential" in the vicinity of the site. The reactor must be designed to shut down safely should this largest expected earthquake occur.

If a reactor can withstand an earthquake more severe than the SSE, then the reactor has a "seismic margin." The extent of the seismic margin is based on a reference earthquake, more severe than the SSE, for which there is a "high confidence of a low probability of failure" (HCLPF).[24] For example, if the SSE corresponds to a peak ground acceleration of 0.3 g at the reactor site, the seismic margin condition might be established by demonstrating fulfillment of the HCLPF condition for an acceleration of 0.45 g. The NRC requirement for new reactors will be that they demonstrate an adequate seismic margin.

Although the safety study NUREG-1150 included estimates of seismic core damage probabilities for two of the five reactors, no estimates were given of large release probabilities for seismic effects. The rationale was that if an earthquake is severe enough to damage a reactor, there will be severe damage to other structures such as buildings and dams [9, p. 1-4]. In the absence of good techniques to estimate the effects of this other damage on the surrounding population, it apparently seemed one-sided to try to calculate the off-site effects of reactor accidents alone.

Overall, while considerable effort has gone into making nuclear reactors "safe" against earthquakes, there has been difficulty in quantifying the level of safety. In a quite different approach, some designs for new reactors incorporate seismic isolation between the reactor and the surrounding ground, with the goal of decoupling the reactor from possible ground motion.

11.4.5 Observed Reactor Performance

Comparison to Predictions of Core Damage

Estimates of core damage frequency based on PSAs cannot be checked against actual experience because there has not been an LWR accident that has caused core damage since the TMI accident in 1979. Thus, the predictions concerning post-TMI reactor performance

---

[23]The definition and use of the safe shutdown earthquake is discussed in Appendix A to Part 100 of Ref. [20]. The SSE had been previously called the *design basis earthquake*.

[24]The HCLPF criterion can alternatively be established by deterministic or probabilistic determination of failure modes. In the latter case, it is assumed to correspond to a greater than 95% confidence that the failure probability is less than 5% [17, p. 5-4].

cannot be tested, although a rough upper limit can be set. In the 1980–1994 period there were roughly 1300 reactor-years of operation in the United States, with no core damage. This record indicates that the core damage frequency was not much greater than 10^{-3} per reactor-year. However, so high an upper limit considerably exceeds both what is acceptable and what is predicted, and therefore it is not very useful.

The data base of course substantially increases if one looks at all LWRs throughout the world, as the number of LWRs in the world is more than three times the U.S. total. Again, there has been no core damage. But it is not appropriate to compare worldwide experience to estimates made in the NRC studies of individual U.S. reactors. The reactors studied by the NRC may be somewhat representative of U.S. reactors as a whole, but reactors in other countries are likely to have different safety performance. The guiding principles are the same, but the regulatory and construction practices are not necessarily similar. Reactors in foreign countries could be more safe or less safe than U.S. reactors,[25] and their performance should be compared to more directly specific studies.

Analysis of Precursor Events

While the PSA predictions on core damage have not, and one hopes will not, be put to the ultimate test, the PSA techniques can be used in a different sort of comparison with actual experience. Short of core damage, there are continuing malfunctions of reactor equipment, spanning a wide range of severity, which the utilities are required to report to the NRC. These events can be considered to be potential precursors of worse accidents.

The implications of the events are studied in the so-called Accident Sequence Precursor Program carried out for the NRC at Oak Ridge National Laboratory. The events are analyzed by incorporating them into the PSA formalism and calculating a conditional core damage probability for each. From the array of events, it is then possible to calculate an overall "inferred mean core damage frequency." This core damage index is an overall measure of reactor performance.

This index suggests that there has been a considerable improvement in reactor safety since the 1970s [21, 22]. The semi-empirical index dropped from an average of 3×10^{-3} per reactor-year (RY) for 1969–1979 to an average of under 5×10^{-5}/RY for 1986–1991, an improvement of more than a factor of 60 [21].

A more detailed analysis shows that the index is dominated by a few events of relatively great severity (high conditional core damage probability) rather than by a large number of relatively minor events [23]. This leads to large year-to-year fluctuations in the index. Nonetheless, there is an overall consistency. In every year in the 1970s, the index exceeded 10^{-4}/RY, and in five of these years it exceeded 10^{-3}/RY. In contrast, in every year of the 1987–1991 period, the index was below 4×10^{-5}/RY.

The NRC attributes the improvement to a variety of changes in equipment and training instituted after the TMI accident [23]. Such a program illustrates one of the main benefits of the PSA approach: It makes it possible to identify and correct potential weak links. Overall progress can be measured in terms of the number of near and not-so-near misses.

---

[25] As an example of the variability, German authorities shut down on safety grounds all reactors in the former East Germany following the reunification of Germany. These were all PWRs supplied by the Soviet Union. There is a fear that other reactors still operating in Eastern Europe and the former Soviet Union, including LWRs, do not meet current international safety norms.

Frequency of Scram Events

Any significant aberration in reactor performance triggers the insertion of the control rods and the shutdown of the reactor. Such an emergency shutdown is called a *scram*. Scrams are a means of avoiding serious accidents, but the need for a scram is an indication of a problem. Therefore, the number of scrams per year suggests how well a plant is operating. This number has dropped in the United States during the 1980s. Measured in terms of the number of scrams per 7000 hr of reactor operation,[26] the rate fell from 7.3 in 1980 to 1.3 in 1991 [4, p. 31] and has since dropped to 0.8 in 1994 [24].

11.5 THE SOURCE TERM

One surprising aspect of the TMI accident, as discussed in the following chapter, was the fact that although the damage to the reactor fuel assemblies was very great, there was very little release of radioactive material to the environment outside the reactor containment. This caused attention to be directed to the accident *source term*. The source term is the full inventory of radionuclides released to the environment in a nuclear reactor accident.

For ^{131}I and other iodine isotopes, the source term at Chernobyl was essentially the total initial core inventory. At TMI it was close to zero: 18 Ci out of 64 MCi, in one estimate [23, p. 1-12]. The RSS had assumed a very much greater release. The explanation of the low release at TMI is obvious in retrospect. There is much more cesium than iodine in the core inventory, because the half-lives of the most abundant isotopes are greater. The iodine remained in the containment because it formed CsI, rather than the volatile gas I_2, and was trapped by dissolution in water or deposition on surfaces (see, e.g., Ref. [25], p. 8-9).

A crucial question is whether this very good performance of the containment system is generic to *all* LWRs or was peculiar to TMI. There have been a number of studies of this matter, for example, studies carried out under the auspices of the American Nuclear Society (ANS) [25] and the American Physical Society (APS) [14]. The ANS study reached an overall conclusion:

> reductions in the source term from estimates reported in the 1975 pioneering Reactor Safety Study (WASH-1400) could range from more than a factor of ten to several factors of ten for the critical fission products in most of the accident scenarios that have been recently considered [25, p. ES-1].

The APS study in effect reached a similar, but more tentatively stated, conclusion:

> In a number of cases, new calculations indicate that the quantity of radionuclides that would reach the environment is significantly lower than calculated in the Reactor Safety Study [14, p. S127].

These appear to be the more likely cases. However, the APS study did suggest a mechanism that might *increase* the release.[27] It suggested that further study was needed to determine the behavior of all radionuclides and of each type of reactor.

Although neither the ANS or APS study provides unequivocal assurance that the source term will always be small in LWR accidents, they suggest that it often may be. Understanding the source term more definitively requires further research. More importantly, a

---

[26]This is equivalent to one reactor-year for a 1000-MWe reactor operating at an 80% capacity factor.

[27]This is the release of nonvolatile radionuclides in case of a core–concrete interaction (see Section 11.3.3). The APS report indicated that the understanding of the matter was inadequate.

better understanding of the source term can in principle lead to designs that will increase the chance that the source term will in fact be small.

REFERENCES

1. International Atomic Energy Agency, *Nuclear Power Reactors in the World*, Reference Data Series, no. 2 (Vienna: IAEA, April 1993).
2. Bengt Pershagen, *Light Water Reactor Safety* (Oxford: Pergamon, 1989).
3. U.S. Environmental Protection Agency, *Accidents and Unscheduled Events Associated with Non-Nuclear Energy Resources and Technology*, report EPA-600/7-77-016 (Washington, D.C.: EPA, 1977).
4. OECD Nuclear Energy Agency, *Achieving Nuclear Safety: Improvements in Reactor Safety Design and Operation* (Paris: OECD, 1993).
5. Uranium Institute, *The Safety of Nuclear Power Plants: An Assessment by an International Group of Senior Nuclear Safety Experts* (London: 1988).
6. International Atomic Energy Agency, *The Safety of Nuclear Power: Strategy for the Future* (Vienna: IAEA, 1992).
7. Ronald Allen Knief, *Nuclear Engineering: Theory and Technology of Commercial Nuclear Power*, 2nd ed. (Washington, D.C.: Hemisphere Publishing, 1992).
8. "Report to the APS by the Study Group on Light-Water Reactor Safety," H. W. Lewis, chairman, *Reviews of Modern Physics* 47, supplement no. 1 (1975).
9. U.S. Nuclear Regulatory Commission, *Severe Accident Risks: An Assessment for Five Nuclear Power Plants*, final report NUREG-1150, vols. 1 and 2 (Washington, D.C.: 1990).
10. U.S. Nuclear Regulatory Commission, *Reactor Safety Study: An Assessment of Accident Risks in U.S. Commercial Nuclear Power Plants*, report WASH-1400 (NUREG 75/014) (Washington, D.C.: 1975).
11. H. Kouts, "The Safety of Nuclear Power," in *The Safety of Nuclear Power: Strategy for the Future* (Vienna: IAEA, 1992), 47–54.
12. U.S. Nuclear Regulatory Commission, *Risk Assessment Review Group Report to the U.S. Nuclear Regulatory Commission*, report NUREG/CR-0400, H. W. Lewis, chairman (Washington D.C.: 1978).
13. David Bodansky, "Risk Assessment and Nuclear Power," *Journal of Contemporary Studies* 5, no. 1 (winter 1982), 5–27.
14. "Report to the APS of the Study Group on Radionuclide Release from Severe Accidents at Nuclear Power Plants," Richard Wilson, chairman, *Reviews of Modern Physics* 57, no. 3, part 2 (July 1985).
15. "World List of Nuclear Power Plants," *Nuclear News* 37, no. 3 (March 1994): 43–62
16. American Nuclear Society, *Report of the Special Committee on NUREG-1150, the NRC's Study of Severe Accident Risks* (LaGrange, Ill.: ANS, 1990).
17. Electric Power Research Institute, *Use of Probabilistic Seismic Hazard Results: General Decision Making, the Charleston Earthquake Issue, and Severe Accident Evaluations*, EPRI report TR-103126, prepared by Risk Engineering Inc. (Palo Alto, Calif.: EPRI, 1993).
18. U.S. Nuclear Regulatory Commission, *Revised Livermore Seismic Hazard Estimates for 69 Nuclear Power Plant Sites East of the Rocky Mountains*, draft report NUREG-1488 (Washington, D.C.: 1993).
19. U. S. Nuclear Regulatory Commission, Recommendations for *Probabilistic Seismic Hazard Analysis: Guidance on Uncertainty and Use of Experts*, report NUREG/CR-6372, prepared for LLNL by the Senior Seismic Hazard Analysis Committee, Robert J. Budnitz, chairman, to be released in 1996.
20. *Energy, U.S. Code of Federal Regulations*, title 10 (1993).
21. T. E. Murley, "Developments in Nuclear Safety," *Nuclear Safety* 31, no. 1 (1990): 1–9.
22. T. E. Murley, "Current Issues in Reactor Regulation," MIT Reactor Safety Course (Cambridge, Mass.: July 1993).
23. "Changes in Probability of Core Damage Accidents Inferred on the Basis of Actual Events," Nuclear Regulatory Commission staff report (Washington, D.C.: April 24, 1992).
24. Nuclear Energy Institute, "1994 Another Stellar Year for U. S. Nuclear Industry," *Nuclear Energy Insight* (April 1995): 4.
25. American Nuclear Society, *Report of the Special Committee on Source Terms* (La Grange, Ill.: ANS, 1984).

Chapter 12

Nuclear Reactor Accidents

12.1 REACTOR ACCIDENTS PRIOR TO THREE MILE ISLAND

Despite the largely successful precautions taken to avoid nuclear reactor accidents, the record is not perfect. We list below the more important known reactor accidents, excluding accidents in submarine reactors and possible accidents in the former USSR and Soviet Bloc countries, other than the Chernobyl accident.[1]

The decision as to which accidents qualify as "major" accidents is somewhat arbitrary. In particular, ordinary industrial non-nuclear accidents are omitted. For example, in 1972 two workers at the Surry Power Station were fatally scalded by steam escaping from a faulty valve. This did not involve the nuclear components of the power station and therefore is not pertinent to the broader issue of nuclear reactor safety. The major past accidents are:

Chalk River, Canada (1952). There was a partial meltdown in a 30-MWt experimental reactor. The reactor was cooled by light water and moderated by heavy water. The accident was initiated by operator errors and a failure of the control rod system. This led to an elevated power output and some boiling and loss of cooling water. In an LWR the accident would have been mitigated by a negative void coefficient (see Section 11.2.2), but in this case the feedback was positive. The reactor was eventually shut down by draining the heavy water moderator. There were no known injuries or deaths, but the reactor core was damaged, and there was escape of an unspecified amount of radioactivity [4, p. 101].

National Reactor Testing Laboratory, Idaho (1955). The 1.4-MWt experimental breeder reactor, EBR-I, suffered a 40% to 50% core meltdown during a test in which the power level of the reactor was intentionally raised but, due, to operator error, was not

---

[1]There are reports of reactor accidents in these countries prior to the much larger Chernobyl accident (see Refs. [1] and [2]), although accounts of their course and magnitude are in dispute. In addition, there was a major release of radioactive material in an accident in 1957 at the Kyshtym nuclear complex in the Urals. The accident was a non-nuclear explosion in tanks of reprocessed radioactive wastes, not a reactor accident. It led to the evacuation of 10 730 people and caused a collective effective dose of about 2500 person-Sv [3, p. 116].

reduced promptly. There was little contamination of the building, no injuries occurred, and the release of radioactive material was "trivial" [4, p. 103].

Windscale, England (1957). Overheating and fire occurred in a graphite-moderated reactor used for plutonium production. The accident began in the course of heating the fuel above normal operating temperatures to release energy stored in the graphite crystal lattice.[2] This energy is a consequence of radiation damage to the graphite, a problem that arises in graphite-moderated reactors if they operate below the temperature necessary for annealing radiation damage. In this case, the heating and the energy release, although intentional, were too rapid. The reactor was shut down with control rods, but the heating had been sufficient to cause a fire in the uranium fuel and eventually in the graphite. The fire smoldered for about five days, until extinguished by flooding with water [6]. The most serious consequence was the release of about 20 000 Ci of ^{131}I ($T=8.0$ d), which was carried by winds over much of central and southern England. The estimated consequences for England and continental Europe are 260 thyroid cancers and 13 thyroid cancer fatalities over a period of 40 yr, plus 7 additional fatalities or hereditary effects [7, p. 24].

National Reactor Testing Laboratory, Idaho (1961). Three army technicians were killed when one of them apparently manually removed a control rod from a 3-MWt test reactor, known as SL-1, on which they were working. Reactors of this type were intended for heating and electricity production at remote sites, and they were so primitive that the control rods could be moved by an operator standing on top of the reactor. There was a rapid increase in reactor output, followed by a steam explosion, leading to lethal levels of radiation within the reactor building. Most, but not all, of the activity was contained within the building [4, p. 109].

Fermi Reactor, Detroit (1966). There was a partial meltdown in a 200-MWt (61-MWe) commercial breeder reactor, which was a one-of-a-kind prototype. The cause was a blockage in the flow path of the sodium coolant. There were no injuries or significant release of radioactivity, and the reactor was briefly put back into operation before final shutdown in 1973 [4, p. 32].[3]

Lucens, Switzerland (1969). There was partial fuel melting in a 30-MWt experimental reactor, due to loss of CO_2 cooling. There was severe damage to the reactor but no radiation release beyond permitted levels [4, p. 121].

Browns Ferry 1, Alabama (1975). A fire in the electrical wiring did extensive damage to the control systems and threatened the reactor, but the reactor was turned off and cooling maintained with no radiation release and no injuries other than one individual receiving a minor burn from the fire. Despite the absence of damage to the reactor itself, this accident was of importance because it was the first major accident in a commercial LWR, and it demonstrated a serious vulnerability in the control systems of that period due to inadequate redundancy.

Three Mile Island, Pennsylvania (1979). This accident is discussed in more detail in Section 12.2.

Chernobyl, USSR (1986). This accident is discussed in more detail in Section 12.3.

---

[2]The storage and release of energy in a graphite moderator is the so-called *Wigner effect* (see, e.g., Ref. [5]).
[3]For two very different assessments of the significance of this accident and the level of hazard it created, see Refs. [8] and [9].

Table 12.1. *Major nuclear reactor accidents.*

Year	Reactor	Purpose	Capacity (MW)	Environmental consequences		
				Radioactivity release	Prompt deaths	Delayed cancers[a]
1952	Chalk River	Experimental	30 (t)	some	0	0
1957	Windscale	Pu production		large	0	≈13–20
1961	Idaho Falls	Test (army)	3 (t)	small	3	0
1966	Fermi I	Demo breeder	61 (e)	very little	0	0
1969	Lucens	Experimental	30 (t)	very little	0	0
1975	Browns Ferry 1	Power	1065 (e)	none	0	0
1979	TMI-2	Power	906 (e)	small	0	≈0–2
1986	Chernobyl	Power	1000 (e)	very large	31[b]	≈50 000[b]

[a]Indicated cancers are calculated cancer fatalities, using standard estimates of effects of low-level radiation.
[b]See Section 12.3.4 for further discussion of Chernobyl fatalities.

Some aspects of these accidents are summarized in Table 12.1.

The only accidents that caused a clearly identifiable loss of life were at Idaho Falls, where three workers died from the effects of the explosion and radiation, and at Chernobyl, where 31 operating and fire-fighting personnel reportedly died within about two months from radiation and from thermal burns. In addition, it is estimated that there will be a small number of eventual, or "delayed," cancer fatalities from the Windscale radiation release and a large number of delayed fatalities from Chernobyl. There already have been reports of additional deaths occurring from Chernobyl, with widely varying numbers quoted (see Section 12.3.4). It may be noted that none of these three reactors were commercial LWRs, and except for Chernobyl, the accidents took place more than 30 yr ago. Their history therefore has only limited pertinence to the present safety of commercial LWRs or of other non-Soviet commercial reactors.

12.2 THE THREE MILE ISLAND ACCIDENT

12.2.1 The Early History of the TMI Accident

The Three Mile Island (TMI) accident occurred in one of two similar reactors at the Three Mile Island site in Pennsylvania.[4] The accident was in the second unit, known as TMI-2. It was a 906-MWe pressurized water reactor built by Babcock and Wilcox, the smallest (in terms of number of units completed) of three U.S. manufacturers of PWRs. It had first received a license to operate at low power in February 1978 and was in routine operation at full power by the end of 1978. A schematic diagram of the TMI-2 facility is shown in Fig. 12.1.

The accident started with a failure of the cooling system of TMI-2 in the early morning of

---

[4]Extensive studies were carried out after the accident. One, referred to below as the "Kemeny Report," was by a commission appointed by President Carter and chaired by John Kemeny, the president of Dartmouth College [10]. The second, the Rogovin Report, was by a special inquiry group instituted by the Nuclear Regulatory Commission and chaired by Mitchell Rogovin, a partner in an independent Washington law firm. The description here is drawn largely from the Kemeny Report [10] and Part 2 of Volume II of the Rogovin Report [11], as well as a further review article [12].

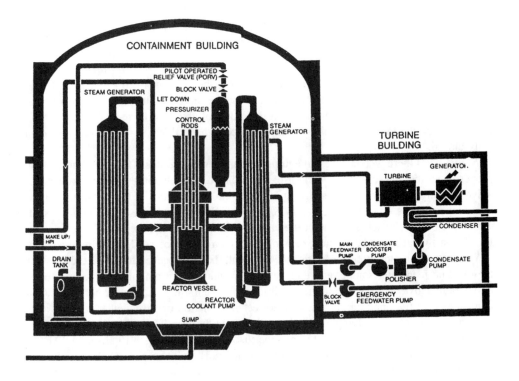

Figure 12.1. Schematic diagram of the TMI-2 facility, including reactor building and turbine building. Piping goes, from right to left in the diagram, through the containment building in to the auxiliary building (not shown); piping also goes, from left to right, through the turbine building to the condensate storage tank and cooling tower (not shown). (From Ref. [10], pp. 86–87.)

March 28, 1979. The initial problem was an interruption in the flow of water to the secondary side of the heat exchanger in the steam generator. This water is the so-called feedwater. In the secondary loop, feedwater enters the steam generator, and steam emerges to drive the turbine. The steam is condensed in a second heat exchanger (the condenser), and water is returned to the steam generator after passing through a "polisher," in which dissolved impurities are removed. The flow of water between the condenser and steam generator is maintained by the condensate pump and the main feedwater pump (see Fig. 12.1).

The chain of events that led to the accident appears to have been initiated by work done to clean the polishers. In a sequence that has not been conclusively established, this operation may have caused one or more of the valves in the condensate polisher system to close, automatically shutting off (tripping) one of the condensate pumps. The tripping of the condensate pump, whatever the cause, in turn tripped the main feedwater pumps.[5] This failure caused the emergency feedwater pumps to start automatically, in order to maintain the flow of water to the steam generator. Maintenance of feedwater flow is essential to cool the water from the reactor; this water flows through the primary side of the steam generator.

Up to this point, everything was "normal," in the sense that reactors are designed to handle occasional equipment failures; protection then comes from backup systems. However,

---

[5]Figure 12.1 does not show the redundancy in the system. There were two main feedwater pumps and three emergency feedwater pumps.

the block valves in the emergency feedwater lines (there were two) were closed; according to proper operating procedures, they were supposed to be open. Indicator lights in the control room showed the closed status, but the operators at first did not notice this. Thus, no water was being fed to the secondary side of the steam generator, because the pumps for the main supply were off and valves in the emergency line were closed. With no flow of water, the pressure in the steam generator rose, and in response the pilot-operated relief valve (PORV) on the "hot" side of the steam generator heat exchanger opened. The pressure excursion also caused the reactor to trip, with automatic insertion of the control rods. With the reactor turned off and the PORV open, the pressure dropped; the PORV should then have automatically closed.

At this point there were additional equipment and design failures. The PORV did not close properly, but the control panel indicator light displayed the status of the control power to the valve (namely, that it was supposedly closed), not the actual status of the valve (namely, that it was open). Thus, the operators had to cope with unusual conditions in the cooling system without knowing the actual status of the valves in it. In particular, the PORV remained open for almost two and one-half hours, causing a very large loss of needed cooling water.

Within two minutes after the start of the accident, the steam generators boiled dry, because they had no feedwater source, and there was a substantial heat output from the reactor core due to radioactive decay. Overall, the conditions of the cooling system were both unusual and confused, with the operators not having correct information or sufficient training to recognize the nature of the evolving anomalies and cope with them. They did recognize that there were serious problems, and by 4:45 am supervisory personnel began to arrive at TMI, only three-quarters of an hour after the start of the accident. By 6:22 am the PORV was closed but the problems were not over. At 7:00 am a "site emergency" was declared because there had been some release of radioactivity.

12.2.2 Evolution of the TMI Accident

Over the next few days the accident continued to unfold, with continued difficulty in establishing proper cooling conditions. There were some small releases of radioactivity outside the plant and one misinterpreted report of radiation levels that led to the incorrect belief that there had been a large release. There was a great sense of emergency both at the site and in the surrounding area, as no one was willing to give unequivocal assurances that matters were under control. This led to a recommended evacuation of pregnant women and preschool children from the immediate vicinity and a large self-initiated evacuation by individuals.

Concern reached a peak on Saturday, March 31, over the possibility of a hydrogen explosion inside the pressure vessel. As described in the subsequent Kemeny Report:

> The great concern about a potential hydrogen explosion inside the TMI-2 reactor came with the weekend. That it was a groundless fear, an unfortunate error, never penetrated the public consciousness afterward, partly because the NRC made no effort to inform the public it had erred [10, p. 126].

Hydrogen is produced by the reaction of steam with the zircaloy cladding at high temperatures. Oxygen is formed by the breakup of water under radiation, so-called radiolysis. Together, hydrogen and oxygen can form an explosive mixture. There was fear that such an explosion could occur within the pressure vessel. Within a day or so, some NRC experts came to the conclusion that a hydrogen explosion was impossible, but this conclusion was not immediately accepted by all the authorities. In the meantime the hydrogen bubble had become a matter of great public concern, a concern not unambiguously dismissed by the NRC. However, by 6:00 pm on April 1 the hydrogen was removed from the bubble by "letdown, leakage, and venting" [11, p. 535]. It never had been the threat that had been believed.

The reason that the problem was not a real one was an insufficient accumulation of oxygen. In a PWR it is normal to have some hydrogen dissolved in the water and to have continued recombination of oxygen and hydrogen. This recombination prevented the amount of oxygen from rising sufficiently to create a danger of explosion. This point is brought out in the Rogovin Report:

> Little or no oxygen was present in the bubble and a very low probability of explosion existed. The incorrect perception of an explosion hazard stemmed from contradiction among supposed experts. This perception was known or should have been known to be false by the afternoon of April 1 [11, p. 535].

It may be noted that a Babcock and Wilcox scientist had given assurances from the first that there was no problem from oxygen production [11, p. 534], but apparently this assurance did not receive much attention.

President Carter visited the site on Sunday, April 1, the hydrogen bubble itself dissipated (although not the perception of a near miss with hydrogen) and the worst of the crisis apparently was over. But it took more than another week for the advisory evacuation of pregnant women and preschool children to be withdrawn by the governor of Pennsylvania.

12.2.3 Effects of the TMI Accident

Core Damage and Radionuclide Releases

In retrospect, several major aspects of the Three Mile Island accident were not fully appreciated at the time and might seem to be in conflict:

- There was very little release of radioactivity and very little exposure of the general population. According to the Kemeny Commission, "the maximum estimated radiation dose received by any one individual in the off-site general population (excluding the plant workers) during the accident was 70 millirems. . . . three TMI workers received radiation doses of about 3 to 4 rems; these levels exceeded the NRC maximum permissible quarterly dose of 3 rems" [10, p. 34]. In essential agreement, the Rogovin Report found that "the maximum off-site individual dose was less than 100 mrem" [11, p. 400].

- The total collective dose to the two million people living within 50 miles of TMI was approximately 2000 person-rem (20 person-Sv).[6] From this, the Kemeny Commission estimated a 50% chance of no fatal cancers from the accident, a 35% of one fatal cancer, and a 15% chance of more than one [10, p. 12]. These results correspond to an average expectation of 0.7 cancer fatalities. If the 1993 NCRP risk estimate of 0.05 per Sv is adopted, then one fatal cancer is calculated for the collective dose of 20 person-Sv (see Section 2.5.4). Among these two million people, it is expected that 325 000 will die of cancers unrelated to TMI, so the TMI impact, if any, will be far below any detectable level.

---

[6]See Ref. [10], p. 34 and Ref. [11], p. 399.

- The core damage was very great. As cleanup and dismantling of the TMI-2 reactor proceeded, it was found that the core damage was greater than originally thought, and some observers have expressed surprise that the pressure vessel itself withstood the molten fuel at its bottom.

Thus, those who thought that the accident was causing, or was about to cause, large releases of radioactive material were proven to be wrong. Those who thought that matters were being exaggerated and that there was relatively little actual damage to the reactor were also wrong. The biggest surprise, however, was having these two outcomes together—great core damage and almost negligible radionuclide releases. Only about 15 Ci of ^{131}I were released [11, p. 358], despite an initial core inventory more than one million times greater. It had been commonly assumed that with core damage of this magnitude a very large fraction of the iodine would escape. Thus, the concrete containment system performed unexpectedly well. In the aftermath of TMI, understanding this performance—now part of what is known as the source term question—became a major issue in reactor safety studies (see Section 11.5).

Studies of Health Effects of TMI

The release of radioactivity from the Three Mile Island plant and the resulting radiation exposures were too small to have produced any observable effects, if one accepts official accounts of the magnitude of the releases and standard dose–response relationships. One or even ten cancer deaths would be lost among a total of over 300 000 "natural" cancer deaths. Nonetheless, there have been persistent claims of health problems from TMI. In response to some of the early concerns, the Pennsylvania Health Secretary stated in a news release: "After careful study of all available information, we continue to find no evidence to date that radiation from the nuclear power plant resulted in an increased number of fetal, neonatal, and infant deaths. That simply isn't the case" [13]. This was based on an examination of death rates near TMI and in Pennsylvania as a whole, before and after the accident.

The Pennsylvania Department of Health carried out a later study of spontaneous abortions, in the face of a rather widespread belief among residents of the TMI area that there had been an increase in stillbirths and miscarriages [14]. The study identified 479 women living within 5 miles of the plant who were pregnant at the time of the accident. For this group, there were 436 live births, 28 spontaneous abortions or stillbirths, and 15 other abortions. The rate of spontaneous abortions and stillbirths was compared to the rate expected from earlier studies of nonexposed populations, and it was found that there was no excess. In the author's words, the TMI incidence rates "compared favorably with the four baseline studies."

In a broader study of cancer rates near Three Mile Island, investigators found a statistically significant increase in cancer incidence, compared to rates at greater distances, during 1982 and 1983 [15].[7] However, this excess did not persist, and in 1985 the cancer incidence rate was slightly lower for the near-TMI group than for the more distant group. Further, the increase was seen only in cancer *incidence*, not in cancer fatalities. This lack of increase was commented on particularly for lung cancer, which progresses rapidly from incidence to fatality, as pointing to possible "screening bias." The authors concluded:

> we observed a modest postaccident increase in cancer near TMI that is unlikely to be explained

---

[7]The authors were from the School of Public Health at Columbia University, with the exception of one from the Audubon Society.

by radiation emissions. The increase resulted from a small wave of excess cancers in 1982, three years after the 1979 accident. Such a pattern might reflect the impact of accident stress on cancer progression. Our study lacked a direct, individual measure of stress, however. The most plausible alternative explanation is that improved surveillance of cancer near the TMI plant led to the observed increase.

These results are consistent with the belief that there is virtually no possibility that there have been or will be observable health effects from radioactivity released in the TMI accident, given the low exposure levels. However, the post-TMI history illustrates the extent of skepticism about official reassurances in situations of possible radiation hazard. This skepticism is fed by anomalies in the data (such as the increase in observed cancer incidence in 1982). Anomalies often cannot be explained in any conclusive fashion, and the ruling out of radiation exposure as the cause may hinge on somewhat indirect arguments, such as comparisons to standard models of the time displacement between radiation exposures, cancer incidence, and cancer fatalities. The families and friends of the "victims" of the anomalies may have little incentive to accept these arguments.

These difficulties may be of only marginal interest in the case of TMI, where the weight of evidence and scientific opinion is strong, but they could assume much greater importance in evaluating the Chernobyl accident, where the exposures were very much greater and the conditions for systematic epidemiological studies are poorer. It is probable that there will be large health consequences from Chernobyl, and it is possible that some will be observable, but it may prove difficult to assess the validity of individual reports and to resolve the disagreements that will arise.

12.3 THE CHERNOBYL ACCIDENT

12.3.1 The Chernobyl Reactors

Among reactor accidents, the Chernobyl accident in 1986 stands alone in its magnitude.[8] The design features of "Chernobyl-type" reactors are unique, and it is the standard assumption of nuclear analysts that a similar accident could not occur in any of the other types of reactors operating today. Nonetheless, Chernobyl demonstrated the seriousness of a near "worst-case" accident. It intensified pre-existing public concern about reactors of all sorts and strengthened the position of those who oppose nuclear power on safety grounds.

The Chernobyl reactor was one of the Soviet RBMK-1000 reactor series, designed to operate at a (gross) capacity of 1000 MWe but since down-rated to 925 MWe. These are graphite-moderated and water-cooled, of a type originally used in the USSR (and with important differences, in the United States) for the production of plutonium. Such reactors were also used for the generation of electricity in the USSR, dating back to a 5-MWe water-cooled, graphite-moderated reactor at Obninsk, put into operation in 1954 [17, p. 9].

At the beginning of 1986 there were four RBMK-1000 reactors at the Chernobyl site in the Ukraine, located about 130 km north of the major city of Kiev. The most recently installed reactors, completed in 1983, were Units 3 and 4, housed in a single building. At the time of the accident all four reactors were operating and two more were under construction. The accident itself occurred in Unit 4 on April 26, 1986, at 1:24 am. Unit 3 was turned off by the operators after almost 5 hours, and the nearby Units 1 and 2 were turned off after about 24 hours. Units 1 and 2 were returned to operation in late 1986 and Unit 3 in December 1987.

---

[8]The account given here is based on parts of Refs. [16] through [22].

Construction was suspended on the two reactors being built, and they have since been cancelled [23].

As of mid-1995 Units 1 and 3 were still in operation at Chernobyl, although there has been Western pressure for a shutdown. Unit 2 was closed, following a fire on October 1, 1991. The fire was in a non-nuclear part of the plant, and there was no release of radioactivity, but there was damage to the engine room [24]. Outside Ukraine, there are 11 RBMK-1000 reactors operating in Russia (4 at Kursk, 4 near St. Petersburg, and 3 at Smolensk) and two larger RBMK reactors (1380 MWe) at Ignalina in what is now Lithuania [25].

Although the USSR in the past exported reactors to its neighbors, these were all PWRs, not RBMKs. Outside the USSR, the only reactor bearing some similarities to the Chernobyl reactors was the not so very similar N reactor at Hanford, which stopped operations in 1988. One of the RBMK-1000 units at Smolensk was put on-line in 1990, so the Chernobyl accident did not stop all building of the RBMK series, although future Russian plans are based on LWRs. In an effort to reduce the chance of another accident, steps have been taken since 1986 to improve operator training, and significant modifications have been made in the RBMK reactors themselves (see below).

Nonetheless, there have been calls for the shutdown of all reactors at Chernobyl, because some people living in the surrounding area, as well as some foreign safety experts, consider them a continuing hazard. Authority over the fate of the Chernobyl reactors lies with the Ukrainian government, and there has been a great deal of oscillation in its decisions, arising from the conflict between the need for electrical power and concern over safety. At the end of 1995 the Ukrainian authorities reached a preliminary, not yet binding, agreement with concerned foreign countries to shut down the Chernobyl reactors by the year 2000 in exchange for aid "to develop Ukraine's energy sector, including completion of two safer nuclear plants" [26].

12.3.2 History of the Chernobyl Accident

Deficiencies in Attention to Safety

There was no crisp single cause of the Chernobyl accident. Even six years later a review by a major international technical body, the International Nuclear Safety Advisory Group (INSAG), reported: "It is not known for certain what started the power excursion that destroyed the Chernobyl reactor" [22, p. 23]. The quotation may suggest more ignorance than is the case. A great deal is known about the conditions before and during the accident, even if the development of a definitive, detailed scenario has been made difficult by the complexity of the reactor conditions, the speed of unfolding of the accident, and the damage itself.

Overall, the accident was the result of a combination of design deficiencies, operator errors, and an unusual set of prior circumstances, all of which put the reactor and the operators to a test that they failed. In the INSAG view, "the accident can be said to have flowed from deficient safety culture, not only at the Chernobyl plant, but throughout the Soviet design, operating and regulatory organizations." The INSAG report lists a whole gamut of weaknesses in institutions and attitudes [22, p. 24]. In this view, with greater vigilance in both design and operations, there would have been no accident.

Design Weaknesses

Two aspects of the design have been particular targets of criticism: a positive void coefficient of reactivity and an improperly configured control rod system. Both contributed positive feedbacks, which turned an initial excursion in reactor performance into the Chernobyl disaster.

The effects of voids are considered in more detail in Section 12.3.6. In brief, the cooling water acts (whether one wants it to or not) as both a moderator and a poison. In an LWR, when water is lost or steam bubbles develop, the reactivity drops, corresponding to the negative void coefficient discussed in Section 11.2.2. However, in a Chernobyl-type reactor most of the moderation is provided by the graphite, and the water acts mainly as a poison. When some of the cooling water is replaced by steam, there is a different balance than in an LWR between the competing feedbacks: increased resonance absorption in ^{238}U (negative void coefficient), increased neutron leakage from the reactor (negative void coefficient), and decreased absorption of thermal neutrons in H_2O (positive void coefficient). The net outcome depends upon the relative amounts and arrangement of the uranium, carbon, and water in the reactor. For the Chernobyl reactor the overall void coefficient was positive, while at the same time a graphite-moderated water-cooled reactor in the United States, the Hanford-N reactor, had a negative void coefficient.

The second major defect, involving the control rods, also is related to the role of water. At the onset of the accident, most of the control rods had been fully withdrawn from the reactor. Remarkably, the first effect of the insertion of the control rods from the full-out position was to *increase* the reactivity. This was due to a peculiarity of the RBMK-1000 control rod system, which has since been corrected. The control rods move vertically through the reactor core and are withdrawn by being lifted upward. To prevent the control rod from being replaced by water, which acts as a poison and lessens the effect of withdrawing the rod, a long graphite "displacer" was attached to the bottom of the rod. When the rod was fully withdrawn, most of the channel in the core was occupied by this graphite, and the control material was above the core. However, the graphite displacer did not completely fill the channel; instead there was still a 125-cm column of water below the graphite displacer [22, p. 4]. The first effect of inserting the control rod, before the absorbing part of the control rod reached the core, was for the graphite to drive out the water column below it, increasing the reactivity by reducing the poison. The full motion of the control rods was slow, and by the time the control rod proper entered the core region, it was too late.

Since Chernobyl there have been a number of changes to correct these defects in the RBMK reactors. These include increasing the enrichment of ^{235}U in the fuel, changing the control rod geometry so that the displacer will not displace water when inserted, and speeding up the control rod insertion.

Reactor Operations Prior to Accident

The accident evolved from a test that disturbed normal operating conditions. Ironically, the test was undertaken to demonstrate a safety feature of the reactor. Power for the pumps and other plant facilities normally comes from the plant's own turbogenerator units or from the off-site power grid. Should there be an off-site power outage and a shutdown of the reactor, standby diesel generators at the plant come on-line to supply power. There could be an interval of several seconds between the loss of normal power and the startup of the diesel generators. The test was to demonstrate that the inertial coasting of the turbogenerator would provide sufficient power to operate pumps during this interval.

The first step taken to perform this test was a reduction of reactor power to one-half of its normal 3200 MWt, beginning at about 1:00 am on April 25. One of the turbogenerators was switched off at 1:06 am, and the power reached 1600 MWt at 3:47 am [22, p. 53]. The remainder of the test, involving a further reduction of power, was to start at about 2: 00 pm. As part of the test, the emergency core-cooling system was disconnected at 2:00 pm.

However, the reactor's power was required for the electricity grid fed by Chernobyl, and instructions were given to postpone the further power reduction. The test did not resume until 11:10 pm on April 25. It was then intended to reduce the power to about 700–1000 MWt, but there was an overshoot in the shutdown and the power level dropped to 30 MWt. By about 1:00 am on April 26, it had been brought back to 200 MWt, but the period of operation at low power caused a buildup of xenon poisoning (see Section 5.5.3), which was compensated for by removal of a large number of the control rods, more than proper under operating guidelines.[9]

At this point there were at least two unusual circumstances: the power level for the test was lower than planned, and the margin of safety, in terms of the ability to shut down the reactor with the control rods, was less than the normal operating limit. In addition, a number of safety systems had been turned off to facilitate the planned test. In hindsight, it is clear that the test should have been terminated at this point, but it was continued.

Initiation and Progress of the Accident

As the test proceeded at low power, water flow conditions were not normal, there was some decrease in steam, and the reduced reactivity caused automatic control rods to withdraw further to restore the reactivity. This was a manifestation of the fact that the Chernobyl reactor operated with a positive void coefficient.[10]

This first manifestation of the positive void coefficient was in itself benign, although compensated for by a raising of the control rods. At 1:23:04 am, despite warning indications of the dangerous control rod configuration, the operators initiated the turbine test by shutting a valve and reducing steam flow to the turbine. The resulting changes in steam pressure and in water flow from the cooling water pumps led to a decreased water flow through the core and some boiling in the core. The displacement of water by steam caused the reactivity to rise.

In response, at 1:23:40 am, an emergency shutdown (scram) was attempted. But the control rods had been withdrawn too far to take immediate effect, and, due to the graphite displacers at their ends, their first effect was to increase rather than decrease the reactivity. Within 3 sec there was a sharp increase in the neutron flux and power output, as the reactor went prompt critical. Within not more than 20 sec, there were two large explosions, one apparently a steam explosion that exposed the reactor fuel to the air and the other explosion due to exothermic reactions, including interaction of liberated hydrogen and carbon monoxide with the air.

---

[9]After power is reduced, decay of ^{135}I ($T=6.61$ hr) continues, but the destruction of ^{135}Xe ($T=9.09$ hr) by neutron capture is decreased. Therefore, the amount of ^{135}Xe increases for several hours.

[10]According to Richard Wilson [19], Russian experts explained to him that this design was adopted for convenience in establishing the graphite configuration. For power levels of 20% of normal or higher, the negative fuel temperature coefficient (the Doppler coefficient) was supposed to provide an adequate margin of safety, and at lower power levels there were to be stringent operating regulations.

Table 12.2. *Estimated radionuclide releases from the Chernobyl accident, for selected radionuclides.*

Isotope	$T_{1/2}$	Core (MCi)	Release[a] (MCi)	Fraction released
^{85}Kr	10.8 yr	0.67	0.67	1.0
^{133}Xe	5.25 d	170	170	1.0
^{131}I	8.02 d	82	46	0.6
^{137}Cs	30.2 yr	6.2	2.4	0.4
^{90}Sr	28.8 yr	4.6	0.2	0.04

Source: Core data from Ref. [20], p. 3.6. Release data from Ref. [20], pp. 3.6, 4.15.
[a]The actual release was less for short-lived isotopes, because of decay before release; these numbers are "corrected" back to the activity at the time of the accident.

These explosions breached the reactor building (there was no true containment) and sent burning fragments into the air, which started fires on the roof of the reactor. Firefighters from the nearby towns of Chernobyl and Pripyat arrived shortly after the accident and put out the building fires by 5:00 am. There had been a threat that the fires would spread to the other units at Chernobyl (Units 1, 2, and 3), but this was prevented by firemen working under extreme conditions of heat and radiation exposure. Remarkably, Unit 3 was not turned off until about 6:00 am.

Although the exterior fires had been extinguished, the problem of heat generation in the reactor continued, due to (chemical) burning of the fuel and graphite in the reactor and to radioactive decay heat. These interior fires were not extinguished until May 6, following a series of attempts to quench them by dropping massive amounts of boron carbide (intended to prevent recriticality), limestone, lead, sand, and clay. In total, Unit 4 was entombed under about 5000 tons of material, and the acute phase of the accident was then over.

12.3.3 Release of Radioactivity from Chernobyl

The initial explosions and subsequent reactor fires caused large amounts of radioactive materials to be released from the reactor. The release was not all immediate, with about 24% the first day, 28% over the next five days, and 48% over the following four days [20, p. 3.9]. The release was mainly of the volatile nuclides, including the noble gases, iodine, and cesium. Much less of the nonvolatile nuclides, such as strontium, escaped. Estimated release fractions and total releases are given in Table 12.2 for some of the most important radionuclides [20].[11]

The cloud of radioactivity from the accident spread generally to the north and west and thus did not severely impact Kiev. The accident was not immediately made public, and the first awareness outside the USSR came from radiation measurements in Sweden and Finland. The cloud reached Sweden at about 2 pm on April 27 and was first detected about 18 hr later by monitors at the Forsmark nuclear power station [17, p. 13]. This was approximately two days after the start of the accident itself.

---

[11]A 1990 review of existing studies of the Chernobyl radionuclide source inventory and releases [27] indicated considerable disagreements as to their magnitudes. The core inventories presented in Table 12.2, from Ref. [20], fall within the range of estimates of other cited studies; the indicated releases of ^{131}I and ^{137}Cs are higher than those given in most of the other studies.

12.3.4 Radiation Exposures at Chernobyl and in Vicinity

Effects on Plant Workers and Firemen

The accident led to the death within several months of 31 people from severe radiation sickness, burns from beta-particle radiation, and burns from the fire. These deaths were all among plant personnel and fire-fighting personnel. The latter appear to have performed in an exceedingly dedicated and self-sacrificing manner. One of the deaths was immediate, 1 within 5 hr, and 29 others after a period of hospitalization of up to 2 months. A total of 203 people were hospitalized, including the 29 who died. The deaths among those hospitalized were strongly correlated with whole-body radiation exposures. There were 21 fatalities among the 22 people who received exposures above 6 Sv, and 8 fatalities among the 181 people exposed to doses below 6 Sv. For exposures below 2 Sv, there were no deaths from acute radiation sickness [20, p. 2.2].

There have been subsequent statements, along with denials, that a large number of additional fatalities and sickness have occurred among workers who participated in the cleanup following the Chernobyl accident. A delegation from the Commission of the European Community visited Ukraine in November 1993 and reported 11 subsequent deaths. Eight of these were among workers at the site (seven firemen and one helicopter pilot) and three were cases of thyroid cancer among children [28]. It is anticipated that many additional deaths will occur from these exposures in the future, as is to be expected given the long latent period for radiation-induced cancer.

Exposure of Population in the "Affected Region"

On the day after the accident about 45 000 people were evacuated from Pripyat, three kilometers from the accident. Eventually, a total of about 135 000 people were evacuated from a zone within 30 km of the reactor, to avoid high acute exposures of the nearby general population. This is the so-called prohibited zone.

In addition, there was an intermediate population for whom further evacuation became the adopted policy goal, in view of the possibility of prolonged exposures from radionuclides in the ground, particularly ^{137}Cs. This was the population outside the initial evacuation zone but still relatively near Chernobyl, in the so-called affected region.

An assessment of the consequences of the Chernobyl accident for this intermediate population was undertaken by an international committee under the sponsorship of the International Atomic Energy Agency, with the assistance of the World Health Organization, the United Nations Scientific Committee on the Effects of Ionizing Radiation, and other international organizations. This study was in response to a October 1989 request from the USSR government, apparently prompted by concerns among people living in the vicinity of Chernobyl but not close enough to have been among the 135 000 originally evacuated from the prohibited zone. The affected area was the region where ground surface concentrations of ^{137}Cs exceeded 5 Ci/km^2. It embraced an area of about 25 000 km^2, with a population of about 825 000.

An overview of the results of this study, known as the International Chernobyl Project (ICP), was published in the spring of 1991 [29]. The chairman of the committee was Itsuzo Shigematsu, of the Radiation Effects Research Foundation in Hiroshima. Given the motivation of the study, radiation exposures within the prohibited region were not examined, nor were exposures considered for workers involved in the immediate cleanup of Chernobyl.

The main conclusion of the study was that the Soviet authorities, presumably in the interests of conservatism, had *overestimated* the levels of radioactivity in the region rather than underestimated them, as some of the local population suspected.

The isotope of greatest concern was ^{137}Cs. The ICP values for ^{137}Cs in the soil agreed reasonably well with earlier Soviet estimates, but the values for ^{137}Cs concentration in milk were lower. Exposure from external radiation is largely due to ^{137}Cs in the soil, and therefore the new estimate of external exposure was in approximate agreement with the earlier one. On the other hand, the estimates of internal dose were substantially reduced, partly because of lower estimates of concentrations of ^{137}Cs and other radioisotopes in foods, and partly because results of whole-body counting of ^{137}Cs showed lower concentrations than suggested by models of the retention rate of ingested ^{137}Cs.

Comparisons of the original Soviet dose estimates with the ICP dose estimates are summarized in Table 12.3. The indicated doses are effective dose equivalents from the accumulated radiation exposures over a period of 70 yr. The values given are ranges of the average doses in various sampled localities. The thyroid dose has not been included in this estimate, in part because the radioisotope responsible for this dose, ^{131}I, has an 8.0-day half-life, and there is no direct way to estimate doses of ^{131}I several years after the Chernobyl accident.

The total dose in the ICP estimates was about one-half the original Soviet estimate. Typical cumulative total 70-yr doses averaged 80–160 mSv (see Table 12.3), and the average dose over 70 yr is not more than about 2 mSv per year, although the dose in the first few years is greater than the average. This is, of course, the incremental dose due to Chernobyl, over and above natural background. As a point of comparison, the average annual dose in the United States from natural background is about 3 mSv, primarily from indoor radon (see Section 2.4.2).

Health Effects Near Chernobyl

In the populations it studied, the International Chernobyl Project found no indications of adverse medical consequences from the radiation. In its conclusions it stated:

> [there were] no health disorders that could be attributed to radiation exposure. The accident had substantial negative psychological consequences in terms of anxiety and stress. . . [29, p. 32].

This refers to health effects to date. Future health effects cannot be excluded, but with the possible exception of thyroid tumors, it was judged that "future increases over the natural incidence of cancers or hereditary effects would be difficult to discern" [*ibid*]. This is another way of saying that whatever effects may occur are expected to be too small to identify, given

Table 12.3. *Comparison of International Chernobyl Project estimates and original Soviet estimates of the 70-yr cumulative dose commitment: range of average doses in sampled localities at intermediate distances from Chernobyl.*

	Estimated dose (mSv)	
	ICP	Orig. Soviet
External dose	60–130	80–160
Internal dose (cesium)	20–30	60–230
Total (including strontium)	80–160	150–400

Source: From Ref. [29], pp. 25–26.

the large number of cancers and hereditary effects in a normal (unexposed) population.

More recent studies have found a large increase in the thyroid cancer rate among children exposed to fallout from Chernobyl, although the relation between the radiation exposure and the cancer incidence is not yet fully understood [30]. A somewhat earlier study of leukemia rates in children, comparing incidence in the 1979–85 and 1986–91 periods, showed no appreciable change [31].

Controversies Over Health Effects from Chernobyl

It is too soon for the health effects of the Chernobyl accident to have fully manifested themselves, for either the workers who participated in the Chernobyl cleanup efforts or the surrounding population. Many cancers have a latent period of 10 years or longer, and it is impossible to attribute a given cancer to a given cause. The ambiguities of statistical evidence, the dramatic nature of anecdotal evidence, and the strong incentives to reach one conclusion or another almost guarantee that there will be very different assessments of the consequences of Chernobyl.

Even at Three Mile Island there has been some controversy over the consequences (see Section 12.2.3), and the situation is likely to be far more difficult in the Chernobyl case, especially in view of the political and economic problems in the area. These may make it difficult to obtain satisfactorily comprehensive and reliable records. It is important that vigorous efforts be made to carry out detailed health surveys and data analyses, in order to improve our understanding of the effects of prolonged exposures to radiation at low and intermediate levels.

The disagreements about Chernobyl could be reduced if there are thorough epidemiological studies by a group whose legitimacy is widely accepted. There is a precedent for this in the Radiation Effects Research Foundation,[12] which has carried out continuing studies of the aftermath of Hiroshima and Nagasaki. It is not clear, however, if the social stability and the internal or international resources will be sufficient to carry out an equivalently convincing set of studies in Ukraine and surrounding regions. Instead, we may face decades of unresolved controversies.

12.3.5 Worldwide Radiation Exposures from Chernobyl

The spread of the radiation cloud around the northern hemisphere led to extensive attempts to measure and calculate worldwide radiation exposures, present and future. The most important paths for radiation exposure are from radionuclides deposited on the ground and from ingested radionuclides. Ingestion is the more important in the first year, but as deposition from the atmosphere ends and the radionuclides are washed off vegetation by rain, the ground exposure becomes the more important. Overall, in terms of a 50-yr dose commitment, including the dose from ingested radionuclides that remain in the body, the external surface dose and the ingestion dose are roughly equal.

The single radionuclide ^{137}Cs is responsible for about 50% of the calculated ingestion dose and 80% of the ground exposure dose. Its importance follows from the fact that aside from noble gases, which stay in the atmosphere rather than on vegetation or ground surfaces, ^{137}Cs

---

[12]This was formerly known as the Atomic Bomb Casualty Commission. It is a joint research activity of the United States and Japan.

Table 12.4. *Collective and mean individual lifetime radiation doses due to the Chernobyl accident, and projected cancer fatalities.*

Region	Population (millions)	Lifetime dose (50-yr)		Fatal cancers, lifetime	
		Collective (kSv)	Individual (mSv)	Natural	Chernobyl[a]
USSR	279	326	1.2	35 000 000	16 000
Europe (non-USSR)	490	580	1.2	88 000 000	29 000
Asia (non-USSR)	1900	27	0.014	342 000 000	1400
U.S. and Canada	250	1.2	0.005	48 000 000	60
Northern hemisphere	2900	930	0.32	513 000 000	47 000

Source: Table 5 of Ref. [32].
[a]These values are about a factor of three higher than those given in Table 5 of Ref. [32]. They are based on a cancer risk factor of 0.05 per Sv (see Section 2.5.4).

is the only long-lived radionuclide among the copiously emitted radionuclides (see Table 12.2).

A detailed study of the projected cancer fatalities from Chernobyl has been reported in *Science* [32]. In a summary table, the calculated fatal cancers extending over 50 yr were listed for the populations of the northern hemisphere, there being very little exposure in the southern hemisphere. Table 12.4 is based on these results.

The individual average radiation doses shown in Table 12.4 are very low. Most of the projected fatalities are in populations where the *lifetime* dose is about 1.2 mSv, or an average of 0.02 mSv per year. This is less than 1% of the average annual dose from natural sources.

Nonetheless, if one uses standard extrapolations of the dose–response relationship, the total number of fatalities, summed over hundreds of millions of people in a 50-yr period, is large. In Table 12.4 the total is found to be 47 000, assuming a cancer risk factor of 0.05 fatalities per person-Sv. As discussed in Section 2.5.4, this risk factor is based on the assumption of a linear dose–response curve at low doses, but it incorporates a dose-rate reduction factor of two.[13] A total of 47 000 fatalities would increase the average cancer rate during this period by about 1 part in 10 000, or 0.01% for the northern hemisphere. In the (former) USSR as a whole, the projected increase is 0.05%.

Such numbers are very uncertain. The largest uncertainty is in the dose–response relationship. The value adopted here (0.05 fatalities per person-Sv) is consistent with the estimates in recent UNSCEAR and NCRP reports and slightly lower than the estimate of BEIR V (see Section 2.5.4). However, it is also possible that at the low doses received by the bulk of the populations of the USSR and Europe, there may be virtually no effect.

Further, as seen above, the studies of the International Chernobyl Project indicate that the radiation exposures in the region near Chernobyl were substantially overestimated. This raises the possibility that the worldwide doses reported in Table 12.4 were also overestimated.

Different perspectives on the impact of Chernobyl may be stated as follows: (a) the accident may lead to about 50 000 cancer deaths; (b) Chernobyl will not increase the cancer rate in the former USSR by as much as 0.1%; (c) the average annual exposure from Chernobyl in the former USSR is less than 1% of the average *annual* radiation exposure of an

---

[13]Anspaugh et al., using standard dose-response estimates of the time, projected a total of 17 400 cancer fatalities but gave a range of uncertainty extending about a factor of three above and below that estimate, also indicating that "the possibility of zero health effects at very low doses and dose rates cannot be excluded" [32].

individual in the United States. Depending upon which of these formulations appears most appropriate, Chernobyl may be considered a major global disaster or no more than a serious accident.

12.3.6 The Void Coefficient in the Chernobyl Reactor[14]

Over- and Under-Moderation

The Chernobyl accident and general issues of reactor safety are often discussed in terms of the void coefficient or, more completely, the void coefficient of reactivity. The void coefficient α_ϕ can be expressed in the case of a steam–water mixture (as at Chernobyl and in a BWR) as[15]

$$\alpha_\phi = \frac{\delta k}{\delta \phi},$$ (12.1)

where k is the effective multiplication factor and the parameter ϕ takes on the value 0 when there is no void (no steam) and 1 when there is all void (all steam). The same qualitative concepts are relevant to other cases when the reactivity is modified by a loss of water or a decrease in water density due to thermal expansion. In such cases, however, there is no steam void and a parametrization in terms of ϕ is not directly applicable. A positive void coefficient corresponds to positive α_ϕ, a negative void coefficient to negative α_ϕ.

The effective multiplication factor k is schematically represented as a function of ϕ in the qualitative sketch of Fig. 12.2. At point A, where the slope is positive, the void coefficient is positive, and a reactor would be *over-moderated*. At point B, where the slope is negative, the void coefficient is negative, and a reactor would be *under-moderated*. An alternative terminology, which might be more appropriate to the Chernobyl case, is to describe the condition at point A as "over-poisoned," because in the graphite-moderated Chernobyl reactor the water is more significant as a poison than as a moderator.

Evaluation of Void Coefficient

Approximating k with the standard four-factor formula (Equation 5.5), $k = \eta_F \epsilon f p$, it follows that for a reactor of infinite size

$$\alpha_\phi = \frac{k}{\eta_F}\frac{\delta \eta_F}{\delta \phi} + \frac{k}{\epsilon}\frac{\delta \epsilon}{\delta \phi} + \frac{k}{p}\frac{\delta p}{\delta \phi} + \frac{k}{f}\frac{\delta f}{\delta \phi}.$$ (12.2)

We will consider each term on the right-hand side of Eq. 12.2:

1. The magnitude of η_F depends on the ratios of neutron cross sections in the fuel. The neutron spectrum and hence the cross sections change as the void fraction changes, but this effect is small and the first term is small.

2. The fast fission enhancement factor, ϵ, depends on the rate of fission of ^{238}U by fast neutrons. A larger void fraction means less efficient moderation, more fission in ^{238}U,

---

[14]This section addresses somewhat detailed issues relating to the void coefficient, and is more technically oriented than most of the preceding sections.

[15]We here use a conventional notation adopted, for example, in Ref. [21]. The reactivity ρ is related to the effective multiplication factor k by the expression $\rho = (k-1)/k$ (see Eq. 5.14). For $k \approx 1$, $\delta k \approx \delta \rho$.

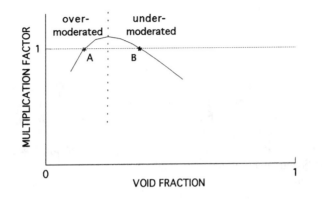

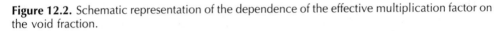

Figure 12.2. Schematic representation of the dependence of the effective multiplication factor on the void fraction.

and a larger ϵ. This term is positive, but not great in magnitude.

3. The resonance escape probability, p, depends on the rate of capture of neutrons in ^{238}U in the resonance region. A larger void coefficient and less efficient moderation mean more capture in ^{238}U and a smaller p. This term is negative. If it is the dominant term, α_ϕ will be negative.

4. The thermal utilization factor, f, depends upon the competition between thermal neutron reactions in uranium, in hydrogen in the water, and in other materials. This term is positive. Its magnitude determines whether the void coefficient α_ϕ is on net balance positive or negative. We will consider it in more detail in the next section.

This discussion, based on the four-factor formula, applies to an infinite reactor. In a finite reactor, leakage of neutrons contributes a negative term to the void coefficient, because an increase in void fraction leads to greater leakage. This effect must be included along with the effects already mentioned to determine the overall sign and magnitude of the void coefficient.

Dependence of Thermal Utilization Factor, f, on Void Fraction φ

If we designate the rates of thermal neutron reactions in uranium, hydrogen, and all other absorbers by u, h, and a, respectively, the thermal utilization factor depends on the ratio of reaction rates:[16]

$$f = \frac{u}{u+h+a}.$$ (12.3)

We assume that the only changes in the reactor conditions are due to changes in ϕ, i.e., in the amount of water. For changes in the rate of reactions in hydrogen, it follows from Eq. 12.3 that

---

[16]Roughly speaking, each of these "rates" can be thought of as the product of a target abundance and the neutron reaction cross section in that target, with an integration over the paths of the neutrons as they move through the complex nonhomogeneous geometry of the reactor core. For simplicity, in the present discussion we treat the rates as applying to a fixed neutron flux, although the competition between processes couples the rates.

$$\frac{1}{f}\frac{\delta f}{\delta h} = -\frac{1}{u+h+a}. \tag{12.4}$$

Obviously, $\delta f/\delta h$ is always negative.

When the void fraction increases, ϕ rises and the reaction rate in hydrogen, h, drops. Thus, $\delta h/\delta\phi$ is negative, and the signs of $\delta f/\delta\phi$ and $\delta f/\delta h$ are opposite. This explains the positive contribution of the $\delta f/\delta\phi$ term to the void coefficient. Of course, the sign of this result can be seen without any equations. Hydrogen is acting as a poison for thermal neutrons, and the less the poison the greater the fission rate.

The magnitude of $\delta f/\delta\phi$ increases as the magnitude of $\delta f/\delta h$ increases. The sensitivity to void fraction will be greater, i.e., the magnitude of $\delta f/\delta h$ will be greater, if u and a are small in Eq. 12.4. Therefore, to avoid a large positive void coefficient, one wants u and a to be large.

These were not the conditions at the time of the Chernobyl accident. The reaction rate in uranium, u, was relatively low, because the initial enrichment in ^{235}U was low and there had been considerable burn-up. The role of other absorbers involved a balance between the low contribution from the control rods, which were nearly withdrawn, and the unusually large contribution from xenon poisoning. Overall, the absorption rate a was small enough for the fourth term in Eq. 12.2 to provide a dominant positive contribution to the void coefficient.

Implications for Modifications of Chernobyl-Type Reactors

The discussion above can help explain the rationale behind some of the steps taken in an effort to make the continued operation of the RBMK-1000 reactors safer [21]. In particular, we consider those that reduce the magnitude of the positive contribution to the void coefficient represented by the $\delta f/\delta\phi$ term:

- The initial enrichment of ^{235}U in the fuel has been increased to 2.4%, raising u in Eq. 12.4.

- Permanent boron-steel absorbers have been inserted into the core, which is equivalent to increasing a in Eq. 12.4.

In qualitative terms, the reactor is less sensitive to the loss of hydrogen poison if it is operating with a higher ^{235}U content and with more poisons from other components. Although these two changes both decrease the positive void coefficient, they have opposite effects on the reactivity, and in that sense they balance each other.

REFERENCES

1. Zhores Medvedev, *The Legacy of Chernobyl* (New York: W. W. Norton, 1990).
2. Grigori Medvedev, *The Truth About Chernobyl* (New York: Basic Books, 1991).
3. United Nations Scientific Committee on the Effects of Atomic Radiation, *Sources and Effects of Ionizing Radiation* (New York: United Nations, 1993).
4. H. W. Bertini *et al.*, *Descriptions of Selected Accidents that Have Occurred at Nuclear Reactor Facilities*, report ORNL/NSIC-176 (Oak Ridge, Tenn.: Oak Ridge National Laboratory, 1980).
5. F. J. Rahn, A. G. Adamantiades, J. E. Kenton, and C. Braun, *A Guide to Nuclear Power Technology* (New York: John Wiley, 1984).
6. U.K. Atomic Energy Office, *Accident at Windscale No. 1 Pile on 10th October, 1957* (London: November 1957).
7. M. J. Crick and G. S. Linsley, *An Assessment of the Radiological Impact of the Windscale Reactor Fire, October 1957*, report NRPB-R135 (Chilton, U.K.: National Radiation Protection Board, 1982).

8. John G. Fuller, *We Almost Lost Detroit* (New York: Ballantine Books, 1975).
9. Earl M. Page, "The Fuel Melting Incident," in *Fermi-1, New Age for Nuclear Power*, edited by E. Pauline Alexanderson (La Grange, Ill.: American Nuclear Society, 1979): 225–254.
10. *Report of the President's Commission on the Accident at Three Mile Island*, John G. Kemeny, Chairman (New York: Pergamon Press, 1979).
11. *Three Mile Island, A Report to the Commissioners and to the Public*, report of Special Inquiry Group, Mitchell Rogovin, director, vol. II, part 2 (Washington, D.C.: U.S. Nuclear Regulatory Commission, 1980).
12. David Okrent and Dale W. Moeller, "Implications for Reactor Safety of the Accident at Three Mile Island, Unit 2," *Annual Review of Energy* 6 (1981): 43–88.
13. Pennsylvania Department of Health, "Health Department Discounts TMI as Connected to Infant Deaths," news release, May 19, 1980.
14. Marilyn K. Goldhaber, Sharon L. Staub, and George K. Tokuhata, "Spontaneous Abortions after the Three Mile Island Nuclear Accident: A Life Table Analysis," *American Journal of Public Health* 73, no. 7 (1983): 752–759.
15. M. C. Hatch, S. Wallerstein, J. Beyea, J. W. Nieves, and M. Susser, "Cancer Rates after the Three Mile Island Nuclear Accident and Proximity of Residence to the Plant," *American Journal of Public Health* 81, no. 6, (1991): 719–724.
16. USSR State Committee on the Utilization of Atomic Energy, *The Accident at the Chernobyl Nuclear Power Plant and Its Consequences*, information compiled for the IAEA Experts' Meeting, Vienna, 25–29 August 1986.
17. C. Hohenemser, M. Deicher, A. Ernst, H. Höfsass, G. Lindner, and E. Recknagel, "Chernobyl: An Early Report," *Environment* 28, no. 5 (June 1986): 6–43.
18. "Chernobyl: The Soviet Report," *Nuclear News* 29, no. 3 (October 1986): 59–66.
19. Richard Wilson, "Comments on the Accident at Chernobyl and Its Implications Following a Visit to the USSR on February 13–24, 1987" (1987).
20. M. Goldman, *et al.*, *Health and Environmental Consequences of the Chernobyl Nuclear Power Accident*, report DOE/ER-0332 (Washington, D.C.: U.S. DOE, June 1987).
21. A. A. Afanasieva, E. V. Burlakov, A. V. Krayushkin, and A. V. Kubarev, "The Characteristics of the RBMK Core," *Nuclear Technology* 103 (July 1993): 1–9.
22. International Atomic Energy Agency, *INSAG-7. The Chernobyl Accident: Updating of INSAG-1*, report by the International Nuclear Advisory Safety Group (Vienna: IAEA, 1992).
23. U.S. Council for Energy Awareness, *Nuclear Power Plants Outside the United States* (Washington, D.C.: 1991).
24. Yuri Kanin, "Chernobyl Fire Chronology," *Nature* 353 (1991): 690.
25. "World List of Nuclear Power Plants," *Nuclear News* 38, no. 3 (March 1995): 27–42.
26. "Ukraine Agrees to Close Chernobyl A-plant," *The New York Times*, December 21, 1995, p. A3.
27. S. A. Khan, "The Chernobyl Source Term: A Critical Review," *Nuclear Safety* 31 (1990): 353–374.
28. "Late News in Brief," *Nuclear News* 36, no. 15 (December 1993): 75–76.
29. International Atomic Energy Agency, *The International Chernobyl Project, An Overview*, report by an International Advisory Committee (Vienna: IAEA, 1991).
30. Michael Balter, "Chernobyl's Thyroid Cancer Toll, *Science* 270 (1995): 1758–1759.
31. E. P. Ivanov, G. Tolochko, V. S. Lazarev, and L. Shuvaeva, "Child Leukemia after Chernobyl," *Nature* 365 (1993): 702.
32. Lynn R. Anspaugh, Robert J. Catlin, and Marvin Goldman, "The Global Impact of the Chernobyl Reactor Accident," *Science* 242 (1988): 1513–1519.

Chapter 13

Reactor Safety and Future
Nuclear Reactors

13.1 OPTIONS FOR FUTURE REACTORS

13.1.1 Are Present Reactors Safe Enough?

In the 1980s, when planning began for a next generation of reactors in the United States, there was no consensus as to the relative merits of two possible approaches, and both have continued to be pursued. On the one hand are those who believe that the existing types of reactors, with incremental improvements, are quite satisfactory for the near future. On the other hand are those who believe that a major expansion in the use of nuclear power calls for substantially different and safer reactors. These reactors have been given names such as *inherently safe* or *passively safe* reactors, to emphasize the nature of the intended changes.

One of the most prominent of the U.S. advocates of such inherently safe nuclear reactors has been Alvin Weinberg. He and colleagues at the Oak Ridge Institute for Energy Analysis have taken the position that while existing reactors are very safe, they are not safe enough for a major future nuclear buildup [1–5].

The Nuclear Regulatory Commission proposed in the early 1980s a criterion that the core melt probability for any reactor be less than 10^{-4} per year [6, p. xiv]. In the opinion of Weinberg *et al.*, "there was no doubt that the majority of LWRs today [1984] meet this criterion" [4, p. 672]. Further, the chance of a catastrophic release of radioactivity was much less than that of core melt because, as shown at TMI, the containment was likely to retain most of the radionuclides. Nonetheless, in Weinberg's thesis, this is not good enough. If one

232

assumes, as a conservative estimate, that the core melt probability is 1 in 10^4 per reactor-year, then:

> In a 500-reactor world, one might expect a meltdown every 20 years in reactors satisfying the safety goal. As the number of reactors increases in a Second Nuclear Era, the likelihood of a meltdown would also increase. Though posing little risk to the public at large, a meltdown of any one nuclear plant would pose many risks to all investors in nuclear power. We propose, therefore, that as the number of reactors increases, the safety goals would be tightened for new reactors [4, p. 672].

Weinberg's solution, to pave the way to a nuclear expansion into a "Second Nuclear Era," was to adopt inherently safe designs. Although this suggestion by Weinberg and others was at first greeted coolly by most in the nuclear industry, it has gradually won favor and now represents a major thrust within the U.S. nuclear world.

Somewhat apart from the choice of reactor type, the NRC has been grappling with the problem of possible standards for probabilities of core damage and large releases. Thus, it has been suggested that the target be raised from the earlier 1 in 10^4 chance of a core melt per reactor-year (10^{-4}/RY) to a new standard:

> the requirements for future plants should be directed towards ensuring a core-damage frequency of no greater than 10^{-5}/reactor-year and a large-release frequency of no greater than 10^{-6}/reactor-year [7, p. 31].[1]

The distinction between core damage and large release stems from the protective barriers provided by the reactor pressure vessel and the reactor containment.

Although there are frequent references in NRC documents to a large release guideline of 10^{-6}/RY, these remained as of 1994 in the form of proposals rather than officially adopted requirements (e.g., Ref. [8]). It is possible that some such guideline will be adopted before new reactors are licensed. The issue may turn out to be somewhat moot, however, as all of the planned new reactors project a core damage frequency of less than 10^{-5}/RY [9, p. 134]. Given the expected performance of the containment, this would imply a large release frequency of less than 10^{-6}/RY. Thus, it is possible that the teeth in the NRC design approval process will come in demanding evidence that the plants will meet the targets set by the manufacturers.

13.1.2 General Considerations in Reactor Safety

Design Goals

As light water reactors have evolved, the guiding principles of reactor safety have been defense-in-depth and redundancy. Components of the safety system, beyond those for the avoidance of accidents, include systems for assured shutdown of the reactor, alternative cooling systems, and a reactor pressure tank and a surrounding concrete containment to limit the escape of radioactive material.

The most complex challenge has been to assure that the flow of coolant to the core is maintained under any accident circumstances, so that the decay heat is removed. Attempts to guarantee this have led to the installation of alternative cooling paths and redundant components. For example, several independent diesel generators provide electric power to the pumps should the outside electricity supply fail. A profusion of pipes, valves, and control

---

[1]An obvious misprint in the quotation has been corrected (i.e., 10^{-5} rather than the stated 10^5).

systems has evolved from the search for greater safety, often introduced by retrofitting an existing reactor. This has added to the complexity and cost of the reactors, and it puts extra burdens upon the operators.

With the hiatus in reactor orders since the 1980s, reactor manufacturers have had the incentive and the opportunity to step back and see how they might design safer and simpler reactors, without being constrained by existing configurations. There have been two general approaches to achieving improved reactor safety. *Evolutionary* reactors build upon the experience gained to date, making incremental improvements without changing the fundamental reactor design. *Advanced* reactors employ designs that are substantially, sometimes radically, different from the designs of existing plants. In the former case, one relies on the benefits of accumulated experience. In the latter case, reliance is placed on a revised reactor safety strategy.

Terminology for "Safe" Reactors

The terms "inherent" and "passive" are intended to suggest that the safety of the reactor will depend on immutable physical phenomena rather than on the proper performance of individual components or correct actions by reactor operators. In the extreme version of the concept, in a passively safe reactor all operators could become incapacitated and all external electricity and water could be shut off, and the reactor would still turn itself off and gradually cool with no damage.

This terminology has been widely used and also widely criticized. The objections have had several strands:

- Inherent or passive safety is a matter of degree rather than a totally new departure. A negative temperature coefficient or a negative void coefficient is a passive safety feature, and therefore most existing reactors already have passive safety features.
- The terms are misleading because they seem to suggest that an accident would be *totally* impossible, while in fact one can find circumstances in which any given reactor might fail, if arbitrarily improbable scenarios are permitted.
- The terms could appear to have a prejudicial aspect, because they could seem to suggest that existing reactors are *not* safe.

The criticisms have had some force, and to defuse them alternative words have sometimes been suggested. Weinberg has introduced the term *transparently passive* as an equivalent to "inherent" [10, p. 476] and has also suggested the use of the terms "passively stable" (adopted from Taylor [11]) and the still more comprehensive "transparently passively stable" [5, p. 137]. Whatever words are used, however, the concept is clear: it is safer to rely on systems that depend only upon basic physical processes (e.g., gravity or thermal expansion) rather than on the consistent good performance of equipment and operators.[2]

13.1.3 Categories of Future Reactors

Specific options for new reactors were considered in a 1992 National Academy of Sciences study on nuclear power [9]. Table 13.1 lists the leading possibilities, based on the classifi-

---

[2]The International Atomic Energy Agency is in the process of codifying the terminology. In the proposed usage, "*inherent* safety is achieved through the elimination or avoidance of hazards." A *passive* system is one that does not require action by people or external mechanical or electrical equipment [5, p. 136].

Table 13.1. *New reactor designs under development (partial listing).*

Reactor Designation	Manufacturer[a]	Size (MWe)	NRC FDA
Large evolutionary LWRs			
Advanced boiling water reactor [ABWR]	GE	1350	July 94
System 80+ (PWR)	ABB-CE	1300	July 94
Advanced pressurized water reactor [APWR]	W	1350	
Mid-size passive LWRs			
Advanced passive 600 [AP600] (PWR)	W	600	late 96[b]
Simplified boiling water reactor (SBWR)	GE	600	late 97[b]
Other LWRs			
Safe integral reactor [SIR] (PWR)	ABB-CE	320[c]	
Process inherent ultimate safety (PIUS)[e]	ABB-Atom	600	
Advanced Non-LWRs			
Gas-turbine modular helium reactor (GT-MHR)[d]	GA	≈250[c]	
Advanced liquid-metal reactor (ALMR)[f]	GE	≈300[c]	
CANDU-3 (HWR)	AECL	450	

[a]ABB=Asea Brown Boveri; AECL=Atomic Energy of Canada, Ltd; CE=Combustion Engineering; GA=General Atomics; GE=General Electric; W=Westinghouse.
[b]These are estimated times for the granting of final design approval by the NRC, as reported in Ref. [12].
[c]Capacity per module; modular deployment assumed.
[d]This reactor is the successor to the modular high-temperature, gas-cooled reactor, previously studied by General Atomics (see text).
[e]This reactor operates as a PWR but has a radically different design from standard PWRs.
[f]This reactor is the successor to the PRISM reactor, previously studied by General Electric (see text).

cations used in the NAS report [9, p. 92]. The list is not exhaustive, but it includes the reactors that the NAS study considered to be the most promising. It is similar to an earlier list prepared by the Nuclear Regulatory Commission [7]. While it includes the main designs in the United States, it leaves out some developments in other countries. Along with indicating the reactor manufacturer and size, Table 13.1 gives actual and expected dates for the NRC to grant final design approval (FDA) (see Section 13.5.1), reflecting developments subsequent to the publication of the NAS report.

The reactors in Table 13.1 have been divided into several categories:

Evolutionary LWRs. These are essentially refined versions of current large light water reactors. As they embody the fewest changes, these will probably be the first available for construction in the United States, although it does not mean that they will be the first ordered. Progress on such reactors has been more rapid in other countries. A brief further discussion of evolutionary LWRs is given in Section 13.2.1.

Mid-sized passive LWRs. The two reactors listed in the table incorporate substantial changes in the reactor configuration, compared to present LWRs, but the basic principles are the same. The chief differences involve increased use of passive safety features, especially in the system for emergency core cooling. These reactors are discussed further in Section 13.2.2, with particular emphasis on the Westinghouse version, the AP600.[3]

Other LWRs. The two reactors listed in this category represent fairly radical departures from standard designs. Neither is a leading contender at present in the United States. The design of reactors is becoming more of an international enterprise than in the past. The original PIUS design was by a Swedish company (which has evolved into ABB-Atom),

---

[3]See Table 13.1 for the fuller designation of this and other reactors referred to here by their acronyms.

and the SIR reactor was originally a United Kingdom–United States project. Both now involve the same parent company, Asea Brown Boveri (ABB).[4] The PIUS reactor is discussed briefly in Section 13.2.3.

Advanced non-LWRs. These include an array of diverse designs. The GT-MHR is a recent variant of the high-temperature, gas-cooled reactors being studied by the General Atomics Company. The ALMR is a sodium-cooled, fast neutron reactor. Both the GT-MHR and the ALMR are successors to earlier versions considered in the NAS report [9]. They are discussed further in Sections 13.3 and 13.4, respectively. CANDU-3 is an advanced and somewhat smaller version of the heavy water reactors developed in Canada.

Of these reactors, the PIUS, GT-MHR, and ALMR reactors have design features that their proponents claim make them inherently safe. It is less common for the proponents of the various other reactors to make equivalently strong claims, although they quote very low accident probabilities and stress passive safety features, especially in the case of the advanced passive LWRs.

As seen in Table 13.1, two of these reactors have received FDA from the NRC and two others are expected to receive FDA within the next several years. This establishes the leading candidates, if new reactors are to be ordered in the United States in the near future. Reflecting their status are front-runner, these four reactors (the ABWR, System 80+, AP600, and SBWR) are the only ones mentioned in a 1994 Electric Power Research Institute review on "Reopening the Nuclear Option" [12]. However, with an eye on a longer time frame, other reactors are also considered in the discussion below.

13.2 NEW LIGHT WATER REACTORS

13.2.1 Evolutionary Light Water Reactors

The three main reactor vendors in the United States are all developing large evolutionary reactors (see Table 13.1), and versions of these have already been ordered abroad as part of international collaborations. Two 1315-MWe advanced boiling water reactors (ABWRs) are under construction in Japan, in a collaboration of General Electric, Hitachi, and Toshiba; the reactors are scheduled to go on-line in 1996 and 1997 [14]. Four 950-MWe Combustion Engineering System 80 reactors are under construction in South Korea [14]; these are a preliminary and smaller version of the System 80+ under development ABB-CE. Westinghouse is designing an advanced PWR (APWR) in collaboration with Mitsubishi.[5]

Although the General Electric ABWR is under construction in Japan, no new reactor has yet received NRC approval for construction in the United States. However, the ABWR and the System 80+ reactor each passed an important milestone towards licensing when they were granted NRC final design approval in July 1994 [15].[6] A further stage, that of "stan-

---

[4]Asea Brown Boveri (ABB) is a major Swedish–Swiss company with two subsidiaries for building reactors: ABB-Atom and ABB Combustion Engineering (ABB-CE). The latter was formed by the purchase in the late 1980s of the American reactor manufacturer Combustion Engineering [13].

[5]The terminology used in describing these reactors may lead to some confusion. In particular, the term "advanced" has been applied in different ways. In recent usage, the term "advanced LWRs" has been used to include large evolutionary LWRs, along with LWRs of more innovative design such as the mid-sized passive LWRs (see Table 13.1) [12].

[6]The timing of these approvals compares reasonably well with an earlier NRC schedule of May 1994 for the ABWR and August 1994 for the System 80+ reactor.

dardized design certification," remains before a plant can be licensed and construction begin (see Section 13.5.1).

As implied by the designation "evolutionary," the differences between the ABWR and previous BWRs are not striking. Nonetheless, the expectations of General Electric for improved performance are striking [16]. On the basis of its probabilistic safety assessment, the company projects a core damage probability of less than 10^{-6}/RY and a probability for an off-site dose of more than 0.25 Sv (25 rem) of 2×10^{-9}. In addition, the projected construction time is only 48 months and the overall cost well below that of recent LWRs.

Of course, it is important to have an independent evaluation of these expectations. Further information on construction time and cost will be provided by the experience in Japan, although there are differences in the two settings. The safety claims will receive further scrutiny in the course of the U.S. design certification process, including quantitative probability safety assessments. However, it would be remarkable if the new ABWR did not represent an improvement over existing BWRs in the United States, the last of which was ordered in 1973.

The United States and U.S. collaborations are not alone in developing large evolutionary PWRs. France is far along, having begun construction of next-generation 1455-MWe PWRs, the so-called N4 series. Four units were under construction in 1995, and these are scheduled to go into operation in 1996 through 1998 [14]. The one new reactor being put into operation in the United Kingdom is the 1188-MWe Sizewell B reactor, a PWR being put into operation in 1995 [14]. In addition, there is a joint French–German program to develop a future 1500-MWe "European Pressurized Water Reactor" (EPR), but the timing of a first order for an EPR is not yet established.

13.2.2 Mid-sized Passive Light Water Reactors

General Considerations

The "advanced" or "innovative" LWRs that are under consideration are smaller and simpler than the current generation of LWRs. In the U.S., two designs have been under development with the support of the DOE and the Electric Power Research Institute (EPRI): the Westinghouse AP600 and the General Electric SBWR. Both reactors are about 600-MWe in size. Special features of this category of reactors, as summarized by Forsberg and Weinberg, include the following [5, p. 140]:

- The emergency cooling systems are simpler and more passive, relying on large pools of water fed by gravity, rather than on flow sustained by pumps.
- Emergency electric power requirements are reduced so that they can be satisfied by batteries rather than emergency diesel generators.
- Reactor power densities are reduced.
- The designs have been simplified to reduce costs and sources of possible operating or maintenance error.

We will discuss below the AP600 as an example of these reactors, although it is premature to say which one, if either, will prove to be more successful commercially.

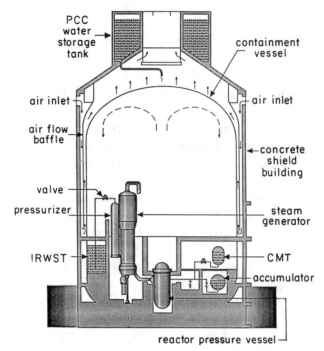

PCC water storage tank

containment vessel

air inlet

air inlet

air flow baffle

concrete shield building

valve

pressurizer

steam generator

IRWST

CMT

accumulator

reactor pressure vessel

Figure 13.1. Sketch of the AP600 containment vessel, showing the reactor pressure vessel, steam generators, and systems for emergency cooling of the containment vessel. (Courtesy of the Westinghouse Electric Corporation.)

Design Features of the AP600

The AP600 is designed to be simple in overall configuration and relatively inexpensive to build.[7] Part of the savings will come from modular construction, in which components will be built and to some extent assembled off-site, substantially reducing the construction time at the reactor itself. When compared to standard PWRs of similar size, the design simplifications reduce the requirements for valves by 50%, for safety grade pipe by 80%, and for cable by 70%.[8] Among the significant improvements are a digitized control and instrumentation system and sealed pumps. All of this is intended to provide reliable and economical operation.

These are incremental improvements. The more significant differences between the AP600 and present-day PWRs are the passive systems to cope with accidents. The AP600's reactor pressure vessel, steam generator, and much of the emergency cooling system are all within the containment vessel, a large steel tank with a 130-foot diameter (see Fig. 13.1). The containment is rated for a pressure of 45 psig ($\approx$3 atm above the outside pressure) and is expected to be able to survive considerably higher pressures. This containment is in turn surrounded, with a not very large gap, by a concrete shield building.

In case of an accident the two urgent goals are to maintain cooling of the reactor core and to avoid a breach in the containment that could allow the escape of radioactive material. The

---

[7]This discussion is based primarily on information from Westinghouse, including published documents and private communications.

[8]The gains are somewhat less if the comparison is made to standard larger plants, normalizing to capacity.

former is accomplished in the AP600 through passive emergency core-cooling systems. The latter is accomplished by cooling the containment vessel from the outside, to prevent a large buildup of internal pressure. The large volume of the containment coupled with the relatively low power rating of the reactor help in this regard.

We first consider a loss-of-coolant accident in the primary loop, which includes the reactor core and the input side of the steam generator. There are three types of tanks (five tanks in all) within the containment that can supply replacement water:

1. **Core makeup tanks (CMT).** There are two such tanks, each containing 2000 ft^3 (5.7×10^4 l) of borated water. When the reactor pressure or water level falls below prescribed safety levels, the reactor scrams and air-operated valves open between these tanks and the pressure vessel. These are "fail-open" valves that open if the air supply is lost. The tank is situated at a greater height than the reactor vessel, providing gravity-fed flow. It does not matter if the reactor is still under pressure, because the piping is such that the same pressure appears at the top of the core makeup tank as in the reactor vessel.

2. **Accumulator tanks.** There are two such tanks, each containing 1700 ft^3 (4.8×10^4 l) of borated water, driven by nitrogen gas at high pressure. They supplement the core makeup tanks and come into play when the reactor pressure falls below the accumulator pressure.

3. **In-containment refueling water storage tank (IRWST).** The tanks cited above suffice to provide cooling water for something on the order of 30 min.[9] A much larger tank, the IRWST (normally used for other purposes), provides longer-term cooling in case of an accident. It has a capacity of 67 000 ft^3 (1.9×10^6 l). In the first instance it provides direct cooling, but it also sets up a closed cooling cycle, as discussed below. A redundant set of valves is opened when the water level drops sufficiently in the core makeup tank, releasing the pressure in the reactor vessel and allowing gravity-fed flow from the IRWST into the reactor vessel.

It is to be noted that no pumps are required in this system, the flow is passively driven, and the valves are opened without operator intervention.

As the cooling process continues, it becomes self-sustaining. The reactor core sits near the bottom of the reactor vessel, well below the inlet and outlet pipes. In the event of the sort of break that requires emergency cooling, water is boiled off and steam escapes to the interior of the containment through the break and through a depressurization system that relieves the reactor pressure in case of cooling system failures. A cycle is set up in which the steam condenses on the inside wall of the steel containment and flows down, either to the IRWST or to a pool at the bottom. Either way, it is returned to the reactor.

To avoid the buildup of high internal pressure in the containment, the outside of the containment must be cooled. This is at first accomplished with gravity-fed water from the passive containment-cooling (PCC) water storage tank, located above the containment. This wets and cools the containment. This tank water supply can suffice for three days. For the longer term, the containment can be cooled by an air flow established by convection in the gap between the containment vessel and the concrete shield building. However, to give a larger margin of safety, it is anticipated that water will be externally supplied to the outside

---

[9]The mass of water in these tanks is approximately 1.0×10^5 kg. The heat of vaporization of water at 1 atmosphere is 2.2×10^6 J/kg. The heat output of a 600-MWe reactor at shutdown is about 120 MW (see Section 11.3.1). At this power output, it would require $(2.2 \times 10^{11})/(1.2 \times 10^8) \approx 1800$ sec$=30$ min to vaporize the water.

of the containment after the three-day period.

Failures in the primary loop were considered above. If instead there is a failure in the secondary loop, so that the steam generator no longer serves as a heat exchanger, the water from the core is diverted through fail-open valves into an alternate heat exchanger inside the containment. Flow in the loop from the reactor to the heat exchanger and back is maintained by convection. The reservoir for this heat exchanger is the IRWST, mentioned above in another role. As above, if water from this reservoir boils off, it condenses on the containment interior and the water is recycled back to the IRWST.

The overall expectations of the designers of the AP600 are high. The core damage frequency is projected to be 3×10^{-7}/RY for internal events and 4×10^{-7}/RY when fire and flood (external events) are included as well. The probability of a large release of activity is still lower. Earthquake risks are addressed by the seismic margin approach (see Section 11.4.4). The safe shutdown earthquake corresponds to a peak ground acceleration of 0.3 g. The seismic margin is established by determining for key reactor components the maximum acceleration at which there is a "high confidence of low probability of failure." The goal was to satisfy this criterion for an acceleration of 0.45 g. The Westinghouse analysis concluded that it was met for various components at levels of 0.5 g or higher.[10]

Costs and construction times are both projected to be well below those of recently completed reactors.

13.2.3 PIUS

PIUS could be called a *very* innovative LWR. The PIUS concept was invented by Käre Hannerz, who was with the main nuclear reactor company in Sweden, ASEA/Atom [3] and then with the combined company Asea-Brown-Boveri (ABB-Atom). Although it has received some very favorable evaluations, e.g., from Alvin Weinberg, it has not been pushed by major American manufacturers. The term PIUS is an acronym for "process inherent ultimate safety," and it is also sometimes referred to as the Secure-P reactor. Brief descriptions of PIUS have been given by Spiewak and Weinberg [3], by Weinberg *et al.* [2], and by Hannerz [17].

The PIUS reactor is a light water reactor in which the reactor core and the primary system are submerged in a pool of cold borated water contained in a large concrete vessel. Typically, this vessel might have a 13-m inside diameter (17-m outside diameter) and a 35-m height.

There are two distinct water supplies to consider. The first is the water in the primary reactor cooling loop. It serves the typical LWR functions of transferring heat from the reactor core to the steam generator and acting as the moderator. It is hot and has a low content of boron. The second water supply is the water of the surrounding pool. This pool, with a much greater volume than the primary loop, contains cold, highly borated water and has no function during normal operation. However, it is the key to the avoidance of serious consequences in the case of a reactor malfunction.

In normal operation, the water in the primary system is isolated from the borated pool by "very clever hydraulic locks" [3, p. 447]. These locks are essentially interfaces between the two water supplies, maintained by careful control of the pressure of the water in the primary loop. There is one such lock below the core and one above it. They are normally closed, and there is no interchange between the water in the primary loop and in the pool.

---

[10]These results are described in Appendix H of a probability risk assessment document submitted to the Nuclear Regulatory Commission on June 26, 1992.

In case of an excursion in reactor performance, the pressure in the primary circuit changes, this delicate balance is disrupted, and the reactor is flooded with borated water from the pool, terminating the chain reaction. Natural convection will maintain the flow of borated water from the pool into the reactor, keeping "the reactor cool for at least a week" [*ibid*].

The operation of PIUS relies on the hydraulic locks, which provide isolation during normal operation and bring in borated water during abnormal conditions. Reduced-scale tests with a non-nuclear source have shown that the locks worked as designed, with excellent agreement between the observed and calculated behaviors of the system ([3, p. 449] and [17]).

Despite the ingenuity of its design and evidence that the hydraulic locks can perform properly, PIUS has not been a leading contender for deployment in the United States within the next decade or two. Perhaps the novelty of the hydraulic locks makes people nervous. Even if the locks provide an unusually high degree of safety, there remains a fear that relatively small system perturbations could cause them to open, contaminating the reactor core with borated water and requiring a lengthy cleanup.

The future of PIUS development is not clear. For a number of years the greatest interest in PIUS appeared to be in Italy, which had terminated all nuclear power use pending the development of a new generation of supposedly safer reactors. But uncertainties in the Italian reactor market and the lack of a major commitment for financing have led ABB-Atom to "mothball" the entire PIUS program [18].

13.3 HIGH-TEMPERATURE GAS-COOLED REACTORS

13.3.1 HTGR Configurations: the MHTGR and the GT-MHR

The high-temperature gas-cooled reactor (HTGR) is a carbon-moderated and helium-cooled reactor. Proposals for inherently safe HTGRs are based on modular systems; hence the designation modular HTGR, or MHTGR. Several modules, not necessarily built simultaneously, could be included in a given facility.

The design of HTGRs has been pursued in the United States by the General Atomics Company. One MHTGR design called for 137-MWe modules, with four such modules comprising a 540-MWe unit having a thermal power level of 1400 MWt. In the MHTGR, as originally conceived, the helium circulating through the reactor core passes through a steam generator, the output of which drives a steam turbine. The system had a relatively high design efficiency (38–39%), stemming in part from the high temperature of the helium and the steam.

More recently, General Atomics has emphasized the alternative of using the helium to drive a gas turbine directly; no steam generator is needed. This variant is termed the Gas-Turbine Modular Helium Reactor (GT-MHR). The module capacity is somewhat higher than in the MHTGR. Among the advantages of the GT-MHR is a thermal efficiency even higher than that of the MHTGR (see below).

More broadly, the GT-MHR is one application of the modular helium reactor (MHR), which General Atomics describes as a "universal heat source for the next century" [19]. The MHR is intended to provide heat and steam for industrial applications, in addition to being used for electricity generation.

In either version, the reactor core sits in a steel reactor vessel. Helium pumped through the reactor vessel cools the core and either transfers heat to a steam generator (in the MHTGR) or drives a gas turbine (in the GT-MHR). Each module is placed in a separate below-grade silo.

Although the primary argument for these reactors is based on passive safety, they also have advantages in terms of unusually low releases of radioactive materials. The helium coolant is not itself made radioactive, nor does it carry many impurities that can become radioactive. Further, the fuel encapsulation is more secure than for oxide fuel in zirconium cladding, because the latter can have small leaks. Thus, there is less escape of volatile fission products. As far as the public is concerned, the release rates in normal operation are too low in either case to be of significance. However, the lowering of exposures for plant workers may become a factor in reactor choice, if standards for worker exposures are made more stringent as has been suggested.

13.3.2 Historical Background of Graphite-Moderated Reactors

The history of carbon-moderated reactors is old and mixed. The world's first reactor was the graphite-moderated reactor developed by Enrico Fermi in Chicago during World War II. Reactors for plutonium production are primarily graphite-moderated, because the conversion ratio is high, and it is relatively easy to change fuel frequently and avoid a large buildup of ^{240}Pu. Graphite-moderated reactors can be either water-cooled, as in the Hanford plutonium production reactors and the Chernobyl-type RBMK reactors, or gas-cooled, as in the British CO_2-cooled reactors and the helium-cooled Fort St. Vrain reactor in the United States.

The claims of inherent safety in the HTGRs might seem to fly in the face of the fact that the *only* accidents resulting in major releases of activity have been in graphite-moderated reactors, namely, the Windscale and Chernobyl accidents. However, it is argued that what happened at these plants has no relevance to the planned HTGRs:

Windscale. The HTGRs will run at higher temperature than did Windscale, and thus there will be no buildup of stored energy in the crystal lattice (the so-called Wigner energy), because the graphite will be continually annealed. The temperature for annealing is about 350 °C [20, p. 441], well below the normal graphite temperature in an HTGR.

Chernobyl. In addition to other major design differences, the use of a helium coolant in the HTGR (rather than water, as at Chernobyl) means that loss of the coolant cannot give a positive feedback. This follows from the fact that helium has a negligible absorption cross section for neutrons and therefore, unlike the water at Chernobyl, cannot be a poison.

It might also be noted that the one electricity-producing HTGR in the United States, a 330-MWe prototype unit at Fort St. Vrain in Colorado, had an unusually trouble-plagued life after going into operation in 1979. It was shut down in 1989 by the operating utility because it was not economical to continue to run it. The difficulties were primarily with the cooling system, and it is believed that these difficulties can be avoided in a next-generation helium-cooled reactor.

In light of the history described above, one might imagine that the nuclear industry would shy away from further attempts to develop HTGRs. However, some are convinced by the argument that past experience is not relevant to future performance, and there are strong believers in the HTGR as a super-safe reactor for the future.

Part of this confidence is based on experience with a series of prototype pebble-bed HTGRs built in West Germany. The first of these to be built was the AVR reactor, which was put into operation in 1967 to test the HTGR concepts. It was a small reactor, only 40 MWt and 15 MWe. A 300-MWe pebble-bed HTGR, the THTR-300, was put into operation in 1987. These reactors have provided experience on the behavior of HTGR fuel. However, in early 1991 further development work on the German HTGR systems was halted, due to lack of commercial interest [21].

13.3.3 Safety Features of the MHTGR

We here summarize some of the safety attributes of the earlier MHTGR designs. Presumably, the same considerations apply to the GT-MHR, because the nuclear parts are essentially the same.

In the MHTGR a negative temperature coefficient of reactivity is provided by the fuel itself. There is no problem analogous to the poisoning properties of the water coolant in the Chernobyl reactor, because the MHTGR's coolant is helium, which does not absorb thermal neutrons.[11] Therefore, there is no danger of a runaway reaction as at Chernobyl.

A particular argument for the HTGR is the ruggedness of the fuel, which is in the form of microspheres less than one millimeter in diameter.[12] The spheres have a small central kernel containing the fissile or fertile material, typically in oxide form. A variety of mixtures of uranium, thorium, and plutonium oxides can be used. The fuel kernel is encapsulated in a shell of graphite surrounded by porous carbon and three concentric protective shells, consisting in succession of pyrolytic carbon, silicon carbide and pyrolytic carbon [23]. These microspheres are called TRISO particles [23]. They were first used in a British reactor in 1967, and further experience has been obtained from tests and use in the Peach Bottom and Fort St. Vrain reactors in the United States and the AVR plant in Germany [22, p. 4].

In the General Atomics HTGR design, the microspheres are compacted with carbonaceous material into fuel rods, which are then placed in holes in graphite blocks, commonly prismatic in shape. These fuel elements are grouped in an annular array to form the reactor core.

The multi-coated TRISO microspheres in principle provide a rugged protection for the fuel. The fuel can withstand very high temperatures. The normal operating temperature is about 1000 °C. In tests in the German reactors, less than 1 part in 10^6 of the cesium, krypton, or strontium in the fuel was released when the fuel was held at 1600 °C for 200 h. There is a greater fractional release of silver, because silver can diffuse relatively rapidly through the fuel, but little silver is produced in fission. What silver does diffuse out of the fuel condenses without escaping to the environment [23]. In a hypothetical accident in which the helium cooling is lost, the maximum calculated temperature reached is slightly above 1600 °C, but none of the fuel remains above 1600 °C for more than 50 h [23].

Nonetheless, there are some problems with the fuel form. In addition to withstanding high temperatures, the coatings have to be able to withstand both physical damage in the manufacturing process and radiation damage under neutron bombardment in the reactor. In 1991 protective coatings were added to prevent physical damage, but those were found to compromise the integrity of the original underlying coatings under high neutron bombardment [22]. Work towards obtaining a still more damage-resistant fuel form continues.

---

[11]If heating of the graphite or loss of the helium has any appreciable effect on reactivity (and such an effect is not commonly mentioned), it is to provide negative feedback. The graphite expands slightly with heating, which tends to make it a less effective moderator, and loss of the helium coolant can only serve to decrease the moderation. Both effects would reduce the reactivity, although probably only by a very small amount.

[12]Planned diameters range from 0.65 to 0.8 mm [22].

Normally, the core is cooled by the flow of helium. If this fails, the temperature rise of the fuel is limited by radiation from the core to the reactor vessel. Heat is removed from the exterior of the reactor vessel by convection or radiation to the surrounding silo. The rate of heat transfer by radiation rises rapidly with temperature, increasing as the fourth power of temperature. Thus, reliance on radiation is possible for the HTGR, which tolerates high temperatures, although not for LWRs.

The high thermal capacity of the core limits the rate at which the temperature rises. Because graphite is a relatively poor moderator and has a low cross section for neutron absorption, the ratio of graphite to fuel in an HTGR is much greater than the ratio of water to fuel in an LWR. This leads to a high heat capacity or, essentially equivalently, to a low power density. For the planned MHTGR, the power density is 5.9 kW/l [24], while typical values are 60 kW/l in a BWR and 100 kW/l in a PWR [2, p. 44].[13]

Overall, the MHTGR is a quite simple system in which the ultimate safety is provided by the thermal characteristics of the fuel and the possibility of passive radiative cooling to the surrounding silo. In consequence, the estimated probability for a significant radiation release is much less than 10^{-8} per year [23].

The major danger would be the entry of water or air into the reactor containment and the resulting interactions with the hot carbon. To the extent that there are doubts about the ultimate safety of these reactors—and these doubts do not appear to be great—they hinge on how confident one can be that air and water will be excluded.[14]

13.3.4 Direct-Cycle Gas-Turbine Reactor

The overall configuration of the GT-MHR is shown in Fig. 13.2. This sort of system, in which hot helium drives the turbine directly, with no steam generator, has been advanced for several years by Lawrence Lidsky at MIT and his collaborators [26]. Lidsky *et al.* point out that this is not a new idea conceptually, but they argue that recent technological developments and the push to modular design have made it practical.

In the GT-MHR, hot gas from the reactor drives the turbine. Gas exiting the turbine at reduced temperature and pressure passes through three heat exchangers and a two-stage compressor before returning to the reactor. The first heat exchanger, the recuperator, transfers heat from helium leaving the turbine to helium entering the reactor. The other two heat exchangers, the cooler and intercooler, further reduce the temperature of the gas in the compressor.[15] Partly because the recuperator is very efficient in heating the gas returned to the reactor, the overall generating efficiency is expected to be about 48%, which is well above that of present reactors [9, p. 11]. The module sizes are expected to range between 450 and 600 MWt [*ibid*], corresponding to a module capacity in the neighborhood of 250 MWe.

---

[13]The specific heat of graphite is about 0.45 calories per gram, roughly half that of water, and the density of reactor-grade graphite (which is somewhat porous) is about 1.6 g/cm³. Thus, graphite's heat capacity per unit volume is slightly less than that of water, but the difference is minor compared to the difference in total mass. The thermal inertia is therefore much greater in an MHTGR than in an LWR.

[14]Arguments that the system is safe against air intrusion are presented in Ref. [25].

[15]The temperatures and pressures at various points in this cycle are: turbine inlet—850 °C, 7.02 MPa; turbine outlet—510 °C, 2.65 MPa; precooler inlet—131 °C, 2.62 MPa; compressor inlet—33 °C, 2.60 MPa; compressor outlet—112 °C, 7.24 Pa; reactor inlet—490 °C, 7.07 MPa [19]. Note: 1 MPa=9.9 atmospheres.

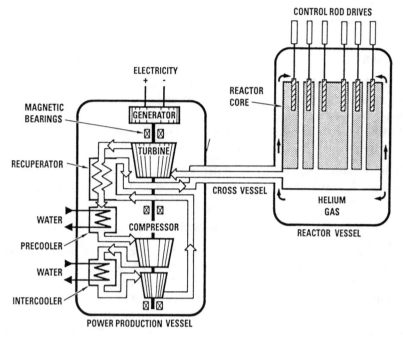

Figure 13.2. Schematic drawing of a GT-MHR power plant, showing the reactor on the right and the turbine on the left. The turbine is driven by helium from the reactor. (Courtesy of General Atomics.)

13.4 LIQUID-METAL REACTORS

13.4.1 Recent United States Programs

The Integral Fast Reactor System

Liquid-metal reactors (LMRs) have had a long history, primarily directed towards the development of breeder reactors. Several breeder reactors are now in operation worldwide (see Section 6.3.3). During the 1980s the United States program became centered upon the Integral Fast Reactor (IFR) system being developed at Argonne National Laboratory. The IFR system consists of the reactor and fuel reprocessing and fabricating facilities located in close physical proximity. Hence the name "integral."

In recent years, with little increase in the demand for nuclear power and some diminution in political support for it, interest in preserving a breeder reactor capability has decreased. At the same time there has been an increased interest in transmutation as a waste disposal option and in finding optimal ways of disposing of plutonium from dismantled nuclear weapons. In response to these trends, the emphasis in the IFR program and the associated reactor program switched from developing a breeding capability to developing an actinide-burning capability. LMRs could be used for either purpose, although a particular reactor would be designed with one or another goal in mind.

In the IFR fuel cycle, the spent fuel from the reactor is reprocessed and the separated actinides, including plutonium and uranium, are fabricated with fresh uranium into new fuel elements, which are returned to the reactor. The overall system has four components: the reactor, a reprocessing plant for spent fuel, a facility to fabricate new fuel elements, and facilities for waste handling and storage. The system is intended to achieve simultaneously a number of goals:

Passive safety. The reactor itself has passive features that should make it unusually safe.

Separation of chemical elements. Spent fuel from the reactor is divided into different output streams through a series of chemical and electro-refining steps, some at high temperatures, in an overall process termed *pyroprocessing*. The actinides can be separated out, leaving a fission-product waste stream from which most of the long-lived components in the spent fuel have been removed.

Actinide burning. The IFR has a closed fuel cycle, in which the actinides are to be returned to the reactor. The fast neutron spectrum of the LMR makes it possible to destroy the actinides by fission, thereby eliminating them from the long-term waste inventory.

Proliferation resistance. The co-located facilities make plutonium diversion difficult.

Despite its de-emphasis, the unique aspect of the IFR in the U.S. nuclear program is its relation to the ultimate development of a breeding capability. There are other avenues to passively safe reactors and coping with actinides. The potential of the IFR in these areas is a plus, but the main long-term motivation for pursuing the IFR was the belief that someday there may be a need for breeder reactors. Even if the first LMRs are not used as breeders, their construction would develop experience for a future breeder program. This argument cuts both ways. The fact that there is no urgent present need for breeders has caused some observers to think the IFR program serves no purpose which outweighs the increased risks of weapons proliferation which they believe would arise were the U.S. to abandon its present rejection of plutonium reprocessing.

Support for the IFR program was removed from the Department of Energy budget for the 1995 fiscal year, effectively terminating (or at least suspending) the IFR program. However, some related activities have continued, in particular work on possible pyroprocessing of the relatively small amount of spent fuel for which the DOE has immediate responsibility.

Reactor Options: PRISM and ALMR

Whatever the current prospects of LMRs, they remain of interest as a technically feasible and potentially important option. Recent U.S. development of LMRs have been based on experience with small test reactors, especially safety tests carried out with the Experimental Breeder Reactor II (EBR-II), a 57-MWt reactor (17 MWe) at Argonne National Laboratory West, in Idaho Falls. A commercial reactor design that builds on this experience was developed by General Electric. The original GE design for a passively safe LMR was based on a 155-MWe reactor module, termed the "power reactor inherently safe module" (PRISM); each module had its own steam generator, and several steam generators would drive a single turbine.

Subsequently, GE decided that it would be more cost-effective to raise the module size to the neighborhood of 300 MWe. In the revised design, two reactor modules comprise a 622-MWe power block operating at a 37% thermal efficiency [27]. The plant size could range up to 1866 MWe, given three power blocks. The reactor is now termed the advanced liquid-metal reactor (ALMR).

The ALMR has a cylindrical reactor vessel (30 ft in diameter and 62 ft long), filled with

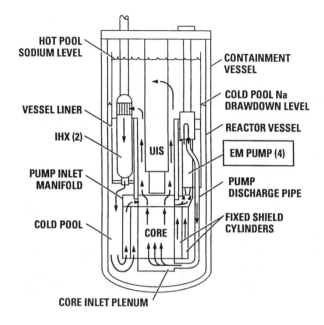

Figure 13.3. Sketch of the LMR system, showing the primary sodium flow. (Courtesy of GE Nuclear Energy.)

liquid sodium in a *pool* design (see Fig. 13.3). The reactor core sits near the bottom of the pool. Liquid sodium is drawn from the pool and circulated with electromagnetic (EM) pumps through the reactor and through two intermediate heat exchangers (IHX), also seated in the pool, and then back into the pool. This cycle constitutes the primary flow through the IHX. The secondary flow, also of liquid sodium, is piped from the IHX to a steam generator outside the containment vessel. The steam generator is a second heat exchanger, with sodium in the hot loop and water (or steam) in the "cold" loop. From this point, the system is functionally very similar to other reactors, with the steam driving a turbine, entering a condenser, and returning as feedwater to the steam generator.

13.4.2 Safety Features of the LMR

Sodium Coolant

The boiling point of liquid sodium at atmospheric pressure is 883 °C [28, p. 344]. The use of sodium, with its relatively high boiling point, gives several safety advantages:

- It is possible to operate at atmospheric pressure and still have a high coolant temperature under normal operating conditions, as desired for high thermal efficiency.[16] The sodium temperature in the primary flow past the core is 500 °C at the outlet [27].
- There is a large margin of safety before the boiling point is reached. The maximum

---

[16]The pressure is near atmospheric at the top of the sodium pool in the reactor vessel. There is a pressure gradient in the pool, with the pumps that drive the sodium circulation creating a pressure head at the bottom of the reactor vessel to drive the sodium upward through the reactor core.

temperature for either a failure of the control rods or a failure of the sodium-flow cooling system is estimated to be about 722 °C, well below the boiling point of sodium. In contrast, LWRs operate with a water or steam temperature of about 300 °C, and if the pressure or temperature shifts too much, one is faced with a complex two-phase system of steam and water.
- In accident situations, heat transfer from the reactor vessel can be primarily through radiation, because of the high temperature that can be safely reached by the sodium coolant. This feature applies also to the HTGR but not the LWR (see Section 13.3.3).
- Liquid sodium has a much higher thermal conductivity than does water, reducing thermal gradients.

Liquid sodium requires care in handling and in the choice of metals used in the system. Sodium (Na) reacts strongly with water and with oxygen in the air; sodium and, more so, sodium hydroxide (NaOH) are corrosive in interacting with some metals. However, LMR proponents argue that experience has shown corrosion by liquid sodium to be less of a problem than is corrosion by water in LWRs [29, p. 147].

Radioactivity is not a long-term problem with a sodium coolant. The main activity produced in the sodium is ^{24}Na from neutron capture in the stable isotope ^{23}Na. The half-life of ^{24}Na is 15 h, so the problem is one of protection during and immediately after operation, rather than of ultimate disposal.

Metallic Fuel

The fuel for the ALMR is expected to be metallic. The leading candidate is an alloy of uranium, plutonium, and zirconium [27, 30]. The alternative of using more familiar oxide fuels, such as mixed UO_2 and PuO_2, remains an option. The fuel would be packed in fuel rods and arranged in assemblies. The core is made up of these fuel assemblies along with assemblies of fertile ^{238}U. The fertile and fissile material of the core is surrounded by other elements to reflect neutrons back to the interior of the core and to shield against escaping neutrons.[17]

Metallic fuels offer the possibility of very high burn-up, because the problem of fuel swelling has been successfully addressed in the alloys now being tested. Fuel swelling is due to fission product gases that cause the fuel to expand and deform, with possible rupture of the cladding and escape of volatile fission products into the coolant. In present LMR fuel, after swelling reaches about 30%, gas bubbles form and escape from the fuel [31]. The fuel pin is designed to accommodate this swelling and has space to contain the escaped gas. With swelling no longer a major problem, it is possible to keep the fuel in the reactor for long periods and achieve a high burn-up of the fissile material. For some fuels tested in EBR-II, the burn-up has exceeded 18.5% of the fissile atoms [31], or about 180 GWd/t.[18]

Metallic fuel has another advantage over oxide fuels in that it has higher thermal conductivity. With better thermal conductivity, the temperature gradients in the fuel are smaller and the temperature of the fuel is closer to the temperature of the sodium coolant. For PRISM, the calculated average operating core temperature was about 1400 °C with mixed-oxide fuel, while with metallic fuel it was about 570 °C. A very high fuel temperature creates a

---

[17]One design for the ALMR core has 108 fuel assemblies and 84 blanket assemblies, as well as 180 assemblies in the reflector and shield and 19 assemblies for control, shutdown, and other purposes [27].

[18]The complete fission of 1 tonne of ^{235}U gives a thermal energy output of 950 gigawatt-days (see Table 7.2). Ignoring the small differences between ^{235}U and ^{239}Pu, this means, for example, that a 10% consumption (by number of atoms) corresponds to a burn-up of 95 GWd/t.

reactivity control problem, because there is *too much* negative Doppler feedback and the reactor is shut down more slowly.[19]

This seems like a paradoxical result. It may be understood as follows: In the shutdown of the reactor, it is important that the sodium not get too hot. From a safety standpoint, the sodium temperature is more crucial than the fuel temperature. As the sodium temperature rises, a number of negative feedbacks come into play (see below) and the reactivity of the reactor drops. At this point, a very large negative Doppler feedback in the fuel is undesirable, because the falling temperature increases the reactivity of the fuel. If this is a strong effect, it impedes the prompt shutdown of the reactor and the termination of heat input to the sodium. The effect is smaller at the comparatively low temperatures of the metallic fuel than it would be at the high temperatures reached with oxide fuel.

Feedback Mechanisms

In case of an abnormal occurrence, such as loss of cooling, the reactor is supposed to be shut down by rapid insertion of the control rods (scram). This requires some controls and is generally considered an "active" system. Should this fail, the safe shutdown of the reactor requires the dominance of negative passive feedback over positive feedback. The feedback mechanisms include:[20]

Thermal expansions (positive and negative). A rise in temperature of the sodium coolant heats other parts of the system, causing them to expand. This brings into play a variety of feedbacks, predominantly negative. The change in the relative position of the fuel elements and control rods causes a positive feedback. Negative feedbacks are provided by the decrease in the fuel density and by the lateral spreading of the fuel elements due to the expansion of the structure which holds them.

Doppler temperature coefficient (negative). This is the most familiar negative feedback, arising from the increase in neutron absorption in ^{238}U as the temperature of the fuel increases. This is an immediate feedback, with less delay than other feedbacks because it depends on the temperature of the fuel itself, rather than that of a somewhat removed part of the system.

Sodium density change (positive). This is closely connected to the "void coefficient." As the temperature of the sodium increases, it expands and its density decreases. This makes it less effective as a partial moderator, increasing the average neutron energy and the reactivity. Partly counterbalancing this is the fact that with a less "thick" sodium blanket around the core, more neutrons will escape the reactor entirely. Overall, however, for usual design configurations, the first of these effects predominates, giving a positive feedback.

Gas expansion modules (negative).[21] The gas expansion module (GEM) is used to exploit the effect mentioned in the previous item, the escape of neutrons when the thickness of the sodium blanket is reduced. A GEM is a gas-filled tube, open at the bottom and closed at the top. Six such modules are placed near the periphery of the ALMR core. Sodium rises in the GEM tube to a level that balances the sodium pressure and the gas pressure in the tube. In case of accident, the sodium pumping system shuts off, the pressure at the bottom of the reactor vessel drops, and the gas in the GEM pushes the

---

[19]The present discussion of these phenomena is based in part on Ref. [32].
[20]This list is based largely on Ref. [32], pp. 43–48, and Ref. [27].
[21]GEMs were not discussed in Ref. [32]. The information on them is from Ref. [27].

sodium downward. A lower sodium level in the GEM allows greater neutron leakage from the reactor.

Taking all the feedback mechanisms into account, the net result is a negative feedback sufficient to terminate the chain reaction ($k<1$) if the sodium temperature becomes too high. The temperature at which this happens is well below the boiling point of sodium.

Removal of Residual Heat

In the event of equipment failure and a reactor shutdown, heat can be removed from the core either by the normal circulation of sodium or, if the pumps fail, by convective circulation. The reactor vessel itself can be cooled, if necessary, by radiation to the surrounding containment vessel, which in turn is cooled by a natural convective air flow.

Safety Assessment

A probabilistic safety (or risk) assessment for the ALMR, carried out by the manufacturer, indicated risks well below NRC safety goals. For example, the chance that an individual in the vicinity of the plant will suffer a delayed fatal cancer as a result of an accident was calculated to be 1×10^{-10}/RY, compared to an NRC goal of 2×10^{-6}/RY (see Section 11.4.4) [27]. The analysis included both internal and external events.[22]

13.4.3 Tests with EBR-II

A series of tests has been carried out with EBR-II to determine if the actual performance of the system matches design expectations. Two tests are particularly cited as demonstrating fail-safe shutdown of this LMR system; these tests were carried out in April 1986 and are termed: (a) loss of heat sink without scram (LOHSWS) and (b) loss of flow without scram (LOFWS). We recapitulate a description given by R. R. Smith of these tests [32, pp. 18–25].

In test (a), the LOHSWS test, the pumps providing secondary flow of sodium to the intermediate heat exchanger (IHX) were turned off. Normally, the control rods would then be inserted, but this was intentionally disabled. Thus, the reactor continued to operate, as did the primary sodium pumping system. With no removal of heat in the IHX, the temperature of the sodium pool began to rise. The negative temperature coefficients associated with the higher sodium temperature caused the reactivity to decrease. The core inlet sodium temperature then continued to rise, but the core outlet temperature dropped, until the two temperatures were almost equal. After 18 minutes the power output of the reactor had dropped to "essentially zero." The inlet sodium temperature had risen about 40 °C, reflecting the rise in the bulk temperature of the sodium, and the outlet temperature had dropped about 50 °C. The temperature changes, in particular the temperature at which the reactor power reached zero, agreed well with predictions, confirming both the general operation and the detailed model. Thermal expansion (see above) was a major factor in the negative feedback.

In test (b), the LOFWS test, the primary and secondary sodium pumps were both turned

---

[22]Protection against earthquakes is provided by a seismic isolation system, in which mechanical bearings isolate the reactor containment from Earth motion. Use of seismic isolation is partly motivated by the fact that the ALMR operates at low pressure and therefore does not have as heavy a reactor vessel and pipe system as does an LWR operating at high pressures [33].

off, as might happen were all power lost. Again, the reactor was purposely not shut down with control rods. In the first instance, the outlet sodium temperature rose quite sharply, by about 110 °C in $1\frac{1}{2}$ min, because the flow through the core (now just convective) was much slower. However, the reactivity also dropped quickly, and after about three minutes (from the start of the test) the coolant outlet temperature began to drop. Due to poorer sodium circulation, the heating was more localized to the core area, than is the case when the primary pumps are operating, and different feedback mechanisms were dominant. In this case, the power level did not drop quite to zero during the time span of the test but was well under 5% within 10 min.

These tests were viewed by the Argonne group, which carried them out at Idaho Falls, as a tremendous success. In the words of Smith:

> In both tests the amount of time needed for remedial action was virtually unlimited. Without touching any control elements the operators could have gone home and returned days, weeks or even years later only to find the system in essentially the same condition as when they left. In the meantime inlet and outlet temperatures would have oscillated with very long periods—days [32, p. 25].

Having a delay between the onset of an accident and the time when remedial measures become urgently needed represents an important advance. Although the description here for the LMR is particularly vivid, it is also characteristic of other reactors with passive safety features. In contrast, at the Three Mile Island accident, the operators began taking remedial action almost immediately and, under confusing circumstances and the urgency to act, made a serious error within the first three minutes.[23] At Chernobyl, it was too late to react once there was unambiguous evidence of a developing accident.

13.5 PROSPECTS FOR NEW REACTORS

13.5.1 Institutional Status

NRC Licensing Schedules

New reactors are to be licensed by the NRC on the basis of a standardized design. This procedure has the advantage that once the design is approved, licensing of individual reactors will be simplified. The manufacturer and the NRC go through a long series of interactive applications, evaluations, and advisory opinions before reaching the culminating stages of *final design approval* and *standardized design certification*. Then, if there is a reactor order, an application can be made for a combined construction permit and operating license.

Four reactors have been going through the process, and their actual or expected dates for NRC's final design approval are given in Table 13.1. Standardized design certification is expected to come about 18 months later, allowing for a period of hearings.[24] The FDA date plus 18 months represents the earliest time at which a reactor might be ordered and construction begin. The first such opportunities would be for the ABWR and System 80+ reactors, which received FDA in July 1994 and, if the process proceeds smoothly, might receive standardized design certification in early 1996. However, it is quite possible that one of the

---

[23]In particular, the flow from high-pressure pumps was mistakenly reduced after $2\frac{1}{2}$ min, under conditions where it appeared urgent to do something quickly and the information available to the operators at the control panel was inadequate [34].

[24]The schedule of procedural stages is discussed, for example, in Ref. [36].

ABWRs under construction in Japan will be in operation before any ABWR is ordered in the United States and perhaps even before the NRC grants standardized design certification for any.

Support for Development

Support for development of new reactors has come from the U.S. Department of Energy, the Electric Power Research Institute (EPRI), and the reactor manufacturers. In the past DOE and EPRI have supported both evolutionary and advanced LWRs, as well as the modular high-temperature, gas-cooled reactor (MHTGR) and the liquid-metal reactor (LMR). However, the base of federal support has been shrinking. In principle, it would be possible for the reactor manufacturers to carry on with development, relying on their own funds or foreign support, but lack of federal help may be a major discouragement, especially for variants of the HTGR or LMR.

A consortium of utilities that goes by the name Advanced Reactor Corporation (ARC) has been established to coordinate private support for selected designs. The ARC consortium announced in early 1993 that it would support further development of the Westinghouse AP600 and the General Electric ABWR under a joint program with the DOE to fund "first-of-a-kind" engineering (FOAKE) [35]. The DOE share was to total $100 million over 5 yr. The award of FOAKE support for these reactors appeared to make the ABWR the front-runner among U.S. evolutionary reactors and the AP600 the front-runner among advanced passive LWRs, but in each case there is a significant competitor (see Table 13.1).

The lack of either FOAKE or direct federal support has put the HTGR and LMR programs in difficulties, with the latter effectively terminated, and there is presently little incentive for reactor manufacturers or utilities to commit major funds to long-range nuclear initiatives. However, the intriguing technical features of these reactors remain, and this could prompt a future revival of federal interest.

Ranking of Reactors in 1992 NAS Study

In response to a congressional request for a National Academy of Sciences study on "technological and institutional options for future nuclear power development," a study was carried out by a committee formed by the National Research Council and chaired by John Ahearne. The resulting report was issued in 1992 [9]. It emphasized the characteristics and prospects of new reactors, without taking a position on the broad issue of desirability of nuclear power. The reactors considered are listed in Table 13.1.[25] Ratings of these reactors on what the study took to be the most important considerations were:

Safety in operation. All of the reactors received a "high" rating, reflecting the committee's belief "that each of the concepts considered can be designed and operated to meet or closely approach the safety objectives currently proposed for future advanced LWRs" [9, p. 9]. These objectives are embodied in design requirements established by the Electric Power Research Institute in 1990. These stipulate a core damage frequency of less that 10^{-5} per reactor-year, as established by probability safety analysis, and a large release frequency of less than 10^{-6} per reactor-year, i.e., a release that causes a radiation dose in

---

[25]The NAS study considered earlier versions of two of these: the MHTGR, not the GT-MHR, and the PRISM reactor, not the ALMR (see Sections 13.3.1 and 13.4.1).

excess of 25 rem (0.25 Sv) at the boundary of the reactor site [9, p. 94].

Cost. The committee believed that the large evolutionary LWRs would be the least costly to build and operate, the other LWRs to be intermediate, and the HTGR and LMR the most expensive. The committee made no absolute cost estimates of its own, but it did report estimates made by the vendors for levelized generating costs that ranged from 3.0 to 3.3 ¢/kWh for the ABWR to 5.5 ¢/kWh for CANDU-3 (in 1989 dollars).

The committee gave encouragement to three categories of reactors:

1. **Large evolutionary reactors.** The committee judged these to be furthest along in the design and licensing procedures and likely to be lowest in cost. Further, in the committee view, "compared to current reactors, significant improvements in safety appear likely" and "significant improvements also appear possible in cost if institutional barriers are resolved" [9, p. 196]. Federal funding for these reactors did not appear essential, although it could speed the final development.

2. **Mid-size passive LWRs.** These, in the committee's judgment, were next closest to deployment, and the committee recommended federal funding to assist in the development of these reactors. The committee took no explicit position on the relative safety of these reactors and the evolutionary LWRs.

3. **Liquid-metal reactors.** The committee recommended an intermediate level of funding for this program, including some test reactor development, but no fuller package of support for the program. It assigned the LMR "the highest priority for long-term nuclear technology development" [9, p. 197]. The proposed support was intended to be sufficient to "maintain the technical capabilities of the LMR R&D community" [9, p. 198].

The committee did not recommend federal funding for any of the remaining options (explicitly naming the MHTGR, SIR, PIUS, and CANDU-3), on the grounds that their designs were further from the point where licensing would be possible and that the costs would be high.

13.5.2 Reactor Choices

In considering the alternative reactor types discussed above, the possible reactors fall into two broad categories: (1) evolutionary reactors, with capacities typically of 1000 MWe or greater; and (2) designs emphasizing passive safety, with capacities typically of 600 MWe or less.

The evolutionary reactors continue a long trend towards larger size, dating to the 1960s and 1970s. For example, 12 of the first 15 commercial reactors for which construction permits were granted in the U.S. had capacities of under 100 MWe, and the median was only about 60 MWe. Starting in 1963 the size jumped to 400 MWe or more. From 1973 onward reactors were all in the 800 to 1300 MWe range, with most in excess of 1000 MWe. The increases in size were motivated by expected economies of scale, which may still favor such reactors, as suggested by both the National Academy of Sciences study (see preceding section) and foreign practice.

However, the other reactors being considered all represent a move to smaller size, in some cases based on modules that can be combined to give a generating facility with a large total capacity. Arguments for small reactors, in preference to large ones, include: a better match to electricity demand, both for slow growth in the U.S. and for developing countries; shorter construction time; less financial risk; and the possibility of greater off-site construction, in assembly-line factories.[26] Further, with a smaller reactor it is easier to prevent accidents, and should there be one, the maximum size of the accident is less.

There is somewhat of a logical coupling between small size and passive safety. Consider the AP600: a relatively low ratio of reactor core power to containment volume makes it possible to use water reservoirs within the tank for emergency cooling and keeps the pressure in the containment from rising excessively. Similarly, in the HTGR and LMR, the peak core temperature reached in an accident would be less with a small core than with a large one, because of better heat transfer from the core to the wall of the reactor vessel and beyond.

Despite the arguments for small reactors, most of the reactors actually brought on-line in the past decade and most of the commercial reactors now under construction are in the 1000-MWe range. This is true for both the U.S. and the world. In particular, as of the end of 1994 there were 66 reactors nominally on order or under construction worldwide, with a total capacity of 54 GWe [14]. This corresponds to an average capacity of over 800 MWe. Thus, the arguments for going to smaller size have not yet proved decisive. Further, in countries that seem particularly determined to continue nuclear power development, e.g., France and Japan, near-term planning appears to be based on large LWRs. Reactors under construction average about 1000 MWe for Japan and about 1400 MWe for France [14].

The degree to which the supposedly ultra-safe reactors will be safer than new versions of the more conventional LWR is a matter of debate. The validity of the distinction can only be evaluated in terms of detailed safety studies of specific designs. In general, the safety features of the "inherently safe reactors" are more transparently obvious than those of the advanced LWRs, but the latter are building upon a better-known technology that is less susceptible to surprises.

It is possible to think of the various reactor possibilities in terms of "generations." Thus, present reactors represent the first generation, the evolutionary and passive LWRs a second generation, and the remainder of the reactors in Table 13.1 a third generation. Given the impetus to do so, the second generation could begin to be constructed in the United States in the late 1990s, with operation beginning about five years later. The third generation would probably lag significantly. Outside the U.S., as in the Japanese construction of the evolutionary ABWR (see Section 13.2.1), initial steps have been taken into the second generation.

13.5.3 Radical Nuclear Alternatives to Present Reactors

Fusion

The traditional nuclear alternative to fission is fusion. There has been continued study of fusion, with significant but slow progress. Fusion research and development has now become a very expensive matter, motivating international collaborations. The next hope for major further progress centers upon the building of the International Thermonuclear Experimental Reactor (ITER), which is being planned in a collaboration involving the United States, Europe, Japan, and Russia. If ITER is built and tests with it are successful, a subsequent step would be to build a demonstration power plant.

It is not possible to say with any confidence when, if ever, this program will come to fruition. Most observers doubt that electric power from a prototype fusion reactor could come before the year 2030, and some would consider this to be an optimistic target date.

---

[26]A comprehensive early review of the argument for small reactors is given in Ref. [37].

Nonetheless, in principle, fusion offers the prospect of virtually unlimited electrical power, probably with fewer concerns about environmental contamination and weapons proliferation than is the case with fission power.

Accelerator-Driven Fission

In recent years a quite different approach has been suggested for obtaining energy from fission. In this approach, a high-energy beam from a proton accelerator produces a large number of neutrons in collisions with heavy nuclei, in a process known as *spallation*. A geometric arrangement is used so that the spallation neutrons irradiate uranium or thorium, initiating fission. The configuration of fissile material is sub-critical and the spallation neutrons are required to maintain the level of neutron flux and reactor output. The sub-critical chain reaction serves to amplify the total number of neutrons above those produced by spallation alone. Thus, many neutrons are produced per initial proton, and the total energy released from fission in a properly designed reactor can considerably exceed the input energy required to run the accelerator.

There are two prominent suggestions for implementing this approach:

Los Alamos proposal. A Los Alamos group has suggested a system in which 1.6-GeV protons irradiate a heavy target (e.g., lead), spallation neutrons from the target are moderated in heavy water, and the resulting thermal neutrons produce fission in the fissile fuel (^{233}U or ^{235}U) [38]. The heavy spallation target would be in the form of molten metal, the fissile fuel in the form of a molten salt, and the fertile fuel (^{232}Th) mixed with heavy water. All of these components would flow in recirculating streams. Given a high-current proton beam and a large spallation yield, the flux of neutrons in the reactor is much higher than in normal reactors. Many of the radioactive products of fission or neutron capture can then be transmuted into products of shorter half-life, reducing the inventories of wastes that require disposal. The system assumes chemical separation of selected radionuclides from the circulating fuel and their return to the reactor for bombardment and transmutation.

CERN proposal. A design developed at CERN is similar to conventional reactors in that the fissile and fertile materials are in solid form and the fuel and moderator are both contained in a reactor vessel [39]. However, the fuel is in a configuration that makes the assembly sub-critical. Again, the system is fed by neutrons produced in proton-induced spallation reactions, either in the fuel and moderator or in a separate target. Actinides are consumed in this cycle, but there is no intent to transmute fission products.

Both systems have little danger of a criticality accident because their fuel is in a sub-critical arrangement. They are envisaged for use in a ^{232}Th–^{233}U breeding cycle, which could greatly extend energy supplies. In this cycle there is less production of long-lived actinides than in ^{238}U reactors, thereby reducing proliferation and waste disposal problems. However, for the case where the reduction is greatest, the Los Alamos approach, it is necessary to compare these gains against the possible hazards created by the chemical handling of the radioactive waste streams from the circulating fuel.

As yet neither of these accelerator proposals has received the critical examination that conventional reactors have received. They clearly represent an ingenious application of accelerator technology to the extraction of energy from nuclear fission. But it is too soon to judge their practicality, the possible hazards that they might create, and their overall advantages or disadvantages compared to more traditional techniques. As yet no major programs have been announced to advance these reactors beyond the present conceptual designs.

REFERENCES

1. Alvin M. Weinberg and I. Spiewak, "Inherently Safe Reactors and a Second Nuclear Era," *Science* 224 (1984): 1398–1402.
2. A. M. Weinberg, I. Spiewak, J. N. Barkenbus, R. S. Livingston, and D. L. Phung, *The Second Nuclear Era: A New Start for Nuclear Power* (New York: Praeger, 1985).
3. Irving Spiewak and Alvin M. Weinberg, "Inherently Safe Reactors," *Annual Review of Energy* 10 (1985): 431–462.
4. A. M. Weinberg, I. Spiewak, D. L. Phung, and R. S. Livingston, "The Second Nuclear Era: A Nuclear Renaissance," *Energy* 10 (1985): 661–680.
5. C. W. Forsberg and A. M. Weinberg, "Advanced Reactors, Passive Safety, and Acceptance of Nuclear Energy," *Annual Review of Energy* 15 (1990): 133–152.
6. U.S. Nuclear Regulatory Commission, *Safety Goals for Nuclear Power Plants: A Discussion Paper*, report NUREG-0880 (Washington D.C.: 1982).
7. Eric S. Beckjord, "Safety Aspects of Evolutionary and Advanced Reactors," in *Proceedings of the Conference on Technology-Based Confidence Building: Energy and Environment*, edited by John C. Allred, Roger C. Eckhart, and Arthur S. Nichols (Los Alamos, N.M.: Center for National Security Studies, LANL, 1989), 29–37.
8. Forrest J. Remick, "Large Release Guideline as Part of Safety Goals," memorandum to J. M. Taylor, March 2, 1993.
9. National Research Council, *Nuclear Power: Technical and Institutional Options for the Future*, report of the Committee on Future Nuclear Power Development, John F. Ahearne, chairman (Washington, D.C.: National Academy Press, 1992).
10. Alvin M. Weinberg, "Nuclear Energy, Carbon Dioxide, and International Cooperation, in *Proceedings of the Conference on Technology-Based Confidence Building: Energy and Environment*, edited by John C. Allred, Roger C. Eckhart, and Arthur S. Nichols (Los Alamos, N.M.: Center for National Security Studies, LANL, 1989): 474–477.
11. John J. Taylor, "Improved and Safer Nuclear Power," *Science* 244 (1989): 318–325.
12. Electric Power Research Institute, "Reopening the Nuclear Option," *EPRI Journal* 19, no. 8 (December 1994): 7–17.
13. C. W. Forsberg, L. J. Hill, W. J. Reich, and W. J. Rowan, *The Changing Structure of the International Commercial Nuclear Power Reactor Industry*, report ORNL/TM-12284 (Oak Ridge, Tenn.: Oak Ridge National Laboratory, 1992).
14. "World List of Nuclear Power Plants," *Nuclear News* 38, no. 3 (March 1995): 27–42.
15. U.S. Nuclear Regulatory Commission, "NRC issues Final Design Approval for GE advanced boiling water reactor" and "NRC issues Final Design Approval for ABB-CE System 80+ standard nuclear plant," NRC press releases (Washington, D.C.: NRC, July 1994).
16. D. R. Wilkins and J. Chang, "GE Advanced Boiling Water Reactors and Plant Systems Design," paper presented at the 8th Pacific Basin Nuclear Conference, San Jose, Calif., April 1992.
17. Käre Hannerz, "Making Progress on PIUS Design and Verification," *Nuclear Engineering International* 33, no. 412 (1988): 29–31.
18. Ariane Sains, "With Italian Prospects Dim, ABB Will Shelve PIUS Design," *Nucleonics Week* 35, no. 4 (January 27, 1994): 14–15.
19. W. A. Simon, J.-D. Wistrom, and K. T. Etzel, "The MHR-A Universal Heat Source for the Next Century," General Atomics preprint GA-A21455 (San Diego: September 1993).
20. F. J. Rahn, A. G. Adamantiades, J. E. Kenton, and C. Braun, *A Guide to Nuclear Power Technology* (New York: John Wiley, 1984).
21. "International Briefs," *Nuclear News* 34, no. 3 (March 1991): 56.
22. *Coated Particle Fuel Technology For Modular HTGR Reactor Systems*, rev. 1, General Atomics report (August 14, 1993).
23. Paul R. Kasten, "The Safety of Modular High Temperature Gas-Cooled Reactors," paper V.A.2 in *Small & Medium Sized Nuclear Reactors: You and Your Environment* (San Diego: General Atomics, 1989).
24. T. Dunn, J. Cardito, and J. Cunliffe, "Perspectives of Modular High Temperature Gas-Cooled Reactor (MHTGR) on Effluent Management and Siting," paper III.6 in *Small & Medium Sized Nuclear Reactors: You and Your Environment* (San Diego: General Atomics, 1989).
25. G. J. Cadwallader, "MHTGR—New Production Reactor, Safety Significance of Air Ingress," General Atomics report (San Diego: March 1988).
26. L. M. Lidsky, D. D. Lanning, J. E. Staudt, and X. L. Yan, "A Direct-Cycle Gas Turbine Power Plant for Near Term Applications: MGR-GT," paper presented at the 10th International HTGR Conference, San Diego, September 1988.

27. P. M. Magee, E. E. Dubberley, A. J. Lipps, and T. Wu, "Safety Performance of the Advanced Liquid Metal Reactor," paper presented at the ARS '94 Topical Meeting—Advanced Reactors Safety, Pittsburgh, April 1994.

28. Yeram S. Touloukian, "Thermal Physics" in *A Physicist's Desk Reference, The Second Edition of Physics Vade Mecum*, edited by Herbert L. Anderson (New York: American Institute of Physics, 1989): 336–347.

29. Charles E. Till and Yoon I. Chang, "The Integral Fast Reactor," *Advances in Nuclear Science and Technology*, 20 (1988): 127–154.

30. B. R. Seidel, L. C. Walters, and Y. I. Chang, "Advances in Metallic Nuclear Fuel," *Journal of Metals* 39, no. 4, (April 1987): 10–13.

31. Y. I. Chang and C. E. Till, "Advanced Breeder Cycle Uses Metallic Fuel," *Modern Power Systems* 11, no. 4 (April 1991): 59.

32. R. R. Smith, "The Quenching Temperature and the Roles of Reactivity Feedback in the Development of Passively Safe Nuclear Powerplants" (1987).

33. Emil L. Gluekler (GE Nuclear Energy), private communication with author, May 9, 1994.

34. *Report of the President's Commission on the Accident at Three Mile Island*, John G. Kemeny, chairman (New York: Pergamon Press, 1979).

35. "$158 Million Contract to Support AP600 Work," *Nuclear News* 36, no. 7 (May 1993): 24.

36. James M. Taylor, "Integrated Review Schedules for the Evolutionary and Advanced Light Water Reactor Projects," Nuclear Regulatory Commission Policy Issue memorandum SECY-93-097, April 14, 1993.

37. Joseph R. Egan, *Small Power Reactors in Less Developed Countries: Historical Analysis and Preliminary Market Survey* (Westmont, Ill.: ETA Engineering, October 1981).

38. C. D. Bowman *et al.*, "Neutron Energy Generation and Waste Transmutation Using an Accelerator-Driven Thermal Intense Neutron Source," *Nuclear Instruments and Methods in Physics Research* A320 (1992): 336–367.

39. R. Carminati, R. Klapisch, J. P. Revol, C. Roche, J. A. Rubio, and C. Rubbia, "An Energy Amplifier for Cleaner and Inexhaustible Nuclear Energy Production Driven by a Particle Beam Accelerator," preprint CERN/AT/93-47(ET) (November 1, 1993).

Chapter 14

Nuclear Energy and Nuclear-Weapons Proliferation

14.1 CONCERNS ABOUT LINKS BETWEEN NUCLEAR POWER AND NUCLEAR WEAPONS

Many observers believe that the most profound problem with using nuclear energy for electricity generation is the connection between nuclear power and nuclear weapons. In this view, the threat of nuclear-weapons proliferation increases if the world relies on nuclear power, because nuclear power capabilities could be translated into nuclear-weapons capabilities.

Assuming that fossil fuels cannot indefinitely remain the world's main source of energy, it will be necessary in the future to rely on renewable energy in one or another of its forms, on nuclear energy, or on both. The relative merits of renewable energy and nuclear fission energy (omitting fusion as speculative) are highly controversial, with unresolved arguments over relative economic costs, environmental impacts, and safety. However, the weapons connection is unique to nuclear power and for some people provides a strong motivation for limiting or abandoning it.

Giving up nuclear power would clearly avert the danger that nuclear power facilities might be modified for nuclear weapons purposes. But it would not avert all dangers of weapons development. It is quite possible to have nuclear weapons without nuclear power as well as nuclear power without nuclear weapons. In fact, most countries that have nuclear weapons or are suspected of having them had those weapons well before they had civilian nuclear power. Other countries that rely heavily on nuclear power, such as Sweden and Switzerland, are rarely perceived to be potential nuclear weapons threats.

Nonetheless, a program in one area can aid a program in the other. The connections will be considered in more detail in this chapter, and the rudiments of bomb technology and the history of weapons proliferation will also be discussed. The goal of this analysis is to understand something of the nature of proliferation dangers and the extent that nuclear energy policies might have an impact on these dangers.

14.2 NUCLEAR EXPLOSIONS

14.2.1 Basic Characteristics of Fission Bombs

Fissionable Materials for Nuclear Weapons

In a nuclear weapon it is necessary to have a rapidly developing chain reaction. Therefore, the time interval between successive fission generations must be short, and the chain reaction is propagated by unmoderated neutrons from fission, which typically have energies E_n in the neighborhood of 1 MeV. It is also necessary to have a high multiplication factor. Therefore the weapons material must have a large neutron fission cross section for $E_n \approx 1$ MeV, and the number of neutrons emitted per fission, ν, must be relatively large. The materials that meet these criteria are the same as those that can serve as the fissile fuel in nuclear reactors:

- **^{235}U.** The bomb dropped at Hiroshima was a ^{235}U bomb. For reactors, uranium is typically enriched to 2% to 5% in ^{235}U. Uranium enriched to more than 20% is termed "highly enriched." For nuclear weapons, it is the norm to use ^{235}U enrichments in the neighborhood of 90% or more, but it is possible to make a bomb with enrichments well below 90% (see Sec. 14.2.7). The enriched material is obtained by isotopic separation, starting with natural uranium.

- **^{239}Pu.** The bomb dropped at Nagasaki was a ^{239}Pu bomb. A plutonium bomb requires about 5 kg of ^{239}Pu, which must not be excessively contaminated with ^{240}Pu if the bomb is to have a reliably high yield. (The problems created by ^{240}Pu are discussed in Sec. 14.2.5.) Both ^{239}Pu and ^{240}Pu are produced in any reactor that uses ^{238}U as a fertile fuel. The plutonium is extracted from the spent fuel by chemical separation; the more demanding step of isotopic enrichment is unnecessary with ^{239}Pu.

- **^{233}U.** No ^{233}U bombs have been made, and there have been no apparent incentives to produce them. ^{233}U could in principle be produced in a reactor that uses ^{232}Th as a fertile material. A bomb based on ^{233}U would require uranium that is highly enriched in ^{233}U. However, the chief interest in ^{233}U is not as a bomb material but as the fissile component in a possible ^{232}Th–^{233}U fuel cycle, where ^{232}Th is the fertile component. This cycle may have advantages in protection against weapons proliferation (see Sec. 14.5.4).

Types of Nuclear Bombs

The minimum mass sufficient to sustain a chain reaction is known as the "critical mass." This is the mass for which the neutrons produced in fission just balance the neutrons that escape. For an effective nuclear weapon, the critical mass must be assembled quickly. Two general approaches are used to achieve this (see Fig. 14.1):

Gun assembly. In the gun-type approach, two subcritical masses of fissionable material are brought together by firing one into the other, or by firing both. This can be done in a straightforward way using a chemical propellant. The assembly speed depends upon the velocity imparted to the mass or masses. As discussed in Sec. 14.2.4, the gun assembly method is too slow for use with ^{239}Pu bombs and is only used with ^{235}U bombs.

Implosion technique. In the implosion approach, a given mass is changed from subcritical to critical by compression, produced by chemical explosives surrounding the mass (see Sec. 14.2.2). The arrangement of the explosives and the timing of their firing require a more sophisticated design than is necessary in the gun assembly. The motivation for making this extra effort is speed of assembly. The implosion technique has been used for both ^{235}U and ^{239}Pu bombs.

Many of the nuclear weapons developed since World War II obtain a major fraction of their explosive energy from the fusion of hydrogen, in conjunction with fission. However, such "hydrogen bombs" are more complicated to build than are simple fission bombs, and while they may be more dangerous in terms of the size of the explosion produced, they are less immediate threats from the standpoint of weapons proliferation.

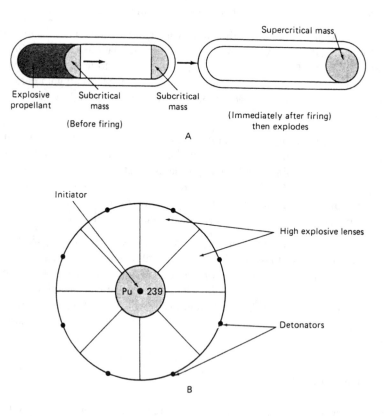

Figure 14.1. Schematic sketches of fission bomb designs. Top: gun-type bomb; bottom: implosion-type bomb [from Paul P. Craig and John A. Jungerman, *The Nuclear Arms Race: Technology and Society*, 2nd ed. (New York: McGraw-Hill, 1990), 213; with permission of McGraw-Hill].

Table 14.1. *Properties of fissile materials for nuclear weapons: fission cross section, neutrons per fission, mean free path at 1 MeV, and critical mass and radius.*

Material	E_n=1 MeV		Density (g/cm^3)	λ (cm)	No Reflector		With Reflector	
	σ_f (b)	ν			M_c (kg)[a]	R_c (cm)	M_c (kg)[b]	R_c (cm)
^{235}U	1.22	2.52	19	17	52	8.7	13–25	≈ 6
^{239}Pu	1.73	2.95	19.6^c	12	10	5.0	5–10	≈ 4
^{233}U	1.9	2.5	19	11	16	5.9	5–10	≈ 4

Sources: σ_f and ν data for ^{235}U and ^{239}Pu are from Ref. [2], p. 20; σ_f and ν data for ^{233}U are from Ref. [8]. M_c data are from Ref. [5], p. 24.
[a]Values of M_c identical to those listed here are given in Ref. [1] for ^{235}U, specified as "weapons-grade" (94% ^{235}U), and for ^{239}Pu, presumably in an isotopically pure form.
[b]The critical mass depends upon the reflector material and configuration. Critical masses as low as 4 kg are cited for ^{239}Pu and ^{233}U in Ref. [7], p. 908. In Ref. [6] the critical masses for 100% pure ^{235}U and ^{239}Pu, in each case surrounded by a thick natural uranium reflector, are given as 15 and 4.4 kg, respectively.
[c]This density is for plutonium metal in the α phase; for the other phase most relevant to plutonium explosives, the δ phase, the density is 15.7 g/cm^3 [1].

Energy Yield of Nuclear Weapons

The energy yield of nuclear weapons is commonly expressed in kilotons (kt) or megatons (Mt) of TNT equivalent, where 1 kiloton of TNT is assumed to release 10^{12} calories (4.18×10^{12} J). The complete fission of ^{235}U in a reactor releases 8.2×10^{13} J/kg (see Table 7.2). About 85% of the energy comes from the fission fragments themselves and 5% from prompt neutrons and gamma rays (see Sec. 4.4.2). Therefore, complete fission of 1 kg of ^{235}U would give a prompt explosive yield of about 7×10^{13} J, or 17 kt. The yield for ^{239}Pu is similar.

Actual yields in nuclear weapons are less than 17 kt/kg, because a bomb will disassemble without complete fissioning of all the material. For example, the world's first nuclear bomb, used in the Trinity test in New Mexico in July 1945, is reported to have had a mass of 6.1 kg [1, p. 127] and a yield of 18.6 kt.[1] This corresponds to a yield of 3 kt/kg, or an efficiency of under 20%.[2] Modern bombs are more efficient, reportedly "approaching 40 per cent" [4, p. 266].

14.2.2 Critical Mass for Nuclear Weapons

Critical Mass with and without Reflectors

For a sphere of fissionable material, the mass will be less than the critical mass if the radius is small compared to the mean free path for fission, λ. A crude initial estimate is that the critical radius R_c is roughly equal to λ. For 1-MeV neutrons in ^{235}U, $\lambda = M/\rho N_A \sigma_f = 17$ cm. (See Sec. 3.1.2 for the expression for λ and Table 14.1 for numerical values of the fission cross section σ_f and the uranium density ρ.)

While λ sets a crude scale for the dimensions, its use to determine R_c corresponds to a grossly oversimplified geometric picture. Furthermore, this description does not take into

---

[1]Richard Rhodes, in the Introduction to Ref. [2].
[2]The yields of the two bombs dropped during World War II were about 15 kt for the Hiroshima bomb and about 21 kt for the Nagasaki bomb [3, p. 15]. The Nagasaki bomb reportedly used 6.2 kg of plutonium, again corresponding to an efficiency of about 20% [4, p. 266].

account two crucial nuclear parameters: the number of neutrons produced per fission, ν, and the cross section for elastic scattering of neutrons. Larger ν means lower R_c, all other things being equal, and elastic scattering on average increases the path length traversed by a neutron before it can escape, and therefore decreases R_c. A somewhat better calculation for ^{235}U gives a critical radius R_c of about 9 cm [2, p. 74], equal to roughly one-half of the mean free path λ. The corresponding uranium mass is $M_c = 4\pi\rho R_c^3/3 \approx 60$ kg.

More accurate values of the critical mass for ^{235}U, along with values for ^{239}Pu and ^{233}U, are given in Table 14.1, assuming the fissile material to be isotopically pure, or almost so, and in metallic form. Results are given for a bare device and for one with a surrounding reflector. The reflector returns escaping neutrons to the fissile volume by one or more elastic scattering collisions; this can reduce the critical mass by a factor of 2 or 3 (see Table 14.1). Many materials can be used for the reflector, including those as diverse as uranium and beryllium. Reflectors cited in discussions of weapons include examples with thicknesses of 4 cm of uranium [5, p. 24], 15 cm of uranium [6, p. 412], and 15 cm of beryllium [7, p. 909]. In addition to reflecting neutrons, the mass of the reflector, also known as a "tamper," acts to slow the dispersion of material after the explosion, thereby sustaining a criticality for a longer period of time.

The critical mass for ^{239}Pu is considerably less than that for ^{235}U, a consequence of the former's higher fission cross section σ_f and higher number of neutrons per fission, ν. The critical mass is also small for ^{233}U, but there is no public record that such bombs have ever been built.

Critical Mass and Implosion

The critical mass is reduced below the values given in Table 14.1 if the density of the material is increased through compression. The mean free path for fission, λ, and therefore approximately R_c, is inversely proportional to the density ρ. The critical mass therefore varies as ρ^{-2}. This is the principle underlying implosion bombs. While the actual masses made possible by this approach remain secret, arms experts often suggest the importance of this high compression. For example, Theodore Taylor, a physicist with extensive experience in the field, cautions:

> It should not be concluded that the minimum amount of ^{235}U from which a fission explosive can be made is in the range of 10–20 kg, corresponding to the critical mass at normal density in a reasonably good neutron reflector. It is well known that "implosion" types of fission explosives achieve supercriticality by compression of the core substantially above normal density. . . . How little fissionable material can be used in a practical fission explosive depends on the knowledge, ingenuity, and skills of the explosive designers and fabricators [6, p. 412].

Of course, the same considerations apply to plutonium bombs. In the same vein, a group of nongovernmental arms experts state: "Most weapons in the U.S. arsenal are believed to use only a fraction of a critical mass (at normal density)." [5, p. 25].

Nominal Numbers for a Nuclear Weapon

The actual critical mass for nuclear weapons cannot be precisely stated. Different designs give different results, and the nature of advanced designs and the resulting sizes are not made public by weapons builders. However, it is useful to have a nominal number to provide a measure of the amount of material needed. One author suggests the following nominal numbers for pure material in metallic form: 10–11 kg for ^{235}U and 5 kg for ^{239}Pu [7, p. 907]. The numbers are useful for purposes of qualititative orientation, even if advanced technology may make possible the building of weapons with less material.

The possibility of making bombs with considerably smaller amounts of material has been suggested in a study published by the National Resources Defense Council (NRDC), assuming that the bomb builders have sufficient technical capability and a willingness to settle for a relatively small bomb [9]. The amounts quoted in the previous paragraph correspond, in the NRDC summary, to the amounts needed for a bomb with a 10-kt yield, built using "low technology." On the other hand, according to the NRDC authors, as little as 1 kg of ^{239}Pu or 2.5 kg of uranium highly enriched in ^{235}U would suffice for a 1-kt bomb if the builders had a "high technical capability." One important factor in reducing the amount of material needed is the achievement of high compression during the implosion of the bomb material.

The masses suggested in the NRDC study are on the low end of published estimates. A somewhat higher mass is implied in a 1994 National Academy of Sciences report on plutonium handling, which states that "[S]everal kilograms of separated weapons-grade plutonium . . . . would be enough to build a nuclear weapon" [10, p. 29].

Settling upon a best estimate of the amount of material needed to produce a bomb of a given size at a given level of technology requires information that is currently classified. Furthermore, designations such as "high" and "low" technology capability are not well defined. Regardless of which of the published numbers are used, it is clear that highly damaging weapons can be made with relatively small amounts of fissionable material. For the present purposes, it suffices to use the nominal numbers cited earlier, i.e., about 10 kg for ^{235}U and 5 kg for ^{239}Pu.

14.2.3 Buildup of a Chain Reaction

The total yield of a nuclear weapon greatly exceeds the energy necessary to blow the material of the weapon apart, and dispersion of the fuel begins before the developing chain reaction builds up to its full design level. For an effective bomb, this buildup should occur in a time that is short compared to the time for appreciable disassembly of the material. This means that the time between successive fission generations must be short, and the chain reaction must therefore rely on fast neutrons, not on the thermal neutrons that propagate the chain reaction in most nuclear reactors.[3]

The time rate of change in the number of neutrons N present in the system due to fission is

$$\frac{dN}{dt} = \frac{k-1}{\tau} N = \alpha(t)N, \tag{14.1}$$

where k is the effective multiplication factor, τ is the mean time between successive fission generations, and $\alpha=(k-1)/\tau$. [This is essentially the same as Eq. (5.16), with some minor changes in notation.] For a reactor, the effective multiplication factor k must be very close

---

[3]The discussion in this section draws to a considerable extent on *The Los Alamos Primer*, which originally appeared as notes of lectures delivered at Los Alamos by Robert Serber during World War II and was reissued in 1992 with explanatory notes and emendations by Serber [2]. It provides a simple introduction to the physics of nuclear weapons.

to unity. For a bomb, it will be larger. Both k and α may vary with time, as explicitly suggested for α in Eq. (14.1).

The time between successive fission generations is $\tau = \lambda/v$, where λ is the mean free path of the neutrons and v is their average velocity. A typical fission neutron has an energy in the neighborhood of 1 MeV and therefore a velocity of 1.4×10^9 cm/sec. As mentioned in Section 14.2.2, $\lambda = 17$ cm for a 1-MeV neutron in ^{235}U. Therefore, the time between successive fission generations in ^{235}U is $\tau \approx 10^{-8}$ sec. Similar calculations for ^{239}Pu also give $\tau \approx 10^{-8}$ sec.

The fission rate in a chain reaction rises rapidly as $e^{\alpha t}$. The logarithm of the number of neutrons present at time T is equal to the integral $\int \alpha(t) dt$, integrated from the start of the chain reaction (by one neutron) to T. Roughly speaking, for a large multiplication factor, this is also the cumulative number of fissions. In 1 kg of fissile material there are about 2.5×10^{24} nuclei. Therefore, for complete fission of 1 kg of material the integral must reach a numerical value close to $\ln(2.5 \times 10^{24}) = 56$. For a case where k and α are constant with a doubling in each generation, this would correspond to 81 generations for the fission of 1 kg. It would take only a few more generations to fission 10 times this amount or to make up for a slightly lower multiplication rate. Thus, under these conditions the chain reaction develops completely in a time on the order of 10^{-6} sec. This time is short compared to the time for the material to disperse physically and become subcritical.

14.2.4 Explosive Properties of Plutonium

Different Grades of Plutonium

Manufacturing a bomb from plutonium extracted from a reactor faces a special difficulty due to the presence of ^{240}Pu. This isotope has a half-life of 6570 years. Its primary decay mode is by alpha-particle emission, but it also sometimes decays by spontaneous fission. This fission produces neutrons that may cause premature initiation of a chain reaction in a bomb before the fissile material has been fully brought together or compressed. This is known as *predetonation*. With predetonation, the weapon gives a much smaller explosion than designed; in other words, it "fizzles."

During the World War II atomic bomb program, the discovery that ^{240}Pu had a high rate of spontaneous fission came as a surprise and appeared briefly to be a major threat to the development of a ^{239}Pu bomb.[4] Part of the solution to this problem was simple. If the fuel is withdrawn from the reactor before a long exposure, there is insufficient time for a large concentration of ^{240}Pu to form; this results in "weapons-grade" plutonium. The plutonium from a power reactor, operated in a normal cycle, has a much higher ^{240}Pu concentration and is termed "reactor-grade" plutonium; it is unsuitable for the production of efficient, reliable weapons. However, it is well accepted that sophisticated designers could make some sort of weapon even with reactor-grade plutonium (see, e.g., Ref. [11], p. 259).

Information on the issues involved has gradually entered the unclassified open literature. A brief review by J. Carson Mark, formerly director of the Theoretical Division at Los Alamos National Laboratory, has put some of the key information in a generally available form [1]. This section draws heavily from his exposition. As a preliminary, some information on different grades of plutonium is summarized in Table 14.2, based on information from Mark's paper [1].

---

[4]See, e.g., Ref. [2], p. 55 for a brief discussion of this surprise.

Table 14.2. *Properties of different grades of plutonium: isotopic abundances, neutron emission from spontaneous fission (SF), and decay heat from radioactive decay.*

	Grade of plutonium		
	Super	Weapons	Reactor[a]
Fraction (by mass)			
^{238}Pu		0.0001	0.013
^{239}Pu	0.98	0.938	0.603
^{240}Pu	0.02	0.058	0.243
^{241}Pu[b]		0.0035	0.091
^{242}Pu		0.0002	0.050
SF neutrons per g sec	20	66	360
Decay heat (W/kg)	2.0	2.3	10.5

Source: Ref. [1].
[a]For LWR fuel with a 33-GWd/t burn-up, reprocessed after 10 years.
[b]Includes ^{241}Am.

The ^{240}Pu arises from neutron capture in ^{239}Pu. Initially, starting with enriched uranium fuel, the ^{239}Pu content rises linearly with fuel burn-up and the ^{240}Pu content rises quadratically. As the burn-up progresses, the description of the rates of rise becomes less simple, but the ratio of ^{240}Pu to ^{239}Pu continues to increase. For PWR fuel with a burn-up of 33 GWd/t, the isotopic abundance of ^{240}Pu is about 24% (see Table 14.2). This is typical of present-day spent fuel.

The neutron emission rate from spontaneous fission is about 910 (g sec)$^{-1}$ for ^{240}Pu [1, p. 115], contributing about 220 neutrons per second per gram of reactor plutonium. ^{238}Pu and ^{242}Pu also have sizable spontaneous fission rates, and the total neutron emission rate is 360 (g sec)$^{-1}$ for reactor plutonium [1, p. 122]. If one assumes a bomb with a total plutonium mass of 6 kg, this means the emission of 2×10^6 neutrons per second. The number of neutrons from spontaneous fission is considerably less in weapons-grade plutonium and still less in "super-grade" plutonium (see Table 14.2), and the probabilities of premature detonation are correspondingly reduced.

Although ^{240}Pu creates problems due to spontaneous fission, it does not greatly increase the required mass of ^{239}Pu for criticality. In this, there is some difference between ^{240}Pu in a ^{239}Pu bomb and ^{238}U in a ^{235}U or ^{233}U bomb. For ^{238}U, the fission cross section even at 1 MeV is low, and a fast neutron chain reaction in ^{238}U is impossible.[5] In contrast, the fission cross section for ^{240}Pu, which is very small at thermal energies, is about 1.5 barns at 1 MeV and a chain reaction is possible in pure ^{240}Pu [1, p. 115]. Because ^{238}U contributes little to the fission yield but absorbs neutrons, a ^{238}U contaminant means that a greater mass of ^{235}U is required for criticality. However, with its substantial fission yield, a ^{240}Pu contaminant raises only slightly the required mass of ^{239}Pu [6, p. 412].

Predetonation of Bombs and Implosion

The second part of the solution to the problem of spontaneous fission was to bring together a critical mass very quickly (see, e.g., Ref. [2], p. 55). This was accomplished by the implosion method, in which a plutonium sphere is symmetrically compressed by a shock

[5]A simple early discussion of this point is reprinted in Ref. [2], p. 21.

wave caused by detonating chemical explosives placed around the sphere. The critical mass is inversely proportional to the square of the density, and compression can change the mass from subcritical to supercritical.

In a typical bomb, the implosion shock wave has a speed of about 5000 m/sec and the implosion proceeds at a rate such that the bomb goes from initial supercriticality to full compression in a period of about 10^{-5} sec [1, p. 118]. This is the vulnerable period during which the bomb is subject to predetonation, before it is fully compressed and the reactivity has reached its maximum value. In a "proper" bomb, ignition is triggered only when the material is fully compressed.[6]

In discussing the progress of the chain reaction in the context of a possible fizzle, it is useful to focus on several times, adopting the notation and general model used in the Appendix to Ref. [1]:

- $t=0$, the time at which the implosion has progressed to the point of first achieving a critical mass ($k=1$).
- t_i, the time at which a chain reaction starts, assuming predetonation.
- t_f, the time at which the chain reaction has progressed sufficiently for the bomb to begin to disassemble.
- t_0, the time when the plutonium is compressed sufficiently for the bomb to give its maximum yield.

The condition for a full bomb explosion, not a fizzle, is $t_f > t_0$.

For purposes of qualitative orientation, it is useful to assume that $k-1$ increases linearly with time, from $t=0$ to $t=t_0$ [1, 2]. Thus, the parameter α can be expressed as [see Eq. (14.1)]:

$$\alpha = (t/t_0)\alpha_0,\qquad (14.2)$$

where $\alpha_0 = \alpha(t_0)$. For an implosion device, $t_0 \approx 10^{-5}$ sec and for $k=2$, $\alpha_0 \approx 10^8$ sec^{-1}.

The increase in k and α continues only as long as the implosion proceeds in an uninterrupted fashion. Predetonation causes the implosion to reverse at t_f. As the system then expands, the multiplication constant drops and the chain reaction eventually ceases. The time interval $t_f - t_i$ is the time during which the chain reaction builds to a magnitude sufficient to begin the expansion. To determine the magnitude of this interval, it is necessary to determine the condition for the expansion to begin. According to Mark, this occurs when there have been about e^{45} fissions (3.5×10^{19} fissions), i.e., when $\int \alpha(t)dt \approx 45$ [1, p. 117]. The corresponding energy release is about 10^9 J, equivalent to 0.2 kg of TNT.

However, the fission rate increases as long as $k>1$. The total fission yield Y_F therefore comes primarily from fission events that occur after the expansion begins, i.e., after t_f. According to a simplified analysis, Y_F varies as the cube of α, evaluated at t_f [1].[7] Thus,

$$Y_F/Y_0 = (\alpha_f/\alpha_0)^3 = (t_f/t_0)^3,\qquad (14.3)$$

where Y_0 is the nominal bomb yield.

---

[6]Triggering can be produced, for example, by breaking a thin barrier between a radioactive source of alpha particles and an adjacent beryllium target. Reactions between the alpha particles and the ^{9}Be nuclei will produce a burst of neutrons.

[7]Mark here relies on a qualitative argument presented by Serber [2], stating "his conclusions, though qualitative, will be adequate for our needs, which are also qualitative" [1, p. 118].

The time t_f will be earliest, and therefore the magnitudes of α_f and Y_F smallest, if the chain reaction starts at $t=0$. Using Eq. (14.2), the condition for the earliest predetonation becomes

$$45 = \int_0^{t_f} \alpha(t)\,dt = (t_f^2/2t_0)\alpha_0. \tag{14.4}$$

Then, from Eq. (14.3) and taking $\alpha_0 = 10^8 \sec^{-1}$ and $t_0 = 10^{-5}$ sec,

$$Y_F/Y_0 = (t_f/t_0)^3 = (90/\alpha_0 t_0)^{3/2} = 0.3^3 = 0.027. \tag{14.5}$$

This describes the smallest fizzle. It occurs when $t_f/t_0 = \alpha/\alpha_0 = 0.3$. The fizzle yield is then 2.7% of the full design yield Y_0. For $Y_0 = 20$ kt, this means a fizzle yield of about 0.5 kt, which is still a sizable explosion.

Dependence of Outcome on ^{240}Pu Content and Implosion Speed

It is to be noted that the minimum yield, as given by Eq. (14.5), is independent of the fraction of ^{240}Pu in the plutonium. However, the ^{240}Pu content does affect the probability of predetonation. In reactor-grade plutonium, with 360 spontaneous fission neutrons per gram-second, there will be 2×10^6 neutrons per second in a 6-kg bomb. This makes it likely that a chain reaction will be initiated within the first microsecond, assuring predetonation.

With these general assumptions, a more detailed calculation shows that there is a 70% chance that the yield with reactor-grade plutonium will be less than 10% of the full design yield and a near certainty that the yield will be less than one-half the full yield (see the Appendix of Ref. [1]). For weapons-grade plutonium with 6% ^{240}Pu, on the other hand, the neutron rate is only 3×10^5 neutrons per second, and there is a 50% chance of achieving more than 40% of the design yield. With 1% ^{240}Pu, there is a 90% chance that the yield will be more than 40% of the design yield and an 80% chance of it being the full design yield.[8] These numbers are for an implosion interval t_0 of 10^{-5} sec.

The discussion above brings out the qualitative factors relating to predetonation, but it is doubtful that it can be used for quantitative estimates of yield. One important consideration is the speed of implosion. For faster implosions, the probability of obtaining the full design yield increases. In addition, the fizzle yield itself will be greater, which may account for typical statements in the literature, suggesting that fizzle yields will be "in the kiloton range" [11, p. 259]. An authoritative National Academy of Sciences report has stated in regard to reactor-grade plutonium:

> even with relatively simple designs such as that used in the Nagasaki weapon—which are within the capabilities of many nations and possibly some subnational groups—nuclear explosives could be constructed that would be assured of having yields of at least 1 or 2 kilotons. Using more sophisticated designs, reactor-grade plutonium could be used for weapons having considerably higher minimum yields [10, p. 4].

For slower implosions, predetonation is more probable and the explosive yield will be less. According to Mark (presumably taking somewhat representative velocities), at the still slower speeds of a gun-type weapon in which the critical mass is created with an assembly velocity of 300 m/sec, the minimum fissile yield would be at least a factor of 30 less than that

---

[8]The plutonium used for the Trinity test has been inferred to have had about 1% ^{240}Pu, on the basis of predictions by Robert Oppenheimer of the yields and the sort of analysis outlined here [1, p. 127].

for an implosion weapon with a shock-wave velocity of 5000 m/sec [1, p. 119]. In addition, the high probability of predetonation with gun assembly makes it unlikely that the yield would much exceed the minimum. That is why plutonium weapons use implosion.

14.2.5 Reactor-Grade Plutonium as a Weapons Material

Difficulties in Use of Reactor-Grade Plutonium

It is clear, from the arguments reviewed above and from a long history of less detailed but authoritative statements about the possibilities, that reactor-grade plutonium can be used to make an explosive device that would release a substantial amount of energy. This would be enough to create a damaging explosion, and the damage would be compounded by the dispersion of plutonium from the weapon.

It does not seem very likely that such a project would be undertaken by a national government. A country with access to reactor-grade plutonium could probably arrange to have a domestic reactor irradiate fuel for a shorter time, yielding weapons-grade plutonium. This would require a deliberate decision by the national authorities, but one could imagine such a situation arising after a change of government or changes in relations with neighboring countries.

A large terrorist organization might be a more likely user of reactor-grade plutonium. Such an organization, if it wanted plutonium, might take whatever was accessible. Furthermore, while uncertainties in the explosive yield could make a weapon unsuitable for national military purposes, such uncertainties would probably be less important for terrorist purposes.

However, while possible in principle, an effort to turn reactor-grade plutonium into an explosive device faces formidable obstacles:

- If the plutonium has not already been separated from the fission products in the spent fuel, a chemical separation must be carried out with very radioactive material. A small reprocessing facility would be needed to accomplish this.
- The plutonium must be carefully machined and shaped, and then assembled in a subcritical geometry together with the surrounding reflector. A mistake at this point could create a microfizzle, probably disastrous to the people making the mistake.
- The explosive must be arranged properly to obtain a rapid and symmetric implosion. If the implosion is not rapid, the yield will be small.
- In assembling the device, precautions must be taken to avoid overheating the surrounding explosive. Reactor-grade plutonium produces heat at a rate of 10.5 W/kg (see Table 14.2). A tight blanket of chemical explosives could cause the system to overheat, with a possible chemical explosion resulting [1].

The barrier provided by the high level of radioactivity of the spent fuel is often taken to be an adequate protection against the use of this material. For example, 15 years after a fuel assembly is removed from a reactor, the activity at a distance of 1 m from its center is about 20 Gy per hour [10, p. 151].[9] A lethal dose is about 4 Gy. Therefore, handling this material without elaborate equipment would be a fatal activity.

A typical assessment has been given by Frank Barnaby, past executive secretary of the Pugwash Conferences on Science and World Affairs:

---

[9]At this time the external radiation is dominated by gamma rays from ^{137}Cs. Therefore, the decrease in radiation dose with time is determined by the decay rate of ^{137}Cs, which has a half-life of 30.2 years.

Spent reactor fuel elements are so radioactive as to be self-protecting. It would [be] extremely hazardous for people to handle them without large remote-handling equipment. However, when the fuel elements have been reprocessed and the plutonium separated from the radioactive fission products, the plutonium is in a form that can be relatively easily handled [12, p. 1].

This implies a high threshold against diversion of spent fuel. The radioactivity is probably a sufficient safeguard against any terrorist group, including a subnational group that lacks a secure geographical base. A national effort could surmount this threshold by developing reprocessing facilities and, if desired, improve the effectiveness of weapons derived from reactor-grade plutonium by isotopic enrichment of the material—although here, as discussed above, it would be simpler to obtain spent fuel that had been irradiated for only a short time.

Reprocessing of Spent Fuel

The difficulties of making nuclear weapons from reactor plutonium are significantly reduced if the reactor fuel has already been reprocessed and the plutonium extracted. This is an argument against reprocessing. Without reprocessing, there is no separated plutonium available for theft or diversion.

Reprocessing has also been opposed in the context of possible nuclear-weapons development by nations, as distinct from terrorist groups. Any country with facilities for reprocessing, gained as part of a normal civilian fuel cycle, has the facilities for reprocessing for weapons production. Civilian reprocessing can be used to conceal clandestine reprocessing for weapons. Even if there is no weapons program initially, having reprocessing creates the opportunity to move quickly into the weapons business.

For these reasons, many informed observers who are not otherwise opposed to nuclear power have opposed the reprocessing of spent fuel and the extraction of plutonium. This contributed to the U.S. decision to forgo reprocessing, as discussed in Sec. 7.4.2.[10]

While proliferation concerns helped to doom commercial reprocessing in the United States, they have not had a decisive effect everywhere. There are substantial reprocessing facilities for commercial spent fuel in France, the United Kingdom, Russia, Japan, and India.[11] It appears improbable that reprocessing will be abandoned in these countries, and in some cases reprocessing facilities are being expanded. According to a 1994 summary by an IAEA staff member, the only countries that have opted for direct disposal of wastes instead of reprocessing are Canada, Sweden, and the United States, while many countries, including China, France, Japan, and the Netherlands, have opted for reprocessing, and others have either deferred decisions or are currently planning for both [13].

The reprocessing in France of spent fuel from Japanese commercial reactors and the shipment of the extracted plutonium back to Japan have been a target of special criticism from groups in the United States. These criticisms have not reversed Japan's policies, and it is unlikely that they are well received. Thus, at a conference in the United States in which proliferation policies were being discussed, one highly placed representative of the Japanese nuclear establishment stated that Japan was tired of outside advice on handling plutonium [14]. In this statement of its position, Japan looks at plutonium as a component of a comprehensive energy program. Although not explicitly stated, it does not appear that Japan will look to the United States for intellectual, much less moral, leadership in matters related to nuclear weapons.

---

[10]Economic factors and difficulties with specific reprocessing facilities contributed to this decision, but these were encountered in an atmosphere where an influential body of opinion welcomed, rather than regretted, the difficulties.

[11]Some details have been given in Table 7.3.

Japan is currently strong in its assurances that the plutonium obtained from reactors will not be used for weapons and to date appears to have scrupulously refrained from weapons development. However, it is uncertain for how many years, or decades, Japan will be content to be protected by a United States nuclear umbrella if it is faced with nuclear weapons in China and, possibly, North Korea. While a plutonium stockpile would speed the pace of a program to develop weapons, even with no prior stockpile, Japan has the personnel and facilities to develop nuclear weapons quite quickly, should it choose to do so.

There is a contrary position on the issue of plutonium reprocessing and proliferation. In this view, the plutonium in spent fuel will remain a permanent danger *unless* the fuel is reprocessed and the plutonium consumed (see, e.g., Ref. [15]). In fact, as the fission products decay, the spent fuel becomes less hazardous and thus more valuable to a potential terrorist or national weapons program. However, even if one were to accept this view, there appears to be no antiproliferation reason to rush into reprocessing as a means of destroying plutonium, as long as the fuel remains self-protected, as it will for many decades.

Even for separated reactor-grade plutonium, the difficulties outlined above make it most unlikely that the high-school student of legend (who could fashion a bomb in the basement) exists or would long exist if he or she made the attempt. However, the necessary equipment and expertise could presumably be brought together by a well-organized terrorist group. It might be argued that even for such a group it would be irrational to proceed with a plutonium bomb when there are simpler alternatives for major destruction and terror. However, it is not prudent to rely on the rationality of terrorist groups. It is therefore very important to have safeguards that will prevent any diversion of separated plutonium, reactor-grade or otherwise.

14.2.6 Production of Plutonium in Reactors

The production of plutonium in a reactor is accomplished most efficiently if the reactor has a high conversion coefficient. This is more readily achieved in a heavy-water- or graphite-moderated reactor than in an LWR (see Sec. 6.3.2), and most weapons programs have in fact obtained their plutonium from graphite-moderated reactors. However, plutonium production is unavoidable in any reactor that uses ^{238}U as a fertile fuel. Thus, even if ostensibly being used for other purposes, any reactor is a present or potential source of plutonium for weapons. For weapons-grade plutonium, the fuel would be removed after a short burn-up period. But plutonium from fuel with a long burn-up also could be diverted for weapons purposes.

For 1-GWe reactors operating with a 70% capacity factor, average annual plutonium outputs (all isotopes) are about 230, 370, and 430 kg for light-water-, heavy-water-, and graphite-moderated reactors, respectively [6, p. 417]. The range here corresponds to 330 to 610 kg of plutonium per GWyr(e) or, in an alternative description, about 0.3 to 0.5 kg of plutonium per GWd(t). The critical mass for an explosive device using reactor-grade plutonium is about 10 kg (see Ref. [6], p. 413). Thus, each year of normal LWR operation at a 70% capacity factor provides enough plutonium for at least 20 such devices.

For reliable weapons, it is desirable to have a low ^{240}Pu fraction. In principle, this could be accomplished by isotopic enrichment of plutonium that has been chemically extracted from the spent fuel. A simpler approach is to reduce the burn-up of the fuel. In this case, the rate of plutonium output per unit energy generated is greater than in a normal fuel cycle, because there is little destruction of the ^{239}Pu in the reactor. At low burn-up in a dedicated reactor—for example, a graphite-moderated reactor—the output is in the neighborhood of 1 kg of plutonium (primarily ^{239}Pu) per GWd(t), or 1 g of plutonium per MWd(t).[12] Thus, even a small reactor can in a few years produce enough weapons-grade plutonium for a modest nuclear arsenal, at a rate of about 1 bomb per 5000 MWd(t) of operation.

In summary, the number of bombs that can be obtained from reactor fuel is large at both low and high rates of burn-up of the fuel. This is the reason for the concern over reactor plutonium as a source of weapons proliferation.

14.2.7 Uranium as a Weapons Material

It is considerably simpler to make a bomb using enriched uranium than to make one using plutonium. The fission cross section σ_f and the average number of neutrons per fission, ν, are both somewhat smaller for ^{235}U than they are for ^{239}Pu, making the critical mass larger (see Table 14.1). However, with uranium there is essentially no problem of premature detonation due to neutrons from spontaneous fission, because the spontaneous fission rates are much lower for ^{235}U and ^{238}U than they are for ^{240}Pu. Therefore, a gun-assembly uranium bomb is practical. Of course, it is also possible to make a uranium implosion weapon. For example, China's first nuclear test weapon was a ^{235}U implosion bomb (see Sec. 14.4.1).

Uranium bombs can be made over a wide range of ^{235}U enrichments, but the mass of uranium required is greater for lower enrichment. For example, with a good reflector, the critical mass for 60% enrichment is 22 kg of ^{235}U and 37 kg of uranium, while only 15 kg of ^{235}U (and U) are required at 100% enrichment [6, p. 412]. Not only is very highly enriched uranium preferable for building a compact bomb, but it requires more separative work to obtain a 37-kg critical mass at 60% enrichment than to obtain the smaller critical mass (roughly 18 kg) at 90% enrichment.[13] Thus, high enrichments are used in ^{235}U bombs. The bomb at Hiroshima was apparently built with uranium enriched to about 89% [2, p. 22]. Nuclear weapons now are reported to use uranium with an enrichment of at least 90% [16, p. 417].

14.3 NUCLEAR PROLIFERATION

14.3.1 The Non-Proliferation Treaty

In an effort to forestall the spread of nuclear weapons, a Treaty on the Non-Proliferation of Nuclear Weapons, commonly known as the Non-Proliferation Treaty (NPT), was first signed in 1968 and went into force in 1970. The original "Depository Governments" were the United Kingdom, the United States, and the USSR, reflecting the parentage of the treaty. The preamble to the treaty[14] included, as motivating purposes, the desire to:

- Prevent the wider dissemination of nuclear weapons.
- Make peaceful applications of nuclear technology widely available.
- Achieve cessation of the nuclear arms race and move towards nuclear disarmament.
- Seek to achieve discontinuance of test explosions of nuclear weapons.

---

[12]This output corresponds to a conversion ratio C of about 0.8. The production of 1 GWd(t) requires the fission of 1.05 kg of ^{235}U and, considering capture reactions, the consumption of 1.23 kg of ^{235}U (see Tables 5.1 and 7.2).

[13]At 60% enrichment, the separative work is 125 SWU/kg, or 4600 SWU for 37 kg (assuming natural uranium feed and 0.003% enriched tails); this may be compared to 193 SWU/kg and about 3500 SWU total for 18 kg of 90% enriched uranium. (See Sec. 7.2.2 for a discussion of separative work and defining equations.)

[14]The text of the NPT is reproduced, for example, in Ref. [17], Appendix A, pp. 352–356, and in Ref. [18], Appendix G, pp. 363–368.

The treaty thus attempted simultaneously to discourage military applications of nuclear energy and to foster peaceful applications.

The NPT starts with a crucial asymmetry between nuclear-weapon states and non-nuclear-weapon states. A nuclear-weapon state (NWS) is defined in the NPT (Article IX, Par. 3) to be one that had "manufactured and exploded a nuclear weapon or other nuclear device prior to January 1, 1967." By this definition, the NWSs are China, France, the USSR, the United Kingdom, and the United States.

Upon becoming a party to the treaty, the key obligation assumed by a NWS is to do nothing, directly or indirectly, to aid nuclear-weapons development in a non-NWS. Each non-NWS party to the treaty is committed not to receive or to make nuclear weapons. The non-NWSs also undertake to accept safeguards against the diversion of nuclear energy from peaceful purposes to weapons purposes. The safeguards are to be set forth for each country in an agreement to be negotiated with the IAEA. Amendments to the treaty can be made by a majority of the parties to the treaty, but this must include the affirmative votes of *all* of the NWSs. The NPT specified (in Article X) that a conference was to be held in 25 years (1995) to "decide whether the Treaty shall continue in force indefinitely, or shall be extended for an additional fixed period or periods."

The asymmetry embodied in the NPT could exist only because it reflected a substantial asymmetry in power as of 1970. The non-weapon states received explicit inducements in terms of a commitment by the weapon states to "pursue negotiations in good faith on effective measures relating to cessation of the nuclear arms race at an early date and to nuclear disarmament..." (Article VI). All parties to the treaty also agreed to "the fullest possible exchange of equipment, materials and scientific and technological information for the peaceful uses of nuclear energy" (Article IV, par. 2).

As is to be expected, some countries have resented the inequalities in status contained in the treaty. However, most would have gained little by abstaining, and by 1994 almost all countries had adhered to the NPT. The most important holdouts, India, Israel, and Pakistan, are not abstaining as a matter of principle, but rather because they have nuclear-weapons programs that would be in violation of the NPT.

The treaty and the associated IAEA safeguards have obviously had an inhibiting effect, because several countries refrained from signing the NPT while they were pursuing their nuclear-weapons programs. But despite this measure of success, the NPT has been an imperfect shield. Iraq, which had a well-advanced nuclear-weapons program by the time of the Persian Gulf war, was an NPT signatory with a full safeguards agreement with the IAEA. North Korea was an NPT signatory but has resisted implementation of full IAEA safeguards.

Whatever its imperfections, the NPT is deemed sufficiently valuable that it was made permanent at the scheduled international conference of its signatories in May 1995 [19]. The conference, held in New York with the participation of 175 nations, was convened to carry out the prescribed 25-year review of the NPT. Although the indefinite extension of the NPT was accepted without explicit dissent, there was sufficient dissatisfaction on the part of a substantial minority that approval was achieved by what amounted to general consent, rather than by a formal vote. In essence there was agreement that a majority favored indefinite extension and this provided the basis for the conference president to declare, without reported objection, that the extension was adopted.

The dissatisfaction stemmed in part from the belief on the part of many of the non-NWSs that there had been insufficient progress towards nuclear disarmament, as was called for in the NPT. In addition, the Arab states objected to the fact that Israel had not acceded to the NPT. However, rather than explicitly oppose the indefinite extension of the NPT, the dissenting nations accepted the extension, and some of their concerns were addressed in companion documents. In particular, again by general consent rather than a formal vote, the conference adopted a set of 20 "Principles and Objectives for Nuclear Non-Proliferation and Disarmament." Among the key principles were the following [19]:

- A call for a comprehensive test ban treaty by 1996 and for a "universally applicable convention banning the production of fissile materials for nuclear weapons." The NWSs were also called upon to undertake "systematic and progressive efforts to reduce nuclear weapons globally, with the ultimate goal of eliminating those weapons."
- A call to encourage the "development of nuclear-weapon-free zones, especially in regions of tension, such as in the Middle East . . . ."
- An affirmation that the peaceful use of nuclear energy is an "inalienable right of all parties to the treaty," together with a call for "the fullest possible exchange of equipment, materials, and scientific and technological information for the peaceful use of nuclear energy."

The NPT is to be reviewed every five years to monitor progress towards the implementation of the treaty objectives.[15]

It may be particularly noted that the principles reaffirmed both opposition to nuclear weapons and encouragement of peaceful uses of nuclear energy. Thus, as far as the NPT is concerned, there is no conflict between the maintenance and expansion of civilian nuclear power and the achievement of a world free of nuclear weapons. The International Atomic Energy Agency retains a central role both in establishing and verifying safeguards against nuclear-weapons proliferation and in furthering the safe development of nuclear power.

14.3.2 Forms of Proliferation

The proliferation of nuclear weapons can occur in a wide variety of ways. Prominent possibilities include the following:

1. A NWS could increase its arsenal.

2. A NWS could publicly transfer part of its arsenal to other states. Normally, this would be a violation of the NPT, but in effect, this happened following the breakup of the Soviet Union and the division of weapons among Russia, Ukraine, Belarus, and Kazakhstan.[16]

3. A NWS could have weapons surreptitiously transferred to other states or groups, by government action, by action of dissident officials, or by theft. Weapons in the countries of the former Soviet Union may be particularly vulnerable to such transfers.

4. A non-NWS could covertly establish a weapons program or intensify an existing weapons program.

5. A subnational group could build weapons using materials obtained by theft or diversion.

---

[15]Within several months after the end of the conference, both China and France carried out nuclear weapons tests, perhaps out of a desire to complete testing programs in anticipation of a 1996 ban.

[16]See the paragraphs on the USSR in Sec. 14.4.1 for a brief discussion of the status of weapons in the former Soviet Union.

6. Individual terrorists could build weapons using materials obtained by theft or diversion.

To date, the most serious dangers have come from activities in categories 1 and 4. More recently, categories 2 and 3 have come to the fore, with the breakup of the Soviet Union and the possible loss of tight control over the Soviet nuclear weapons arsenal. While there is no firm evidence as yet of any major activity in categories 5 and 6, there have been reports of discoveries in Western Europe of stolen nuclear materials, presumably originating in countries of the former Soviet Union. So far, the reported amounts have been small, but the appearance of any such material highlights the potential dangers.

14.3.3 Means for Obtaining Fissile Material

A potential proliferator has several options for obtaining fissile material:

- Isotopically separating natural uranium to obtain uranium highly enriched in ^{235}U.
- Irradiating uranium fuel in a reactor and separating the plutonium formed in neutron capture in ^{238}U.
- Obtaining it—by purchase, friendly transfer, or theft—from a country with fissile material.

Methods for enrichment were discussed in Sec. 7.2.2. These require a moderately sophisticated and expensive effort, but the procedures are not particularly hazardous. Extracting plutonium from spent fuel, on the other hand, is hazardous because it is necessary to handle highly radioactive materials. Otherwise, the chemical separation of plutonium is simpler than the isotopic separation of uranium. Once the plutonium is extracted, the level of radioactivity is much reduced, and further handling of the material is easier. Thus, any group intent on building a nuclear weapon would prefer to start with separated plutonium rather than spent fuel.

Proliferation has often been treated as a matter of safeguarding plutonium. Hence, there has been a substantial focus on reprocessing. However, consideration of the technological options and of the actual history of countries undertaking weapons programs suggests that ^{235}U may be equally attractive, making it important to restrict enrichment technology.

In fact, one authoritative observer has hazarded the thought that North Korea may be one of the *last* countries to try the plutonium route, with the next proliferators opting for ^{235}U if they have uranium available from domestic resources or open purchase [20]. (A ^{235}U bomb requires more natural uranium than does a ^{239}Pu bomb, and the priority equation changes in favor of ^{239}Pu if only small amounts of natural uranium are available.) Another observer has stressed that the "examples of Pakistan, South Africa, Argentina, and Iraq have demonstrated that enrichment is not the exclusive province of the technological elite" [15].

It may be noted in this connection that only a comparatively modest amount of separative work is required to obtain enriched uranium for weapons. For example, the separative work for the 3% enriched uranium for 1 GWyr of normal reactor operation is enough to produce highly enriched uranium for more than 20 bombs. Perhaps more surprisingly, more natural uranium and total separative work are required for a 10-kg plutonium bomb made with reactor-grade plutonium from a LWR than for a 20-kg uranium bomb made with highly enriched uranium [21, p. 66]. (However, the relative requirements are reversed if weapons-grade plutonium from a graphite reactor is used.[17])

14.4 HISTORY OF WEAPONS DEVELOPMENT

14.4.1 Nuclear-Weapon States

As of 1995 five countries officially acknowledged possession of nuclear weapons: China, France, Russia, the United Kingdom, and the United States. These are the nuclear-weapon states as defined in the NPT. Several other countries are suspected of having, or almost having, nuclear weapons, but these are not technically NWSs. In addition, the breakup of the USSR has added some new elements to the situation, as discussed below in the paragraphs on the USSR.

In the immediately succeeding sections, we briefly outline the early history of weapons development in the NWSs. The programs of these countries have progressed well beyond their early form, and in each case there are now extensive facilities for the reprocessing of plutonium and the separation of uranium. Fusion weapons (hydrogen bombs) have been added to fission weapons. However, the early history is pertinent to the question of weapons proliferation, because new entries may find it easiest to retrace some of the same initial steps.

United States

Development of U.S. nuclear weapons was accomplished in the massive Manhattan Project during World War II. The program culminated in the successful Trinity test carried out at Alamogordo, New Mexico, in July 1945. This was the world's first nuclear explosion. It was quickly followed by the dropping of bombs at Hiroshima and Nagasaki in Japan in August 1945.

The Trinity test and the Nagasaki bomb used ^{239}Pu, produced in reactors at Hanford, in Washington. The fissile material for the Hanford reactors was natural uranium. The reactors were moderated with graphite and cooled with water, and plutonium was extracted from the spent fuel discharged from these reactors. The Hiroshima bomb used ^{235}U, produced in electromagnets from uranium partially enriched by thermal diffusion and gaseous diffusion. The thermal diffusion process was soon abandoned, electromagnetic separation fell by the wayside, and after World War II gaseous diffusion became the sole U.S. enrichment method [22, p. 15].

The USSR

The Soviet Union, although allied with the United States during World War II, was excluded from the Manhattan Project. Nonetheless, it became aware of the U.S. program through surmise and through information from spies. A Soviet nuclear bomb program was initiated during the war. Exploration of a reactor fueled with natural uranium and moderated with graphite began in 1944, culminating in the explosion of a plutonium bomb in August 1949 [23, p. 88].

---

[17]With weapons-grade plutonium the bomb can be smaller (nominally 5 kg), and with a graphite reactor it is not necessary to use enriched uranium.

This is just the path the United States had taken a few years earlier. The Soviets reached the point of a critical chain reaction at the end of 1946 and exploded a bomb less than four years later. There is some dispute as the extent to which the information obtained from spies sped the progress of this program, but even with no information other than knowledge of the U.S. use of the bomb in 1945, it is highly likely that Soviet scientists would have been able to develop a bomb without any great delay.

As part of this effort it was necessary for the Soviets to develop reprocessing facilities. A parallel program to produce enriched uranium was also undertaken in the early years. Andrei Sakharov, who went on to lead the Soviet hydrogen bomb effort, reports that as a student in 1945 he had actively speculated about techniques for isotope separation, and one can assume that senior scientists were more seriously pursuing the matter [24, p. 92].

Following the breakup of the Soviet Union, nuclear weapons remained in a number of the newly formed states, namely, Russia, Ukraine, Belarus, and Kazakhstan. Of these, only Russia appears firmly determined to keep nuclear weapons, and only Russia is viewed by the United States as the legitimate heir of the USSR as a NWS. The other three countries have acceded to the NPT, Ukraine completing the process in December 1994 [25]. These countries are now committed to giving up their nuclear weapons, although whatever the official commitments it is possible that some weapons may be surreptitiously retained.

United Kingdom

The United Kingdom was a partner of the United States in the Manhattan Project, albeit a junior partner in the end. After the war British scientists had all the necessary information to proceed, although they were excluded from further U.S. weapons developments. The first British bomb, which used plutonium from specially built plutonium-production reactors, was tested in 1952 [26, p. 120]. The British program was also based on graphite-moderated natural-uranium reactors.

France

France, overrun by Germany in World War II, was not officially a participant in the Manhattan Project, but individual French scientists had joined an Anglo-Canadian project in Canada during the war. This program worked in loose collaboration with the Manhattan Project. Its main wartime project was the building of a heavy-water reactor. After World War II these scientists returned home and were able to contribute to a newly started French atomic energy program despite somewhat ambiguous commitments to secrecy [23, p. 65].

With a strong base of nuclear scientists, including those who had remained in France during the war, development of low-power research reactors began by 1946, starting with a natural-uranium heavy-water reactor. France gradually developed plutonium-production capabilities, also based on natural-uranium graphite-moderated reactors, but there was considerable political ambivalence about weapons development. By the late 1950s momentum towards bomb development had developed, and the decision to proceed with nuclear weapons was confirmed when General Charles de Gaulle assumed power in 1958. The first French ^{239}Pu bomb was set off in the Sahara Desert on February 13, 1960 [23, p. 139]. Since then, following policies established by de Gaulle, France has maintained an independent nuclear-weapons capability.

Table 14.3. *Comparison of years of achieving nuclear weapons and civilian nuclear electric power, for acknowledged nuclear-weapons countries.*

| Country | Year of achieving | | First power reactor |
	Weapon	Elec. power	
United States	1945	1957	Shippingport (60 MWe)
Former USSR	1949	1958	Troisk A (100 MWe)
United Kingdom	1952	1956	Calder Hall 1 (50 MWe)
France	1960	1964	Chinon A1 (70 MWe)
China	1964	≈1992[a]	Quinshan 1 (300 MWe)

[a]Quinshan 1 achieved criticality in 1991 and produced some power in 1992, but in some tabulations it was still not considered to be in "commercial operation" until 1994 [28].

China

Initial Chinese efforts to develop nuclear weapons were facilitated by a military collaboration agreement in October 1957 between China and the Soviet Union. Under this agreement, China obtained atomic bomb design information. This agreement lasted only until 1959, but it gave China a valuable head start.

The original plan was to develop both ^{239}Pu and ^{235}U weapons. However, a decision was made in 1960 to give priority to the ^{235}U program [27, p. 113], perhaps because with the loss of Soviet help, the plutonium program was temporarily deemed more difficult. Gaseous diffusion plants for uranium enrichment were built and put into operation in the early 1960s. The first uranium bomb was an implosion bomb. After prior practice, the enriched uranium was machined into a sphere, with tight specifications, by a single person in a single night, according to a dramatic account of the event [27, p. 167]. The machining was accomplished in May 1964, and the bomb itself was exploded on October 16, 1964, at the Lop Nur test site in a desert region of northwest China [27, p. 185]. Although China soon moved on with plutonium-producing reactors, and very quickly to hydrogen weapons, it was unique among the NWSs in testing a ^{235}U bomb before testing a ^{239}Pu bomb.

Lag Between Nuclear Weapons and Commercial Nuclear Power

For each of these countries, the development of civilian nuclear power lagged behind weapons development. Table 14.3 compares the year in which each of these countries obtained nuclear weapons and the year in which the country first obtained power from a reactor designed for electricity production for civilian use.

14.4.2 Suspected Nuclear-Weapons Countries

India

India began a rudimentary nuclear program in 1948, shortly after achieving independence from Great Britain.[18] It built a small research reactor in 1956 using enriched uranium from

---

[18]In the discussion of India and other countries that do not admit to nuclear weapons but are suspected of having them or having tried to obtain them, we rely extensively on the writing of Leonard Spector of the Carnegie Foundation for International Peace (Refs. [29] and [16]).

Britain, and in 1960 it obtained a larger research reactor under a joint Canadian–Indian–U.S. program [29, p. 24]. The reactor was given the acronym CIRUS in recognition of its parentage. This was a natural uranium reactor moderated with heavy water. Part of the initial fuel was provided by Canada and the heavy water by the United States. By 1962 India had indigenous sources of uranium and a small plant for producing heavy water.

India then proceeded to develop a reprocessing facility to extract plutonium from CIRUS. Whatever its initial motivations, India was able to use this plutonium to build a nuclear explosive device, which it set off on May 18, 1974. It is estimated that the device used about 15 kg of ^{239}Pu [29, p. 34].[19] It is termed a "device" rather than a bomb, because it is thought to have been too bulky to be used in a deliverable weapon. India claimed that the purpose of the explosion was peaceful, to study the effects on rocks. However, outside India the explosion was considered a first step in an Indian weapons program, undertaken as a potential counter to China and, more recently, as part of a nuclear competition with Pakistan.

While India does not acknowledge the possession of nuclear weapons, it is widely believed to have them. It clearly has the facilities that would permit a large nuclear program. These include several heavy-water reactors from which fuel can be continuously removed, facilitating the extraction of low-burn-up, weapons-grade plutonium and reprocessing. Its reactors and reprocessing facilities were estimated by Leonard Spector in 1990 to be able to produce enough plutonium annually for 15 weapons, taking into account difficulties in operating these facilities at full capacity. India was estimated to have had enough plutonium on hand as of mid-1990 for 40 to possibly 60 weapons, and it has tested missiles that could be used to deliver nuclear weapons [16, pp. 72, 74].

India is flanked by China, which avowedly has nuclear weapons, and by Pakistan, which has established a nuclear program in response to India's program. It would probably take a complex three-way agreement to terminate what seems to be a low-level nuclear arms race in that area, and as of late 1995 there was no indication that such an agreement was forthcoming. In carrying out its nuclear program, India has been able to blur the distinction between peaceful research and electricity generation on the one hand and a weapons program on the other. Perhaps this is the leading case of a positive link between possession of civilian nuclear power and the development of nuclear weapons. The start appears to have come from using a research reactor, rather than a power reactor, but an expanded weapons program can be partly hidden under the power generation program, if India is choosing to do so.

Israel

Since the 1960s there have been reports that Israel was or might be developing nuclear weapons. There has been no official acknowledgment from Israeli sources that this indeed has been done, but somewhat ambiguous statements have been interpreted as confirmation that Israel could take whatever components it has and make bombs in very short order. This is probably more a semantic issue than a technical issue, and most observers assume Israel has a substantial number of bombs. In a 1990 summary, Spector estimated that Israel's arsenal "probably consists of 60 to 100 devices" [16, p. 149].[20] Overall, it appears that Israel

---

[19]A mass of 15 kg is above the critical mass specified in Table 14.1, even for a bare ^{239}Pu sphere. It could have been subcritical before implosion if, for example, it was in the form of a spherical shell with a sizable hollow core.

[20]Spector leans substantially on information originating from Mordechai Vanunu, an Israeli who reported details of the nuclear program in 1986. Vanunu, who had worked as a technician at Dimona, made his statements about the program in London and shortly after was abducted and taken back to Israel, where he was convicted of espionage and treason. Many features of the program were surmised before Vanunu's statements, which have not been

has a well-developed nuclear weapons program, started with the aid of France and continuing with at least the acquiescence of the United States.

A key early step in this program was the construction of a reactor at Dimona; the reactor was fueled by natural uranium and moderated with heavy water. Constructed with the aid of France starting in the late 1950s, the reactor began operation in 1963. The reactor was nominally a 24-MWt research reactor, but a series of changes have apparently raised its capacity to 40 MWt or higher. The initial source of heavy water was Norway, under nominally strict controls to restrict use to peaceful purposes. Israel completed a plutonium reprocessing facility at Dimona in the 1960s, with a capacity estimated to be 15 to 40 kg/yr. It has been suggested that Israel has centrifuge or laser uranium enrichment capabilities for producing 2 to 3 kg of highly enriched uranium per year [16, p. 173]. It may have developed fusion weapons as well.

North Korea

North Korea illustrates how a country with a relatively small technical base may be able to go it alone in weapons development, given sufficient determination and some small initial help. North Korea first received a small research reactor from the USSR in 1964. The start of its potential weapons program is thought to have come later, with the construction in the 1980s of a gas-cooled, graphite-moderated reactor at Yongbyon (about 100 km north of Pyongyang), which went into operation in the late 1980s. This reactor is estimated to have a capacity of about 25 MWt [30]. It was of an elementary design, similar to those developed in the 1940s, and it may have been built without significant outside help [16, p. 123]. North Korea has also developed a reprocessing facility.

Although North Korea ratified the NPT in 1985, it managed for many years to keep this reactor free of continuous, comprehensive IAEA inspections. In the absence of adequate inspections, there could only be speculations about the purposes for which this reactor was being used and the history of its operation. An IAEA team concluded from a 1992 inspection that plutonium had been separated from spent reactor fuel in 1989, 1990, and 1991, in conflict with the North Korean claim that the only plutonium separation had been of 100 g in 1990.[21]

Assuming that the reactor was used to produce plutonium for nuclear weapons, there have been attempts to estimate the amount of plutonium extracted. According to the rule of thumb cited in Sec. 14.2.6, a dedicated reactor can produce about 1 g of plutonium per MWd(t) of operation. Therefore, continuous operation of a 25-MWt reactor for one year would yield about 9 kg of plutonium. In one plausible estimate, which does not appear to differ markedly from others that have been published, David Albright suggests that enough fuel might have been unloaded from the reactor in 1989 to provide about 7 to 14 kg of plutonium [30]. The higher number could have been achieved by operating the reactor at a little over 50% efficiency during a three-year period.

While North Korea does not admit to the removal of so large an amount of fuel before 1994 or to the construction of any nuclear weapons, estimates of this sort led to suggestions that North Korea had probably built one or two bombs by 1994, and perhaps more. Further-

---

independently confirmed in publicly available information.

[21]This conclusion was based on the relative amounts of ^{241}Am and ^{241}Pu in material samples taken by swabbing the insides of North Korean plutonium-handling facilities. The ratio of ^{241}Am to ^{241}Pu is a measure of the elapsed time since plutonium was extracted from the fuel, because the ^{241}Am in these samples comes from the radioactive decay of ^{241}Pu ($T = 14.35$ yr).

more, the capacities of North Korea for plutonium production were increasing. Most immediately, there was an announced refueling of the 25-MWt Yongbyon reactor in 1994, creating
the potential for the extraction of a substantial amount of plutonium from the reactor's spent
fuel.

As of the end of 1994, this fuel had not been removed from the reactor cooling ponds and
hence had still not been reprocessed. A still greater eventual threat was posed by two larger
graphite-moderated, gas-cooled reactors that were under construction in 1994: a 50-MWe
(200-MWt) unit at Yongbyon and a 200-MWe (600–800-MWt) unit at Taechon [30]. These
reactors could have been completed by about 1996 and would have given North Korea the
potential ability to produce roughly 200 kg of weapons-grade plutonium per year (assuming
70% efficiency for the reactors).

The prospect of a significant North Korean nuclear-weapons program led to protracted
negotiations involving North Korea, the IAEA, the United States, and other countries. These
negotiations took a surprising turn in the autumn of 1994, with an agreement between the
United States and North Korea under which North Korea agreed to halt construction of the
two graphite reactors and allow international inspection of its nuclear facilities [31, 32]. In
return, financial aid from Japan, South Korea, and the United States was to be given North
Korea to obtain two new LWRs with a total capacity of about 2000 MWe, plus fuel oil to
provide energy pending completion of these reactors in about 2003. There has been difficulty
in working out details of the implementation of this agreement, including the part to be
played by South Korea, and negotiations continued through 1995.

It is striking that the U.S. government, at a time when it showed no interest in promoting
nuclear power in the United States, found it prudent to help provide nuclear reactors for
North Korea. Proponents have argued that this was the best solution under circumstances that
were growing increasingly dangerous and that North Korea's acceptance of inspection
provides assurance against, or at least a warning of, violations of the agreement. Opponents
point out that North Korea, with 2 GWe of nuclear capacity, would be able to produce close
to 500 kg of reactor-grade plutonium per year or, with frequent refueling of the reactor,
several hundred kilograms of weapons-grade plutonium. Overall, the usefulness of the agreement appears to depend more upon the intentions and the resolve of the various parties than
on any decisive change in the technological options.

Pakistan

Following its defeat in a miniwar with India in 1971, Pakistan in 1972 reportedly decided to
develop its own nuclear weapons [16, p. 90]. It has taken a dual approach towards obtaining
fuel, developing facilities for both uranium enrichment and plutonium extraction. Much of
this has been done clandestinely to circumvent export control laws. Thus, an enrichment
plant was obtained by "smuggling" it from Germany [16, p. 91]. Pakistan has also received
help from China. Its weapons program is believed to be based on enriched uranium produced
in ultracentrifuges. As of 1990 Spector concluded that "Pakistan probably could deploy five
to ten nuclear bombs for delivery by aircraft" [16, p. 89].

Relation of Suspected Weapons States to the NPT

Of the suspected weapons states, India, Pakistan, and Israel are not parties to the
NPT. Each has some facilities under IAEA safeguards, prompted by agreements with

countries that helped with these facilities, but there are large gaps in the applicability of the safeguards. Thus, each of these countries is able to pursue a nuclear program free of decisive outside control or monitoring.

The existence of the NPT and of comprehensive safeguards has served to turn the spotlight on those who do not accept these controls. Each of these countries has found it useful to be less than forthcoming in describing its program, and there are charges of outright deception. Nonetheless, world pressures have not had a decisive effect on these programs, perhaps because each of these countries has had powerful protectors and a certain amount of sympathy for its situation.

India is threatened by China and Pakistan, Pakistan is threatened by India, and Israel is threatened by the Arab world. Countries otherwise friendly to one or another of them have a difficult time in condemning their possession of nuclear weapons. Thus, the USSR provided political and technical help to India in the past, while the United States to one degree or another has tolerated the Pakistani and Israeli nuclear programs and would prob- ably prevent sanctions against them. The same arguments of security put forth to justify nuclear weapons in the NWSs can be advanced on behalf of other fearful states.

North Korea has come under particular scrutiny, perhaps because it has no powerful protector but also because it is suspected of having violated commitments made under the NPT. It engaged in a long duel with the IAEA, permitting some inspections in accord with the NPT but apparently attempting to conceal the true nature of its programs. There is no clear public information on the extent to which the NPT and IAEA pressures inhibited the North Korean weapons program, but they may have been important if only in strengthening the position of the United States in opposing the North Korean nuclear program.

14.4.3 Countries that Have Abandoned Nuclear-Weapons Programs

Argentina

Argentina obtained a small research reactor in 1958 and a 320-MWe heavy-water reactor in 1974. It made its first public moves towards a weapons program in 1978, with the announce- ment of a planned plutonium-reprocessing plant and the start on construction of a gaseous diffusion plant. Work on the reprocessing plant was terminated in 1990, apparently before it went into operation, but the gaseous diffusion plant went into operation in 1988. The plant is designed for a 20% enrichment in ^{235}U but could be used for higher enrichments, with a theoretical capacity of 100 kg/y of weapons-grade uranium [16, p. 228].

Although this program appeared originally to be directed towards a weapons program, in competition with Brazil, there has been a change in the political climate in South America, and in 1990 both Argentina and Brazil renounced nuclear weapons [33, p. 177]. In December 1991 they signed a joint agreement with the IAEA for mutual inspections [34]. In January 1994 Argentina officially joined the Treaty of Tlatelolco, the regional South American agree- ment barring nuclear weapons [35], and it has since signed the NPT.

Brazil

Partly in response to Argentina's nuclear program, Brazil took an important step towards nuclear-weapons capabilities in 1979, with the start of an ultracentrifuge uranium enrichment plant. It was put into operation in 1988, with the ostensible purpose of producing uranium enriched to 5% or 20% in ^{235}U [16, p. 250]. Brazil has also had a very small plutonium-reprocessing program. At the same time a 626-MWe PWR was put into commercial operation in 1985. Whatever its original aspirations, Brazil has now given up its nuclear-weapons program, paralleling the decision by Argentina.

South Africa

South Africa originally built a weapons program based on the enrichment of uranium using a variant of the jet nozzle technique developed in Germany. Reportedly, the initial pilot plant was built with clandestine German help and began operation in about 1975 [16, p. 271]. It ran during the 1980s, with a suspected capacity of about 50 kg per year. This gave South Africa the ability to build a substantial arsenal.

In 1990, along with many other changes in political direction, South Africa decided to abandon its nuclear-weapons program. It shut down its plant for producing highly enriched uranium and suggested that it would join the NPT. Culminating its renunciation of nuclear weapons, South Africa signed the Non-Proliferation Treaty and entered into a safeguards agreement with the IAEA in 1991 [36]. The scope of the South African nuclear program was revealed by President F. W. de Klerk in an address to the country's parliament in March 1993, when he reported that six nuclear weapons had been built by 1990 and later dismantled [37].

14.4.4 Suspected Nuclear-Weapons Aspirants

Iran

Iran has for many years attempted to develop a nuclear power program with foreign assistance. Under the last Shah, who was driven from power in 1979, this was with the cooperation of Western countries, starting with a research reactor from the United States. As part of a larger long-term program, construction began on two West German reactors for power generation [16, p. 204]. Iran also made contracts for obtaining less-enriched uranium for its reactors. After the fall of the Shah this program was aborted due to domestic opposition, foreign unhappiness with Iran, and bomb damage during the Iraq–Iran war.

However, a large group of trained nuclear scientists remained in Iran. Efforts to establish an ostensibly civilian nuclear program have progressed subsequently, with the aid of a number of countries, including Argentina, China, and perhaps Pakistan. More recently, in April 1993 Iran ratified agreements with Russia and China for nuclear cooperation that was expected to lead to the purchase of two 300-MWe reactors from China and two 440-MWe reactors from Russia [38, p. 20]. IAEA safeguards were to be applied to these reactors, but this did not ease the concerns of people who assumed that Iran was striving for weapons capabilities.

The U.S. government has opposed both initiatives. There have been conflicting reports of its success in dissuading China from continuing its commitment to the reactor project. As of the end of September 1995, neither the fate of this project nor the effectiveness of U.S. pressures was clear [39]. The cooperation between Russia and Iran appears to be moving forward in more definite fashion, and a contract was signed in early 1995 under which Russia will complete one, and possibly both, of the reactor facilities started and abandoned by West Germany [40]. Instead of an intermediate-size reactor, as earlier contemplated, a 1000-MWe light-water reactor of Russian design is to be placed in the structures built previously for the German reactor. Although U.S. attempts to induce Russia to give up this plan apparently failed, Russia did agree in May 1995 not to provide a centrifuge uranium enrichment facility, which had also been part of its agreement with Iran.

There have been other suspected Iranian efforts to obtain uranium enrichment facilities or enriched uranium itself. In an action to thwart one such an attempt, the United States in 1994 bought a supply of highly enriched uranium from Kazakhstan to forestall a possible purchase by Iran [40].

Iraq

Iraq's nuclear ambitions have expressed themselves in two stages. It obtained a large research reactor from France in 1976, the Osirak reactor, and purchased in 1980 and 1981 large amounts of uranium, including 11 tonnes of depleted uranium from Germany [16, p. 187]. This uranium could be placed around the reactor to produce ^{239}Pu using neutrons escaping from the reactor core, reportedly in amounts sufficient for one or two bombs [16, p. 187]. The Osirak reactor was destroyed by an Israeli bombing raid in June 1981, an action that was widely, although not unanimously, criticized at the time. Iraq was a signatory to the NPT, and its facilities were under IAEA supervision, supposedly averting the danger that prompted the Israeli action. The failure of IAEA controls to prevent Iraq's later weapons efforts gave plausibility to Israel's concerns.

Following the destruction of the Osirak reactor, a standard view was that Iraq was "unlikely to join either the civil or military nuclear club during the rest of this century" [41, p. 42]. However, Iraq was not so easily discouraged. It apparently made an unsuccessful effort to purchase plutonium from Italian arms smugglers and later began efforts to produce enriched uranium. By late 1989 Iraq's efforts to develop centrifuge enrichment facilities had been widely recognized [16, p. 192].

After Iraq's military defeat in early 1991, IAEA inspections revealed a previously unreported major electromagnetic separation program, an ambitious but incomplete centrifuge enrichment program, and the separation of small amounts of plutonium from a research reactor [42]. Iraq had been helped in this program, presumably unwittingly, with equipment and information from many countries, including West Germany, Switzerland, and the United States. It is apparent from a 1990 Iraqi Progress Report, obtained by IAEA inspectors and reprinted in a report by Peter Zimmerman [42], that Iraqi scientists had very considerable scientific understanding and technical expertise.

14.4.5 Summary of Pathways to Weapons

We summarize in Table 14.4 the ways in which fissile material has been sought or obtained by the various countries aspiring to nuclear weapons. A striking feature of the table is the reminder it provides that many of the countries that built or aspired to build nuclear weapons have first used enriched uranium rather than plutonium. This complicates antiproliferation efforts, because plutonium is only half of the story.

Countries seeking fissile material for weapons need amounts that are small compared to the amounts handled in nuclear reactors. A single 1000-MWe reactor each year produces enough plutonium for 20 or more bombs, albeit perhaps not very efficient bombs. Producing the enriched uranium to fuel a reactor for one year requires in the neighborhood of 100 000

Table 14.4. *Initial methods used to obtain fissile material for countries that have nuclear weapons or are believed to have attempted to develop nuclear weapons.*

Country	Material	Initial source of fissile material
Nuclear weapons states		
United States	^{239}Pu	Natural U reactor, graphite moderated
	^{235}U	Diffusion and electromagnetic separation[a]
Former USSR	^{239}Pu	Natural U reactor, graphite moderated
United Kingdom	^{239}Pu	Natural U reactor, graphite moderated
France	^{239}Pu	Natural U reactor, graphite moderated
China	^{235}U	Gaseous diffusion
Relinquished weapons		
South Africa	^{235}U	Jet nozzle technique
Probably built weapons		
India	^{239}Pu	Natural U research reactor, D_2O moderated
Israel	^{239}Pu	Natural U research reactor, D_2O moderated
North Korea	^{239}Pu	Natural U reactor, graphite moderated
Pakistan	^{235}U	Ultracentrifuge
Tried to build weapons		
Argentina	^{235}U	Gaseous diffusion[b]
Brazil	^{235}U	Ultracentrifuge[b]
Iraq	^{239}Pu	Enriched U reactor with ^{238}U in external blanket[c]
	^{235}U	Electromagnetic separation[d]

[a]See text (Sec. 14.4.1).
[b]These programs were terminated voluntarily.
[c]The reactor for this program was destroyed before it became operational.
[d]The enrichment program was terminated in 1991, following the Persian Gulf War.

kg of separative work. This amount of separative work could produce 500 kg of 90% uranium, enough fuel for about 30 implosion weapons. Thus, the size of the facilities needed to obtain material for a significant number of nuclear weapons is on a much smaller scale than that required for the generation of a significant amount of nuclear electricity.

14.5 NUCLEAR POWER AND THE WEAPONS THREAT

14.5.1 Peaceful Uses of Nuclear Energy

Starting with the Atoms for Peace program of the Eisenhower administration and continuing through the formulation of the Non-Proliferation Treaty, it was believed by the U.S. government that it would be possible to keep the military and civilian aspects of nuclear energy separate and that the promise of aid with peaceful nuclear activities could help induce countries to forgo military uses. Articles IV and V of the NPT in fact obligate NWSs to aid non-NWS parties to the treaty in pursuing peaceful applications of nuclear power. The result of these policies has been the spread of research reactors throughout the world.

It can be debated whether this policy was wise or quixotic. However, the good or the damage has been done.[22] In examining the links between commercial nuclear power and nuclear weapons, it is necessary to keep this history in mind. A wide diffusion of nuclear

---

[22]We here are focusing on the negative aspects of nuclear energy, namely, the potential to produce weapons. There are positive practical aspects as well, quite aside from electricity production. Examples include the use of radio-nuclides in medical diagnosis and therapy and myriad industrial applications.

knowledge and technology has already taken place, extending to many countries that as yet have no nuclear power. Further, to the extent that peaceful nuclear energy has been involved in helping start weapons programs, research reactors, not power reactors, have to date been the main culprit.

14.5.2 The Role of Nuclear Power in Weapons Proliferation

Countries with Nuclear Weapons, Admitted or Suspected

The most immediate and perhaps most serious nuclear-weapons threats come from countries that already possess weapons:

- The NWSs, which admit to nuclear weapons and have no plans to renounce them: China, France, Russia, the United Kingdom, and the United States.
- Countries that admit to nuclear weapons and plan to relinquish them: Belarus, Kazakhstan, and Ukraine.[23]
- Countries that are believed to have nuclear weapons but do not explicitly admit having them: India, Israel, Pakistan, and possibly North Korea.

Our reaction to the existence of these weapons depends in part on our assessment of the likelihood that the holders of the weapons may have aggressive intent.

A major related concern is that weapons from one of these countries might fall into the hands of other countries or of internal or external terrorist groups. The weapons of the former Soviet Union, divided among Russia, Ukraine, Belarus, and Kazakhstan, may be the most vulnerable. There is the continuing danger that some of these new holders of weapons will in the end decide not to relinquish all weapons, as well as the possibility that some nuclear weapons may be surreptitiously retained or diverted. There is undoubtedly an eager market for these weapons.

Commercial nuclear power has played virtually no role in the weapons buildup of these countries, although small reactors built ostensibly for research purposes have been of importance in some cases. The first eight states in the set listed above obtained bombs through weapons programs that preceded nuclear power. Of the last four, Israel and North Korea have no nuclear power, and for India and Pakistan the chief early connection with peaceful uses of nuclear energy was with research reactors, not power reactors. There is no known case of plutonium having been diverted for weapons purposes from a civilian reactor used for electricity generation.

Aspiring Weapons States

Iraq has made a substantial effort to obtain nuclear weapons, although it has been thwarted in this, at least temporarily. Again, commercial nuclear power was irrelevant to this program, because Iraq has neither nuclear power nor any announced plans to obtain it.

Iran is another matter. Steps to help it with civilian nuclear power are viewed with suspicion, because it is feared that in the end Iran will use whatever nuclear facilities it has to make weapons. Civilian power can be just a pretext for obtaining reactors and, further along in the process, perhaps enrichment or reprocessing facilities. Many people would breathe easier if Iran did not have nuclear power, even those who would favor nuclear power

---

[23]As noted in Sec. 14.4.1, these countries have acceded to the NPT.

in countries they view as benign. The feared Iranian program might be undertaken surreptitiously, evading IAEA inspections, or openly, with or without withdrawal from the NPT.

Overall, we would feel more secure if we could deny nuclear power to countries that we deem to have aggressive intents. The United States has taken a particularly active role in promoting such a policy, but the it has only limited influence and is constrained by pragmatic considerations—as in its decision to help provide reactors to North Korea while attempting to deny them to Iran. There is no recognized international mechanism for implementing a policy of barring nuclear power selectively, especially in view of the difficulty of getting agreement as to the intent of the suspected nation. In particular, there is no international body that is likely to have both the will and the means to prevent major powers, such as Russia and China (or the United States, for that matter), from providing nuclear facilities to other nations as they see fit.

Countries with Nuclear Power but No Weapons

Many countries exhibit the other side of the coin: They have nuclear power but do not have nuclear weapons. Some have never shown any inclination towards weapons—for example, Canada, Germany, and Japan. All have strong, comprehensive nuclear power programs and could easily develop weapons with no external aid. Sweden, which also has a strong civilian nuclear power program and technological base, embarked on a very sophisticated program of weapons development, but the effort was abandoned before attracting much outside public attention [43]. Beyond these, Argentina and Brazil had moved towards obtaining nuclear weapons, and South Africa succeeded, but all three have given up these programs.

Virtually all of these countries could obtain nuclear weapons should they desire to do so. There are different motivations for abstaining: matters of principle, the belief that a nuclear weapons race would be counterproductive in terms of security, and the feeling that they are already under a protective umbrella established by the United States. Economic pressures also play a part. The action of Argentina in renouncing nuclear weapons was explained in terms that have a broad relevance to other "third-world" countries. As stated by the Argentinian Under Secretary of Foreign Affairs:

> We found we were blacklisted by the international community for our aggressive policies and in the end found we had to cooperate with the netherworld of third-world countries. . . . The paradox was that Argentina was trying to reach high tech through its nuclear program, but because of a lack of openness Western countries made sure it would not get there. I think we have learned our lesson [35].

Here the premise of the atoms-for-peace policy seems to have been vindicated. For countries seeking weapons, the stick is technological denial. For those who renounce them, the carrot is the technological reward, such as help in obtaining commercial nuclear power. It would be a safer world if it could be demonstrated that solar power is more of an economic reward than is nuclear power. But at the moment, such a claim may not be believed by many countries.

Countries with Neither Nuclear Power Nor Nuclear Weapons

Countries that lack both nuclear power and nuclear weapons tend to be countries without a highly advanced technological base. As such, they might not be thought to pose a significant threat. However, the "aspiring" nuclear-weapons countries listed above (Iran and Iraq), as well as North Korea, which may already have built weapons, all could be placed in this category, and they are threats. Many other countries, such as Indonesia, Turkey, and Venezuela, to pick more or less at random, could mount similar efforts. Those that have abstained often have lacked the desire more than the technical ability.

Peter Zimmerman has suggested the concept of a "bronze-medal" technology for weapons development [44]. To win the bronze medal, a country need only achieve the technological level reached by the United States 50 years earlier. The nuclear weapon thus developed might not win plaudits from sophisticated weapons experts, but it could wreak great actual or threatened damage. South Africa exemplifies a success of this approach, building its six bombs using a gun design similar to that of the Hiroshima bomb, without any very great investment of money or personnel [44]. With differences in details, Iraq and North Korea have also adopted bronze-medal approaches, and other countries could do likewise.

It is not clear whether commercial nuclear power would help towards that end. The guiding principle is probably the same as the unsatisfactory principle suggested above in the case of Iran: Deny nuclear power to suspected enemies and support it for your friends. Obviously, such a policy for the United States cannot be defended in terms of any general principles of international equity. At best, it is a pragmatic response of only incomplete and temporary applicability, requiring that the United States have more influence and economic power than it probably will possess in most situations. Of course, this influence and power would be amplified if it could be exerted through the United Nations or a strengthened IAEA, but here there is likely to be difficulty in obtaining an effective consensus.

Subnational Groups

The main premise of the preceding sections was that it is relatively easy for many countries to build nuclear weapons, including countries labeled as "developing." Conversely, it would be a difficult matter for any subnational group that does not have a secure geographical base. The difficulty is greatest if it is necessary to employ plutonium from nuclear wastes or spent fuel. In this case, the subnational group would face an almost insuperable task unless the host country acquiesces in the construction of reprocessing or isotopic enrichment facilities. This would in effect turn the effort into a national, rather than subnational, program.

The difficulties are reduced if it is possible to obtain separated plutonium. Even then, it might be difficult to assemble the people and facilities to make a bomb without the knowledge of the host country. However, if separated plutonium is available, the threshold is lowered enough to make it essential to guard separated plutonium, including reactor-grade plutonium, from potential diversion.

14.5.3 Nuclear Power and Moderation of Weapons Dangers

Economic and Resource Considerations

So far we have stressed ways in which nuclear power might contribute to nuclear-weapons dangers. There are also contrary aspects. Conflicts between nations, including conflicts that might escalate into nuclear conflicts, are more likely given severe competition for resources or domestic unrest due to extreme economic difficulties. Were nuclear power able to alleviate these sources of conflict, it could serve to lessen the risks.

The most obvious connection of this sort has been between nuclear weapons and oil. One of the precipitating factors in Japan's entry into World War II was the tension between the United States and Japan over oil. In an attempt to restrain what it saw as a program of aggressive expansion, the United States sought to limit Japan's access to oil. The sequels were the attack on Pearl Harbor and the bombs at Hiroshima and Nagasaki. Looking to the future, competition for oil could again become desperate enough to lead to dangerous military confrontations, possibly including nuclear war.

The Persian Gulf War with Iraq is a case in point, although Iraq had not yet succeeded in building nuclear weapons and the United States had no need or impulse to use them. But the war serves as a reminder of the high stakes involved in Persian Gulf oil. Oil gave the United States a motivation to go to war, and it gave Iraq the resources to build a strong military machine and to begin an ambitious nuclear weapons program.

To some extent nuclear power can serve as an alternative to oil, directly as a substitute in electricity generation and less directly through increased electrification. It can thereby lessen some of the pressures arising from the central role of Persian Gulf oil in the world's energy economy. Although even with the largest plausible expansion in its use, nuclear power could not alone eliminate the world's heavy dependence on this oil, it could reduce the level of dependence and thereby dampen the intensity of the competition and the associated risks.

Disposition of Fissile Material from Dismantled Weapons

The planned reduction of the nuclear-weapons stockpiles of the former Soviet Union and the United States will make available large quantities of enriched uranium and weapons-grade plutonium. The uranium can be "denatured" by mixing it with natural or depleted uranium. This would provide relatively harmless supplies of slightly enriched uranium, suitable as a fuel for reactors.

The plutonium problem is less tractable. The amount of plutonium from dismantled weapons is large. One estimate places the eventual amount at 150 to 200 tonnes [45]. A 1994 National Academy of Sciences (NAS) study indicates that "50 or more metric tonnes of plutonium on each side are expected to become surplus to military needs" [10, p. 1]. Even at 50 tonnes each from the United States and Russia, there is enough plutonium for about 20 000 small bombs, assuming a nominal 5 kg per bomb. There are several alternatives for keeping this material out of circulation:[24]

- The plutonium could be stored in well-guarded facilities. This is the most immediate solution, requiring the least handling. It is an economical solution in that it would save a potentially useful resource, but also a dangerous one should there be any breach of the integrity of the storage. For these security reasons, the NAS study cautions against extending such storage "indefinitely" [10, p. 226].
- The plutonium could be consumed in a nuclear reactor. With some choices of reactors and fuel types, the spent fuel discharged from the reactor would be similar to LWR spent fuel and would have a significant plutonium component. However, the spent fuel would be protected by its high level of radioactivity and would be less attractive for weapons purposes due to an increased ^{240}Pu contamination. With other reactors and fuel cycles, the amount of plutonium remaining could be substantially reduced. In the NAS study, these approaches are termed the *spent fuel option*.
- The plutonium could be mixed with highly radioactive wastes and stored as waste, for

---

[24]This summary is based on Refs. [10], [45], and [46], which give overviews of the problem of dealing with separated plutonium.

example, in glass placed deep underground. This would put the plutonium in a relatively inaccessible form, making it difficult to reclaim for malign or benign purposes. In the NAS study, this is termed the *vitrification option*.

- The plutonium could be placed in deep boreholes, at depths of several kilometers. Recovery would be difficult, and perhaps prohibitive, if the boreholes were sealed with clay and concrete.

The NAS study also mentioned an array of other possibilities, including subseabed disposal, dilution in the oceans, launching into space, and transmutation in nuclear reactors or accelerators.[25] However, the study's preferred options were the spent fuel option, the vitrification option, and, contingent upon further study, possibly the deep borehole option [10, p. 143].

Most of these solutions have nothing to do with nuclear power, except to the extent that they might spur disposal programs for civilian wastes. However, the spent fuel option is connected to nuclear power and has given reactor manufacturers a glimmer of hope. Nuclear power, in this scenario, would be used not only to generate electricity but also to destroy plutonium. Many different types of reactors can be used for this, although not at equal levels of effectiveness.[26]

An irony in this range of options has been the suggested conversion of plutonium-producing breeder reactor programs into plutonium-destroying fast reactor programs. This has been suggested for the Superphenix reactor in France and was advanced as an argument for the U.S. integral fast reactor program before that program was terminated (see Sec. 13.4.1). Reactors operating with fast neutrons use a higher inventory of fissile material in the core than do thermal reactors, and thus for the same electrical output the throughput of plutonium would be greater in a fast reactor than in a thermal reactor.[27]

However, LWRs are far more numerous than fast reactors, and they offer the more immediate prospect for large-scale plutonium consumption. The plutonium would be used in the form of a *mixed-oxide fuel* (MOX), with 3% to 7% PuO_2 and the remainder UO_2. At the higher ^{239}Pu enrichments, a burnable poison would be added to the fuel to reduce its initial reactivity. Most LWRs are limited to using only about a one-third fraction of MOX in the reactor core, with the remainder being ordinary uranium fuel.[28]

If a full load of MOX fuel is used, the ^{239}Pu consumption rate is roughly the same as the rate of consumption of ^{235}U, namely, about 1 tonne per GWyr. The precise consumption rate varies with reactor operating conditions. With a high initial plutonium concentration in the fuel, the consumption could be well over 1 tonne per GWyr. In very approximate terms, a U.S. inventory of 50 tonnes of plutonium could be consumed by use of full MOX loads for something in the neighborhood of 40 reactor-years or one-third loads for 150 reactor-years [10, pp. 158–159].

It is not as yet clear that weapons-grade plutonium will be disposed of in reactors of any

---

[25]There is no sharp boundary between the spent fuel and transmutation options, because in either case there would be *some* transmutation and *some* residual plutonium.

[26]Summaries of alternatives are presented in Refs. [10] and [47].

[27]The number of fission events will be about the same, but both the input of separated plutonium and the output of contaminated plutonium would be greater.

[28]The limitation to a one-third fraction of MOX fuel stems from differences in the properties of ^{239}Pu and ^{235}U as nuclear fuels. The delayed neutron fraction β is smaller for ^{239}Pu than for ^{235}U (see Sec. 5.3.3), and the neutron mean free path for fission is shorter [45, p. 177]. As a result, it is necessary to have more control rods if a full load of MOX fuel is used in place of uranium fuel. This cannot be readily accomplished in most LWRs. However, it is possible in the so-called System-80 PWRs. Three such reactors are operating at the Palo Verde nuclear plant in Arizona. It should be noted, however, that at present no U.S. LWR is licensed by the NRC to operate with MOX fuel.

sort, i.e., that the spent fuel option will be adopted. If it is adopted and commercial LWRs are used, the electrical output would be useful, and the swords-into-ploughshares aspect might provide an important psychological or public relations dividend for nuclear power. But the real benefit would be in the destruction of weapons-grade plutonium, not in the gaining of an additional source of reactor fuel.

14.5.4 Measures to Reduce Proliferation Dangers

Technological Measures

There is no technological fix to the problem of nuclear proliferation. There are many routes to nuclear weapons for determined countries that possess even a modest industrial and scientific base. However, as we have seen, the threshold of difficulty would be raised for relatively weak countries, and certainly for subnational groups, if there is no easy access to separated plutonium. Several measures have been suggested to limit this access:

- Restrict nuclear power to a once-through fuel cycle, as is now the practice in the United States. However, as discussed in Sec. 14.2.5, this approach is rejected by other important countries, in particular, France and Japan, who are unwilling to accept what they consider to be a long-term economic penalty.
- Adopt a reprocessing fuel cycle in which the extracted plutonium is quickly returned to the reactor, greatly reducing the possibility of diversion. For example, this was a motivating consideration in the integral fast reactor program at Argonne National Laboratory (see Sec. 13.4.1).
- Adopt a breeding fuel cycle in which none of the material is readily usable for weapons. An oft-suggested approach is the ^{232}Th–^{233}U cycle, in which ^{233}U is the fissile fuel and ^{232}Th is the fertile fuel. ^{233}U extracted from the spent fuel could be mixed with natural uranium to obtain uranium fuel of low ^{233}U enrichment to be returned to a reactor. The ^{239}Pu, which is eventually formed from neutron capture in ^{238}U, could be left as part of the highly radioactive wastes. There may be interest in pursuing such an option in the future if the perceived need for breeder reactors increases, but at the moment there are no active programs to implement this cycle.

Institutional Measures

The most effective means of restraining nuclear proliferation, at least in the short run, is a combination of rigorous inspections, presumably by the IAEA, backed by economic pressures, presumably exerted through the United Nations or, if necessary, unilaterally by the United States and its allies. Such mechanisms can be faulted for their initial failure to stop the Iraqi and North Korean nuclear programs, which proceeded vigorously despite the NPT, IAEA inspections, and intelligence gathered by the United States and others. However, Iraq was eventually brought to its nuclear knees and will remain there as long as the IAEA, the United Nations, and the United States persevere. The situation with North Korea is more ambiguous, and it may take some years to determine whether the agreements discussed in Sec. 14.4.2 succeed in preventing the buildup of an expanded nuclear-weapons arsenal.

However, the inhibiting power of inspections and economic coercion should not be totally deprecated. Three countries (Argentina, Brazil, and South Africa) have given up nuclear weapons, and while their reasons may have differed it would appear that international pressures played an important role. India, Israel, and Pakistan have paid the NPT the compliment of not adhering to it. It may be too late, or even undesirable, to try to roll back their programs, but the overall experience suggests that international safeguard measures are not completely toothless. One solution may be to strengthen the IAEA's teeth.

In the case of Iran, the conflict between the IAEA's obligation to encourage peaceful nuclear applications and its obligation to forestall weapon diversions puts the agency under a severe strain. The strain is compounded by the fact that the IAEA governs by the consent of the governed.

14.5.5 Policy Options for the United States

No attempt will be made here to outline a nonproliferation policy for the United States. The issues are too complex and in many cases too removed from nuclear power to be appropriately treated. However, it is worth making a few observations:

- The coupling between nuclear power and nuclear weapons is weak. The issue is important and profound because the stakes are great. But in the end, nuclear power policies, and more particularly U.S. nuclear power policies, can do little to prevent proliferation and are unlikely to contribute importantly to proliferation.
- However, little is not zero. In a few cases, for example, Iran, there may be a significant coupling, and it would be useful from a weapons proliferation standpoint to discourage nuclear power in Iran and similar countries, if that is possible.
- The United States could play a role in securing greater powers for the IAEA in carrying out inspections and in making the United Nations more determined in dealing with countries that pose nuclear threats. Support of these international bodies does not preclude unilateral action on the part of the United States and its allies, especially in terms of sanctions and trade limitations.
- It would accomplish little for the United States to abandon nuclear power in the name of nonproliferation. Many other countries, particularly Japan and France, would decline to follow suit. Further, abandoning nuclear power without entirely giving up nuclear weapons might be seen as an empty gesture.[29]
- In attempting to influence nuclear programs elsewhere, economic pressure may be the most effective tool available to the United States. Economic strength is here a prerequisite. Therefore, even from the standpoint of weapons proliferation, our energy policy should seek to increase U.S. economic strength and economic and political freedom of action. Arguments will still remain as to the extent to which nuclear power can contribute to overall economic strength and to reduced dependence on energy exporters.

In the end, however, it should be recognized that success in preventing the use of nuclear weapons and in limiting nuclear proliferation depends on finding prudent and effective policies in areas that have little to do with nuclear power. Specific key matters include the degree and manner of reductions in the U.S. and Russian nuclear arsenals, means of encouraging responsible control over weapons and nuclear materials, in the former Soviet Union, mechanisms for bringing pressure to bear on potential proliferators, and the future role for nuclear deterrence.

---

[29]It is here assumed that the United States will not give up all nuclear weapons, if for no other reason than fear that another country would then seek nuclear dominance through a few illicit nuclear weapons.

REFERENCES

1. J. Carson Mark, "Explosive Properties of Reactor-Grade Plutonium," *Science and Global Security* 4, no. 1 (1993): 111–124. Frank von Hippel and Edwin Lyman, "Probabilities of Different Yields," *ibid*, 125–128.
2. Robert Serber, *The Los Alamos Primer, The First Lectures on How To Build An Atomic Bomb* (Berkeley: University of California Press, 1992).
3. J. Malik, E. Tajima, G. Binninger, D. C. Kaul, and G. D. Kerr, "Yields of the Bombs," in *U.S.–Japan Joint Reassessment of Atomic Bomb Radiation Dosimetry in Hiroshima and Nagasaki, Final Report*, edited by William C. Roesch (Hiroshima: Radiation Effects Research Foundation, 1987): 26-36.
4. Frank Barnaby, "Types of Nuclear Weapons," in *Plutonium and Security*, edited by Frank Barnaby (New York: St. Martins Press, 1992), 264–279.
5. Thomas B. Cochran, William M. Arkin, and Milton M. Hoenig, *Nuclear Weapons Data Book, Vol. I of U.S. Nuclear Forces and Capabilities* (Cambridge, Mass.: Ballinger Publishing, 1984).
6. Theodore B. Taylor, "Nuclear Safeguards," *Annual Review of Nuclear Science* 25 (1975): 407–421.
7. F. J. Rahn, A. G. Adamantiades, J. E. Kenton, and C. Braun, *A Guide to Nuclear Power Technology, A Resource for Decision Making* (New York: John Wiley, 1984).
8. Victoria McLane, Charles L. Dunford, and Philip F. Rose, *Neutron Cross Section Curves*, Vol. 2 of *Neutron Cross Sections* (New York: Academic Press, 1988).
9. Thomas B. Cochran and Christopher E. Paine, "The Amount of Plutonium and Highly-Enriched Uranium Needed for Pure Fission Nuclear Weapons" (Washington, D.C.: National Resources Defense Council, 1994).
10. National Academy of Sciences, *Management and Disposition of Excess Weapons Plutonium*, report of the Committee on International Security and Arms Control (Washington, D.C.: National Academy Press, 1994).
11. D. Albright and H. A. Feiveson, "Plutonium Recycling and the Problem of Nuclear Proliferation," *Annual Review of Energy* 13 (1988): 238–265.
12. Frank Barnaby, introduction to *Plutonium and Security*, edited by Frank Barnaby (New York: St. Martin's Press, 1992), 1–9.
13. F. Takáts, "International Status and Trends for Spent Fuel Management," *Journal of Nuclear Materials Management* XXII, no. 3 (April 1994): 38–44.
14. Ryukichi Imai, comments at Global '93, Future Nuclear Systems: Emerging Fuel Cycles and Waste Disposal Options, Seattle, Wash., September 15, 1993.
15. Myron B. Kratzer, "Demythologizing Plutonium," paper given at Global '93, Future Nuclear Systems: Emerging Fuel Cycles and Waste Disposal Options, Seattle, Wash., September 15, 1993.
16. Leonard S. Spector, *Nuclear Ambitions: The Spread of Nuclear Weapons 1989–1990* (Boulder: Westview Press, 1990).
17. Stockholm International Peace Research Institute, *Nuclear Energy and Nuclear Weapon Proliferation* (London: Taylor and Francis, 1979).
18. National Academy of Sciences, *Nuclear Arms Control: Background and Issues*, report of the Committee on International Security and Arms Control (Washington, D.C.: National Academy Press, 1985).
19. William Epstein, "Indefinite Extension—with Increased Accountability," *The Bulletin of the Atomic Scientists* 51, no. 4 (July/August 1995): 27–31.
20. Peter D. Zimmerman, private communication with author, March 1994.
21. Karl P. Cohen, "The Front End of the Fuel Cycle," in *The Nuclear Connection*, edited by Alvin Weinberg, Marcelo Alonso, and Jack N. Barkenbus (New York: Paragon House, 1985), 55–86.
22. A. S. Krass, P. Boskma, B. Elzen, and W. A. Smit, *Uranium Enrichment and Nuclear Weapon Proliferation*, Stockholm International Peace Research Institute (London: Taylor and Francis, 1983).
23. Bertrand Goldschmidt, *The Atomic Complex: A Worldwide Political History of Nuclear Energy*, translated by Bruce M. Adkins (La Grange, Ill.: American Nuclear Society, 1982).
24. Andrei Sakharov, *Memoirs* (New York: Knopf, 1990)
25. "Ukraine: Country Joins NPT in December Ceremony," *Nuclear News* 38, no. 1 (January 1995): 45–46.
26. William Sweet, *The Nuclear Age: Atomic Energy, Proliferation and the Arms Race*, 2nd ed. (Washington, D.C.: Congressional Quarterly, 1988).
27. John Wilson Lewis and Xue Litai, *China Builds the Bomb* (Stanford: Calif.: Stanford University Press, 1988).
28. "World List of Nuclear Power Plants," *Nuclear News* 38, no. 3 (March 1995): 27–42
29. Leonard S. Spector, *Nuclear Proliferation Today* (New York: Vintage Books, 1984).
30. David Albright, "How Much Plutonium Does North Korea Have?," *The Bulletin of the Atomic Scientists* 50, no. 5 (1994): 46–53.
31. Michael R. Gordon, "U.S.–North Korea Accord Has a 10-Year Timetable," *The New York Times*, October 21, 1994: 4.
32. Alan Riding, "U.S. and North Korea Sign Atom Pact," *The New York Times*, October 22, 1994: 7.
33. Peter A. Clausen, *Nonproliferation and the National Interest, America's Response to the Spread of Nuclear Weapons* (New York: HarperCollins, 1993).
34. "Late News in Brief," *Nuclear News* 35, no. 1 (January 1992): 18.
35. Nathaniel C. Nash, "Sequel to an Old Fraud: Argentina's Powerful Nuclear Program," *New York Times*, January 18, 1994, p. A6.

36. "Late News in Brief," *Nuclear News* 34, no. 13 (October 1991): 26.
37. "De Klerk Tells World South Africa Built and Dismantled Six Nuclear Weapons," *Nuclear Fuels* 18, no. 7 (March 19, 1993): 6.
38. "Late News in Brief," *Nuclear News* 36, no. 7 (May 1993): 20.
39. Christopher S. Wren, "Mixed Signals Over Status of Iran Deal," *The New York Times*, September 23, 1995: 4.
40. David Albright, "An Iranian Bomb?" *The Bulletin of the Atomic Scientists* 51, no. 4 (July/August 1995): 21–26.
41. Peter Auer, Marcelo Alonso, and Jack Barkenbus, "Prospects for Commercial Nuclear Power and Proliferation," in *The Nuclear Connection*, edited by Alvin Weinberg, Marcelo Alonso, and Jack N. Barkenbus (New York: Paragon House, 1985), 19–47.
42. Peter D. Zimmerman, *Iraq's Nuclear Achievements: Components, Sources, and Stature*, CRS Report for Congress, report 93-323 F (Washington, D.C.: Library of Congress, 1993)
43. Peter D. Zimmerman, private communication with author, May 1994.
44. Peter D. Zimmerman, "Proliferation: Bronze Medal Technology is Enough," *Orbis* 38, no. 1 (1994): 67–82.
45. F. Berkhout, A. Diakov, H. Feiveson, H. Hunt, E. Lyman, M. Miller, and F. von Hippel, "Disposition of Separated Plutonium," *Science and Global Security* 3 (1993): 161–213.
46. U.S. Congress, Office of Technology Assessment, *Dismantling the Bomb and Managing the Nuclear Materials*, report OTA-O-572 (Washington, D.C.: U.S. Government Printing Office, 1993).
47. Carl E. Walter and Ronald P. Omberg, "Disposition of Weapon Plutonium by Fission," in *Future Nuclear Systems: Emerging Fuel Cycles and Waste Disposal Options, Proceedings of Global '93* (La Grange Ill.: American Nuclear Society, 1993), 846–858.

Chapter 15

Costs of Electricity from Nuclear Power

15.1 ELECTRICITY COSTS: QUALITATIVE CONSIDERATIONS

15.1.1 Direct Costs and External Costs

The "costs" of electricity, or energy in general, can be defined either narrowly or broadly. In the narrow definition, the costs correspond to the expenses of the utility that provides the electricity. The basic components are the costs of constructing the plant and related facilities, fuel costs, the costs of maintenance and operation of the plant, and the costs of distribution and sale of electricity. All of these are tangible costs, which are clearly defined by accounting practices and go into determining the price of electricity as established by utility regulators (see Section 15.1.3).

However, there are other costs of energy. These include, for example, the costs to society of environmental pollution and of hidden or explicit government subsidies. If one takes a very broad view, one could also include the costs of providing energy security through a strategic oil reserve or military action. These are sometimes called "social costs" or "hidden costs," but the more common term is *external costs*.

In some models of a desirable approach, a system of taxes and fees would be created in which all of the external costs are recovered and are reflected in the price of energy. However, this has not been the practice, and determining these costs is fraught with ambiguity. What, for example, is the cost of carbon dioxide emissions? To answer this, we would have to know the resulting climate changes and their impact, and then attach a price to this impact.

There can be extensive debate about such matters, with no standard accounting rules to resolve the disagreements. In practice, some of these costs (only a small fraction) are addressed by assessing charges against utilities or by compelling utilities to take steps to address undesirable impacts. Thus, utilities that produce nuclear power are charged a fee of 0.1 ¢/kWh to cover the costs of a (future) nuclear waste repository. Similarly, under the Price–Anderson Act, utilities are required to contribute to a fund maintained to reimburse victims of a hypothesized nuclear accident. Coal-burning utilities have been required to install controls on effluents. There also have been recurring suggestions that a "carbon dioxide tax" be imposed on fossil fuel use, or a "BTU tax" on all energy use.

Such fees, equipment expenditures, and taxes, when paid by the energy producer, automatically become a part of the producer's expenses and thus are included in the narrow definition of cost. The issue of the appropriate magnitude to assign to an external cost is thereby resolved by law or regulation, even if the decision is sometimes arbitrary. When an external cost is not defined by law or regulation, attempts to include it in costs carry the discussion into regions of substantial uncertainty. Therefore, for the purposes of this chapter, we will usually not go beyond the somewhat narrow definition, which does not include costs for which specific charges have not been assigned.

However, especially in the context of utility planning, efforts are being made to take a broader view of costs. An approach of growing importance is *integrated resource planning*, discussed briefly in Section 15.1.3. Further, as discussed briefly in Section 15.4.4, external factors can override the impact of conventional costs, whether or not they are codified as "external costs".

15.1.2 Who Provides Electricity?

The main suppliers of electricity are the investor-owned utilities, commonly referred to as private utilities. In addition, in many parts of the country federal and local entities have created publicly owned utilities for electricity generation, including large regional organizations such as the federal Tennessee Valley Authority in the southeast and small local bodies such as the city-run Seattle City Light in Washington.

In the early days of electricity generation, a large fraction of the national electricity supply was generated by industry for its own use. But by 1980 all but 3% of U.S. electricity was being produced by electric utilities, reflecting the increase in overall consumption but also the efficiencies of central power stations [1]. A reversal of this trend eventually came from the effects of the Public Utilities Regulatory Policy Act of 1978 (PURPA), which mandated the purchase of electricity by utilities from "qualifying facilities" at the utility's avoided cost.[1] Qualifying facilities are limited to relatively small size and must use renewable resources or, if the efficiency is high enough, co-generation.

Such entities are one type of *non-utility generator* (NUG). With the impetus of PURPA, the NUGs have assumed increased importance in the past decade. The NUG contribution to electricity sold by utilities grew from 1 TWh in 1980 to 189 TWh in 1993, amounting in 1993 to about 6% of the total distributed by the utilities [3, p. 229]. Overall, about 9% of electricity generated in 1991 was by NUGs, including industrial plants that generated electricity for their own use, for example, in the forest-products industry. The largest energy

---

[1]The avoided cost is the cost that the utility nominally avoids by not having to produce the electricity itself. According to P. L. Joskow, these costs have often been set by contract at prices much higher than the true avoided costs [2, p. 220].

sources for generation by NUGs were natural gas (54%), coal (16%), and wood (11%), [3, p. 231].

It is anticipated that the role of NUGs will grow in future years, with the entry of companies set up specifically for the generation of electricity, the so-called independent power producers (IPPs), which would sell electricity to the utilities for distribution. These may or may not be qualifying facilities under the terms of PURPA. There has been a great deal of controversy over the desirability of IPPs. Proponents cite the advantages of increased competition in power generation. Critics argue that an increased role for IPPs could undermine the ability of utilities to meet their commitments to provide reliable electricity service. In either case it is not clear whether IPPs would differ from standard utilities in their willingness to undertake nuclear projects.

15.1.3 The Role of Costs

Regulation of Electricity Rates

The sale of electricity has traditionally been a monopoly in a given locale. This monopoly structure has the obvious attractions of efficiency and simplicity, and is controlled by having the utility rates determined by governmental regulatory bodies. In particular, the bulk of electricity in the United States is provided by the investor-owned utilities, whose rates are controlled by state public utility commissions (PUCs). As governmental entities, the state PUCs are subject to political pressures and must find a middle path between the consumer's desire for low rates and the utility's desire for high ones.

An appropriate electricity rate in this framework, intended to be fair to both the utility and the consumer, is one that covers the cost of producing and delivering the electricity plus a reasonable rate of return upon the capital invested. The total investment is called the rate base, and the costs attributable to the costs of capital are the product of the rate base and an average rate of return. The utility commissions have discretion in both establishing the rate of return and determining what capital expenditures are to be included in the rate base.

As might be expected, disputes arise in these decisions. For example, the issue of "prudence" has been of particular pertinence to the nuclear industry. We quote one description of how this consideration enters:

> Before according rate base treatment to a new plant, regulators now question the prudence of the decision to build it, its cost, and its usefulness. In some instances, they do find serious misman-agement; but since most utility managers were careful in planning and spending funds for plants, many of these "prudency" reviews amount to second guessing with perfect hindsight and discourage utilities from undertaking future, needed projects. A finding of imprudence leads to rate base reduction, shifts expenses from ratepayers to stockholders, increases risks to investors, calls for new accounting practices to deal with growing uncertainties, and raises the cost of capital to utilities [4, p. 180].

Nuclear power facilities have been particularly hard hit by prudency reviews. As reported in a National Academy of Sciences study:

> During the 1980s, rate base disallowances by state regulators totaled about $14 billion for nuclear plants, but only about $0.7 billion for non-nuclear plants [5, p. 182].

This disproportionate impact could be attributed to the relative newness of nuclear power and attendant errors, to the difficulty in forecasting demand given the long time interval between the start of construction of a nuclear plant and the beginning of its commercial operation, or to nuclear power's political unpopularity. In any event, this history provides cautionary signals to any utility contemplating a nuclear project.

Utility Planning and Integrated Resource Planning

Traditionally, utilities have chosen among energy-generation options on the basis of their own estimate of costs. As long as their decisions were deemed prudent by the PUCs, they could usually count on electricity rates that provided an adequate rate of return on their investment. As part of this arrangement, the utilities were obligated to provide reliable service to all customers.

However, for more than a decade there has been a growing movement towards *integrated resource planning* (IRP) or *least-cost planning* in guiding utility decisions.[2] This has been gradually mandated in individual states and was adopted nationally in the federal Energy Policy Act of 1992, which amends PURPA with the stipulation that "each electric utility shall employ integrated resource planning" [6, Sec. 111(a)]. The IRP proposals are to be submitted to state authorities and updated regularly, with the requirement that the utilities are to "provide the opportunity for public participation and comment."

The act provides the following definition of integrated resource planning:

> The term "integrated resource planning" means, in the case of an electric utility, a planning and selection process for new energy resources that evaluates the full range of alternatives, including new generating capacity, power purchases, energy conservation and efficiency, cogeneration and district heating and cooling applications, and renewable energy resources, in order to provide adequate and reliable service to its electric customers at the lowest system cost. . . . [6, Sec. 111(d)]

The IRP process marks a significant change. In traditional planning, utilities thought in terms of central power stations and themselves decided which type of plants they should use and own.[3] With IRP, plans receive input from a broader group than the utility management alone, and the utilities are required to look at a wide array of electricity sources beyond their own plants. As expressed by authors who favor this approach, writing before the passage of the 1992 act:

> Key characteristics of this planning paradigm include: (a) explicit consideration of energy-efficiency and load management programs as alternatives to some power plants, (b) consideration of environmental factors as well as direct economic costs, (c) public participation, and (d) analysis of the uncertainties and risks posed by different resource portfolios and by external factors [7, p. 91].

This description of IRP gives strong emphasis to both *demand-side management* and consideration of external costs, or "externalities." Demand-side management has two facets: the reduction of overall demand by using electricity more efficiently (or more sparingly) and the reduction of peak electrical demand by encouraging the shift of some consumption to times when the demand is less. In concept, IRP provides the framework for a broad, forward-looking approach to electricity planning.

---

[2]The two terms are used more or less interchangeably, although the term "integrated resource planning" may be replacing the earlier "least-cost planning."

[3]Differences between the IRP approach and traditional planning are discussed in Ref. [7], especially p. 92.

It may be too soon to assess the full impacts of this approach. It gives more influence to planners outside of the utility structure itself, which will bring broader perspectives but may dilute responsibility. The inclusion of externalities is difficult to make complete, and it is not required in all states. It is obviously difficult to establish the proper external cost of the possibility of nuclear accidents, the effects of carbon dioxide emissions, or, still harder, the problems of assuring the security of energy supplies. Such factors can be either omitted or assigned essentially arbitrary values.[4]

However, in the context of nuclear power, poorly defined externalities may play a critical unacknowledged role, to the extent that they represent considerations that influence an individual's overall attitude towards nuclear power. In a planning process as open as the IRP process, in which it is necessary to make estimates of future costs that are often little more than educated guesses, these attitudes may color the projections. Thus, important but nonquantified external factors can be reflected in the seemingly analytic results that come out of an IRP study. Of course, the IRP process is not unique in this respect, and conscious or unconscious biases can tilt cost projections whatever the mechanism for decision making. But the IRP process is new in that a more diverse set of players participates in making the projections.

A striking outcome of an IRP evaluation was the decision in early 1993 to shut down the Trojan nuclear plant [8]. Various groups participated in the overall process and reached differing conclusions as to the relative costs of continuing to operate the plant for several years and of immediate closure in favor of "a mix of demand-side and renewable resources, cogeneration and combined-cycle combustion turbines [gas turbines]" [8, p. 4A.1]. Pro-nuclear observers believe that the final verdict stemmed from the desire of the utility to avoid further conflict with a sizable fraction of the public, the Nuclear Regulatory Commission, and possibly the Oregon Public Utility Commission. The IRP process, in this view, provided the utility with a seemingly "objective" cover under which to retreat. Nuclear opponents or skeptics believe that the verdict represented a prudent balancing of economic and environmental factors, in which the utility finally began to rectify a past mistake. It is difficult to know the extent to which Trojan represents a precedent, because it had a troubled history and was operating in a region where antinuclear feeling was relatively strong.[5]

The Role of Government in Utility Actions

Other levels of government beyond the public utility commissions can also have important impacts on electricity generation. This can be done through subsidies, hidden or direct, regulations designed to encourage or discourage a given form of generation, or taxes and fees.

For example, coal has been the beneficiary of federal incentives to railroads dating to the 1800s, and nuclear power has been the beneficiary of technology developed for nuclear weapons and ship propulsion. For many years these were the cheapest sources of new electricity capacity, although in many cases hydroelectric dams, also beneficiaries of past federal policies, provided still cheaper power.

---

[4]In the past such externalities have been ignored by utilities, except when forced to consider them, as in the payment by utilities of an accident insurance fee under the terms of the Price–Anderson Act.

[5]Trojan's lifetime capacity factor, through the end of 1991, was only 53%. It recently had undergone steam generator repairs and was out of operation at the time the decision was reached due to the need for further steam generator repairs. The operating utility, Portland General Electric, had been forced to fight several Oregon ballot initiatives calling for its closure.

However, all three of these previous beneficiaries of federal assistance now face environmentally based objections. Coal plants are being required to adopt emission controls, nuclear plants are subject to very strict safety regulation, and hydroelectric output is being curtailed in some areas to preserve stream flow. This climate encourages electric utilities and NUGs to consider other sources.

Environmental groups have long advocated the use of renewable resources beyond hydroelectric power for electricity generation, and there has been increasing interest in renewable resources on the part of the Electric Power Research Institute and some utilities. However, for the near future it would appear that natural gas will be preferred by most utilities as a more immediately practical and economical alternative. It is becoming the compromise fuel of choice, aided by low prices of natural gas and technological advances that have improved the efficiency of gas-fired plants.

This represents a striking reversal in policy. In 1978, through the Power Plant and Industrial Fuel Use Act, the federal government banned the construction of new power plants using natural gas or oil. This action was motivated by fears of eventual fuel shortages. This ban was lifted in 1987, although there was no important change in the prospects for future natural gas supplies.

Future increases in the price of natural gas would not create critical problems for utilities, as long as state regulatory commissions look with favor on gas generation and authorize rates that pass on to the consumer the cost of any such increases. However, plants generating electricity using natural gas are also subject to prudency reviews, and it is conceivable that the utilities might face difficulties in such reviews if natural gas prices rise far above the levels predicted at the time the plants were constructed.

An extreme example of the negative impact that government policies can have, in this case those of local and state government, is the Shoreham nuclear power plant, built for the Long Island Lighting Company (LILCO) in New York. The plant was completed in the mid-1980s, and after a variety of battles in the courts and in NRC deliberations, it was given an NRC license in 1989 to operate at full power. However, local and state opposition prevented it from going into operation, and it is now being dismantled (see, e.g., Ref. [9]). The resulting loss is borne, in proportions that are hard to assign, by LILCO, the New York state government, and individual electricity ratepayers.

Government can also express its preferences or concerns through financial penalties or incentives. The charge of 0.1 ¢/kWh on nuclear-generated electricity for nuclear waste disposal, is a modest example of what can be done; in this case the fee is a charge against expected expenses rather than a punitive measure or a significant discouragement. A carbon dioxide tax, which at the present does not exist in the United States, could be an explicit discouragement to the use of fossil fuels. On the other side of the coin, the federal government is now providing strong incentives for the development of renewable energy. Under the Energy Policy Act of 1992, a subsidy of 1.5 ¢/kWh will be paid for up to 10 years to a "qualified renewable energy facility" that generates electricity using solar or wind energy or some forms of biomass or geothermal energy [6, Sec. 1212].

An Increased Role for Competition?

A possible looming change in the utility industry is the much increased role for competition, which could give consumers more choice among electricity suppliers. The California Public Utilities Commission has announced a plan to foster competition that is scheduled to go into effect by 2002 [10]. The final form of this plan, the extent to which it will be emulated by other states, and its full range of impacts are not known at this time. The intent is, in the end, to reduce electricity prices to the consumer without sacrificing reliability of supply.

These developments are still in a formative stage, and it is too soon to incorporate them in the discussion of costs. Thus, they will be ignored in the remainder of the chapter. Nonetheless, it is worth remembering that major changes in utility structure may be in the offing. It remains to be seen whether the changes, if they materialize, will increase or decrease the relative competitive positions of the various fossil fuels, of nuclear power, and of renewable sources.

15.2 THE DIRECT COST OF ELECTRICITY

15.2.1 Components of Electricity Cost

Restricting consideration to the traditional formulations, the total electricity cost can be expressed as the sum of the following components: (a) the generation cost, (b) transmission and distribution costs, and (c) other costs such as expenses of customer relations and administration.[6] The generation cost is the main component of the total cost. It is the cost of electricity as it leaves the plant, sometimes referred to as the "busbar cost."[7] The generation cost is the sum of two components:

1. *Production cost.* This includes operations, maintenance, and fuel costs. The production cost can be divided into "non-fuel operating costs," often abbreviated to "operating costs," and fuel costs. The operating costs include the normal operation and maintenance (O&M) costs and sometimes include post-operational capital expenditures, known also as *capital additions* [13].

2. *Carrying charges.*[8] These reflect primarily the construction costs, including the return on the capital expended in constructing the plant plus taxes, where the return on capital is the product of the investment in the plant and the rate of return. This return serves to pay depreciation and interest charges and shareholder dividends.

Capital expenditures for modifications or additions made after the plant is completed are sometimes included in the carrying charges, if these expenditures are added to the total investment in the plant, or they can be included in the operating costs. These additions made after the plant has been completed can exceed the original construction cost if extensive retrofitting is deemed necessary, as was common for nuclear power plants in the 1980s.

15.2.2 Recent Trends in Electricity Costs

Price of Electricity

The history of electricity prices in the United States since 1960 is shown in Fig. 15.1, which gives average prices expressed in both current (or nominal) dollars and constant (or real)

---

[6]For the most part, we follow here the terminology of the Department of Energy publication *Electric Plant Cost and Power Production Expenses 1991* [11]. External costs are included only to the extent that they result in actual equipment expenses (e.g., for pollution control) or in fees paid (e.g., for nuclear waste disposal).

[7]The *busbar* is the conductor that connects the output of the generator to the input of a transformer that feeds the transmission lines. Thus, the busbar cost is the cost of electricity as it leaves the plant [12, p. 11].

[8]The terms "fixed charges," "capital-related charges" and "capital charges" are sometimes used in place of "carrying charges." They all mean the same thing.

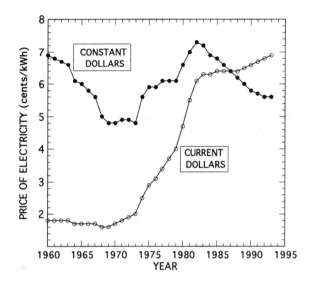

Figure 15.1. Average U.S. retail electricity prices, 1960–1993, expressed in current dollars and in constant 1987 dollars.

dollars [3, p. 249].[9] Overall, real electricity prices have dropped during this three-decade period, although the price in current dollars has consistently risen. There were large electricity price changes, especially the rise in the decade following 1973, but the variations were less than those for oil and natural gas.

Different classes of customers are charged at different rates. Thus, in 1994 the overall average price was 6.9 ¢/kWh, while residential customers paid an average of 8.4 ¢/kWh and industrial consumers 4.7 ¢/kWh [14, p. 122]. There are also large regional differences. For example, in regions where "old" hydroelectric power is important, as in the Pacific Northwest, electricity prices are lower than the national average.

Costs of Fossil Fuels

For electricity generation with fossil fuels, fuel costs are usually the largest component of total costs. Average costs of fossil fuels used by U.S. utilities from 1973 through 1993 are plotted in Fig. 15.2, using data from Ref. [14]. The translation of the cost of the fuel in dollars per MBTU to a contribution to the cost of electricity in ¢/kWh depends upon the efficiency of conversion, η:[10]

$$\text{(electricity cost in ¢/kWh)} = (0.3412/\eta) \times \text{(fuel cost in \$/MBTU)}.$$

The rate of conversion of fossil fuel energy into electric energy (net output) in steam electric power plants since 1963 has been between 10 280 and 10 520 BTU per net kilowatt-hour [3]. The 1994 rate was 10 280 BTU/kWh, corresponding to a thermal efficiency of 33.2% or 0.332 [14]. At this efficiency, the cost of electricity is 1.03 ¢/kWh for each

---

[9]See Section 15.5.1 for further discussion of these terms. The conversion from current to constant dollars is made using an implicit price deflator to correct for inflation.
[10]1 MBTU = 10[6] BTU; 1 kWh = 3412 BTU.

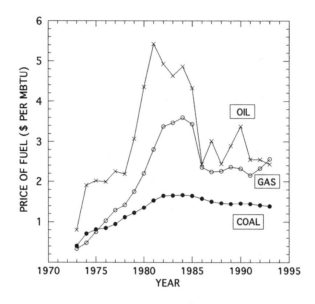

Figure 15.2. Average cost of coal, oil, and natural gas used by U.S. steam electric utilities, 1973–1993, in current dollars per million BTU.

$/MBTU cost of fossil fuels. Thus, for example, the 1993 coal cost at steam electric plants averaged $1.385 per MBTU, corresponding to 1.42 ¢/kWh [14].[11] The "fuel cost" is somewhat higher, because it includes the costs of handling the coal once it reaches the plant.[12]

It is seen in Fig. 15.2 that oil and natural gas prices have fluctuated considerably. Oil prices rose sharply from 1973 to 1974 and from 1978 to 1981, with a large subsequent drop. Gas prices have roughly followed. For the most part, oil has been the most expensive of the three and coal the least expensive. As a result of price changes and restrictions on permitted sources of new generation in force from 1978 to 1987 (see Section 15.1.3), coal's share of U.S. electricity generation grew in the 1973–1994 period from 46 to 56%, natural gas's share dropped from 18 to 10%, and oil's share dropped from 17 to 3%, while nuclear's share rose from 4.5 to 22% [14].

Coal and Nuclear Generating Costs, 1975–1991

Nuclear generation of electricity in the United States had its first beginnings in the late 1950s, entered a period of rapid growth in the 1960s, and became a significant factor in U.S. electricity generation in the mid-1970s. Coal has long been the main source of electricity generation in the United States, and comparisons to the cost of coal-generated electricity have been a standard for gauging economic attractiveness. By 1975 nuclear electricity was significantly cheaper than electricity from coal and much cheaper than electricity from oil or natural gas. There was a widespread expectation that the cost advantage of nuclear power would increase with time, as nuclear experience grew and fossil fuel prices rose.

---

[11]This calculation ignores any small difference between the efficiency of coal-fired steam plants and the average efficiency of all fossil fuel steam plants, the very large majority of which are coal-fired.

[12]For example, for 1991 the quoted average "fuel cost" for coal plants was 1.70 ¢/kWh [11, p. 92], while the cost of the coal itself was 1.49 ¢/kWh [14].

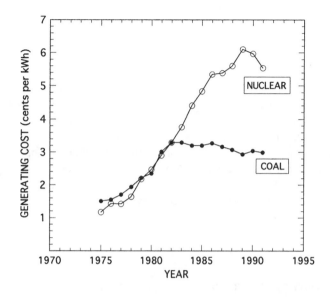

Figure 15.3. Comparison of U.S. coal and nuclear mean generating costs, 1975–1991. Generating costs are taken primarily from Department of Energy reports and are expressed in current dollars.

The actual course of events has been very different, especially in the United States. Nuclear power remains economical in some countries, for example, France. But mean nuclear generation costs in the United States in 1991 were roughly 85% higher than the costs for coal. Thus, in addition to the problems that the nuclear industry faces in surmounting public concerns about safety and environmental issues, it no longer can make a claim on the basis of present average costs.

The dramatic change in the economic status of nuclear power in the period following 1975 is exhibited in Fig. 15.3, where generating costs are compared for coal and nuclear power for 1975–1991.[13] Nuclear generating costs rose by over a factor of five in the 1975–1991 period, more than twice the rate of inflation. They went from being significantly less than those for coal to almost double those for coal. Factors in the change have included a rapid escalation in nuclear costs and a drop in the cost of coal to utilities (in constant dollars).

Natural Gas as an Alternative to Coal and Nuclear Power

Even after the recent decreases in oil prices, oil does not appear attractive as a source of electricity generation from the overall standpoint of cost, resources, and CO_2 production. Although natural gas suffers from somewhat the same liabilities, they are less severe, and natural gas is now preferred by many utilities and NUGs for new construction. Gas-fired gas turbines can be built quickly, reducing the lead time between recognition of increased

---

[13]For the periods 1976–1980 and 1986–1991, costs were taken from Department of Energy tabulations in Refs. [15], p. 18, and [11], p. 92, respectively. (The 1976–1980 data were converted from constant 1980 dollars to current dollars, using the implicit price deflator [3].) For the remaining years, cost figures were obtained from surveys of the Atomic Industrial Forum [16], renormalized to fit the DOE series using quoted results for 1975 and 1980. In effect, the normalization reduced the AIF coal-to-nuclear cost ratios by roughly 10%.

electricity demand and the time when the power can be provided. They can achieve efficiencies of 50% or more in combined-cycle operation, in which the high-temperature gas-turbine exhaust is used to drive a steam turbine.

According to the data of Fig. 15.2, the 1993 cost per BTU of natural gas exceeded that of coal by about 85%. However, comparing a 50% efficient gas-fired combined-cycle system to a coal-fired steam plant with a 35% efficiency, the excess cost of the fuel itself is only 29%. The competitiveness of gas-fired plants is further improved by lower construction and O&M costs (see Table 15.4 below).

However, the volatility of natural gas prices remains an issue (see Fig. 15.2). It is to be expected that gas prices will rise in the future, assuming that consumption increases and that no major low-cost resources are found. A return to 1984 prices, in constant dollars, would mean doubling the 1993 price. It is difficult to know either the timing or the magnitude of future increases, but the more the energy economy turns to natural gas, the greater will be the pressures on its price. Natural gas consumption in the U.S. rose about 27% between 1986 and 1994, following a drop in the preceding decade, and it is to be expected that the increase will continue.

15.3 PAST AND PRESENT NUCLEAR POWER COSTS

15.3.1 Components of Generating Costs: Nuclear Power and Coal

Table 15.1 shows a breakdown of generation expenses for coal and nuclear power plants [11, p. 92]. Data are limited to those for investor-owned utilities. Expenses are shown for both 1986 and 1991 to exhibit some of the trends with time. The ratio of mean nuclear costs to mean coal costs was 1.85 in 1991 and the ratio of median costs was 1.62. The main cost component for nuclear power has been capital-related costs (59% in 1991), while for coal the main component has been fuel costs (57% in 1991).

The ranges in costs are great, especially for nuclear power. Thus, although the average

Table 15.1. *Average generation expenses for privately owned U.S. coal and nuclear power plants, 1986 and 1991 (in cents per kilowatt-hour).*

	Nuclear		Coal	
Category	1986	1991	1986	1991
Mean of all Plants				
Operation and maintenance	1.25	1.57	0.44	0.54
Fuel	0.75	0.67	1.85	1.70
Capital related[a]	3.34	3.29	0.99	0.79
Total	5.34	5.54	3.27	2.99
Distribution of Total Costs				
Lowest (total)		1.55		1.35
First quartile		3.24		2.39
Median (total)		4.77		2.94
Third quartile		8.04		4.05
Highest (total)		25.05		6.79

Source: Mean data are from Ref. [11], Table A2. The total generation expense is taken from this table; the small discrepancy between the quoted total (1991) cost for coal and the sum of the components is not explained. Distribution data are from Ref. [11], Table A1.
[a]These are the same as the "carrying charges" or "fixed charges."

costs for nuclear and coal differ widely, there is considerable overlap in the distributions, with the best nuclear plants considerably lower in cost than the worst coal plants, or even the average coal plants.

15.3.2 Variations in Nuclear Costs Among Different Units

Extreme Cases

While it is possible to define a mean or median cost of nuclear power, there is no such thing as *the* cost of nuclear power. Differences among plants are too great to make any single number a suitable characterization of performance. As seen in Table 15.1, the lowest- and highest-cost nuclear producers had generation expenses that differed by a factor of 16 in 1991.

The lowest-cost nuclear power plant in 1991 was a 920-MWe plant that operated in 1991 with a capacity factor of 78%.[14] It went into operation in 1972 and has a plant capital cost of about $400 per kilowatt (including subsequent plant additions but not depreciation). The highest-cost facility in 1991 was apparently a newly commissioned 1215-MWe reactor that operated in 1991 with a capacity factor of 50%. It went into operation in 1990, after a construction time of over 15 yr and at a plant capital cost of about $5900 per kilowatt. This comparison illustrates the role of low capacity factors and long construction delays in raising costs.

Capacity Factor Variations

The capacity factor for a given year is a measure of the extent to which the plant operated at full capacity during that year. More specifically, it is the ratio of the actual generation to the generation had the plant operated for 8760 hours at its full capacity rating.[15] While the capacity factor alone cannot explain all the differences between plants, it plays a significant part. This is illustrated in Fig. 15.4 where production expenses (the sum of O&M costs and fuel costs) are plotted together with the capacity factor, using data from Ref. [11]. Data are excluded for single-reactor plants that began operation before 1970, as well as for three plants that had capacity factors below 20%.[16] As is to be expected, there is a clear trend to lower production expenses for plants with higher capacity factors. With one exception, plants with capacities of 80% and greater had production costs below 2 ¢/kWh, while the average was well above this level for plants with capacity factors below 60%.

The dependence of production costs per kWh on the capacity factor stems from the fact that over two-thirds of the O&M costs are fixed, independent of the generation achieved [17, p. 34]. Capital-related costs continue at the same rate whether the plant operates or not,

---

[14]The identifications of the lowest- and highest-cost plants are not stated explicitly in Ref. [11] but can be inferred with reasonable certainty from the data presented.

[15]There are disparities between values of the capacity factor for a given plant quoted in the literature. The capacity on which it is based may be the nameplate capacity, the actual gross capacity, or the net capacity; generation can be gross or net generation. The capacity factors in Fig. 15.4 are slightly lower than the more commonly quoted values; they are taken from Ref. [11], where they are calculated from the nominal installed gross generating capacity, also known as the nameplate capacity, and the net generation. The capacity factors used elsewhere in this book are taken from Ref. [14], which uses the net summer capability and net generation. Net generation typically is about 5% less than gross generation.

[16]These are Yankee Rowe, Big Rock Point, Indian Point, Haddam Neck, and Oyster Creek in the first group and Trojan, Turkey Point (2 units), and Browns Ferry (3 units) in the second.

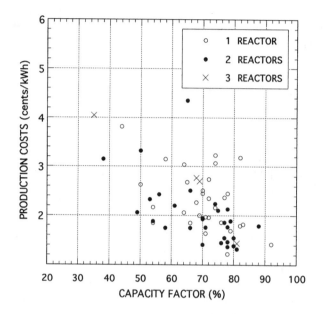

Figure 15.4. Nuclear power production costs per kWh vs. capacity factor in 1991, for U.S. plants with one, two, and three reactors. Plants entering operation before 1970 and those with capacity factors below 20% are omitted. (From Ref. [11].)

leading to an even more sensitive dependence of costs per kWh on the capacity factor.

Number of Reactors per Site

There are also systematic differences between nuclear plants that have one reactor at the site and those with two reactors. In 1991 there were 31 plants with one unit that had gone into commercial operation in 1970 or later and 30 plants with two units (i.e., 60 reactors). Production costs for these groups are displayed in Fig. 15.4. Their average capacity factors were essentially the same, but the production expenses differed significantly. For the two-unit group the median was 1.9 ¢/kWh, while for the one-unit group it was 2.3 ¢/kWh, reflecting economies of scale for the former.[17]

One of the suggestions for a second-generation nuclear era is to have many nuclear plants concentrated at a single site. Some countries have already gone a good way in this direction. For example, Canada has two sites with eight reactors at each, France has one with six, and Japan has one with six reactors and is scheduled to have another with seven reactors by 1997 [18]. The United States has no site with more than three reactors, reflecting the decentralized structure of the U.S. nuclear power industry.

The decentralization of the U.S. nuclear industry is part of the overall decentralization of U.S. electric utilities. The U.S. situation can be contrasted to the highly centralized French system. As of 1994 the 109 U.S. reactors were operated by 47 different utilities, while 54 out of 55 French reactors were operated by a single utility (Electricité de France)[18] [18]. Canada

---

[17]There are also five plants with three units each, but these constitute too small a sample for meaningful averages.

[18]The one exception among operating reactors is the Phenix breeder reactor.

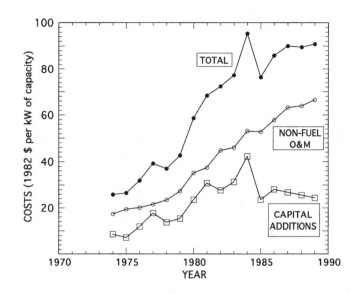

Figure 15.5. Increase in U.S. nuclear plant operating costs, 1974–1989.

and Japan are intermediate in centralization, while South Korea has only a single utility for all its nuclear plants. With centralization comes various possible benefits of standardization, including greater opportunity to create large nuclear parks. Of course, it carries the risk of magnified errors as well.

15.3.3 Escalation in Operating Costs

Total Operating Costs

At the time of the rapid increase in orders in the late 1960s and early 1970s, it was believed that the main cost of nuclear power would be the initial cost of construction, and that once construction was completed, the real cost of nuclear power would decrease as the capital costs were paid off in gradually inflating dollars. This has not proven to be the case, partly because operating costs have risen rapidly with time.

The term *operating costs* as used here is equivalent to *non-fuel operating costs*. It is the sum of O&M costs and capital additions made after the plant goes into operation.[19] Capital additions have turned out to be a large factor in the rise of operating costs, although they were not an important consideration in the estimates made originally.

The period of rapid rise in operating costs, from 1974 through 1989, was studied in the DOE report *An Analysis of Nuclear Plant Operating Costs: A 1991 Update.* In this period these costs (in 1982 dollars) rose from $26 per kWe of (gross) capacity in 1974 to $95 in 1984, corresponding to a rate of increase of 14% per year [13, p. 4] (see Fig. 15.5). This rapid rise has since reversed, and in 1989 these costs were down to $91 per kWe (in 1982 dollars), of which 27% was for capital additions. This corresponds to a cost of about 1.76 ¢/kWh

---

[19]The operating costs, as defined in Ref. [13] and used here, differ from the production costs, as used in the preceding section, by the inclusion of capital additions and the exclusion of fuel costs.

(net).[20] The capacity factor has since improved from 62% in 1989 to almost 74% in 1994 [14, p. 105], further reducing operating costs per kWh.

O&M Component of Operating Costs

Non-fuel O&M costs rose rapidly from 1974 to 1984 and more slowly, especially in fractional terms, from 1984 to 1989 (see Fig. 15.5). The 1989 costs were equivalent to about 1.3 ¢/kWh (1982 dollars). The study of Ref. [13] did not reach a definitive conclusion on the reasons for the increases in O&M costs, although more demanding NRC requirements have been identified as a large factor. Other factors included increased wage rates and increased prices of materials, but their individual contributions were not quantified [13, p. 27].

The nuclear industry has made a determined effort to keep O&M costs from continuing to rise, and the cost increase did not continue after 1989. From 1989 to 1991, helped by higher capacity factors, the O&M costs per kWh dropped about 3% in current dollars and over 10% in constant dollars [11, p. 92], and further drops were achieved in the following three years.

Capital Additions Component of Operating Costs

Modifications made after the plant was completed have been an important contributor to increased operating costs. These modifications have been undertaken to meet more stringent requirements on plant equipment imposed by the Nuclear Regulatory Commission, which led to extensive retrofitting of nuclear power plants, especially after the Three Mile Island accident in 1979 [13]. Costs of capital additions dropped after 1984, resulting in an overall decrease in average operating costs from 1984 to 1989 (in constant dollars).

Utilities have the option of including the cost of these modifications in their operating costs ("expensing" them) or adding them to the total plant capital cost, resulting in a higher base on which rates are calculated.[21] In the latter case, they do not appear as part of the operating costs, but nonetheless add to the total generating costs.

15.3.4 Increase in Construction Costs

Overall Cost Increases

Construction costs of nuclear power plants have increased dramatically over the past two decades. The median cost, in current dollars per unit gross capacity, was about $190/kWe for reactors going into operation in 1970–1974, $470/kWe for 1975–1979, $1000 for 1980–1984, and $2800 for 1985–1989.[22] This is an increase of roughly a factor of 15 in a period of 15 years. Part of the increase was due to inflation, but the implicit price deflator rose by only a factor of 2.6 from 1972 to 1987 (see Section 15.5.1), leaving a factor of six to explain otherwise.

---

[20]This value is based on a 62% capacity factor for 1989 and a difference of about 5% between gross and net capacity.

[21]The choice is determined, at least in large part, by tax considerations.

[22]The cost per kWe is taken from Ref. [19]. It includes allowance for funds used during construction, or AFUDC (see Section 15.5.4), and is based on gross capacity.

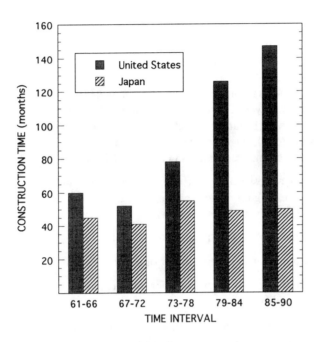

Figure 15.6. Construction times for reactors in the United States and Japan, 1961–1990. The indicated year refers to the date of completion.

Increased Construction Time

One factor in this increase in costs has been the stretching out of the time between the start of construction and commercial operation. Longer lead times mean higher interest expenses and increased costs due to escalation of prices. At an interest rate of 10%, a delay of 7 yr means a doubling of the investment cost.[23] For reactor starts after 1968, well over one-half of the total actual costs (in mixed current dollars) were attributed to these "time-dependent" components.

Fig. 15.6 shows the construction times of nuclear power plants for six-year intervals from 1961 through 1990.[24] Through 1978 average construction times for U.S. reactors were under 7 yr. No reactors came on-line in 1979, the year of the Three Mile Island accident. Subsequently, construction times averaged over 10 yr for 1980–1984 and over 12 yr for 1985–1990. Plants completed in the later years were the stragglers, all having been ordered by 1974. For contrast, construction times for Japanese reactors are also shown in Fig. 15.6. These have consistently averaged under 5 yr.

Reducing construction times requires good planning on the part of the utility and contractor as well as a licensing process that moves in a prompt and predictable fashion. Steps being taken towards this end for future reactors are discussed in Section 15.4.1.

---

[23]This formulation overstates the case if the construction schedule is readjusted so that the expenditures are also delayed. However, if a nearly completed plant stands idle, the costs escalate as quickly as indicated.

[24]These data are taken from an IAEA compilation [20] in which construction time is defined as the time "from the first pouring of concrete to the connection of the unit to the grid." There is usually a further delay, lasting from one month to over a year, between the time of grid connection and the time of commercial operation, i.e., the date at which the utility takes over the plant from the contractor responsible for building it.

Increase in Overnight Costs

There was also a very large increase in the actual construction costs, often referred to as the overnight costs (see Section 15.5.4). Some of these factors have been discussed in a detailed, unpublished Department of Energy study [21, p. III-15], in which comparisons were made between the requirements for the early generation of nuclear power plants and later ones. The early plants considered were Dresden 2 and 3 and Indian Point 2 and 3 (all in the 770 to 970-MWe range), which began construction in 1966 and 1967 and went into operation by 1976 [22, p. 14]. Later plants were those required to conform to standards adopted in 1976 and 1980. Changes, normalized to the same capacity in MWe, include:

- An increase of almost a factor of two in amounts of concrete, due to more stringent demands on shielding protection (e.g., against tornadoes) and the need for greater separation between components enclosed within the concrete shielding.

- An approximate doubling in the amount of steel.

- More than a doubling in the linear feet of electrical cable.

- An increase in the cost of special components such as valves by factors of four to eight due to "rigid quality assurance."

Associated with these equipment changes were very large increases in the number of hours of craft labor, from 3.5 million hours by 1967 standards to 19 million hours by 1980 standards (for a nominal increase in plant size from 1000 MWe to 1139 MWe) [21, III-19]. For many of the later reactors, changes in requirements occurred during construction, leading to further cost increases.

Failure of the Learning Curve

In the introduction of new equipment, costs commonly drop according to a learning curve, as experience is gained in constructing and using the equipment. U.S. nuclear reactors have been a striking exception: Instead of costs dropping with time, they have risen. The failure to learn is partly related to the attempted pace of expansion. Among U.S. reactors actually completed, the last was ordered in 1973 (see Section 1.2.1), and the last construction start was in 1977. Yet the first reactor with a capacity of over 500 MWe (Haddam Neck) was not connected to the grid until 1967 [20]. Thus, there was no large base of accumulated experience to draw upon in the design of the reactors that followed.

Further diluting the benefits of experience, U.S. light water reactors were built by four different manufacturers, with individual units tailored to the demands of the many different utility customers. Again, France stands at the other extreme, with only one manufacturer for the one customer and a succession of standardized designs.

15.4 PROJECTIONS FOR THE FUTURE

15.4.1 Nuclear Power Costs in the United States

General Considerations for Future Reactors

Unless the construction cost of power plants can be reduced, nuclear power will be too expensive to make further expansion economically acceptable. Reductions can be achieved through the related factors of simpler plant designs, standardized designs, and shorter periods of construction. The new light water reactors discussed in Section 13.2 are intended to meet these criteria. If, as is to be expected, the next-generation plants have a longer operating lifetime than existing plants that too will reduce the electricity cost attributable to plant construction costs.

Licensing reform is a crucial matter. An important step was taken in 1989 when the Nuclear Regulatory Commission (NRC) adopted a "one-step" licensing policy, under which a utility or other power producer can obtain a combined license that includes both the construction permit and the operating license.[25] At the time this combined license is issued and construction can start, all fundamental issues of safety would have been addressed. Subsequent regulatory review would be limited to verifying that the construction was consistent with the criteria set at the time the license was issued. Two further regulatory measures allow potential reactor builders to proceed with fewer uncertainties, even before a reactor is actually ordered: (a) early site permits, which give a decision on the general suitability of a site for the installation of a nuclear power plant, and (b) standard design certification, which provides a decision on the detailed design features of a class of standardized reactors.

These provisions are embodied in the *U.S. Code of Federal Regulations* (Part 52 of Title 10). Since adoption of these new rules, there has been no application for a combined license, so there is no experience as to how this process works in practice and the extent to which it forestalls delays caused by court challenges and extended NRC reviews.

Intrinsic to all these cost considerations are rationalized and standardized designs. Nuclear power plants in the United States have not been consistently designed and operated with a single, stable set of safety features. Instead, there has been a succession of changes and retrofits. This may have been understandable in a rapidly evolving new industry, but it has led to waste and excessive costs. The future development of nuclear power depends upon how well the lessons from past experience are put into effect.

Costs for New U.S. Nuclear Reactor Designs

As discussed in Chapter 13, the two most developed approaches to new reactors in the United States are the *evolutionary* advanced light water reactors, usually well over 1000 MWe, and the *passive* advanced light water reactors, in the neighborhood of 600 MWe. As the name suggests, the evolutionary 1200-MWe reactors are similar to present reactors but take advantage of the experience in design and construction gained over the past three decades. The passive 600-MWe reactors embody more substantial changes in design. It is anticipated that costs for either will be less than those for the present generation of reactors.

We present in Table 15.2 estimates of future power plant costs, prepared by a group at Oak Ridge National Laboratory [23, p. 90] as part of a series of cost studies. Five cases are presented from the Oak Ridge study, each for a midwestern site:

1. Present generation 1100-MWe PWRs that represent median experience.

2. Present generation 1100-MWe PWRs that represent best (or better) experience.

3. Future 1100-MWe improved or evolutionary PWRs (EPWRs).

4. Future advanced passive PWRs, assumed here to be a pair of 550-MWe units (APWR).

5. Coal plants, assumed to be a pair of 550-MWe units.

---

[25]The combined license approach is incorporated into Title XXVIII of the Energy Policy Act of 1992 [6].

Table 15.2. *Projected costs of electricity generation for plants going into operation in the year 2000, in cents per kWh.*

| | Oak Ridge Study | | | | | USCEA Study |
Plant Type	Capital	O&M	Fuel	Decomm.	Total	Total
Present Plants (1100 MWe)						
PWR (median experience)	5.64	1.30	0.72	0.05	7.7	
PWR (better experience)	2.83	0.91	0.64	0.05	4.4	
Future Plants						
EPWR (1100 MWe)[a]	2.20	0.91	0.64	0.05	3.8	3.8
APWR (2×550 MWe)[a]	2.21	1.04	0.64	0.07	4.0	4.1
Coal (2×550 MWe)[a]	2.34	0.75	2.09	0.01	5.2	4.6
Natural gas (4×250)[b]						4.2

Sources: Oak Ridge data are from Ref. [23], p. 90; costs are levelized costs in 1987 dollars. USCEA data are from Ref. [24], pp. 21–22; costs are levelized costs in 1992 dollars.
[a]For the USCEA study, the plant sizes were 1200 MWe and 2×600 MWe.
[b]The natural gas plants are based on combined-cycle combustion turbine units.

All costs are expressed as levelized costs (see Section 15.5.3) in 1987 dollars, with cost estimates based on plants that would begin operation in the year 2000.

The Oak Ridge study, as seen in Table 15.2, projects that future reactors will give only a modest improvement over the "better experience" reactors with present generation designs but a quite substantial improvement over the "median experience" cases. There is little difference, in terms of cost, between one large evolutionary reactor and two smaller advanced reactors with the same total capacity. Both are projected to give electricity at a somewhat lower cost than coal.

Results of a study published by the U.S. Council for Energy Awareness (USCEA), a pro-nuclear advocacy organization, are also presented in Table 15.2. This study, carried out with the aid of a panel of industry, government, and financial representatives, reflects an overall assessment by the nuclear industry [24]. The numerical results are quite similar to those of the Oak Ridge study, although one is expressed in 1987 dollars and the other in 1992 dollars.[26] There is still a cost advantage for nuclear power over coal in the USCEA report, but it is less marked than in the Oak Ridge study. Natural gas in the USCEA study lies between coal and nuclear power in terms of cost.

The results of Table 15.2 should be viewed with some caution, in part because past projections of costs for all energy sources, and nuclear energy in particular, have been faulty and in part because at least one of the projections is from an explicitly pro-nuclear organization. Nonetheless, the results are plausible, because with the benefit of hindsight and planning, it should be possible for the nuclear industry to equal or improve upon its past construction and operating performance.

Of course, comparisons with coal and natural gas are only part of the story. In the past, renewable sources of electricity, other than hydroelectric and geothermal power, have not been competitive in cost with fossil fuels and nuclear power. However, if one takes a partially symmetric view and adopts estimates from "pro-renewable" authors, the future competition could be a close one. For example, in one major review of renewable energy sources, the costs of electricity from wind and biomass are projected to be under 5 ¢/kWh by

---

[26]The implicit price deflator for 1992 is 120.9, with the year 1987 at 100.

the year 2000 or sooner, with further decreases anticipated [25]. Solar-thermal and photo-voltaic sources will remain more expensive for a period of time, but these authors and other solar advocates project eventual decreases to 5 ¢/kWh or less. Thus, if the nuclear optimists and the renewable energy optimists are both correct, either option could provide an economical substitute for fossil fuels. (However, even if costs are not excessive, problems of intermittence remain for some of the renewable sources.)

15.4.2 International Cost Projections

The OECD Cost Projections

Results of a series of studies of projected electricity costs have been published by the Nuclear Energy Agency (NEA) of the Organization of Economic Co-operation and Development (OECD). Successive studies were published in 1983 [26], in 1986 [27], in 1989 [28], and in 1993 [29], the last two done jointly with the International Energy Agency. The earlier studies emphasized comparisons between nuclear power and coal, while the 1993 study gave comparable emphasis to natural gas as well. Fragmentary attention was given to renewable resources in the 1989 and 1993 studies.

These studies report electricity cost projections for individual countries. The basic estimates are provided by energy experts from each country, typically "utilities (at the request of governments) or in some cases by government agencies" [29, p. 13]. To put these separate estimates on a common basis, some of the data were adjusted to common parameters—for example, the same discount rate and plant lifetime.

The reported costs are intended to be comprehensive. Thus, for example, decommissioning costs are included for nuclear plants and costs of emission controls are included for coal plants. Here, as in other analyses, some important external costs are omitted, but the OECD study was atypically explicit in pointing out the omissions. The impact of carbon dioxide emissions was not included except for the few cases where countries had imposed "carbon taxes." Further, the costs related to maintaining the security of fuel supplies were not included. This is an important, if poorly defined, issue for nations that rely heavily on imports. It was omitted because "it is difficult if not impossible to set a meaningful value to the benefit of energy supply diversity" [29, p. 50].[27] We will return briefly to the issue of externalities in Section 15.4.4.

Comparison of Costs of Nuclear Power, Coal, and Natural Gas

Results of the OECD projections are presented in Table 15.3 for some of the larger energy producers among the OECD countries, as well as for several major non-OECD countries [29, Table 23]. The projections are for plants "that could be commercially available for commissioning in the year 2000 or shortly thereafter" [29, p. 9]. A breakdown of the costs into their major components is shown in Table 15.4 for the United States and Japan. Results are given for two values of the real discount rate d_0: 5% and 10% (see Section 15.5.2). In the reports

---

[27]To take a specific case, it would be virtually impossible to assign a cost to the U.S. side of the war with Iraq in 1990–1991. Among many other difficulties, there is no recognized way of determining how important oil was either as a motivating cause in U.S. engagement in the war or in the building up over decades of the capability to wage war effectively.

Table 15.3. *Projections of relative electricity-generation costs for future nuclear, coal-fired, and gas-fired power plants for selected countries. Cost are levelized costs at different discount rates, d_0.*

Country	Ratio: Nuclear to Coal[a,b]			Ratio: Nuclear to Gas[c]		
	d_0=5%	d_0=10%	Average	d_0=5%	d_0=10%	Average
OECD Members						
France	0.65	0.77	0.71	0.60	0.78	0.69
Germany[d]	0.72	0.90	0.81			
Japan	0.85	0.94	0.90	0.69	0.92	0.80
United Kingdom	1.02	1.27	1.14	1.11	1.69	1.40
United States						
Midwest	0.96	0.99	0.98	0.90	1.19	1.04
Northeast	0.85	0.93	0.89	0.86	1.15	1.00
West	1.19	1.17	1.18	0.86	1.14	1.00
Other						
China	0.86	1.05	0.96			
India	0.86	1.00	0.93			
South Korea[e]	0.75	1.00	0.88			

Source: Ref. [29].

[a]For OECD countries, nuclear plant capacities are assumed to be ≥1200 MWe, with 1 to 4 reactors at the site. Costs are calculated for a standardized capacity factor of 75% and a lifetime of 30 yr.
[b]For OECD countries, coal plants are assumed to have emission controls for SO_x, NO_x, and particulates.
[c]Gas-fired plants are combined-cycle plants, with thermal efficiencies ranging from 45 to 50%. Costs are calculated for a standardized capacity factor of 75% and a lifetime of 30 yrs.
[d]For Germany, the coal data assume 50% domestic and 50% imported coal. (At present, Germany uses primarily domestic coal, which is the *more* expensive [30].)
[e]For South Korea, data correspond to PWRs; indicated nuclear costs are lower for heavy water reactors (PHWR).

from individual OECD countries the assumed discount rates ranged from 5 to 10%, with 5% the most common value.[28]

As seen in Table 15.3, in most countries the projected cost for nuclear electricity is less than that of electricity from coal. The nuclear advantage is greater at the lower discount rate, due to the high initial investment in a nuclear plant. The price of fuel is also an important variant. For example, coal-fired power is substantially less expensive in the U.S. west than in the northeast, because the projected price of coal is less than half as great [29, Table 20].

The comparison between nuclear power and natural gas shows wide variations among countries, with the overall result somewhat evenly balanced. Comparing the two extremes in Table 15.3, France and the United Kingdom, there are differences in both the capital costs of the nuclear plant (with the UK costs more than twice as great) and in the projected price of natural gas (with the UK costs roughly 20% lower).

Differences between countries are in part real and in part consequences of differences in assumptions used in the calculations or difficulties in establishing a suitable conversion factor between currencies. Coal costs represent a real difference. For example, for Germany there is a 1.3 ¢/kWh difference between using imported coal and the unusually expensive domestic coal. For most other countries, costs are lower using domestic coal and higher using imported coal. For example, coal costs are about 1.7 ¢/kWh less for the United States, a coal producer, than for Japan, a coal importer (see Table 15.4).

---

[28]The U.S. rate was 7%. The poorer OECD countries (Portugal and Turkey) used values at the top of the range, 10%. The cited non-OECD countries reported high discount rates, sometimes exceeding 10%.

Table 15.4. *Components of projected electricity-generation costs for future nuclear, coal-fired, and gas-fired power plants for the United States (midwest) and Japan. Costs are levelized costs expressed in U.S. centers per kWh (1991 dollars).*

	Cost (¢ per kWh)					Percentage of Total			
	Capital	O&M	Fuel	Total		Capital	O&M	Fuel	Total
U.S. midwest									
Discount rate: 5%									
Nuclear	2.11	1.64	0.52	4.27		49	38	13	100
Coal	1.74	1.02	1.71	4.47		39	23	38	100
Gas	0.62	0.25	3.90	4.77		13	5	82	100
Discount rate: 10%									
Nuclear	3.77	1.64	0.55	5.96		63	28	9	100
Coal	3.28	1.02	1.71	6.01		55	17	28	100
Gas	1.10	0.25	3.67	5.01		22	5	73	100
Japan[a]									
Discount rate: 5%									
Nuclear	2.44	1.09	1.83	5.37		45	20	18	100
Coal	2.06	0.79	3.45	6.30		33	12	55	100
Gas	1.27	0.69	5.77	7.73		16	9	75	100
Discount rate: 10%									
Nuclear	4.65	1.11	1.71	7.46		62	15	23	100
Coal	3.82	0.80	3.34	7.96		48	10	42	100
Gas	2.29	0.70	5.14	8.13		28	9	63	100

Source: Ref. [29].

[a]Japanese cost estimates converted to U.S. currency at the July 1, 1991 rate of 137.84 yen per dollar [29, p. 62].

Although the results suggest that nuclear power has an overall cost advantage over coal, the situation is not clear-cut. Projected costs have been changing fairly rapidly with time, and experience has shown that actual costs often differ substantially from projected ones. Further, these estimates are dependent upon parameter choices that are somewhat arbitrary. For example, when a 10% discount rate is assumed rather than a 5% discount rate, the capital costs of nuclear plants go up substantially and the ratio of nuclear costs to coal costs rises.

Whatever credence such projections deserve, it is not clear that they play a decisive part in determining national energy policy. The advantage for nuclear power for France is mirrored in its heavy dependence on nuclear power. However, while the estimated nuclear cost advantage for Germany is greater than that for Japan, Germany has had a *de facto* moratorium on new nuclear power plant construction, while Japan is continuing an active expansion.

15.4.3 Renewable Sources

Given the concerns about nuclear power and coal, there is increased interest in alternatives, including renewable sources. The 1993 OECD report discussed present costs of renewable sources but did not present many projections for future costs. There were some estimates for future costs of wind power for several countries in the 1989 OECD study [28]. These are summarized in Table 15.5, with comparisons to projections made at that time for other sources. For the United States, the 1993 study cited a projection of 4 ¢/kWh for windpower by the year 2000, down from 8 ¢/kWh in 1992 [29, p. 157]. The estimates of windpower

Table 15.5. *Projected electricity generation costs for future nuclear, coal-fired, natural-gas, and wind-turbine power plants for several OECD countries. Costs are levelized costs, in U.S. cents per kilowatt-hour (1987 dollars).*

	Nuclear[a]	Coal[a]	Gas[a]	Wind[a]
Canada	2.62	2.15	5.60	
Denmark		3.77	4.74	4.48
Italy[b]	4.04	5.04	5.62	
Japan	4.33	5.57	7.07	
Netherlands	3.46	3.28	4.84	4.02
United Kingdom	3.58	3.81		4.06

Source: Ref. [28], pp. 77–81.

[a]For commissioning in the period 1995–2000, assuming a 5% real discount rate.
[b]For Italy, coal cost is average of high and low estimates.

costs do not include the costs of backup systems, which would be required were wind (which is intermittent) to account for a large fraction of a grid's electricity supply.

15.4.4 Will Costs Be Decisive?

Cost estimates of the sort shown in Tables 15.2 and 15.3 do not point unambiguously to a decision among electricity sources for most countries, and even where there appears to be a significant difference, cost may not be decisive. Thus, despite the quoted advantage of nuclear power over coal in Germany, there is no move in Germany to expand the role of nuclear power.[29]

Under some circumstances, cost can certainly be a major consideration. The rapidly rising cost of oil in the 1970s led to sharp reductions in its use for electricity generation, and the high cost of photoelectric power makes it essentially prohibitive at present for any large-scale use. However, projected prices of many sources are converging enough toward common values that it is plausible to make choices on grounds other than calculated costs alone.

Thus, as already discussed, external factors often play a decisive role, whether or not they are incorporated into a framework of external costs. Important considerations, which have different weights in different countries, include safety, environmental protection, preservation of resources, reduction of dependence on imports, and preservation of jobs (particularly in coal mining). These factors can lead to bans on additional use of natural gas (as imposed temporarily in the United States) or to effective barriers against the construction of new nuclear power plants (as now exist in many countries). The resulting energy choices may not be the most economical in terms of minimizing direct expenditures, but they correspond to a judgment as to what is best for society as a whole.

Can a society afford to ignore the hard numbers of actual costs and take a more expensive option? The answer may be yes, within limits. The real cost of electricity was less in the United States in 1993 than in 1973 (see Fig. 15.1), and electricity is used more efficiently now than in 1973. Further efficiency improvements are possible in many applications, for example, lighting, refrigeration, home heating, and motors. If the same task can be

---

[29]The last German reactor to be brought on-line was in 1989. There were none under construction as of late 1995 and no apparent plans for new orders.

Table 15.6. *The implicit price deflator (IPD), 1960–1994 (the 1987 IPD=100).*

Year	IPD	Year	IPD	Year	IPD	Year	IPD
1960	26.0	1970	35.2	1980	71.7	1990	113.3
1961	26.3	1971	37.1	1981	78.9	1991	117.6
1962	26.9	1972	38.8	1982	83.8	1992	120.9
1963	27.2	1973	41.3	1983	87.2	1993	123.5
1964	27.7	1974	44.9	1984	91.0	1994	126.1
1965	28.4	1975	49.2	1985	94.4		
1966	29.4	1976	52.3	1986	96.9		
1967	30.3	1977	55.9	1987	100.0		
1968	31.8	1978	60.3	1988	103.9		
1969	33.4	1979	65.5	1989	108.5		

Source: Ref. [3], Table E.

performed with less electricity, the savings can be pocketed either in the form of lower prices or in the use of more socially benign electricity sources.

Thus, the United States and other relatively wealthy countries are not forced to adopt the lowest-cost source of electricity. With growing competition a possibility, individual utilities may become increasingly cost conscious. But society, acting formally through government actions or more informally through a variety of legal and other pressures, can "tilt the playing field" to increase or decrease the cost of a given source, or in some cases ban it outright. When direct cost differences are very large, they may be the decisive factor. When the apparent disparity is of moderate proportions, external considerations are likely to play an important role in determining which forms of electricity are encouraged and which are confronted with major obstacles.

15.5 SPECIAL TERMS AND CONCEPTS

15.5.1 Current and Constant Dollars

In comparing the cost of any item in one year—for example, the price of oil or the cost of building a generator—to the cost in another year, it is useful to correct for the intervening inflation (or deflation). The cost in *current* dollars is the amount actually expended in a given year. The effects of inflation are removed by converting to *constant* dollars, where some arbitrarily chosen time period serves as the reference for the constant dollar. Thus, if T_0 is the reference year and T is the year in which the cost is incurred, the constant dollar cost (C_0) is related to the current cost (C) by the relation $C_0 = C/(1+i)^n$, where i is the average annual rate of inflation and $n = T - T_0$. For example, if $i=0.07$ per year and $n=10$ yr, the constant dollar cost is about one-half the current dollar cost. Sometimes the constant dollar cost is termed the *real cost* and the current dollar cost the *nominal cost*.

The concept of an unvarying rate of inflation is an idealization. It is sometimes employed, however, for forecasts. For interpreting the past record, it is possible to use actual inflation rates, although there is some ambiguity as to how to define and determine these rates. One measure, often referred to in the newspapers, is the consumer price index (CPI). Another measure, used by the Department of Energy in many of its analyses, is the implicit price deflator (IPD). Values for the IPD for the period 1960–1993 are presented in Table 15.6. The two indices do not give precisely the same results. For example, comparing 1993 to 1973 the IPD ratio is 3.01 while the CPI ratio is 3.25 [31, p. 22].

In considering capital costs of nuclear and coal plants, the construction extends over a period of time, and the utilities report their costs in "mixed current dollars," namely, the sum of the nominal costs incurred over the years. To correct current costs to constant dollar costs, a special inflation index is sometimes used for power plant costs to reflect inflation rates specific to the equipment and materials used for power plant construction.[30]

15.5.2 Present Value and Discount Rate

Even ignoring inflation, there is a different economic significance in an expenditure today and an equal expenditure at a future date, because an alternative to making the expenditure is to invest the funds. By the future date, the invested funds would have increased and the planned expenditure would require only a part of these funds. Thus, the value of a given amount of money spent in the future is less than the value of the same amount today.

The change in the value of money is expressed in terms of a *discount rate* per year.[31] For a constant discount rate, the present value of a sum of money, V_P, is related to the value of a sum, V, at n years in the future by the expression

$$V_P = \frac{V}{(1+d)^n},$$ (15.1)

where d=discount rate (or nominal discount rate)=$(1+i)(1+d_0)-1 \doteq i+d_0$; i=inflation rate; and d_0=real discount rate (or real interest rate).

Conversely, an investment today of an amount V_P is equivalent to an expenditure in n years of

$$V=(1+d)^n V_P.$$ (15.2)

The discount rate has direct relevance to the cost of facilities where there are high initial investments in the plant, to be paid off later through the sale of electricity. The higher the discount rate, the greater will be the costs incurred due to the initial investment.

It is sometimes appropriate to replace the inflation rate by the *escalation rate*, e, as a quantity distinct from the inflation rate, although they are sometimes taken to be identical. The escalation rate is the rate of increase of particular cost components and may be greater or less than the overall inflation rate. For example, the capital-related costs of paying interest do not change with time. For these, the escalation rate is zero even if inflation is not. Similarly, fuel costs would be expected to outstrip inflation rates if the fuel becomes scarce. The effect of inflation is sometimes removed by introducing the *real escalation rate*, e_0, where $e_0=e-i$.

15.5.3 Levelized Cost

In making cost comparisons between different generation choices, it is useful to couch the predicted cost in terms of a *levelized cost*.[32] Levelized costs are an economic abstraction. It

---

[30]In analyses of power plant costs, it is common to use the Handy–Whitman Index of Public Utility Construction Costs to convert to constant dollars (see Refs. [32], p. 6, and [33], p. 313).

[31]The choice of discount rates for use in cost estimation is discussed, for example, in Ref. [28], Annex 4.

[32]Levelized costs are defined, with slight differences, in Refs. [33], p. 279, [21], p. C-19*ff*, and [26], p. 19.

is anticipated that actual costs will usually vary over the lifetime of a plant, probably increasing from year to year because of inflation and real escalation. The levelized cost is a nominal, or fictitious, constant cost. It remains the same from year to year (in current dollars). Subject to assumptions about future inflation and other economic factors, the levelized cost is set such that the total present value of payments required to cover the levelized costs, summed over the lifetime of the plant, is equal to the total present value of the payments to cover the actual expected annual costs. The levelized cost is therefore, very roughly speaking, an annual cost in current dollars averaged over the lifetime of the plant.

15.5.4 Base Construction Cost or Overnight Cost

The capital costs of a plant upon completion—that is, the total investment costs—can be expressed in mixed current dollars and consist of four components:[33]

1. The base construction cost in constant dollars (for the start date of construction), which is the sum of the annual real expenditures for construction.

2. The increase in the construction cost due to inflation or escalation.

3. The interest paid during construction on money borrowed to finance construction. This is referred to as the "allowance for funds used during construction" (AFUDC).

4. Contingency funds.

For a protracted construction period and high interest rates, the total capital costs can greatly exceed the base construction cost.

The base construction cost is sometime designated by the descriptive terms *overnight cost* or *initial cost*. It provides a means of comparing costs of construction, independent of inflation or changing interest rates.

15.5.5 Fixed-Charge Rate

The *fixed-charge rate* provides a convenient formulation for relating the capital component of electricity costs of the carrying charge (C_C) to the capital investment. The fixed charges, which are the same as the carrying charges, include the cost of capital, depreciation and amortization, taxes, and insurance. The fixed-charge rate is the ratio of the total of these costs to the initial capital investment.[34] Thus, the fixed charge rate (F) is:

$$F = \frac{C_C}{C_I},\tag{15.3}$$

where C_C and C_I are the total annual carrying charges and the total capital investment, respectively. If the carrying charges are expressed in cents per kilowatt-hour (c_c), the investment cost in dollars per kWe (c_i), and F in percent per year, then

$$F = 8760\,\eta \times \frac{c_c}{c_i},\tag{15.4}$$

---

[33]We are here following the formulation of Ref. [21], p. II-7, which appears to be reasonably standard.
[34]We are here following the definition of Ref. [34], p. 25. Some authors may replace the initial capital investment with the net investment in the plant, which includes subsequent capital additions as well as depreciation and amortization.

where η is the capacity factor and 8760 is the number of hours per year. The usual practical application of these relations is for calculating the carrying charges, assuming a given fixed-charge rate:

$$c_c = \frac{F}{8760\eta}\, c_i. \tag{15.5}$$

For example, for $c_i = \$1000$ per kWe, $F = 10$ percent, and $\eta = 0.80$, Eq. 15.5 gives carrying charges: $c_c = 1.4$ ¢ per kWh.

REFERENCES

1. Marc Gervais, *Electricity Supply and Demand Into the 21st Century* (Washington, D.C.: U.S. Council for Energy Awareness, 1991).
2. P. L. Joskow, "The Evolution of Competition in the Electric Power Industry," *Annual Review of Energy* 13 (1988): 215–238.
3. U.S. Department of Energy, *Annual Energy Review, 1994*, Energy Information Administration report DOE/EIA-0384(94) (Washington, D.C.: U.S. DOE, 1995).
4. L. S. Hyman and E. R. Habicht, Jr., "State Electric Utility Regulation: Financial Issues, Influences, and Trends," *Annual Review of Energy* 11 (1986): 163–185.
5. National Research Council, *Nuclear Power: Technical and Institutional Options for the Future*, report of the Committee on Future Nuclear Power Development, John F. Ahearne, chairman (Washington, D.C.: National Academy Press, 1992).
6. 102d Congress, *Energy Policy Act of 1992*, Public Law 102–486, October 24, 1992.
7. Eric Hirst and Charles Goldman, "Creating the Future: Integrated Resource Planning for Electric Utilities," *Annual Review of Energy and the Environment*, 16 (1991): 91–121.
8. Portland General Electric, *1992 Integrated Resource Plan, Update* (Portland, Ore: Portland General Electric, February 2, 1993).
9. R. M. Bleiberg, "Legacy of Shoreham," *Barron's*, October 9, 1989, p. 9.
10. Mary O'Driscoll, "Utilities' Future in California Includes Retail Wheeling, CPUC Says," *The Energy Daily* 22, no. 75 (April 21, 1994): 1.
11. U.S. Department of Energy, *Electric Plant Cost and Power Production Expenses 1991*, Energy Information Administration report DOE/EIA-0455(91) (Washington, D.C.: U.S. DOE, 1993).
12. U.S. Department of Energy, *Historical Plant Cost and Annual Production Expenses for Selected Electric Plants 1986*, Energy Information Administration report DOE/EIA-0455(86) (Washington, D.C.: U.S. DOE, 1988).
13. U.S. Department of Energy, *An Analysis of Nuclear Plant Operating Costs: A 1991 Update*, Energy Information Administration report DOE/EIA-0547 (Washington, D.C: U.S. DOE, 1991).
14. U.S. Department of Energy, *Monthly Energy Review*, March 1995, Energy Information Administration report DOE/EIA-0035(95/03) (Washington, D.C.: U.S. DOE, 1995).
15. U.S. Department of Energy, *Projected Costs of Electricity from Nuclear and Coal-Fired Plants*, Vol. 2, Energy Information Administration report DOE/EIA-0356/2 (Washington, D.C.: U.S. DOE, 1982).
16. "Nuclear Power Facts and Figures," *Info* and *Info Data* (Bethesda, Md.: Atomic Industrial Forum, March 1984 and July 1987).
17. C. Braun, *U.S. Nuclear Plants Operating and Maintenance Costs—Increased Costs and Controls*, presentation to Commonwealth Edison Company, May 4, 1990.
18. "World List of Nuclear Power Plants," *Nuclear News* 38, no. 3, (March 1995): 27–42.
19. U.S. Council for Energy Awareness, "Nuclear Plant Construction Costs and Duration" (Washington, D.C.: U. S. Council for Energy Awareness, 1993).
20. International Atomic Energy Agency, *Nuclear Power Reactors in the World*, Reference Data Series, no. 2 (Vienna: IAEA, April 1993).
21. U.S. Department of Energy, *A Review of the Economics of Coal and Nuclear Power*, draft report 9/30/81.
22. U.S. Department of Energy, *Nuclear Power Plant Construction Activity, 1987*, Energy Information Administration report DOE/EIA-0473(87) (Washington, D.C.: U.S. DOE, 1988).
23. J. G. Delene, K. A. Williams, and B. H. Shapiro, *Nuclear Energy Cost Data Base*, report DOE/NE-0095 (Washington, D.C.: U.S. Department of Energy, 1988).
24. U.S. Council for Energy Awareness, *Advanced Design Nuclear Power Plants: Competitive, Economical Electricity, An Analysis of the Cost of Electricity From Coal, Gas, and Nuclear Power Plants* (Washington, D.C.: U.S. Council for Energy Awareness, 1992).
25. T. B. Johansson, H. Kelly, Amulya K. N. Reddy, and R. H. Williams, "Renewable Fuels and Electricity for a Growing World Economy," in *Renewable Energy*, edited by T. B. Johansson, H. Kelly, Amulya K. N. Reddy, and R. H. Williams (Washington, D.C.: Island Press, 1993), 1–71.

26. Organization for Economic Co-operation and Development, *The Costs of Generating Electricity in Nuclear and Coal Fired Power Stations*, report by an Expert Group of the Nuclear Energy Agency (Paris: OECD, 1983).

27. Organization for Economic Co-operation and Development, *Projected Costs of Generating Electricity from Nuclear and Coal-Fired Power Stations For Commissioning in 1995*, report by an Expert Group of the Nuclear Energy Agency (Paris: OECD, 1986).

28. Organization for Economic Co-operation and Development, *Projected Costs of Generating Electricity from Power Stations For Commissioning in the Period 1995–2000*, report of the Nuclear Energy Agency and International Energy Agency (Paris: OECD, 1989).

29. Organization for Economic Co-operation and Development, *Projected Costs of Generating Electricity: Update 1992*, report of the Nuclear Energy Agency and International Energy Agency (Paris: OECD, 1993).

30. Organization for Economic Co-operation and Development, *Energy Balances of OECD Countries 1990–1991* (Paris: OECD, 1993).

31. U.S. Department of Energy, *Monthly Energy Review, April 1994*, report DOE/EIA-0035(94/04) (Washington, D.C.: U.S. DOE, 1994).

32. U.S. Department of Energy, *An Analysis of Nuclear Power Plant Construction Costs*, Energy Information Administration report DOE/EIA-0485 (Washington, D.C.: U.S. DOE, 1986).

33. C. Komanoff, *Power Plant Cost Escalation: Nuclear and Coal Capital Costs, Regulation and Economics* (New York: Komanoff Energy Associates, 1981).

34. American Physical Society, *Principal Conclusions of the American Physical Society Study Group on Solar Photovoltaic Energy Conversion*, Henry Ehrenreich, chairman (New York: American Physical Society, 1979).

Chapter 16

The Prospects for Nuclear Power

16.1 THE NATURE OF THE NUCLEAR DEBATE

The nuclear power debates of the 1970s and 1980s sometimes appeared to have the aspect, and occasionally even the intensity, of a religious war. Nuclear power was pictured as intrinsically good or evil, and for many of the protagonists it became an article of faith that it should be opposed or supported.

That debate has since become relatively muted in most of the world. At least temporarily, the nuclear side has either won, as in France and probably Japan and South Korea, or lost, as in most of the rest of the "western" world. This does not mean that all discussion is at an end or that the basic issues are settled. Rather, the public and most of the participants have grown less interested. Fossil fuel energy is at the moment plentiful and relatively inexpensive, energy problems are not at the center of the world's attention, and nuclear energy is not at the center of whatever consideration is given to energy. In discussions of U.S. energy policy, nuclear power is sometimes just ignored.

Of course, the issue was never properly an ideological one. There is no moral or immoral content to nuclear power *per se*. Moral issues are involved in using nuclear weapons or in using nuclear power in a way that endangers people or the environment. But if the technical claims of nuclear advocates are correct, the moral case against nuclear power evaporates. Nuclear power becomes benign, and "morality" lies in providing people with its benefits. The root issues are thus technical, not moral.

In this spirit, nuclear power is most appropriately viewed simply as one option for generating electricity. The key questions are the following:

- How much electricity is needed? The answer depends on the amount of energy needed, with the potential for conservation taken into account, and on the best balance between the direct use of fuels and the use of electricity.
- What are the relative merits of nuclear power and alternative means of generating

322

electricity, considering their environmental, economic, and national security impacts?

In answering these questions, the full range of costs and benefits of the competing paths should be considered, including both short-term and long-term consequences.

16.2 ELECTRICITY IN THE ENERGY ECONOMY

16.2.1 The Progress of Electrification

The Increasing Role of Electricity

In the twentieth century, electrification has in many ways been almost synonymous with modernization. Electricity has changed the mechanics of the home and has broadened the options within—with convenient lighting, refrigeration, and motor-driven appliances plus greatly expanded entertainment, cultural, and informational resources. In industry, electrical motors allow machines to be used where and when they are needed. Electric equipment can deliver heat in highly controllable forms, for example, with electric arcs, laser beams, and microwaves. Medical diagnosis and treatment have been transformed by the use of equipment ranging from x rays and high-speed dental drills to lasers and magnetic resonance imaging.[1] It has changed the nature of both local and global communication and has made possible the development of computers. In virtually all spheres of technological life, with the exception of transportation, electricity plays a central role, and even in transportation the rapid trains of Japan and Europe may be pointing to future advances.

There can be philosophical arguments as to the gains and losses to society from technological modernization and from electrification in particular. But there is no question that the world has sought it. In the industrialized countries, as represented by the Organization for Economic Co-operation and Development (OECD), electricity generation rose by 120% in the 1970–1993 period.[2] This far outstripped the growth during the same period in population (23%) and in total primary energy supply, including electricity (43%). It also outstripped the increase in gross domestic product (GDP), which rose 87% (in constant dollars) [1].

The rise in electricity generation was even greater in some countries. For France, generation more than tripled, with an average growth of 5.2% per year from 1970 to 1993. There have been still more rapid rises in recent years in the energy and electricity consumption of developing countries. For China, electricity consumption rose from 32.0 GWyr in 1980 to 79.7 GWyr in 1993, an average annual increase of 7.3% [2]. The growth rate for India in this period averaged 7.8% per year, although with a lower absolute level.[3]

It appears inevitable that the demand for electricity will increase on a worldwide basis, even if conservation restrains the growth in some countries. This increase will be driven by (a) increasing global population; (b) increased per capita use of energy, at least in developing countries that are rapidly modernizing; and (c) continued electrification, due to the convenience of electricity in some applications, its uniqueness in others, and the cleanliness of electricity in end-use. For example, the push in the United States for electric vehicles,

---

[1]The magnetic resonance technique was originally called "nuclear magnetic resonance" because magnetic properties of nuclei were being exploited. The word "nuclear" was dropped because of what were felt to be unfavorable connotations.

[2]Gross generation increased from 395 GWyr in 1970 to 869 GWyr in 1993, corresponding to an average annual growth rate of 3.5%.

[3]The data from Ref. [2] are for consumption of electricity, not generation. However, unless there are substantial energy imports or exports, the rates of rise are the same for the two.

despite their present unattractiveness from a cost standpoint, is motivated by the pollution-free character of electric cars. (There may be pollution at the generating plants, but this is often far removed from areas where cars are concentrated, and it is easier to control or eliminate the pollution from the generating facilities than from millions of automobiles and trucks.)

Sources of Electricity

Most electricity is now being generated by combustion of fossil fuels. The 1993 world shares, from utility and nonutility sources, were fossil fuels, 62%; hydroelectric power, 20%; and nuclear power, 18% [2]. The amount generated by renewable sources other than hydroelectric power is not always reliably reported, because it is in large measure produced by entities other than utilities, and so the accounting is not as reliable as that for utility generation. These renewable sources include biomass energy (for example, wood), geothermal energy, wind energy, and direct forms of solar generation. Of these, at present only biomass makes an appreciable contribution, and that varies greatly from country to country.

Table 16.1 shows the electricity generation rates and sources for the OECD's five largest producers of electricity in 1993. Much of the rest of the world wishes to emulate the industrial and economic accomplishments of the OECD countries, including their extensive use of electricity. Thus, even if the OECD countries are not representative of the world today, they may be indicative of the world of the future.

The patterns of electricity generation differ substantially even among the OECD countries listed in Table 16.1, although as advanced industrial nations, they are economically similar. Canada, with ample hydroelectric power, uses the most electricity per capita in this group. Germany and the United States each obtain over 65% of their electricity from fossil fuels, while Canada and France each obtain more than 75% from nuclear and hydroelectric power. Renewable resources other than hydroelectric power play a negligible role in these countries, with the exception of the United States, where it provided in 1993 close to 3% of the electricity, mostly from biomass. The nuclear component for these countries now ranges from about 18% in Canada (some other OECD countries have no nuclear power) to well over 70% in France.

Table 16.1. *Sources of electricity generation in OECD countries as a whole and in selected OECD countries from utility and nonutility sources 1993.*

	OECD	Canada	France	Germany	Japan	U.S.
Population (millions)	964	29	58	81	125	258
Generation						
Annual (gross TWh)	7609	527	468	522	899	3390
Annual (gross GWyr)	869	60	53	60	103	386
Per capita rate (kW)	0.90	2.09	0.93	0.73	0.82	1.50
Fraction (%)						
Coal	39	15	5	58	19	54
Gas	12	3	1	7	21	13
Petroleum	8	2	1	2	20	4
Nuclear	24	18	79	29	28	19
Hydroelectric	16	61	14	3	11	8
Biomass, geothermal, solar	1.9	0.8	0.3	1.1	1.3	2.7

Sources: All data from Ref. [1].

16.2.2 Alternatives to Power from Nuclear Fission

Fossil Fuels

The long-standing impetus for the development of nuclear power has been the eventual need to replace fossil fuels. Their supply is finite, and in a few centuries we will exhaust the store built up over millions of years. There can be disagreement as to the rate at which this will happen. Some warnings of the imminent exhaustion of supplies have been premature, and concern over oil has been less visible in the early 1990s than it had been in the late 1970s. But if oil shortages have been deferred, they cannot in the long run be avoided. The ultimately available oil is heavily concentrated in the Persian Gulf region, and unless substitutes for oil are found, the world will face a series of economic and political crises as it competes for the dwindling supplies. Already there have been three crises directly or indirectly involving this oil: (a) the oil embargo of 1973–1974, following the October 1973 war in which Egypt and Syria fought Israel; (b) the rapid rise of oil prices in 1979–1980 (by more than a factor of 2), following the fall of the Shah of Iran in January 1979; and (c) the Persian Gulf War in 1990–1991, following Iraq's invasion of Kuwait.

Natural gas is about as abundant as oil, measured in terms of total energy content, and the present world consumption rate of gas is less. Therefore, global shortages of gas are somewhat less imminent. Furthermore, there have been suggestions that "unconventional" sources of natural gas exist, in less accessible forms than the "conventional" supplies, which could conceivably stretch the availability of natural gas. It is also possible to obtain gas, and for that matter oil, by conversion from coal. Nonetheless, gas is also a limited resource, with reliance on unconventional resources speculative.

Coal is much more plentiful than either oil or natural gas, but coal is the least environmentally desirable. Its use was banned in England in the 13th century by King Edward I due to the "intolerable smell" and the injury to the health of "magnates, citizens, and others" [3, p. 5]. With no alternatives other than dwindling supplies of wood, coal became important again in England by the seventeenth century, and in many countries it is now the leading fuel. It has not had a clean history, with chronic pollution punctuated by severe incidents such as the 4000 deaths caused by the London smog of 1952 [4, p. 297].

The output of chemical pollutants from coal, particularly sulfur, can be greatly reduced by "cleaner" burning of coal. The more intractable problem is that of carbon dioxide, which is produced in the combustion of fossil fuels in amounts probably much too large for sequestering. The resulting increase in the concentration of carbon dioxide in the atmosphere carries with it the possibility of significant global climate change. Among fossil fuels, the production of carbon dioxide per unit energy is greatest for coal, because the ratio of carbon to hydrogen in the fuel is the highest.[4] If the greenhouse effect is a threat, then use of coal is undesirable, whether in direct combustion or in producing synthetic gas or oil.

Among the fossil fuels, natural gas has significant advantages. It is the least environmentally damaging in terms of both chemical pollutants and carbon dioxide production. If the hypothesized supplies of "unconventional" natural gas live up to optimistic projections, then supply difficulties may be postponed for many decades. Furthermore, natural gas can be used in highly efficient combustion turbines operating in a combined-cycle mode, in which much of the (otherwise) waste heat from the combustion turbine is used to drive a steam turbine. Nonetheless, reliance on natural gas as more than a short-term stopgap involves two

---

[4]Approximate release rates, in tonnes of carbon per terajoule (10^{12} J) of combustion energy, are coal, 23.8; oil, 19.2; and natural gas, 13.7.

Table 16.2. *Sources of electricity generation in the United States from utility and nonutility sources, 1993.*

Source of power	Net generation		Fraction from source (%)
	TWh	GWyr	
Fossil fuels	2231	255	70
Nuclear	610	70	19
Hydroelectric	276	32	8.6
Wood and waste	55	6.3	1.7
Geothermal	17	2.0	0.5
Wind	3	0.3	0.1
Solar thermal	0.9	0.1	0.03
Other[a]	3	0.4	0.1
Total[b]	3197	365	100

Sources: All data are from Ref. [6].
[a]Includes, for example, hydrogen and sulfur.
[b]Nonutility sources provided 10% of this total, primarily from fossil fuels but also in substantial amounts from wood and waste.

significant difficulties. First, gas supplies may be limited to the conventional resources, advancing the time at which the availability of gas will become a problem and prices will rise substantially. Second, natural gas (while preferable to coal) is still a source of greenhouse gases, both carbon dioxide from combustion and methane from leaks.[5]

Renewable Sources

Renewable sources are the alternative to nuclear power for replacing fossil fuels.[6] Renewable energy embraces solar power in both its direct and indirect forms. For electricity generation, the direct forms of solar power either use photovoltaic cells or raise a fluid to high temperatures with concentrated sunlight from mirrors. Indirect forms of solar power include hydroelectric power, wind power, biomass as fuel, wave power, and exploitation of ocean thermal gradients. Other renewable sources, not based on solar energy, are geothermal power and tidal power.

The most important of these, particularly from the standpoint of electricity generation, is hydroelectric power. It was long considered environmentally benign, with the auxiliary benefits of creating lakes and controlling floods. However, some of the impacts of hydroelectic power on land use and water flow are adverse—for example, damage to fish populations in rivers. The growth in environmental concerns and the fact that the best sites for development have been taken have slowed the development of hydroelectric power throughout the world and particularly in the United States.

Renewable sources other than hydroelectric power make only a minor contribution to electricity generation in industrialized countries, as seen in Table 16.1. Among industrialized countries, the United States is a leader in their use. A detailed breakdown of electricity sources in the United States is given in Table 16.2 for 1993.[7] (The contribution from photo-

---

[5]Natural gas is primarily composed of methane (CH_4). Per unit volume (or equivalently, on a per molecule basis), methane is considerably more effective than carbon dioxide as a greenhouse gas.

[6]An extensive review of renewable energy sources is given in Ref. [5].

[7]Table 16.2 is based on data from Ref. [6]. The results in Table 16.2 are expressed in terms of net generation, while those in Table 16.1 are expressed in terms of gross generation.

voltaic power was too small to be listed.)

Although the contribution of the renewable sources is presently small, there may be possibilities for substantial increases. There is a more or less inverse correlation between the extent to which a renewable source is now used and the extent to which its use might be expanded. The rapid expansion of hydroelectric power is nearly ended especially in the United States. The United States has significant electrical generation from biomass and wastes (about 1.7% of total supply), but most of this is for local use in the forest-products industry. It is not clear if biomass supplies can be expanded by establishing large biomass energy plantations. Geothermal power has long been used, but with present techniques there are not many suitable sites for its exploitation, and in some cases there is concern about pollution from effluents and noise.

Prospects may be more open-ended for wind power. It now provides small amounts of electricity in a few places, notably Denmark and California, and the potential resource is very large. Photovoltaic and solar thermal electric power are even less important in terms of present generation, but if methods can be found to reduce the cost of power from these sources, the potential resource is also very large.

It is too soon to gauge the long-term prospects of new renewable energy sources for the large-scale generation of electricity. There has been significant technological progress in reducing costs and increasing the reliability of wind turbines and photovoltaic cells, and there are suggestions that it would be possible to develop dedicated farms for large-scale biomass production. But the input energy resource is dilute, and these sources require large land areas. Some forms, particularly hydroelectric and geothermal power, face well-defined environmental objections that are likely to limit their expanded use. It remains to be seen whether the other forms will be judged environmentally benign or excessively intrusive, if their use is expanded. Cost is also a problem, especially for photovoltaic power. Further, the expanded use of intermittent sources, such as wind and photovoltaic power, will have to confront difficult problems of matching supply and demand.

Thus, the future role of renewable energy sources, even for the United States alone, is highly uncertain. In 50 years they could dominate U.S. electricity production or they could be a minor factor, depending upon the extent to which costs are reduced, the dedication of land to renewable sources is accepted, and the problems of intermittence are surmounted.

Fusion

Fusion energy remains in principle a long-term alternative to fission energy and renewable energy. In the long run, success in developing fusion would have profound effects on the need for the other options. However, as suggested in the brief discussion in Sec. 13.5.3, it is too soon to base energy planning on any specific timetable for the successful deployment of fusion, or even to say with absolute confidence that it will become an important energy source in the predictable future.

Conservation

Conservation, especially in the form of higher efficiency in energy use, has played an important part in restraining all forms of energy consumption, including electricity consumption. Further advances—for example, wider use of highly efficient lighting and motors—may make major future contributions. But conservation is not a "source" of electricity. It can reduce or delay the need for new generating facilities, but it cannot eliminate the need.

Dilemma with Alternatives

In summary, there is no ideal alternative to nuclear fission. Coal is relatively plentiful but highly polluting. Oil and gas are somewhat "cleaner" than coal, but they too produce carbon dioxide, and their supply is limited. A switch to new sources of renewable energy, such as wind or direct solar generation, would require a tremendous expansion over present output. It takes considerable optimism to be confident of the success of an expansion of such magnitude. Comparable optimism is needed to count on fusion for the predictable future.

Overall, in the search for energy sources, we are caught between the fossil devil we know and the nonfossil angel we don't know. To reduce their dependence on either, a few countries are making a major commitment to nuclear power, a source they find environmentally acceptable and practical. Most countries, however, are opting for some mix of the possibilities, with fossil fuels the overwhelmingly dominant component for the present.

16.3 WORLD STATUS OF NUCLEAR POWER DEVELOPMENT

16.3.1 Overall World Picture

Current Plans

Worldwide nuclear power generation grew at an annual average rate of 13% in the period from 1973 to 1993 (see Sec. 1.3). But this expansion tapered off in the later years, and there is no prospect for an immediate resumption of this rapid growth. According to announced plans for individual countries, as compiled in *Nuclear News*, 66 reactors were nominally on order or under construction as of December 31, 1994, with a total capacity of 54 GWe [7]. If all of these are completed, this would mean that the then-existing capacity of 338 MWe would grow by 16%. Were this accomplished within five years, the average annual rate of increase would be about 3%.

It is very unlikely that even this modest increase will be achieved. Many of the plants involved are listed with no definite completion date, reflecting the uncertainties surrounding their ultimate fate. Consistent with this pessimistic picture, the International Atomic Energy Agency estimated in 1993 that worldwide nuclear energy production will grow at an average rate of only about 0.7% to 2.6% per year for the period until 2015 [8]. Similarly, a 1993 U.S. Department of Energy study projected a rise in world capacity from 329 GWe in 1992 to levels in 2010 of 351 GWe in the "lower reference case" and 427 GWe in the "upper reference case" [9, p. 25]. These represent average annual growth rates of 0.4% and 1.5%, respectively.

Figure 16.1 shows some details of the current status of announced plans for the 11 largest users of nuclear power [7]. The figure displays the operating capacity as of December 31, 1994, along with the capacity of plants then on order or under construction, divided between those scheduled for completion before 2000 and those with later or indefinite completion dates. Figure 16.1 brings out the small magnitude of future development plans. There appear to be firmly established programs of nuclear expansion only in France, Japan, and South Korea—and France may be close to saturation.

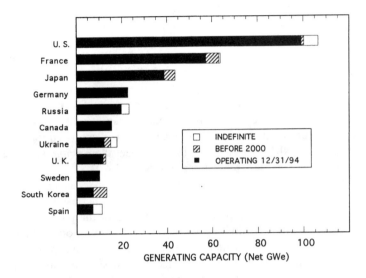

Figure 16.1. Nuclear power generating capacity: actual capacity as of December 31, 1994, added capacity scheduled for operation before the year 2000, and additional plants on order or under construction with more distant or indefinite operation dates.

16.3.2 United States

Present Status of U.S. Reactor Construction

The first phase of nuclear growth is coming to an end in the United States. In the 1990–1994 period only four new reactors went into commercial operation[8] and three operating reactors were shut down (see Table 1.2). In total, there were 109 operable reactors at the end of 1994, including several undergoing long-term shutdowns for safety improvements. There were six more reactors still in some sense "under construction," but only one was actively being worked on, with the others at least tentatively slated for cancellation.

But it makes relatively little difference in the end whether none of these is added or all are added. At most they would represent an increase of less than 10% over present capacity. There is no prospect of new plants being ordered soon enough to go into operation before the year 2000, and it is possible that additional reactors will be decommissioned in this time, reducing the total U.S. nuclear capacity. A major thrust of the nuclear power industry in recent years has been to improve the reliability of reactor performance. A further increase in capacity factors, which in the 1991–1994 period were well above earlier levels, may offer the best hope for increased nuclear power generation in the near future.

Institutional Issues

Some observers, including most advocates of nuclear power, believe that the crucial issues in the United States are more institutional than technical. The division of authority and initiative among many levels of government makes it difficult to adopt and implement

---

[8]These were Limerick 2, Seabrook, and Comanche Peak 1 and 2, all LWRs in the 1050-MWe to 1150-MWe range.

policies that would permit rapid development of nuclear power (or even its prompt curtailment). Important roles are played by the President, Congress, the courts, and a host of federal agencies, including the Nuclear Regulatory Commission, the Department of Energy, and the Environmental Protection Agency. There is also a confused division of power among the federal government, the states, and in some cases individual cities or counties. With many opportunities for *de facto* vetoes, smooth progress in any direction requires a very strong positive consensus. This does not now exist with respect to nuclear power.

Two specific suggestions have been made to address institutional difficulties. One is the streamlining of the licensing process, so that once a plant is approved the only remaining requirement, in terms of NRC procedures, will be to meet the *original* specifications. This change is already reflected in new regulations (see Sec. 15.4.1).

Another, more radical, change is the establishment of independent power producers (IPPs), as mentioned in Sec. 15.1.2. At present, utilities take the responsibility for both the development of the generating facility and the distribution of the power. Prices charged to consumers are heavily regulated, limiting the prospect of large profits while retaining the prospect of high losses (as at Three Mile Island or Shoreham). In some plans, an independent power producer could choose to build whatever facility it wishes, including a nuclear plant, and then market the power to utilities. This might allow for greater financial gains. It is difficult to predict whether this would provide investors with an adequate incentive to take the political and economic risks that have characterized nuclear power to date.

The immediate economic incentive depends strongly on the price of natural gas, which at present is low (the real price in 1994 was less than half the 1984 level) but which may eventually be expected to rise under pressures of growing demand and limited supplies.

16.3.3 France

Status of the French Nuclear Program

France is widely cited as the leading success story for nuclear power development. Since 1986 nuclear power has each year accounted for over 70% of electricity generation in France. Another large fraction, averaging about 15% but with considerable year-to-year variations, has come from hydroelectric power. Therefore, France is virtually saturated in terms of the replacement of fossil fuels. Taking advantage of its ample nuclear capacity, France now exports substantial amounts of electricity to its neighbors—nearly 15% of its gross output in 1993, with expectations of future increases [10].

The chief avenues for further major nuclear growth are through increased exports or a greater electrification of the French energy economy. Recent projections are for continued but relatively slow growth in electricity generation. As of December 31, 1994 France had 54 PWRs in operation plus a small breeder reactor (the 233-MWe Phenix), adding to a total capacity of 57 GWe [7]. Four large PWRs were under construction, with a combined capacity of 5.8 GWe.[9] These new PWRs are all scheduled to begin commercial operation by the end of 1998. Thus, while some nuclear construction is continuing, the expansion is at a rate of one reactor per year, a much lower rate than that of the previous decade. It may even stop for a period after the reactors now on order are completed, because France now depends on exports to provide a market for all its existing nuclear capacity, with only slow growth

---

[9]In addition, France has the Superphenix reactor, which was originally built as a 1200-MWe breeder reactor but has had a troubled record. Its main purpose will be to burn plutonium and other actinides, with electricity generation secondary. Its role as a demonstration breeder reactor has been at least temporarily dropped [11].

expected in the French economy and in domestic demand for electricity [10]. Despite talk in the United States of turning to smaller reactors, the new French reactors under construction are large units, each about 1450 MWe. For the future, a still larger reactor—the 1500-MWe "European Pressurized Water Reactor"—is being designed by French and German utilities, with construction of the first unit possibly beginning by the year 2000.

The Impact of Nuclear Power in France

Over the past two decades, nuclear power has greatly changed the structure of the French energy supply. It was an almost negligible contributor in 1970, but by 1993 nuclear power provided 40% of all primary energy in France and 79% of electricity [1]. During this period, in which the primary energy supply rose by 59%, the fossil fuel energy supply dropped 4%, with decreases of 4% for petroleum (the most used of the fossil fuels) and 60% for coal, and a rise of 251% for natural gas [1]. Carbon dioxide production dropped about 12%, which exceeded the decrease in fossil fuel use because natural gas produces less CO_2 per unit energy than do coal and oil.

Over this same period total electricity generation more than tripled. The chief sources in 1970 were fossil fuels (57%) and hydroelectric power (39%). By 1993 this had changed to a dominant reliance on nuclear power (79%), with hydroelectric power rising slightly in absolute output and its fractional contribution dropping substantially (to 14% in 1993).[10] The French experience has been widely cited as demonstrating the impact that nuclear power can have in curbing dependence on fossil fuels and on the emission of greenhouse gases.

Why Is France Different?

There have been a number of attempts to explain why French nuclear history has developed so differently from that of other major countries. Several factors have been advanced, although many of them need not have been unique to France:

- France is poor in fossil fuels, and nuclear power was seen as the most expeditious way of reducing French dependence on oil and coal imports.
- There has been a concentration on a single type of reactor (a PWR originally modeled on Westinghouse designs), operated under the aegis of a single utility, Electricité de France (EDF), and built by a single reactor manufacturer, Framatome. This has allowed for standardization to a few PWR types, with resulting economies in design, construction, and operation.
- Although France has open political debate, there are few mechanisms whereby opposition to nuclear power can impede its development, short of changing the policy of the central government. Intervention through the courts is much less effective than in, for example, the United States.
- The French Communist party, which in the early years of nuclear power was still an important political force, supported nuclear power. In most other European countries, the political left opposed nuclear power.

---

[10]Comprehensive energy data from the OECD for years after 1993 were not available at the time of writing. For nuclear generation in France, there was a 2% drop from 1993 to 1994. This trend was reversed in the first six months of 1995, when nuclear generation was 2% higher than in the comparable 1993 period and 9% above the 1994 level [12].

- With the initial success of the French nuclear program, the maintenance of French leadership in this area has become a matter of national pride.

Despite these factors, there exists some opposition to nuclear power in France. Although there is no suggestion that this opposition will reverse the French use of nuclear power, it may inhibit France from increasing its electricity exports to other European countries.

Critique of the French Nuclear Program

The record of the French nuclear program has prompted suggestions that it be emulated elsewhere. A comprehensive response to such suggestions was embodied in a 1991 report sponsored by the environmental organization Greenpeace and authored by Francois Nectoux, a French-born economist working in London [13]. The intent of the report was in part to influence the nuclear debate in Great Britain, but it has obvious relevance elsewhere. The report described an extensive list of difficulties and problems, without necessarily succeeding in demonstrating its central thesis that the "French nuclear programme is now in the midst of an economic and industrial crisis" [13, p. 1]. Nuclear opponents could find in the report support for their concerns and objections, while nuclear supporters could find reasons to dismiss many of the arguments.

Nevertheless, there would probably be broad agreement that the report correctly identified several difficulties:

- The reactor manufacturing arm of the French nuclear industry, Framatome, is troubled by overcapacity in light of the slow pace of new reactor orders.
- Efficient use of nuclear reactors calls for continuous operation of the reactors at full power. However, there are daily and seasonal fluctuations in energy demand, creating some difficulties if almost all electricity comes from nuclear power.
- The breeder reactor program, embracing the Phenix and Superphenix reactors, ran into greater difficulties than expected, with numerous technical problems and a lack of an urgent mission.

On the positive side of its ledger, the Greenpeace report indicates that "the development of nuclear power reduced heavy oil and coal import requirements for electricity generation . . . [and] played an important role in the reduction both in volume and in value of the energy trade deficit" [13, p. 89].

Specific mention is made here of the Greenpeace report because of the importance of the French "example." The report argued that the French record is one of accumulating and, by extrapolation, possibly eventually crippling problems, thereby confirming the fears of nuclear opponents. Alternatively, the French program as the report describes it can be interpreted as having had great if not perfect success, confirming for nuclear proponents that even a searching, somewhat hostile inquiry leaves the French program looking successful. In any event, the difficulties anticipated for the French nuclear program in this 1991 study do not appear in 1995 to have materialized.

16.3.4 Japan

Japan is another country with a continuing strong nuclear program, although a smaller one than that of France, especially when adjusted for Japan's larger population and economy. It is particularly driven by Japan's dependence on energy imports. Virtually all of Japan's fossil fuels are imported. The chief domestic resources are nuclear energy and hydroelectric power. In 1993 nuclear power was responsible for 77% of Japan's indigenous energy production and hydroelectric power for 10% [1].[11] An increase in the nuclear energy share of the total energy budget is the most direct way of moving towards greater energy independence. As part of this general strategy, Japan is taking steps towards a self-contained fuel cycle, which could eventually rely on breeder reactors.

Although the Japanese program appears to be vigorous and determined, it is not a crash program, and Japan's efforts to increase electrical generation do not focus on nuclear power alone. For the 1992 to 2002 period it has been projected that nuclear generation will rise from 29% of total utility generation to 34%, while the coal share will rise from 11% to 20% and oil will drop from 25% to 10% [14].

The government approved a plan in October 1990 that called for the construction of 40 nuclear power plants by 2010, roughly doubling the number of reactors and raising nuclear power's share of electricity generation to 43% [15, p. 51]. This target was reaffirmed in 1993 [16]. As of December 31, 1994 four LWRs were under construction, with a total capacity of 4.5 GWe [7]. These include two 1315-MWe advanced BWRs, in a joint program with the General Electric Company. In December 1993 plans were announced for the construction of two units of a new reactor type, a 1420-MWe PWR termed an "advanced PWR," but no construction schedule was set [17].

In addition, construction of the 280-MWe Monju LMFBR has been completed. The Monju reactor is one step in a program pointing to the reprocessing of spent fuel, the use of recycled mixed-oxide fuel in LWRs, and a possible eventual reliance on breeder reactors. (Some delay in the breeder program may be created by the sodium leak at Monju that occurred in December 1995 (see Section 6.3.3).) To date, Japan has depended for reprocessing on a small domestic plant and reprocessing contracts with facilities in France and Great Britain. Japan's own reprocessing facilities will be greatly expanded with the construction of the Rokkasho-mura plant, started in 1993 and originally scheduled for completion in the year 2000 [18, p. 36]. It will have a capacity of 800 tonnes per year, sufficient to handle the output of about 25 1000-MWe reactors each discharging about 30 tonnes/yr of fuel.

A uranium enrichment plant at Rokkashomura began operation in 1992 and is expected to reach an annual capacity of 1500 tonnes SWU, enough to produce about 260 tonnes per year of uranium enriched to 4% in ^{235}U [18, p. 36]. This would still leave Japan partially dependent on foreign suppliers of enriched uranium, and the installation of additional, more advanced facilities is being explored.

Reactor construction times are relatively short in Japan, averaging under five years [19, p. 43], so that fairly rapid changes in the program are possible. Therefore, it is possible that this pace of expansion could be speeded up if circumstances change. Conversely, political opposition to nuclear power in Japan could interfere with present plans. This opposition has not stopped what appears to be a steady and orderly nuclear buildup, but there have been intermittent protests over aspects of Japan's nuclear program, and in some cases these have slowed the rate of development. Nevertheless, except for a dip in 1988, there has been steady growth in nuclear generation (see Fig. 1.5), increasing 38% from 1989 to 1994 [12].

---

[11]In this accounting, use of imported oil to generate electricity is not counted as indigenous production, but electricity generated from imported uranium is counted as indigenous. This is not symmetric, but it is not unreasonable given that the cost of the fuel in the case of nuclear energy is a small fraction of the total cost.

16.3.5 Other Countries

The Former Soviet Union (FSU)

The breakup of the Soviet Union left nuclear reactors in a number of the new states. The resulting number of power plants as of the end of 1994 was Russia, 25; Ukraine, 14; Lithuania, 2; and Kazakhstan, 1 [7]. In addition, two 440-MWe PWRs in Armenia were taken out of service in 1989, after serious earthquakes in the region, but in response to energy shortages the Armenian government has undertaken to restart the reactors, after completion of improvements and inspections [20].

The 1986 Chernobyl accident was a serious setback to the nuclear program in the entire USSR, and its effects were strongly felt in Russia, leading to a moratorium on reactor construction. Nonetheless, five reactors were put into operation in Russia between the time of the accident and the end of 1994, four of which were 950-MWe PWRs and one a 925-MWe RBMK reactor (Smolensk 3 in 1990). In this period 11 older and smaller (50–336 MWe) reactors were taken out of operation, and a considerable number of planned reactors were either cancelled or put on an indefinite schedule. As of the end of 1994 there were four reactors listed as under construction, all with indefinite completion dates [7]. This is a sharp drop from earlier projections.

It is difficult to anticipate the future of nuclear power in Russia, given the unsettled economic and political conditions. Nonetheless, Russia retains a strong industrial capacity in nuclear energy and could have an additional pool of scientific talent stemming from a slowing of its weapons programs. Were a decision made to emphasize nuclear power, which would free more fossil fuels for the export market, it would be possible for Russia to build a greatly expanded nuclear power capacity.

On the other hand, Russia is rich in coal, and the construction costs of coal plants are considerably less than those of nuclear reactors built to high safety standards. Furthermore, geography and differences in scientific assessments may make Russia less concerned than most countries about the greenhouse effect. For these reasons, the magnitude and even the direction of change in Russian nuclear power use are quite uncertain.

Ukraine was the site of the Chernobyl accident, and public opinion there has been hostile to the continued operation of nuclear reactors, especially the remaining RBMK reactors at Chernobyl. Nonetheless, six new PWRs went into operation between 1987 and 1989, before Ukraine separated from the USSR [7]. Since then, the Ukrainian government has oscillated between the need for energy supplies and the fear of nuclear power, leading to a succession of policy reversals. In October 1993 the Ukrainian parliament decided to keep the two remaining Chernobyl reactors in operation and suspended a moratorium on the completion of six additional PWRs that were in varying stages of construction [21].

Eastern Europe

The situation in some of the countries of Eastern Europe is similar to that in the FSU, in that there is continued interest in nuclear power but also considerable uncertainty. Most of the operating reactors are PWRs, with capacities in the neighborhood of 410 MWe, modeled after the Soviet VVER-440 series. Construction and performance of these reactors have varied greatly among countries, and some of these reactors have been widely viewed as constituting safety hazards. However, problems of air pollution from coal burning are severe in Eastern Europe, and this may favor the development of nuclear power. Furthermore, a number of these countries are heavily dependent on nuclear energy for their electricity, so it would be difficult to abandon it. For example, the share of electricity from nuclear power in 1994 was 76% in Lithuania, 49% in the Slovak Republic, 44% in Hungary, and 46% in Bulgaria (see Table 1.4).

Each of these countries is following its own path, and we describe these briefly (with 1994 net generation shown in parentheses):

Former East Germany (0 GWyr). The most draconian remedial measures were taken here following the merger of East and West Germany, with the removal from service of all five East German reactors as not meeting Western safety standards. These were all PWRs of Soviet design, including four VVER-440 reactors and an older 70-MWe reactor [7].

Former Czechoslovakia (2.8 GWyr). The division of Czechoslovakia into the Czech Republic and Slovakia left four operating reactors in each, with two reactors under construction in the Czech Republic (each 890 MWe) and four in Slovakia (each a little over 400 MWe) [7]. These are Soviet-designed PWRs. Western help, from France, Germany, and the United Kingdom, is being extended to upgrade reactor operation and continue construction [22], and the United States has joined through Westinghouse's participation in the completion of two Czech reactors [23].

Bulgaria (1.8 GWyr). There were four VVER-440 reactors in operation in Bulgaria at the time of Chernobyl, and two new 953-MWe Soviet-type PWRs were brought on-line in 1988 and 1992, respectively. Bulgaria has undertaken a succession of phased temporary shutdowns in an effort to bring the reactors up to international safety standards.

Hungary (1.5 GWyr) had four VVER-440 reactors in operation at the end of 1994. No new nuclear plants were under construction.

Romania (0 GWyr) is an anomaly in Eastern Europe, and for that matter a rarity in the world as a whole, in opting for Canadian-type heavy water reactors. Five 625-MWe pressurized heavy-water reactors of the CANDU design were in various stages of construction as of the end of 1994 [7]. One was scheduled to go into operation in early 1996 and another in 1998.

Lithuania (0.8 GWyr) led the world in dependence on nuclear electricity in 1994, obtaining 76% of its electricity from two 1380-MWe RBMK reactors, which went into operation in 1985 and 1987, respectively.

Western Europe

Aside from France, nuclear power development is in the doldrums in Western Europe, although some countries import significant amounts of electricity from France. The status in these countries is indicated briefly (with 1994 generation shown in parentheses):

Germany (17 GWyr) has a *de facto* moratorium on nuclear reactor construction, reflecting the political strength of the Green movement there.

United Kingdom (9 GWyr), an early leader in nuclear power (based on gas-cooled reactors), decided after a long hiatus to build its first PWR, the 1188-MWe Sizewell B reactor, which is completed and achieved criticality in December 1994 as a step towards commercial operation in 1995 [7].

Sweden (8 GWyr) obtains close to half of its electricity from nuclear power. A decision was made in a 1980 referendum to terminate all nuclear power by 2010, and the schedule was stepped up following Chernobyl [24]. However, the decommissioning of nuclear plants was predicated on the availability of alternative sources, and no attractive alternatives have emerged. Thus, the shutdown has been postponed. Sweden is the first country to have a credible, and reasonably widely accepted, program for disposal of spent nuclear fuel.

Spain (6 GWyr) has suspended what had been a rather ambitious nuclear program, halting construction on all its new reactors, including four that were more than 50% completed.

Belgium (4 GWyr) and *Switzerland (3 GWyr)* obtained large fractions of their electricity from nuclear power in 1994, and *the Netherlands (0.4 GWyr)* obtained a negligible fraction. None have announced plans to increase or decrease their nuclear programs in the near future.

Italy (0 GWyr) abandoned its very small nuclear program in 1986, but has undertaken some design studies for a next generation of reactors.

Western Hemisphere (Other Than United States)

Canada completed a significant expansion of its nuclear facilities with the installation of three 881-MWe PHWRs in 1992 and 1993. This completes a program based entirely on the CANDU reactors, which has made Canada a leading country in nuclear capacity and generation. Although the utility, Ontario Hydro, had called for further expansion, no new reactors were in the pipeline as of the end of 1994. Instead, the utility planned to shut down the Bruce-2 plant in 1995, ahead of schedule, in response to a surplus of power capacity and difficulties specific to the history of this reactor [25].

There are also nuclear reactors in use and under construction in Argentina, Brazil, and Mexico, but these are small programs, with no imminent expansion plans.

Asia

Other than Japan, the only countries in Asia that at present derive a substantial fraction of their electricity from nuclear power are South Korea and Taiwan. South Korea has a particularly dynamic program. As of the end of 1994 it had nine reactors in operation and seven on order [7], as a stage in a planned larger expansion. All of the seven reactors ordered are scheduled for operation by the end of 1999. Four of them are 950-MWe to 1000 MWe PWRs, originally undertaken in a joint project with the ABB/Combustion Engineering company using a design that evolved from the 1250-MWe reactors at Palo Verde, Arizona. The other three on order are PHWRs with capacities near 700 MWe, to be built in a joint project with Atomic Energy of Canada, Ltd. Four additional reactors are scheduled for completion by the year 2004 [26]. In this program Korea has sought to achieve "nuclear technology self-reliance," and an increasing role has been taken over by Korean designers and builders.

Taiwan had a relatively early start in nuclear power, and by 1985 six reactors were in operation. There has been no subsequent nuclear construction, but in 1992 the government decided to add two large reactors, and the Taiwan Power Company has invited bids from possible foreign suppliers. If costs can be held to an acceptable level, the utility would like to move ahead with these and additional reactors, but it is not yet clear how this program will evolve [27].

India and Pakistan are large countries with small nuclear power programs. India's total nuclear capacity is still under 2 GWe and nuclear generation in 1994 was only 0.5 GWyr. Pakistan's program is even smaller. Obstacles in these countries include the high capital cost of nuclear units and the reluctance of some foreign countries to enter into cooperative agreements as long as India and Pakistan decline to accept broad IAEA safeguard provisions that might reveal the status of their nuclear weapons programs [28].

China was very late in embarking upon a nuclear power program. Its first three reactors went into commercial operation in 1994: a 300-MWe PWR built domestically (Qinshan) and two 900-MWe reactors (Guangdong 1 and 2) supplied by the French company, Framatome. Although the present program is still very small, China's economy is growing rapidly, and its heavy use of coal is contributing to serious pollution. Negotiations are under way with nuclear suppliers in a number of countries—including France, Russia, South Korea, and Canada—in what may mark the start of a significant nuclear expansion. "Official projections" anticipate 50 GWe of nuclear power by 2020 [29]. Even with this substantial expansion, the nuclear capacity in China in 2020 would be smaller than that in France today—despite more than a 20-fold larger population.

At present, nuclear energy is just beginning to emerge in Asia, aside from Japan and South Korea. However, Kunihiko Uematsu, a Japanese scientist who served as Director-General of the Nuclear Energy Agency of the OECD, described in 1993 an expansive future for nuclear power in Asia. Citing the existing programs in Japan, Korea, Taiwan, and China, and the planning and studies underway in Indonesia, Thailand, and the Philippines, he suggested:

> nuclear power generation in this region will soon reach the level of the OECD's European and North American regions. This is a striking example of the general shift in the world's energy pattern from the traditionally developed countries of the OECD to other parts of the world . . . with the increasing importance of nuclear power in this part of the world, **the future development of this energy source may no longer be spearheaded by the traditionally developed countries of Europe and North America** [30, p. 20].

The boldface type, as shown, was used in the published paper, indicating the significance the author attached to the point being made.

16.4 THE FUTURE PROSPECTS OF NUCLEAR POWER

16.4.1 Developments that Will Impact the Role of Nuclear Power

The future of nuclear power will be influenced by both external factors and developments intrinsic to it. These factors will have different impacts on different countries, but to one degree or another they will influence all countries.

Key considerations external to nuclear power include:

Global warming. If the greenhouse effect looms larger in the public consciousness as a threat to the future environment, then pressures will increase to limit the use of fossil fuels. This will make nuclear power a comparatively more attractive, or at least comparatively less unattractive, option.

Renewable energy. Renewable and nuclear energy compete as alternatives to fossil fuels. The competition has technological, economic, and environmental dimensions. If advances in wind power, photovoltaic power or other power forms are achieved on a rapid time scale and with few deleterious consequences, the perceived need for nuclear power will decrease. Conversely, if technological progress in renewable energy is slow or the land impacts appear to be excessive, nuclear power may be seen as essential.

The world oil and natural gas markets. Higher prices of oil and gas, with or without increased fears of long-term shortages, will make nuclear power more attractive as an alternative, both to reduce present costs and, in some countries, to reduce dependence on increasingly expensive imports.

Electricity demand. The pressure to increase any sort of electricity generating capabilities will depend on the level of the demand for electricity. This demand is influenced by the efficiency of electricity use, the rate of general economic growth, and shifts among energy sources.

Key considerations internal to nuclear power include:

Nuclear accidents. Any major nuclear accident, even on a scale smaller than Chernobyl, will heighten world fears of nuclear power. Likewise, each year of accident-free operation tends to calm the fears.

Reactor designs. The development of new reactors will increase the attractiveness of nuclear power if they are manifestly safe, reliable, and economical. These could be either large evolutionary reactors, of the sort now being built in France, Japan, and South Korea, or smaller reactors that place more explicit emphasis on inherent safety.

Waste disposal. The completion of integrated and fully explained waste disposal plans will encourage people to believe that the problem is "solved," even if final disposal is intentionally delayed to allow the wastes to cool. Conversely, continued delay in the promulgation of convincing plans fosters the belief that the problems may be intractable.

Perceptions of radiation hazards. Most "experts" believe that public fears of radiation exposures from nuclear power are out of proportion to the actual risks. The success of nuclear advocates in imparting their sense of perspective to the general public will be an important factor in winning acceptance of nuclear power, although here cause and effect may be complex. If the need for nuclear power is seen to be great, there is likely to be a greater tendency to accept assurances as to its safety.

These are a complex array of issues, with uncertain outcomes. In some cases, the verdict will not emerge for decades. In the meantime, individual countries are unlikely to all follow the same paths.

16.4.2 Differences Among Countries

The differences in the nuclear policies of different countries can arise from basic aspects of their environment or, at least in the short run, from the political and economic character of their societies.

In terms of its environment, Japan is in a particularly difficult situation. It is poor in fossil fuel resources, and its island location makes it difficult for Japan to import natural gas by pipeline. Further, it has a population per unit area that is roughly 12 times that of the United States, limiting its options for use of renewable energy. For Japan, nuclear power offers a path to at least partial energy independence, which cannot be obtained in any other way. In contrast, the United States and Canada are much richer in fossil fuels and have large areas that could in principle be used for renewable energy. Thus, they are not under the same pressures as Japan.

The effects of differences in physical circumstances are illustrated by a comparison between Norway and Sweden, which in many ways are similar in attitudes and sociology. Sweden, with somewhat limited hydroelectric resources, has been reluctantly relying on nuclear power for roughly one-half of its electricity. Norway, in contrast, uses its abundant hydroelectric power to provide over 99% of its electricity and has no nuclear power. Its per capita electricity consumption is about 60% greater than that of Sweden [1].

Differences in political mood, economic opportunities, and national institutions can also play an important role. The more legal and political avenues that nuclear opponents have for contesting nuclear development, the more difficult it is to proceed with nuclear power. As several observers have pointed out, the difficulties are greater in a country with a federal government than in a country with a centralized national government (e.g., Ref. [30]). A federal government offers many opportunities to raise objections, and the objections are put forth in an atmosphere in which local concerns are likely to take precedence over national priorities. The United States is quite vulnerable in this regard, with important prerogatives held by the states and with a system of checks and balances within the federal government.

It is to be expected that both basic differences in their objective situations and institutions and transient differences in popular attitudes will continue to lead individual countries to different choices. Thus, even were the United States to believe it could do without nuclear power, there is no reason to expect that Japan would follow suit. Other countries, such as China and India, might seek to accelerate their use of nuclear power but be held back by a lack of capital for rapid expansion. At least in the near term—despite common technology, world fuel markets, global environmental concerns, and mutual example—decisions on energy policy will be largely national decisions.

16.4.3 The Future of U.S. Nuclear Power

The United States and the World

The United States was the world pioneer in nuclear energy and, by virtue of the size of its economy, is still the world leader in total nuclear power generation. However, it is not the leader either in the fraction of electricity that comes from nuclear power nor in the rate of growth. The U.S. share of world nuclear generation was 51% in 1975 [2] but had dropped to 30% in 1994 [12], and all indications are that this share will continue to drop. Light water reactors of U.S. design provided the model for reactors in most countries, including France and Japan, but many countries now have strong and confident reactor design and construction capabilities of their own. The world is no longer dependent on the United States for enriched uranium, and after abandoning reprocessing the United States has taken itself out of the near-term use of mixed-oxide fuels.

This is not to say that nuclear power is unimportant in the United States or that the technical capabilities of the industry are gone, but the United States is losing the clear primacy it once had. One assessment of the situation was given by an American participant at an international nuclear conference held in the United States in 1993, who suggested that the conference might be remembered as marking the passing of the "mantle" of nuclear power leadership from the United States to France and Japan [31]. This impression could easily have been gained by considering the number of participants from various countries at the conference, as well as the tenor of their presentations.

A transfer of leadership away from the United States may be viewed with dismay by those interested in American progress in nuclear energy, but it also provides a measure of insurance. In the unlikely event of a crippling diminution of U.S. nuclear design and manufacturing capabilities, the existence of flourishing nuclear programs elsewhere could help facilitate a revival if and when the development of nuclear power again were to become a national goal.

Projections for Future Growth

The uncertainty in the future of nuclear power in the United States is illustrated by alternative Department of Energy projections for the growth of nuclear power up to the year 2030 [9, p. 9]. Three DOE scenarios are presented in Fig. 16.2: (a) a "no-new-orders" scenario, in which nuclear capacity decreases as existing reactors are phased out; (b) a lower-reference case, in which there is a cautious resumption of nuclear expansion; and (c) an upper-reference case, which assumes a vigorous economy and use of nuclear power *and* coal for new generation. U.S. nuclear capacity in the year 2030 is projected to be 5, 119, and 168 GWe in the three scenarios, respectively. These numbers may be compared to the 1994 capacity of 99 GWe.

The actual future for the United States remains a matter of public choice. Evolutionary large reactors, of the general type being built in Japan and South Korea, are moving towards NRC approval and probably could soon be ordered in the United States if the market exists. But it is questionable whether any utility will be ready in the near future to take the implied financial gamble.

This is not only a matter of the economic and engineering attributes of the reactor. Even if the political atmosphere were temporarily supportive at the time a reactor is ordered, there would be a fear that changes in the policies of the federal or state government could make it expensive or impossible to put the reactor into operation when it is completed. Precedents

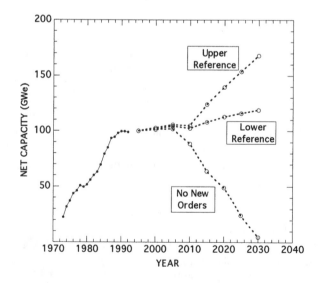

Figure 16.2. Projected United States nuclear power capacity (in net GWe) for 1995–2030 in three DOE scenarios, together with actual capacity for 1973–1992.

such as the Shoreham reactor in New York are not forgotten. Hence, a prudent utility is likely to be hesitant about ordering a large nuclear reactor.

The same arguments apply, but with somewhat less force, to intermediate-size reactors, such as the 600-MWe passive LWRs now being developed (see Chap. 13). The investment in such a unit is less than that for a larger reactor, thereby decreasing the financial risk, and the lead time between an order and reactor operation is expected to be shorter.

The Scale of Possible Expansion

The no-new-orders and upper-reference cases of Fig. 16.2 are by no means limiting cases. A serious nuclear accident in the United States or a failure to grant extensions to operating reactors conceivably could lead to a virtual termination of nuclear power well before 2030. Conversely, a strong commitment to nuclear power could lead to growth that is much more rapid than the quite modest growth hypothesized in the upper-reference scenario.

There is no question that nuclear power can provide the lion's share of electricity for an industrial nation, as demonstrated by France. In a period of roughly 25 years, France installed a nuclear capacity of 57 GWe. The United States has a much larger industrial base than does France. For example, in one rough measure of the relative industrial capabilities, energy use in the industrial sector in 1993 was nine times greater for the United States than for France [1]. If France could "afford" to build 57 GWe of capacity at a time when its annual industrial output was below the 1993 level, it is reasonable to believe that the United States could "afford" to install something like ten times that nuclear capacity in a future 25-year period.

This suggests that, if deemed desirable, the U.S. nuclear capacity could reach the neighborhood of 500 GWe by the year 2025 (allowing for a hiatus until the year 2000). With an 80% capacity factor, this corresponds to an annual output of 400 GWyr. Total U.S. electricity generation was about 365 GWyr in 1993. Were it to rise at a modest 2% annual rate from 1993 to 2025, total generation would be about 690 GWyr in 2025, and nuclear energy could be providing almost 60% of U.S. electricity.

This is an arbitrary scenario and is not put forth as either a target or a ceiling for nuclear growth in the United States, much less a prediction. A level of 500 MWe in 2025 is about three times the level of the upper reference case in Fig. 16.2 and may be a gross overestimate of what is probable. However, it indicates what is possible, were nuclear expansion to become a national goal.

Possible Constituencies for Nuclear Power

In reaching a national decision as to the future of nuclear power, the role of a constituency is important. At present there is a determined and effective constituency against nuclear power, including most of the organized environmental movement.

There has been the image of a comparably active and determined constituency for nuclear power, namely, the nuclear industry. But with the decrease of nuclear reactor construction, the nuclear industry is shrinking, and this is no longer a valid image. To be sure, there is continuing activity in the operation and improvement of existing power plants and in the completion of a few others, plus some prospect of possible future reactors. This sustains interest on the part of both utilities and manufacturers. But the total scale of equipment development is relatively small, and the utilities are more interested in trouble-free operation of existing reactors than in building new ones. Overall, there is no powerful and vocal constituency for further development of nuclear power.

There are two potential enlarged constituencies: the technical community and the environmental community. For the most part, engineering and scientific organizations and their members support nuclear power, and if energy issues become pressing, there might be a greater sense of urgency in this support.

However, the emotional drive behind any position in the nuclear controversy is heightened when there are important environmental concerns. At present the "environmental movement" is largely opposed to nuclear power, although with different degrees of finality in the opposition. The movement is not monolithic, and there are many strands. From a somewhat extreme standpoint, the fundamental difficulty with nuclear power, or any technology that facilitates increased use of energy, is that it increases the potential impact of humans upon the natural environment—impacts that are likely, in this view, to be undesirable. Those who share this fear will always oppose nuclear power.

Other parts of the environmental movement would welcome a truly clean energy source to replace fossil fuels. Over the next years, some environmentalists might turn to nuclear power in preference to fossil fuel combustion, if renewable energy appears unlikely to achieve present goals, if there are no nuclear accidents, and if there is an apparent solution to waste disposal. If such a revisionist view of environmental priorities takes hold, it could provide the impetus for a nuclear revival, which may not come from industry initiatives alone.

The possibility that reactors may be used to burn plutonium from dismantled nuclear weapons could also lead to changes in attitudes towards nuclear power (see Sec. 14.5.3). In the past, nuclear power has suffered from its institutional associations with nuclear weapons. The two enterprises shared some of the same technology and were initially developed by some of the same people and institutions. This contributed to antipathy by association, over and above concrete concerns about proliferation. If in the future nuclear reactors are seen in a new light, as a destroyer of weapons plutonium, the psychological mood may change enough to ease the path to the reassessment suggested above.

16.4.4 Predictions and Their Uncertainty

It would be a mistake to take any predictions of future developments as reliable guides. Presently, nuclear power is being pursued with vigorous, confident programs in France, Japan, and South Korea but is in limbo in the United States and most of Western Europe. Developing countries vary in their interest in nuclear power and in their opportunity to pursue it. The political and economic situation is too fluid in the former Soviet Union and Eastern Europe to warrant confident predictions. But while there is no hint that sudden changes are in the offing, there is no assurance that countries, wherever they now lie in the spectrum, will maintain their present policies over prolonged time periods.

It is appropriate to look back 20 years and examine predictions made then. Conveniently for this purpose, a conference was held in Paris in 1975, with the complacent title "Nuclear Energy Maturity." The underlying premise of the conference was that nuclear power had arrived and that it remained to consider how to proceed so that nuclear power could "represent a long-term solution, that is for thousands of years rather than the few decades set by the uranium supply required by the 'proven' reactors" [32, p. x].

This long-term issue was addressed by a panel on the role of breeders. One speaker gave projections for future generation in the "Western World" (for this purpose; much the same as the OECD countries). In a variety of scenarios, western capacity was projected to be

700–1000 GWe in 1990 and 2000–4000 GWe in 2005 [32, p. 328]. In actuality, total *world* capacity was only 320 GWe in 1990, and it now appears unlikely that it will reach even 500 GWe by 2005.

These were not atypically optimistic projections, with similar projections presented by other speakers [32, pp. 319, 322]. There was at least one dissenting voice [32, p. 324], but it seems to have been a voice in the wilderness. The clear consensus was that the world was moving into a period of very substantial nuclear expansion. The people involved had looked into the recent past and had seen the future.

There is a possibility that we are repeating the same mistake today, and misinterpreting the sluggishness of recent years as a guarantee of future sluggishness. There are no absolute barriers to a return to rapid growth in nuclear power. There are nuclear suppliers in the United States, Europe, and Japan who are eager and able to build reactors if the demand develops. The question is not whether a major expansion of nuclear power is possible, but whether it is desirable. Predicting what will appear to be desirable ten years hence, or even five years hence, is very problematical. It is especially difficult if one attempts to make generalizations that will embrace all countries.

The nuclear future may be very different for different countries, as is the nuclear present. To one extent or another, nuclear power prospects will improve if oil and gas markets appear precarious and if concern about the greenhouse effect rises. They will be damaged if there is another major reactor accident, if progress on waste disposal remains slow, or if there is rapid progress in renewable energy or fusion. Depending upon these developments, nuclear power may shrink over the next several decades and remain significant in only a few countries; at the other extreme, nuclear power may begin a major new expansion in much of the world.

REFERENCES

1. Organization for Economic Co-operation and Development, *Energy Balances of OECD Countries* 1992–1993 (Paris: OECD, 1995).
2. U.S. Department of Energy, *International Energy Annual*, Energy Information Administration reports DOE/EIA-0219(79/93) (Washington, D.C.: U.S. DOE, 1980, 1995).
3. Harold H. Schobert, *Coal: The Energy Source of the Past and Future* (Washington, D.C.: American Chemical Society, 1987).
4. Michael Allaby, *Dictionary of the Environment*, 2nd Ed. (New York: New York University Press, 1983).
5. Thomas B. Johansson, H. Kelly, A. K. N. Reddy, and R. H. Williams, eds., *Renewable Energy: Sources for Fuels and Electricity* (Washington, D.C.: Island Press, 1993).
6. U.S. Department of Energy, *Annual Energy Review, 1994*, Energy Information Administration report DOE/EIA-0384(94) (Washington, D.C.: U.S. DOE, 1995).
7. "World List of Nuclear Power Plants," *Nuclear News* 38, no. 3 (March 1995): 27–42.
8. *IAEA Newsbriefs* 8, no. 4 (July/August 1993): 6.
9. U.S. Department of Energy, *World Nuclear Capacity and Fuel Cycle Requirements 1993*, Energy Information Administration report DOE/EIA-0436(93) (Washington, D.C.: U.S. DOE, 1993).
10. Ann MacLachlan, "EDF Sets 72-Billion KWh as Its Export Goal From 1977 On," *Nucleonics Week* 34, no. 45 (November 11, 1993): 12.
11. "International," *Nuclear News* 37, no. 11 (September 1994): 88–92.
12. U.S. Department of Energy, *Monthly Energy Review, September 1995*, Energy Information Administration report DOE/EIA-0035(95/09) (Washington, D.C.: U.S. DOE, 1995).
13. Francois Nectoux, *Crisis in the French Nuclear Industry: Economic and Industrial Issues of the French Nuclear Power Programme* (Amsterdam: Greenpeace, 1991).
14. Central Electric Power Council of Japan, *Long-Term Electric Power Facilities Development Plan, 1993* (April 1993).
15. "Japan: Nuclear Need Stressed in Three Government Acts," *Nuclear News* 33, no. 15 (December 1990): 51.
16. Yoshinori Ihara, "Future Nuclear Systems: Japanese Fuel Cycle Strategy," in *Future Nuclear Systems: Emerging Fuel Cycles & Waste Disposal Options*, Proceedings of *Global '93* (La Grange, Ill.: American Nuclear Society, 1993).

17. Naoaki Usui and Ann MacLachlan, "Japco Unveils Specifications for its Advanced PWR Project," *Nucleonics Week* 34, no. 49 (December 9, 1993): 14.
18. "Nuclear Energy and its Fuel Cycle in Japan: Closing the Circle," *IAEA Bulletin* 35, no. 3 (1993): 34–37.
19. International Atomic Energy Agency, *Nuclear Power Reactors in the World*, Reference Data Series, no. 2 (Vienna: IAEA, April 1993).
20. "Armenia: Power Reactor Restart on the Horizon," *Nuclear News* 37, no. 15 (December 1994): 39–40.
21. "Late News in Brief," *Nuclear News* 36, no. 14 (November 1993): 19.
22. "International Briefs," *Nuclear News* 37, no. 8 (June 1994): 52.
23. Douglas Frantz, "U.S. Backing Work on Czech Reactors by Westinghouse," *New York Times*, May 22, 1994, p. 1.
24. Carl-Erik Wikdahl, "Sweden: Nuclear Power Policy and Public Opinion," *IAEA Bulletin* 33, no. 1 (1991): 29–33.
25. Ray Silver, "Hydro to Prematurely Close 18-Year-Old Bruce-2 in 1995," *Nucleonics Week* 35, no. 7 (February 17, 1994): 1.
26. Bo Hun Chung, "Nuclear Power Development in South Korea," *Nuclear News* 38, no. 8 (June 1995): 34–37.
27. Walter W. L. Shen, "An Overview of Taiwan's Nuclear Program," *Nuclear News* 38, no. 8 (June 1995): 44–46.
28. Simon Rippon, "Asian Subcontinent: Nuclear Programs in Pakistan, India," *Nuclear News* 38, no. 8 (June 1995): 42–43.
29. Simon Rippon, "China: Ready For More Nuclear Power," *Nuclear News* 38, no. 8 (June 1995): 32–33.
30. Kunihiko Uematsu, "The Outlook for Nuclear Power," in *Future Nuclear Systems: Emerging Fuel Cycles and Waste Disposal Options*, Proceedings of *Global '93* (La Grange, Ill.: American Nuclear Society, 1993), 18–23.
31. Unidentified floor discussant at *Global '93, Future Nuclear Systems: Emerging Fuel Cycles and Waste Disposal Options*, Seattle, Wash., September 1993.
32. Pierre Zaleski, ed., *Nuclear Energy Maturity, Proceedings of the European Nuclear Conference* (Oxford: Pergamon Press, 1976).

Appendix A

Elementary Aspects of Nuclear Physics

A.1 SIMPLE ATOMIC MODEL

A.1.1 Atoms and Their Constituents

Before 1940 scientists had identified 92 elements. These were commonly arranged in the classical *periodic table*, which organized the elements into groups with similar chemical properties. The "last" element in this table was uranium, and for many years it seemed as if this table provided a full representation of matter. Then, with new facilities and insights, attempts were made to produce elements beyond uranium, the so-called *transuranic* elements. These efforts eventually proved successful, starting with the production and identification of neptunium in 1940. Subsequently, more than a dozen additional elements have been produced and identified.

There is no fundamental distinction between the "original" 92 elements and the later "artificial" elements. All of them, and others, were made in the original cosmic processes of nucleosynthesis. The very heaviest elements were unstable and changed relatively quickly into lighter components. Therefore, they are not found on Earth, although some of them can be recreated in the laboratory, lasting for times ranging from a fraction of a second to many thousands of years.

The smallest possible amount of an element is a single *atom*. Each atom is made up of a central *nucleus* surrounded by one or more electrons. The electrons are much smaller in mass than the nucleus, and their distances from the nucleus are much larger than the radius of the nucleus. This is somewhat analogous to the configuration of the solar system, with a heavy central sun orbited by a number of relatively light and distant planets.

The nucleus has two constituents of roughly equal mass, each much more massive than the electron: the *neutron* and the *proton*. Generically called *nucleons*, they differ in that the neutron is neutral while the proton carries a positive charge equal in magnitude to the negative charge of the electron. This equality has been established with great precision from the neutrality of bulk matter. An un-ionized atom is neutral because it contains equal numbers of electrons (outside the nucleus) and protons (inside the nucleus).

It is now recognized that neutrons and protons are not fundamental "building blocks" of matter, but are themselves composed of still more elementary entities, called quarks. We will not discuss quarks nor most of the other so-called *elementary particles*. Despite their importance to the understanding of the earliest origins and ultimate structure of matter, their existence can be ignored in considering the processes important in nuclear reactors, such as radioactivity and nuclear fission.

A.1.2 Atomic Number and Mass Number

The chemical properties of an element are determined by the number of electrons surrounding the nucleus in an un-ionized atom, which in turn is equal to the number of protons in the nucleus. This number is the element's *atomic number*, Z. Each element can be identified in terms of its atomic number. Thus,

Z=atomic number=no. of protons in nucleus=no. of electrons outside.

The "natural" elements range from hydrogen ($Z=1$) to uranium ($Z=92$). Beyond that, the next elements are neptunium ($Z=93$) and plutonium ($Z=94$), both produced in nuclear reactors. A list of elements through $Z=105$ (hafnium) is given in Table B.3 of Appendix B.

Nuclei with a given number of protons need not have the same number of neutrons, although for the most part the spread in the number of neutrons for a given element is rather narrow. The total number of nucleons in a nucleus is the *mass number* of the nucleus and is customarily denoted by A:

$$A=\text{mass number}=N+Z,$$

where N denotes the number of neutrons in the nucleus.

A.1.3 Isotopes and Isobars

For a given element (i.e., same Z), nuclei with different numbers of neutrons (i.e., different A) are called *isotopes* of the element. Nuclei with the same mass number A but different atomic number Z are called *isobars*. Different isotopes of an element are virtually identical in chemical properties (although small differences may arise from their different masses and therefore their different mobilities). In nature, the relative abundances of different isotopes are usually closely the same for different samples of an element, because it is mainly the chemical properties of the atom that determine the atom's history on Earth.

To specify fully a nuclear species, both Z and A must be given. For example, the most abundant isotope of carbon ($Z=6$) has mass number 12. It is called "carbon twelve" and is conventionally written ^{12}C. In this notation, it is not necessary to specify that $Z=6$, because the symbol C, by identifying the element as carbon, defines the atomic number to be 6. Nonetheless, sometimes the redundant notation $^{12}_{6}$C provides a useful reminder. The term *nuclide* is used to denote a particular atomic species, as characterized by its atomic number and atomic mass number; the term *radionuclide* denotes a radioactive nuclide.

A.2 UNITS IN ATOMIC AND NUCLEAR PHYSICS

A.2.1 Electric Charge

In the International System of Units (SI), the unit of charge is the coulomb, itself defined in terms of the unit of current, the ampere. In atomic and nuclear physics, it is usually more convenient to express charge in terms of the magnitude of the charge of the electron, e, where[1]

$$e=1.6022\times10^{-19} \text{ coulombs.}$$

The charge of the electron is $-e$, and the charge of the proton is $+e$. The atomic number of carbon is 6, and therefore the charge on the carbon nucleus is $6e$, or 9.61×10^{-19} coulombs.

A.2.2 Mass

The SI unit of mass is the kilogram (kg). It also remains common in atomic and nuclear physics to express mass in grams (g)—i.e., in the centimeter-gram-second (cgs) system— when mass must be expressed in macroscopic terms. For most purposes in nuclear physics, however, it is more convenient to express mass in terms of the *atomic mass unit* (u), which is defined so that the mass of a hydrogen atom is close to unity.

More precisely, the atomic mass unit is defined in the so-called unified scale by the stipulation that the mass of the neutral ^{12}C atom is precisely 12 u. (Earlier, a scale based on oxygen, not carbon, had been used, and this can occasionally cause confusion, especially as the numerical values for the two scales are close.) With this definition:

$$1 \text{ u}=1.66054\times10^{-24} \text{ g}=1.66054\times10^{-27} \text{ kg.}$$

Although on this scale, the masses, or atomic weights, of the ^{1}H atom (M_H), the proton (m_p), and the neutron (m_n) are all close to unity, they are not exactly unity nor equal to each other (see Table A.1 for actual values). The mass of the electron is $m_e=m_p/1836$.

A.2.3 Avogadro's Number and the Mole

It is convenient in chemistry and physics discussions to introduce the *gram-molecular weight* or *mole* as a unit for indicating the amount of a substance. For any element or compound, a mole of the substance is the amount for which the mass in grams is numerically equal to the atomic (or molecular) mass of the substance expressed in atomic mass units. For example, the mass of one mole of ^{12}C is exactly 12 g, and the mass of one mole of isotopically pure atomic hydrogen (^{1}H) is 1.0078 g.

---

[1]Numerical values for general physical constants and for nuclear masses and half-lives are taken from Ref. [1], and information on nuclear decay schemes is largely from Ref. [2].

Table A.1. *Masses and rest mass energies.*

Quantity or Particle	Symbol	Mass or energy	
		u	MeV
Atomic mass unit		1 (exact)	931.4943
Electron	e	0.0005486	0.5110
Proton	p	1.0072765	938.2723
Neutron	n	1.0086649	939.5656
H atom (A=1)	^{1}H	1.0078250	938.7833
C atom (A=12)	^{12}C	12 (exact)	11177.9

The number of atoms (or molecules) per mole is termed *Avogadro's number*, N_A. It has the numerical value

$$N_A = 6.02214 \times 10^{23} \text{ per mole.}$$

The mass of an atom (in grams) must equal the mass of one mole of the substance (in grams) divided by N_A. Therefore,

$$1 \text{ u} = \frac{1}{N_A} \text{ g} = 1.66054 \times 10^{-24} \text{ g}$$

as stated above. It is experimentally more practical to determine an accurate value for N_A than to make an absolute determination of the mass of an individual atom. For that reason, N_A is the primary experimental number, and the mass corresponding to 1 u, expressed, for example, in grams, is a derived result.

A.2.4 Energy

Energy is expressed in joules (J) in SI units, but it is much more common in atomic and nuclear physics to express energy in electron volts (eV), kilo-electron volts (keV), or mega-electron volts (MeV), where 1 eV is the energy gained by an electron in being accelerated through a potential difference of 1 volt.[2] This energy is $q\Delta V$, with $q = e$ and $\Delta V = 1$. Thus,

$$1 \text{ eV} = 1.6022 \times 10^{-19} \text{ J.}$$

In atomic physics problems, the convenient unit is usually the eV. In nuclear physics problems, where the energy transfer per event is much higher, the more convenient unit is the MeV.

The average energy of an assembly of particles is sometimes expressed in terms of temperature. Thus, the neutrons responsible for initiating fission in typical nuclear reactors are termed "thermal neutrons." Their kinetic energy distribution is the distribution characteristic of a gas at the temperature of the reactor core. In particular, the mean translational kinetic energy of a molecule in a gas at temperature T is $\frac{3}{2}kT$, where T is expressed in degrees Kelvin (K) and k is the Boltzmann constant:

$$k = 1.381 \times 10^{-23} \text{ J/K} = 0.862 \times 10^{-4} \text{ eV/K.}$$

---

[2] 1 keV $= 10^3$ eV; 1 MeV $= 10^6$ eV.

The product kT is a characteristic temperature for a wide range of phenomena. A convenient reference point, or memory aid, is that at room temperature (taken to be $T=20\,°C=293$ K), $kT=0.0253$ eV $=\frac{1}{40}$ eV.

In a gas at temperature T, there is a broad distribution of kinetic energies, and a small fraction of the particles have kinetic energies well in excess of the $\frac{3}{2}kT$ average. Fusion in stars and in (still experimental) fusion reactors is initiated by positively charged nuclei in the high-energy tail of the energy distribution. High energies are important in fusion because the higher the energy the better the prospect of overcoming the coulomb repulsion between the positively charged interacting particles.[3] For neutron-induced fission, on the other hand, there is no coulomb repulsion acting on the neutron.

A.2.5 Mass–Energy Equivalence

The equivalence of mass and energy is basic to nuclear and atomic physics considerations. An energy E is associated with the mass m of a particle by $E=mc^2$, where c is the velocity of light. Expressing mass in kilograms and velocity in meters/sec, the energy equivalent of 1 atomic mass unit is

$$E=(1.66054\times10^{-27})\times(2.9979\times10^{8})^{2}/(1.6022\times10^{-13})=931.5 \ \text{MeV}.$$

Expressed in terms of energy, the mass of the electron is

$$m_e c^2=0.511 \ \text{MeV}.$$

Often it is simply stated that the electron mass is 0.511 MeV, with no distinction made between "mass units" and "energy units." In Table A.1, the particle masses are expressed in both MeV and atomic mass units (u).

A.3 ATOMIC MASSES AND ENERGY RELEASE

A.3.1 Atomic Mass and Atomic Mass Number

In describing a nuclear species—for example, ^{238}U—one could specify either the mass of the nucleus or the mass of the atom (the nucleus plus the electrons). It is virtually universal practice to specify atomic mass. The mass M of an atom (expressed in atomic mass units) is not exactly equal to the mass number A of the atom, except for ^{12}C, where the equality is a matter of definition. However, the numerical difference between M and A is small. This is because the constituents of the atom—the neutrons, protons, and electrons—have masses quite close to 1 u or quite close to zero. Although the atomic mass is slightly less than the sum of the masses of the constituent particles, it is not very different. Tabulations of atomic masses are often expressed in terms of the *mass excess* Δ, where $\Delta=M-A$.[4] For ^{238}U, for

---

[3]Coulomb repulsion is the name commonly given to the repulsion, governed by Coulomb's law, between objects that carry charges of the same sign.

[4]In this expression, although A is ordinarily defined as a dimensionless integer, we attach to it the same units as the units of M, rather than further encumber the notation.

example, $\Delta=47.305$ MeV or 0.05078 u.[5] Correspondingly, the atomic mass of ^{238}U is 238.05078 u.

A.3.2 Isotopes and Elements

It is necessary to keep in mind the distinction between the atomic mass M_E of the element, as it occurs with its natural mixture of isotopes, and the mass M_i of any particular isotope. The atomic mass of an element is given by

$$M_E = \Sigma(f_i M_i), \qquad (A.1)$$

where f_i is the fractional abundance of the isotope, by number of atoms. For example, carbon has two stable isotopes: ^{12}C and ^{13}C, with atomic masses M_i equal to 12.00000 u and 13.00336 u, respectively. For every 10,000 C atoms on Earth, 9890 are ^{12}C and 110 are ^{13}C, giving fractional abundances f_i of 0.9890 and 0.0110, respectively.[6] By Eq. A.1, it follows that $M_E=12.0110$ u.

It is possible to specify relative abundances of the constituents of a sample of matter in terms of the relative number of atoms (or molecules) or in terms of the relative masses of the constituents. It is usual to specify isotopic abundances f_i in terms of the relative number of atoms. However, an important exception in the context of nuclear power occurs in describing the isotopic abundances of uranium isotopes. These are often specified in terms of the fraction by mass. (This distinction is discussed further in Section 7.2.2.)

Elemental abundances, on the other hand, are usually specified in terms of relative masses. For example, the abundance of uranium is commonly specified in terms of parts per million (ppm). An abundance of 2 ppm means that there are 2 μg of uranium per gram of rock. Again there is an exception. For gases, relative abundances are sometimes specified in fraction by volume, which is equivalent to specifying the relative number of molecules.

A.3.3 Binding Energy, B

The mass of a nucleus is almost, but not exactly, equal to the sum of the masses of the constituent neutrons and protons. The difference is the *nuclear binding energy*, B. Thus, for a nucleus characterized by mass number A and atomic number Z,

$$M_{nuc} = Zm_p + (A-Z)m_n - \frac{B}{c^2}. \qquad (A.2)$$

The binding energy B represents the total energy that would be required to dissociate a nucleus into its constituent neutrons and protons.

Neglecting the binding energy of the electrons in the atom,[7] the atomic mass M is equal to the sum of the masses of the nucleus and the surrounding electrons: $M=M_{nuc}+Zm_e$. Similarly, the mass of Z hydrogen atoms is $ZM_H=Zm_p+Zm_e$. Therefore, adding Zm_e to

---

[5]It might be argued that Δ has been defined in units of mass, not energy. These quantities, however, are physically equivalent, and it is common to use the symbols Δ and M to represent either mass or the equivalent energy, as context requires.

[6]We here ignore small variations that may occur in fractional isotopic abundances. These can be caused, for example, by differences in the temperature at the times when different samples of a given material were formed.

[7]Although this neglect inserts some error in Eq. A.3, if B is strictly interpreted as a *nuclear* binding energy, the error is small because electron binding energies in the atom are much smaller than nucleon binding energies in the nucleus.

both sides of Eq. A.2, the atomic mass can be expressed as

$$M = ZM_H + (A-Z)m_n - \frac{B}{c^2}. \tag{A.3}$$

Rewriting Eq. A.3, the nuclear binding energy is given by

$$B = [ZM_H + (A-Z)m_n - M]c^2. \tag{A.4}$$

Eq. A.4, or the equivalent preceding equations, can be taken to be the definition of the binding energy.

For calculational purposes, in view of the manner in which nuclear data are tabulated, it is useful to rewrite Eq. A.4 in terms of the mass excess Δ, in effect subtracting Z, $A-Z$, and $-A$ in successive terms within the bracket of Eq. A.4. It is also convenient to ignore the difference between mass units and energy units and drop the factor c^2 in Eq. A.4. Then the binding energy can be written

$$B = Z\Delta_H + (A-Z)\Delta_n - \Delta, \tag{A.5}$$

where $\Delta_H = 7.2890$ MeV, $\Delta_n = 8.0713$ MeV, and Δ is expressed in MeV.

The average binding energy or binding energy per nucleon, B/A, provides a measure of the stability of a nucleus. In general, more stable configurations have higher values of B/A. As an example, we calculate the binding energy per nucleon for one of the more tightly bound of the nuclei, ^{56}Fe ($Z=26$ and $\Delta=-60.603$ MeV). From Eq. A.5,

$$\frac{B}{A} = \frac{(26 \times 7.2890) + (30 \times 8.0713) - (-60.603)}{56} = 8.79 \text{ MeV/nucleon.} \tag{A.6}$$

A.3.4 Energy Release in Nuclear Processes

The total energy of the system remains unchanged in any nuclear process, in accord with the principle of conservation of energy. Energy will be released in the process if the total binding energy is greater for the final nuclei than for the initial nuclei. Equivalently, energy is released if the total mass of the final nuclei is less than the total mass of the initial nuclei. The energy release, often denoted by the symbol Q, is given by the mass difference:

$$Q = (\Sigma M_i - \Sigma M_f)c^2, \tag{A.7}$$

where the summations are taken over all initial masses M_i and final masses M_f. Eq. A.7 is used in Section 4.4.1 to calculate the energy release in fission.

A.4 ENERGY STATES AND PHOTONS

One of the great breakthroughs of early 20th-century physics, embodied in the Bohr model of atomic structure, was the recognition that atoms can exist only in certain states or configurations, each with its own specific energy. In the simple Bohr picture of the hydrogen atom, different states correspond to electron orbits of different radius, each with a well-defined energy equal to the sum of the kinetic and potential energies of the electron in its orbit. This very simple mechanical picture has been subsequently modified by quantum mechanics, but the basic point remains that the possible configurations of an atom correspond to a limited set of states. The discrete energies associated with these states are the allowed energy levels of the atom. The same rule holds for nuclei, although we are not yet able to account theoretically for the exact energy levels of nuclei with as great precision as we can for atoms.

Thus, in broad terms, each atom or nucleus can exist in a state of lowest energy, the so-called *ground state*, or in one or another state of higher energy, the so-called *excited states*. With a few exceptions, the excited states are short-lived; that is, they quickly emit their excess energy and the system (atomic or nuclear) reverts to its ground state. The energy lost by the atom in a transition from one excited state to a lower one (or to the ground state) is commonly carried off by electromagnetic radiation.[8] For nuclei, a typical time for a transition from an excited state to a lower state is in the neighborhood of 10^{-12} sec, although very much longer and somewhat shorter lifetimes are also possible.

When an atom (or nucleus) in a state of initial energy E_i makes a transition to a final state of lower energy E_f, the energy carried off in electromagnetic radiation is

$$E_{rad} = E_i - E_f. \tag{A.8}$$

Throughout the 19th century, light and (when recognized) other forms of electromagnetic radiation were thought to be properly described by waves. One of the revolutionary new insights of early 20th-century physics was the recognition that light also has particle-like properties. In particular, the electromagnetic radiation corresponding to a single atomic (or nuclear) transition is carried in a single discrete packet, called a photon.

The energy of the photons for a given transition is simply related to the wavelength, or frequency, of the associated radiation:

$$E = h\nu = hc/\lambda, \tag{A.9}$$

where λ is the wavelength, ν is the frequency, and h is a universal constant known as Planck's constant: $h = 6.626 \times 10^{-34}$ joule-sec. Thus, in the transition of Eq. A.8, the photon energy is

$$h\nu = E_i - E_f. \tag{A.10}$$

Visible light is associated with transitions involving the outer electrons of atoms or molecules, with photon energies in the neighborhood of several eV (3 eV corresponds to $\lambda = 4130$ Å $= 0.413 \times 10^{-6}$ m). X rays correspond to transitions involving the inner electrons of atoms, with typical energies of 1 to 100 keV, depending upon the atomic number of the atom. Radiative transitions between nuclear levels typically involve energies in the neighborhood of 100 to 10 000 keV. The photons from nuclei are called gamma rays. There is no difference between these groups of photons other than their energy. In fact, it is possible, although not common, to have gamma rays with energies lower than those of typical x rays. The names "x ray" and "gamma ray" date to the times of the original discovery of the then-mysterious radiations. In principle, there is no need for different terms to distinguish between photons from atomic transitions (x rays) and photons from nuclear transitions

---

[8]There are two classes of exceptions to this: (a) sometimes the energy can be transferred to an electron in a process known as internal conversion, still leaving the nuclide unchanged; and (b) in some cases an excited state can decay by emitting a particle, such as a beta particle or neutron, thereby changing the atomic number or the mass number of the nuclide.

(gamma rays), but the terminology is retained, perhaps because it provides a reminder of their physical origin.

A.5 NUCLEAR SYSTEMATICS

Nuclei are categorized as being stable or unstable. Loosely speaking, stable nuclei are those that remain unchanged forever. Unstable nuclei decay spontaneously into lighter nuclei on a time scale characteristic of the particular nuclear species. This time scale can be expressed in terms of the half-life of the species, defined to be the time interval during which one-half of an initial sample will decay (see Section A.7.2). If the half-life for decay is greater than some (undefined) small fraction of a second, the process of decay is called radioactivity (see Section A.6). Half-lives of different species vary from much less than a second to many billions of years.

The concept of "stability" is not an absolute one. The heaviest "stable" nuclide is bismuth 209 (^{209}Bi), with $Z=83$ and $A=209$. There is some evidence, however, that it decays with a half-life in the neighborhood of 10^{18} years. (This is stable enough for most purposes, as the age of the universe is only on the order of 10^{10} yr.) It should also be noted that some current theories suggest that the proton itself is not stable, but if the proton does decay at all, it decays at an extraordinarily slow rate (a half-life of more than 10^{31} yr). Of course, such slow decays have no relevance to the processes of radioactivity that are of interest here.

Most of the nuclides found in nature that are lighter than ^{209}Bi are stable. However, there are exceptions, such as potassium 40 (^{40}K) and rubidium 87 (^{87}Rb), which are both long-lived residues of stellar nucleosynthesis processes, as well as carbon 14 (^{14}C), which has a relatively short half-life ($T=5730$ yr) but is produced continuously in the atmosphere due to cosmic rays.

Above ^{209}Bi, continuing up to ^{238}U ($Z=92$, $A=238$), the nuclei found in nature are not stable. The reason that some are still here is either that they have very long half-lives themselves, as in the case of ^{232}Th, ^{235}U, and ^{238}U, or they are progeny of these nuclei. Above $Z=92$, a considerable number of nuclei have been artificially produced, and the properties of some of them are well established. As one goes higher and higher, the half-lives of the nuclei tend to decrease. The heaviest firmly identified nuclide, as of major 1993 evaluation, is at $Z=109$ and $A=266$, with a half-life of the order of milliseconds [3]. There have also been reports of the production and identification of nuclides of higher atomic number, up to $Z=112$.

Most of the mass of the universe is concentrated in hydrogen (mostly ^{1}H), in helium (mostly ^{4}He), and in nuclei with even values of Z and with $A=2Z$, starting with carbon ($Z=6$) and continuing through calcium ($Z=20$), i.e., ^{12}C, ^{16}O, ...^{40}Ca. At higher atomic numbers, the stable isotopes have $A>2Z$, i.e., more neutrons than protons in the nucleus. At each mass number through $A=209$ (other than masses $A=5$ and $A=8$, where there are no stable nuclei), there are one or more stable nuclei and a host of nuclei that are unstable against beta-particle emission (see Section A.6.1). The stable nuclei are clustered about a trajectory, depicted in Fig. 2.1, which follows the line $A=2Z$ up to $A=40$, and continues up to $Z=83$, $A=209$. At higher A, all nuclei are unstable for decay by either beta-particle or alpha-particle emission, but some of the alpha-emitting nuclei have long half-lives and are included in Fig. 2.1.[9] Most of the naturally radioactive isotopes lie between $Z=82$ and $Z=92$, with ^{40}K the most important exception.

The binding energy per nucleon (see Eq. A.6) is close to 8 MeV over most of the range of stable nuclei. It is zero for ^{1}H and is small for the very lightest atoms, but is above 7.4 MeV for all stable nuclei from ^{12}C to the top of the periodic table. It rises from values below 8 MeV for the lighter nuclei to a broad peak at about 8.8 MeV near $A=60$ and then falls gradually to 7.57 MeV at ^{238}U. A plot of the binding energy per nucleon, B/A, as a function of mass number A is presented in Fig. A.1.

The lightest and heaviest nuclei are less tightly bound, while intermediate nuclei are more tightly bound. This suggests two paths for liberating energy: form intermediate nuclei by combining very light nuclei or by splitting very heavy ones. These are the respective processes of *fusion* and *fission*. The difficulty, to the extent there is one, is in accomplishing these processes in a controlled fashion.

A comprehensive 1990 compilation of stable and unstable nuclides includes well over 2500 entries [1], with many different entries for most values of atomic number (many isotopes) and for most values of atomic mass number (many isobars). The number of listed entries can be expected to grow as investigations continue of nuclei with extreme combinations of A and Z. However, only a limited number of these nuclides are of interest from the standpoint of nuclear energy. The most important of these are: light nuclei ($A \leqslant 12$), which may be targets or products in fusion or may serve as moderators in fission reactors, neutron-rich medium-mass nuclei formed as fission products ($76 \leqslant A \leqslant 160$), and heavy nuclei used as fuels for fission or produced by neutron capture ($232 \leqslant A \leqslant 246$) in a fission reactor.

A.6 RADIOACTIVE DECAY PROCESSES

A.6.1 Particles Emitted in Radioactive Decay

Many nuclei are stable in the sense that they normally remain unchanged for exceedingly long periods of time—in fact, "forever" as far as it has been possible to study them. Other nuclei are unstable. In its most common form, the instability is exhibited by the phenomenon of *radioactivity*, in which the nucleus spontaneously emits an *alpha* (α) particle or *beta* (β) particle, often accompanied by the emission of one or more *gamma* (γ) rays.

The phenomenon of radioactivity was discovered in the 1890s, first in uranium and later in other natural elements. Beginning in the 1930s a host of additional radioactive elements have been produced in nuclear accelerators and nuclear reactors. All three of the radiations can cause the blackening of a photographic plate or the discharge of an electroscope. This is accomplished by ionization of the medium, and alpha particles, beta particles, and gamma rays are referred to as *ionizing radiations*, distinguishing them from other radiations such as radio waves. Other than the shared ionizing property, the three rays are very different. We will not examine the history by which the properties of these rays were established, but the properties themselves will be briefly described.

Alpha particles are nuclei of helium-4 atoms (^{4}He). Thus, an alpha particle has a mass of about 4 atomic mass units (u) and is positively charged, with a charge of magnitude $q = Ze = 2e$. Compared to the other radiations, alpha particles can penetrate only a small distance

---

[9]The trajectory depicted in Fig. 2.1 is sometimes referred to as the "valley of beta stability," because if a three-dimensional picture is envisaged, with Z and A as axes in the horizontal plane and atomic mass as a vertical axis, this trajectory would follow a valley of minimum mass at each value of A.

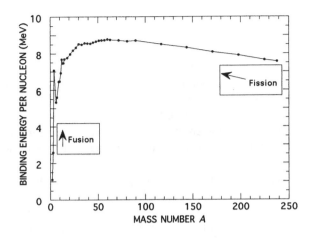

Figure A.1. The binding energy per nucleon of stable nuclei, as a function of mass number. At very high A, the plotted nuclei are not stable against alpha-particle decay but are stable against beta decay.

in matter. The most energetic of the alpha particles emitted from radioactive nuclei are stopped after passing through less than 10 cm of air, i.e., about 10 mg/cm^2 of matter or about 0.1 mm of a material whose density is 1 g/cm^3 (such as water). The penetration distance is strongly dependent on the initial alpha-particle energy, and for lower-energy alpha particles, the penetration will be even less.

Beta particles are electrons. As such, they have a mass far smaller than that of the alpha particle. Except for a very few cases, the beta particles emitted in natural radioactivity are negatively charged; they are more completely designated as β^- particles. The β^- particles are identical to the ordinary electrons surrounding the nuclei of atoms. In the 1930s positive electrons, called positrons or β^+ particles, were discovered. These are emitted from artificial radionuclides produced when positive particles, such as protons or alpha particles, combine with a nucleus to form an unstable "proton-rich" nucleus. β^+ emitters are very rare in natural terrestrial material and among the radioactive nuclei produced in fission. Typical beta-particle penetration distances are on the scale of 0.1 to 1 g/cm^2, increasing with increasing beta-particle energy.[10]

Gamma rays are photons or quanta of radiation. They have neither mass nor charge and are the most penetrating of the trio, with penetration distances typically on the scale of 5 to 20 g/cm^2, depending upon the gamma-ray energy and the atomic number Z of the absorbing material.[11]

---

[10]In specifying typical penetration distances for alpha particles, beta particles, and (below) gamma rays, we are ignoring their very different penetration profiles. To take the extremes, alpha particles of a given energy have a quite well-defined penetration distance, while gamma rays of a given energy are very broadly spread in penetration distance. Further, the processes responsible for the particles being stopped are quite different. Nonetheless, the concept of "typical" distances is useful for a qualitative understanding of the behavior of these particles.

[11]For a high-Z material such as lead (Z=82), the typical penetration distance is only 1 g/cm^2 at 200 keV and becomes much smaller at still lower energies, making, for example, thin sheets of lead very effective for stopping x rays.

A.6.2 Alpha-Particle Emission

Heavy Nuclei as Alpha-Particle Emitters

Within a nucleus, the repulsive electric force (or *coulomb* force) between the protons competes with the attractive force that holds a nucleus together—the *nuclear force*. As one goes to heavier nuclei, with higher nuclear charge and higher nuclear radius, the coulomb force becomes relatively more important, because it has a longer range than does the nuclear force. When it wins out by a sufficient margin, there can be nuclear fission or, less spectacularly and more commonly for nuclei in their ground state, the nucleus can decay by alpha-particle emission.

A requirement for alpha decay, or any decay, is that there be a final system whose total mass is less than the mass of the initial system. In that case, emission can in principle occur, because there will be net energy left over to provide the kinetic energy of the emitted particles. Other things being equal, the greater the available energy, the shorter will be the half-life for alpha-particle decay. The natural radionuclides that emit alpha particles are all heavy nuclei, with atomic numbers Z well over 80. For these nuclei, alpha particles will be emitted at a fast enough rate to be observed—i.e., with a short enough half-life—only if the available energy is not much below 4 MeV. Typically, alpha-particle energies lie between about 4 and 8 MeV.

Conservation Rules in Alpha-Particle Emission

A typical example of alpha-particle decay is the decay of uranium 238 (^{238}U) to form thorium 234 (^{234}Th):

$$^{238}\text{U} \rightarrow {}^{234}\text{Th} + {}^{4}\text{He} .$$

The initial mass number (238) equals the sum of the final mass numbers (234+4), and the initial atomic number (92) equals the sum of the final atomic numbers (90+2). These equalities are dictated by two important rules that apply in radioactive decay:

Charge: The sum of the charges of the final products equals the charge of the original nucleus.

Number of nucleons: The total number of nucleons in the final products equals the total number of nucleons in the original nucleus.

These rules are closely related to the general conservation laws that apply in all nuclear processes.

Energy Relations in Alpha-Particle Emission

In the decay of ^{238}U, about 77% of the alpha particles are emitted with a kinetic energy of 4.20 MeV and 23% with an energy of 4.15 MeV. For the 4.20-MeV alpha particle, which corresponds to a transition to the ground state of ^{234}Th,

$$E_\alpha + E_{Th} = [M(^{238}\text{U}) - M(^{234}\text{Th}) - M(^{4}\text{He})]c^2, \qquad (A.11)$$

where the mass M is the atomic mass of the species in question, and E_α and E_{Th} are the kinetic energies of the alpha particle and thorium nucleus, respectively. The 4.15-MeV alpha particle corresponds to a transition to an excited state of ^{234}Th, with an excitation energy of 0.05 MeV. The excited state then decays to the ground state with the emission of a 0.05-MeV gamma ray.

In each case the total decay energy, summed over two successive steps in the latter case, is 4.20 MeV, plus a small additional kinetic energy of the recoiling ^{234}Th nucleus. From conservation of momentum, when a stationary ^{238}U nucleus decays, the momenta p of the alpha particle and of the ^{234}Th nucleus are equal in magnitude and opposite in direction. Noting that $E = p^2/2M$, it follows that $E_{Th} = (4/234) \times E_\alpha = 0.07$ MeV. Thus the total decay energy is 4.27 MeV, as also could be found from the mass differences if numerical values are substituted in Eq. A.11.

A.6.3 Beta-Particle Emission

Neutrinos and Anti-Neutrinos

In contrast to alpha decay, where only a few discrete decay energies are possible for each nuclear species, the spectrum of emitted beta particles is continuous. All beta-particle energies are possible from zero to a fixed maximum, called the *endpoint energy*. The endpoint energy in beta decay corresponds to the mass difference between the parent atom and the residual product, as would be expected from conservation of energy. However, the average energy of the beta particles is less than one-half the endpoint energy. When this was first discovered, it was thought that there might be a conflict with the demands of energy conservation.

It was eventually realized, by the early 1930s, that the electron shared the available energy (i.e., the energy corresponding to the mass difference between the initial and final constituents) with an elusive partner, which was termed the *neutrino* (ν). In present usage, the "neutrino" emitted in β^- decay is more precisely termed an *anti-neutrino* ($\bar{\nu}$). The neutrino itself is emitted in the analogous process of β^+ decay. When the distinction between the neutrino and anti-neutrino is not important, they are both generically termed neutrinos.[12] In standard physics terminology, the β^- and the ν are called "particles," and the β^+ and the $\bar{\nu}$ are termed "anti-particles." According to very general considerations, in each beta decay one particle and one anti-particle are emitted.[13]

It used to be common to say that the mass of the neutrino is zero. It is now recognized that the neutrino may have a mass, but the results of measurements to date indicate that, if not zero, the neutrino mass is very small. Recent measurements suggest that the mass is probably less than about 5 eV, although uncertainties in the experimental results suggest that there should be some caution in accepting this as a firmly established upper limit [4, p. 153]. Even were the upper limit placed "as high" as 10 eV, it would still correspond to only about 10^{-8} atomic mass units.

---

[12]Unless otherwise indicated, we will use "neutrino" as a generic term and indicate the specific species by ν or $\bar{\nu}$.

[13]There are two issues here. One is the trivial matter of establishing the (arbitrary) convention for choosing which is called the particle and which the anti-particle. We do not here consider the second and much more significant issue: Why are there both particles and anti-particles and what are the relationships between the particles and anti-particles? Ordinary matter is made up of particles and is termed *matter*, as distinct from *anti-matter*. There is no evidence for the existence of bulk anti-matter outside of science fiction, although individual anti-particles are observed in cosmic rays and can be created and observed under spacial laboratory conditions.

The neutrino has zero charge and can typically pass through very large amounts of material without stopping. For example, a flux of neutrinos is not appreciably depleted in passing through the Earth.[14]

Beta Decay and Beta-Particle Energy

β^- decay occurs when a nuclide has "too many" neutrons, i.e., when the neighboring isobar of higher atomic number has a lower atomic mass.[15] In that case, it is energetically profitable for a nucleus to change a neutron into a proton, with the emission of a β^- and an $\bar{\nu}$. A typical process of this sort is

$$^{234}\text{Th} \rightarrow {}^{234}\text{Pa} + \beta^- + \bar{\nu}.$$

Here the initial and final number of nucleons is 234, the β^- and the $\bar{\nu}$ not contributing to the nucleon count. For thorium (Th), the nuclear charge is +90, while for protactinium (Pa), it is +91, so charge is conserved with the emission of a β^- particle. For the case of a nucleus with "too many" protons, β^+ emission will occur; in this case, the β^+ is accompanied by a ν.

The maximum beta-particle kinetic energy, or endpoint energy, can be calculated from the mass differences. The sum of the kinetic energies of the beta particle and the neutrino is equal to the endpoint energy (ignoring the small, and usually negligible, kinetic energy of the residual nucleus). The mean energy of the beta particles is typically about one-third of the endpoint energy, and the mean energy of the neutrinos is typically about two-thirds of the endpoint energy.

Electron Capture

The processes of β^- and β^+ emission are identical, aside from the signs of the charges. However, there is an important asymmetry. In the atoms of ordinary matter, the nuclei are surrounded by (negative) electrons. The nucleus undergoes the same change as in β^+ decay if it captures an electron from outside the nucleus (usually an electron from one of the inner shells of the atom). Thus, *electron capture* exists as an alternative to β^+ emission and in fact can take place with a lesser mass difference between the nuclei. There is no comparable alternative to β^- emission, in the absence of anti-matter atoms with positrons surrounding an anti-nucleus.

Some radionuclides decay by both β^+ emission and electron capture. For example, sodium 22 (^{22}Na) decays about 90% of the time by β^+ emission and about 10% of the time by electron capture. The two processes can be written

$$\beta^+ \text{ emission: } {}^{22}\text{Na} \rightarrow {}^{22}\text{Ne} + \beta^+ + \nu$$

---

[14]Nonetheless, neutrinos are not "infinitely" penetrating, and if sufficiently large numbers of neutrinos pass through material, some will stop. For example, with a high-flux nuclear reactor and a moderately large nuclear detector, it is possible to observe some neutrino interactions with the material of the detector. This observation was first accomplished by Reines, Cowans and collaborators in 1953. The detection of neutrinos and anti-neutrinos— including neutrinos produced in nuclear processes near the center of the Sun—is now routine in some classes of experiments.

[15]A *free* neutron, i.e., a neutron that is not part of a heavier nucleus, is itself unstable due to β^- decay, with a half-life of 10 min. This decay is possible because the mass of a neutron exceeds that of a hydrogen atom. However, for many nuclides the mass of the product of β^- decay would be greater than the mass of the initial atom, making β^- decay impossible.

Electron capture: $^{22}\text{Na}+e^-\rightarrow\,^{22}\text{Ne}+\nu$.

In both cases, the residual nucleus is neon 22 (^{22}Ne).

It is also possible to have cases where the nuclear masses are so close together that electron capture is possible, while β^+ emission is energetically impossible (e.g., ^7Be). For either process to occur, the initial energy of the system must exceed the final energy. The electron rest mass energy adds to the initial energy in electron capture, while it adds to the final energy in β^+ emission.

A.6.4 Gamma-Ray Emission

Ordinarily, gamma-ray emission is not a primary process in radioactive decay but follows alpha-particle or beta-particle emission. It occurs in those decays where the transition is to an excited state of the product nucleus rather than to the ground state. The excited state then gives off its excitation energy by gamma-ray emission, either to the ground state or to a lower-lying excited state. When the transition is to another excited state, the de-excitation sequence continues with further gamma-ray emission. As mentioned in Section A.4, typical half-lives for gamma-ray emission are on the order of 10^{-12} sec.

There are several reservations that should be made in partial modification of the description given above. First, some nuclei have long-lived excited states, known as *isomeric states*, with half-lives ranging from an appreciable fraction of a second to many years. The gamma-ray emission in these cases appears as a primary radioactive process, rather than as a sequel to alpha-particle or beta-particle emission, although ultimately it can be traced back to such initiating processes. Second, not all of the excitation energy is carried off by the gamma ray, because, as in the other decay modes discussed above, the nucleus carries off a small amount of kinetic energy as dictated by conservation of momentum. Third, as an alternative to gamma-ray emission, decay of an excited state can occur by *internal conversion*, a process in which the excitation energy is transferred to one of the inner electrons of an atom. Typically in such cases, gamma-ray emission and internal conversion are competing de-excitation processes. The internal conversion electrons will have an energy equal to that of the competing gamma ray, minus the energy required to remove the electron from the atom.

A.7 RATE OF RADIOACTIVE DECAY

A.7.1 Exponential Decay

The number of nuclei of a given radioactive species that will decay in any time interval is proportional to the number of nuclei present. The constant of proportionality is termed the *decay constant*. Thus,

$$-\frac{dN}{dt}=\lambda N, \tag{A.12}$$

where λ is the decay constant, N the number of nuclei, and dN the change in the number of nuclei in the time interval dt. The negative sign corresponds to the decrease in number with time. Each decay channel, defined in terms of a given initial nucleus and decay mode, has its own decay constant λ. If a nucleus has more than one decay mode—for example, alpha-particle transitions to different states of the residual nucleus—the overall decay constant λ is the sum of the individual decay constants.

The number of nuclei remaining after time t is given by integration of Eq. A.12:

$$N(t)=N_0e^{-\lambda t},\tag{A.13}$$

where N_0 is the initial number (at $t=0$). As seen from Eq. A.13, radioactive decay is an exponential decay. The rate of decay, from combining Eqs. A.12 and A.13 or from differentiating Eq. A.13, is

$$-\frac{dN}{dt}=\lambda N(t)=\lambda N_0e^{-\lambda t}.\tag{A.14}$$

A.7.2 Mean Life and Half-Life

One way of specifying an "average" time before a nucleus decays is in terms of its *mean life*, τ. Following the standard definition of the mean value of a quantity, the mean life is given by

$$\tau=\frac{\int_0^\infty t\lambda N(t)dt}{N_0}=\lambda\int_0^\infty te^{-\lambda t}dt,\tag{A.15}$$

where $\lambda N(t)dt$ is the number of decays occurring in the time interval dt, and $N(t)$ is found from Eq. A.13. Carrying out the integration of Eq. A.15, it follows that

$$\tau=\frac{1}{\lambda}.\tag{A.16}$$

Another way of specifying an "average" time before a nucleus decays is to specify the *half-life*, T. The half-life is the time required for one-half of an original sample to decay. It is defined by the relation

$$e^{-\lambda T}=\frac{1}{2}.\tag{A.17}$$

Evaluating Eq. A.17 and substituting from Eq. A.16, we have the following relations among T, λ, and τ:

$$T=\frac{\ln 2}{\lambda}=\tau\ln 2=0.693\tau.\tag{A.18}$$

It is more common to characterize radionuclides by their half-lives than by their mean lives. Sometimes the ambiguous term "lifetime" is used; it usually means the half-life.

A.7.3 Nuclei Remaining after a Given Time Interval

Given the half-life, or equivalently, the decay constant, it is a simple matter to calculate the fraction of nuclei remaining after the time t. This can be done either on the basis of the definition of the half-life or by using Eq. A.13. It is particularly simple when the time t is an integral number of half-lives. For example, if $t = 3T$, $N/N_0 = 1/8$. More generally, the fraction remaining after time t is given by

$$\frac{N(t)}{N_0} = \left(\frac{1}{2}\right)^{t/T}. \tag{A.19}$$

We also have, from Eq. A.13

$$\frac{N(t)}{N_0} = e^{-\lambda t} = e^{-t \ln 2/T}. \tag{A.20}$$

By comparing the natural logarithms of the right-hand terms of the two equations, it can be seen that Eqs. A.19 and A.20 are equivalent.

Eqs. A.12 and A.13 are examples of statistical equations, describing the average behavior of an assembly of radionuclides. As long as N is large, a statistical description gives accurate results. When N becomes small, the fluctuations about the expected statistical average become more significant, and the time of the last decay, when N goes from 1 to 0, cannot be specified with any precision. It also is a time that has no significance, except perhaps as part of a conceptual game of asking when the activity is "all gone."

It is rarely of importance to consider the statistical variations in radioactive decay, but they can nonetheless be simply described. The expected number of decays in a given time interval can be designated as $\bar{n}$. If $\bar{n}$ is a reasonably large number, the probability that the actual number of decays will be within the interval from n to $n + dn$ is given by the *normal* or *gaussian* distribution:

$$P(n)dn = \frac{1}{\sqrt{2\pi\sigma^2}} e^{-(n-\bar{n})^2/2\sigma^2} dn, \tag{A.21}$$

where the parameter σ is the *standard deviation*. If $\bar{n}$ is sufficiently large, say, greater than 100, then to a good approximation

$$\sigma = \sqrt{\bar{n}}. \tag{A.22}$$

In 68% of the trials, n will lie within 1 standard deviation of $\bar{n}$, and in 95% of the trials it will lie within 2 standard deviations. For example, if $\bar{n} = 1000$, then $\sigma = 31.6$, and in 95% of the cases n will fall within 6% of $\bar{n}$. In most problems of interest in radioactive decay, the relevant $\bar{n}$ is much greater than 1000, and the fluctuations about $\bar{n}$ are of no interest. Ignoring them is rarely a significant omission.[16]

A.7.4 Decay Chains

In many cases, the product of a radioactive decay is another radioactive nucleus. In this case, the decay of the first nucleus is still given as in Eq. A.13:

$$N_1(t) = N_{10} e^{-\lambda_1 t}. \tag{A.23}$$

For the next generation nucleus,

---

[16]On the other hand, statistical fluctuations can be very important in epidemiological studies of the health effects of radioactivity.

$$dN_2/dt = \lambda_1 N_1 - \lambda_2 N_2 . \tag{A.24}$$

The solution of Eqs. A.23 and A.24 is

$$N_2(t) = N_{10} \frac{\lambda_1}{\lambda_2 - \lambda_1} (e^{-\lambda_1 t} - e^{-\lambda_2 t}) . \tag{A.25}$$

We can consider several limiting cases for Eq. A.25, applicable after a time t sufficient for $\lambda_i t \gg 1$, where λ_i is the larger of the two decay constants:

$$\lambda_1 \gg \lambda_2 (T_2 \gg T_1): \quad N_2(t) = N_{10} e^{-\lambda_2 t} , \tag{A.26}$$

and

$$\lambda_2 \gg \lambda_1 (T_1 \gg T_2): \quad N_2(t) = \frac{\lambda_1}{\lambda_2} N_1(t) . \tag{A.27}$$

The second of these limiting cases, given in Eq. A.27, is of importance in the consideration of the naturally occurring radioactive decay chains (see Section 2.3.2). Each of these chains is headed by a very long-lived radionuclide. After its decay, a chain of faster decays follows. By an extension of the previous analysis, the asymptotic condition is approached where

$$\lambda_1 N_1 = \lambda_2 N_2 = \lambda_3 N_3 = \ldots \tag{A.28}$$

The situation described by Eq. A.28 is known as *secular equilibrium*. At secular equilibrium, for any individual species, the same number of nuclei are added per unit time as decay, aside from the very slow variation governed by the original parent.

REFERENCES

1. Jagdish K. Tuli, *Nuclear Wallet Cards* (Upton, N.Y.: Brookhaven National Laboratory, 1995).
2. C. Michael Lederer and Virginia S. Shirley, eds. *Table of Isotopes*, 7th Ed. (New York: John Wiley, 1978).
3. "Discovery of the Transfermium Elements," report of the Transfermium Working Group of IUPAC and IUPAP, D. H. Wilkinson, chairman, *Pure & Applied Chemistry* 65, no. 8 (1993): 1757–1814.
4. R. G. H. Robertson, "Neutrino Mass and Mixing, and Non-Accelerator Experiments," in *Proceedings of the XXVI International Conference on High Energy Physics*, edited by James R. Sanford (New York: American Institute of Physics, 1993), 140–156.

Appendix B

General Tables

Table B.1. *Selected physical constants.*

Quantity	Symbol	Value (in SI units)
Speed of light in vacuum	c	2.9979×10^8 m/sec
Elementary charge	e	1.6022×10^{-19} C
Planck's constant	h	6.6261×10^{-34} J-sec
Avogadro's number	N_A	6.0221×10^{23}/mol
Electron mass	m_e	9.1094×10^{-31} kg
Atomic mass unit	u	1.6605×10^{-27} kg

Table B.2. *Selected conversion factors and approximate energy equivalents.*

Length		
1 foot (ft)	0.3048 m	
1 mile (mi)	1.609 km	
1 micron (μ)	10^{-6} m	
Area		
1 square mile (mi^2)	640 acres	2.590 km^2
1 hectare (ha)	10^{-2} km^2	2.471 acres
1 acre	4.047×10^{-3} km^2	4.356×10^4 ft^2
1 barn (b)	10^{-24} cm^{-2}	
Volume		
1 gallon (gal)	231 in^2	3.785 l
1 liter (l)	10^{-3} m^3	0.2642 gal
1 cubic foot (ft^3)	28.32 l	7.481 gal
Mass		
1 pound (lb)	0.4536 kg	0.0005 ton
1 tonne	1.102 ton	2205 lb
Pressure		
1 bar	10^5 N/m^2 [1 Pa]	0.9869 atm
1 atmosphere (atm)	1.013×10^5 N/m^2	14.7 lb/in^2
Energy		
1 electron-volt (eV)	1.6022×10^{-19} J	
1 million electron-volts (MeV)	1.6022×10^{-13} J	
1 British thermal unit (BTU)	1055 J	
1 kilowatt-hour (kWh)	3.600×10^6 J	3412 BTU
1 gigawatt-year (GWyr)	3.1536×10^{16} J	8.76×10^9 kWh
Energy equivalents		
1 atomic mass unit (u)	931.49 MeV	
1 electron mass *(m_e)*	0.5110 MeV	
1 kWh (e) at 33% efficiency	10,340 BTU (thermal)	
1 fission event (^{235}U)	200 MeV	
1 tonne ^{235}U	≈ 1 GWyr(e)	

Table B.3. *List of elements, Z=1–105.*

Z	Symbol	Name	Z	Symbol	Name	Z	Symbol	Name
1	H	Hydrogen	36	Kr	Krypton	71	Lu	Lutetium
2	He	Helium	37	Rb	Rubidium	72	Hf	Hafnium
3	Li	Lithium	38	Sr	Strontium	73	Ta	Tantalum
4	Be	Beryllium	39	Y	Yttrium	74	W	Tungsten
5	B	Boron	40	Zr	Zirconium	75	Re	Rhenium
6	C	Carbon	41	Nb	Niobium	76	Os	Osmium
7	N	Nitrogen	42	Mo	Molybdenum	77	Ir	Iridium
8	O	Oxygen	43	Tc	Technetium	78	Pt	Platinum
9	F	Fluorine	44	Ru	Ruthenium	79	Au	Gold
10	Ne	Neon	45	Rh	Rhodium	80	Hg	Mercury
11	Na	Sodium	46	Pd	Palladium	81	Tl	Thallium
12	Mg	Magnesium	47	Ag	Silver	82	Pb	Lead
13	Al	Aluminum	48	Cd	Cadmium	83	Bi	Bismuth
14	Si	Silicon	49	In	Indium	84	Po	Polonium
15	P	Phosphorus	50	Sn	Tin	85	At	Astatine
16	S	Sulfur	51	Sb	Antimony	86	Rn	Radon
17	Cl	Chlorine	52	Te	Tellurium	87	Fr	Francium
18	Ar	Argon	53	I	Iodine	88	Ra	Radium
19	K	Potassium	54	Xe	Xenon	89	Ac	Actinium
20	Ca	Calcium	55	Cs	Cesium	90	Th	Thorium
21	Sc	Scandium	56	Ba	Barium	91	Pa	Protactinium
22	Ti	Titanium	57	La	Lanthanum	92	U	Uranium
23	V	Vanadium	58	Ce	Cerium	93	Np	Neptunium
24	Cr	Chromium	59	Pr	Praseodymium	94	Pu	Plutonium
25	Mn	Manganese	60	Nd	Neodymium	95	Am	Americium
26	Fe	Iron	61	Pm	Promethium	96	Cm	Curium
27	Co	Cobalt	62	Sm	Samarium	97	Bk	Berkelium
28	Ni	Nickel	63	Eu	Europium	98	Cf	Californium
29	Cu	Copper	64	Gd	Gadolinium	99	Es	Einsteinium
30	Zn	Zinc	65	Tb	Terbium	100	Fm	Fermium
31	Ga	Gallium	66	Dy	Dysprosium	101	Md	Mendelevium
32	Ge	Germanium	67	Ho	Holmium	102	No	Nobelium
33	As	Arsenic	68	Er	Erbium	103	Lr	Lawrencium
34	Se	Selenium	69	Tm	Thulium	104	Rf	Rutherfordium[a]
35	Br	Bromine	70	Yb	Ytterbium	105	Ha	Hahnium[a]

[a]These are preliminary designations that are in international dispute and may be changed.

Table B.4. *Properties of selected radionuclides.*

Radio-nuclide	Half-life[a]	Main decay mode	Chief source(s) and special aspects
^{3}H	12.3 yr	β^-	cosmic rays; neutron reactions[b]
^{14}C	5730 yr	β^-	cosmic rays; neutron reactions with ^{14}N, ^{17}O
^{40}K	1.277×10^9 yr	β^-	natural (0.0117%)
^{60}Co	5.271 yr	β^-	neutron capture in ^{59}Co
^{85}Kr	10.76 yr	β^-	fission product
^{87}Rb	47.5×10^9 yr	β^-	natural (27.83%); fission product
^{90}Sr	28.8 yr	β^-	fission product
^{99}Tc	0.211×10^6 yr	β^-	fission product
^{129}I	15.7×10^6 yr	β^-	fission product
^{131}I	8.02 d	β^-	fission product
^{135}Xe	9.14 h	β^-	fission product (reactor poison)
^{137}Cs	30.1 yr	β^-	fission product
^{210}Pb	22.3 yr	β^-	^{238}U decay chain
^{222}Rn	3.82 d	α	^{238}U decay chain (gas)
^{226}Ra	1600 yr	α	^{238}U decay chain
^{232}Th	14.05×10^9 yr	α	natural
^{233}U	0.159×10^6 yr	α	neutron capture in ^{232}Th and β^- decay
^{234}U	0.245×10^6 yr	α	natural (0.0055%); neutron capture in ^{233}U
^{235}U	0.704×10^9 yr	α	natural (0.720%)
^{238}U	4.468×10^9 yr	α	natural (99.2745%)
^{239}U	23.45 m	β^-	neutron capture in ^{238}U
^{237}Np	2.14×10^6 yr	α	(n,2n) in ^{238}U and β^- decay; α decay of ^{241}Am
^{239}Np	2.36 d	β^-	β^- decay of ^{239}U
^{238}Pu	87.7 yr	α	neutron capture in ^{237}Np and β^- decay
^{239}Pu	24110 yr	α	neutron capture in ^{238}U and β^- decay
^{240}Pu	6564 yr	α	neutron capture in ^{239}Pu
^{241}Pu	14.35 yr	β^-	neutron capture in ^{240}Pu
^{242}Pu	0.373×10^6 yr	α	neutron capture in ^{241}Pu
^{241}Am	432.7 yr	α	β^- decay of ^{241}Pu

Sources: Half-lives and isotopic abundances from Jagdish K. Tuli, *Nuclear Wallet Cards* (Upton, N.Y.: Brookhaven National Laboratory, 1995).

[a] m=minutes; h=hours; d=days; yr=years.

[b] Tritium is produced, for example, in neutron reactions with ^{6}Li and ^{10}B and as a third fission product in a small fraction of fission events.

Acronyms and Abbreviations

ABB	ASEA/Brown Boveri
ABWR	Advanced boiling water reactor
ACNW	Advisory Committee on Nuclear Wastes [U.S.]
AEC	Atomic Energy Commission [U.S.]
AECL	Atomic Energy of Canada Limited [Canada]
AFR	Away from reactor
AFUDC	Allowance for funds used during construction
AGCR	Advanced gas-cooled graphite-moderated reactor (also AGR)
ALARA	As low as reasonably achievable
ALI	Annual limit on intake
ALMR	Advanced liquid metal reactor
ANS	American Nuclear Society
APS	American Physical Society
APWR	Advanced pressurized water reactor
AVLIS	Atomic vapor laser isotope separation
AVR	Arbeitsgemeinschaft Versuchs Reaktor
BEIR	Biological Effects of Ionizing Radiations
BNL	Brookhaven National Laboratory
BTU	British thermal unit
BWR	Boiling water reactor
CANDU	Canadian deuterium uranium (reactor)
CCDF	Complementary cumulative distribution function
CE	Combustion Engineering
CERN	Centre European de Recherche Nucleaire
CF	Capacity factor
CFR	Code of Federal Regulations
CPI	Consumer price index
DOE	Department of Energy [U.S.]
EBR	Experimental Breeder Reactor
ECCS	Emergency core cooling system
EIA	Energy Information Administration [U.S. DOE]
EPA	Environmental Protection Agency [U.S.]
EPR	European pressurized water reactor
EPRI	Electric Power Research Institute
ERDA	Energy Research and Development Administration [U.S.]
eV	Electron volt
FDA	Final design approval
FOAKE	First-of-a-kind engineering
FSU	Former Soviet Union
GA	General Atomics
GAO	General Accounting Office [U.S.]
GCR	Gas-cooled reactor
GDP	Gross domestic product

GE	General Electric
GeV	Billion (giga) electron volts
GT-MHR	Gas turbine modular helium reactor
GWd/t	Gigawatt-days (thermal) per tonne
GWDT	Gigawatt-days thermal
GWe	Gigawatt (electric)
GWt	Gigawatt (thermal)
GWyr	Gigawatt-year
HCLPF	High confidence of low probability of failure
HLW	High-level (radioactive) waste
HTGR	High temperature gas-cooled reactor
HWLWR	Heavy-water-moderated light-water-cooled reactor
HWR	Heavy water reactor
IAEA	International Atomic Energy Agency
ICP	International Chernobyl Project
ICRP	International Commission on Radiological Protection
IFR	Integral fast reactor
IHX	Intermediate heat exchanger
IIASA	International Institute for Applied Systems Analysis
INEL	Idaho National Engineering Laboratory
INPO	Institute of Nuclear Power Operations
INSAG	International Nuclear Safety Advisory Group [IAEA]
IPD	Implicit price deflator
IPP	Independent power producer
IRP	Integrated resource planning
ITER	International Thermonuclear Experimental Reactor
JTEC	Japanese Technology Evaluation Center [Loyola College]
keV	Thousand (kilo) electron volts
kWe	Kilowatt (electric)
kWt	Kilowatt (thermal)
kWh	Kilowatt-hour
LET	Linear energy transfer
LGR	Light-water-cooled graphite-moderated reactor
LLNL	Lawrence Livermore National Laboratory
LLW	Low-level (radioactive) waste
LMFBR	Liquid metal fast breeder reactor
LMR	Liquid metal reactor
LOCA	Loss-of-coolant accident
LOFT	Loss-of-fluid test
LWR	Light water reactor
MBTU	Million British thermal units
MeV	Million (mega) electron volts
MHR	Modular helium reactor
MHTGR	Modular high-temperature gas-cooled reactor
MOX	Mixed oxide [uranium and plutonium oxides]
MPa	Megapascal
MPC	Maximum permissible concentration; Multi-purpose canister
MRS	Monitored retrievable storage

MTIHM	Metric tons of initial heavy metal
MWDT	Megawatt-days thermal
MWe	Megawatt (electric)
MWt	Megawatt (thermal)
NAS	National Academy of Sciences [U.S.]
NCRP	National Council on Radiation Protection and Measurements [U.S.]
NEA	Nuclear Energy Agency [OECD]
NPT	Treaty on the Non-Proliferation of Nuclear Weapons
NRC	Nuclear Regulatory Commission [U.S.]
NRC	National Research Council [U.S.] (acronym not used in this book)
NRDC	National Resources Defense Council
NUG	Non-utility generator
NWPA	Nuclear Waste Policy Act [of 1982]
NWS	Nuclear-weapon State
NWTRB	Nuclear Waste Technical Review Board [U.S.]
O&M	Operation and maintenance
OCRWM	Office of Civilian Radioactive Waste Management [U.S. DOE]
OECD	Organization for Economic Co-operation and Development
ORNL	Oak Ridge National Laboratory
OTA	Office of Technology Assessment [U.S. Congress]
PHWR	Pressurized heavy-water reactor
PIUS	Process inherent ultimate safety
PORV	Pilot operated relief valve
PRA	Probabilistic risk assessment [or analysis]
PRISM	Power reactor inherently safe module [or innovative small]
PSA	Probabilistic safety assessment [or analysis]
PURPA	Public Utility Regulatory Policy Act of 1978
PWR	Pressurized water reactor
R&D	Research and development
RBMK	Reaktor bolshoi moshchnosti kanalnyi [high-power channel reactor]
RSS	Reactor Safety Study [WASH-1400]
SBWR	Simplified boiling water reactor
SIR	Safe integral reactor
SSD	Sub-seabed disposal
SSE	Safe shutdown earthquake
SWU	Separative work unit
TMI	Three Mile Island
TRU	Transuranic waste
TSPA	Total-system performance assessment
TWh	Terawatt-hour
UNSCEAR	United Nations Scientific Committee on the Effects of Atomic Radiation
USCEA	U.S. Council for Energy Awareness
USSR	Union of Soviet Socialist Republics
WANO	World Association of Nuclear Operators
WDV	Water dilution volume
WIPP	Waste Isolation Pilot Plant [New Mexico]

Glossary

Note: In many cases, the definitions given below are specific to usage in the context of nuclear energy, rather than general definitions.

Absorbed dose. The energy deposited by ionizing radiation per unit mass of matter (also known as the *physical dose*); the common units of absorbed dose are the *gray* and the *rad*. (See Section 2.4.1.)

Absorption, neutron. An inclusive term for neutron-induced reactions, excluding elastic scattering. (See Section 3.1.2.)

Abundance. *See* **isotopic abundance** and **elemental abundance**.

Accelerator. A facility for increasing the kinetic energy of charged particles, typically to enable them to initiate nuclear reactions.

Actinides. Elements with atomic numbers from $Z=90$ (thorium) through $Z=103$ (lawrencium); this group includes the fertile and fissile nuclei used in nuclear reactors and their main neutron capture products.

Activity. The rate of decay of a radioactive sample.

Alpha-particle decay (or alpha decay). Radioactive decay characterized by the emission of an alpha particle.

Alpha particle. A particle emitted in the radioactive decay of some heavy nuclei; it is identical to the nucleus of the ^{4}He atom. (See Section A.6.1.)

Antineutrino. A neutral particle, with zero or small mass, emitted together with an electron in beta-particle decay; sometimes referred to simply as a *neutrino*. (See Section A.6.3.)

Aquifer. A permeable layer of rock through which water can travel.

Assembly. A bundle of nuclear fuel rods, serving as a basic physical sub-structure within the reactor core. (See Section 6.2.3.)

Asymmetric fission. Fission events in which the masses of the two fission fragments are significantly different. (See Section 4.3.1.)

Atom. The basic unit of matter for each chemical element; it consists of a positively charged nucleus and surrounding electrons; the number of electrons in the neutral atom defines the element. (See Section A.1.1.)

Atomic energy. Used as a synonym for *nuclear energy*.

Atomic mass (M). The mass of a neutral atom, usually expressed in atomic mass units. (See Section A.2.2.)

Atomic mass number (A). The sum of the number of protons and the number of neutrons in the nucleus of an atom. (See Section A.1.2.)

Atomic mass unit (u). A unit of mass, defined so that the mass of the ^{12}C atom is exactly 12 u; 1 u$=1.6605\times10^{-27}$ kg. (See Section A.2.2.)

Atomic number (Z). An integer which identifies an element and gives its place in the sequence of elements in the periodic table; equal to the number of protons in the nucleus of the atom or the number of electrons surrounding the nucleus in a neutral atom. (See Section A.1.2.)

Avogadro's number (or Avogadro constant) (N_A). The number of atoms or molecules in one mole of a substance, e.g., the number of ^{12}C atoms in 12 g of ^{12}C; N_A=6.022 $\times 10^{23}$ per mole. (See Section A.2.3.)

Back end (of fuel cycle). In the nuclear fuel cycle, the steps after the fuel is removed from the reactor. (See Section 7.1.)

Backfill. Material placed in the cavities of a nuclear waste repository, after emplacement of the waste canisters.

Barn. A unit used to specify the cross section for a nuclear reaction, equal to 10^{-28} m^2 (or 10^{-24} cm^2).

Becquerel (Bq). A unit used to specify the rate of decay of a radioactive sample, equal to 1 disintegration per second.

BEIR Report. One of a series of reports prepared by the National Research Council's Committee on the Biological Effects of Ionizing Radiations.

Bentonite. A rock formed of clays which swell on absorbing water; often suggested as a backfill material for use in nuclear waste repositories.

Beta particle. A particle emitted in the radioactive decay of some nuclei; negative beta particles are electrons and positive beta particles are positrons. (See Section A.6.1.)

Beta-particle decay (or beta decay). Radioactive decay characterized by the emission of a beta particle.

Binding energy. *See* **nuclear binding energy**.

Biomass fuel. Fuel derived from organic matter, excluding fossil fuels; e.g., wood, agricultural crops, and organic wastes.

Biosphere. The zone where living organisms are present, extending from below the Earth's surface to the lower part of the atmosphere.

Boiling water reactor (BWR). A light water reactor in which steam formed inside the reactor vessel is used directly to drive a turbine. (See Section 6.2.1.)

Boron (B). The 5th element (Z=5); used as a control material because of the high cross section of ^{10}B for thermal-neutron absorption.

Borosilicate glass. A glass with high silicon and boron content, used for encapsulating reprocessed nuclear wastes.

Breeder reactor (or breeder). A reactor in which fissile fuel is produced by neutron capture reactions at a rate which equals or exceeds the rate at which fissile fuel is consumed.

Burnable poison. A material with a large thermal-neutron absorption cross section that is consumed during reactor operation; inserted in reactor to compensate for the decrease in reactivity as fissile fuel is consumed and fission product poisons are produced. (See Section 5.5.2.)

Burnup. Denotes the consumption of nuclides in a nuclear reactor; also used to specify the energy output from nuclear fission per unit mass of initial fuel. (See Section 7.3.1.)

Cadmium (Cd). The 48th element (Z=48); used as a control material because of the high cross section of ^{113}Cd for thermal-neutron absorption.

CANDU reactor. A reactor design developed in Canada, using heavy water as the coolant and moderator; also known as the *pressurized heavy water reactor* (PHWR).

Canister. *See* **waste canister**.

Capacity. The power output of a reactor under the designed operating conditions, commonly expressed in megawatts or gigawatts. *See also* **net capacity** and **gross capacity**.

Capacity factor. The ratio of the actual output of electricity from a reactor during a given period (usually a year) to the output that would have been achieved had the reactor operated at its rated capacity for the full period.

Capture. *See* **neutron capture**.

Carbon (C). The 6th element ($Z=6$); used as a moderator in some types of nuclear reactors because of its relatively low atomic mass and the very low thermal-neutron absorption cross section of its stable isotopes.

Carbon dioxide (CO_2). A gas present naturally in the atmosphere; the atmospheric concentrations of carbon dioxide are being substantially increased by the combustion of fossil fuels, perhaps leading to changes in the Earth's climate.

Chain reaction. A sequence of neutron-induced nuclear reactions in which the neutrons emitted in fission produce further fission events.

China syndrome. A term applied to a hypothetical sequence of events in which molten nuclear fuel penetrates the bottom of a reactor vessel and works its way into the Earth.

Cladding. The material surrounding the pellets of nuclear fuel in a fuel rod, isolating them physically, but not thermally, from the coolant.

Collective dose. The radiation dose to an entire population, equal to the sum of the individual doses.

Commercial reactor. A nuclear reactor used to generate electricity for sale to consumers.

Committed dose. *See* **dose commitment**.

Compound nucleus. An unstable nucleus formed when an incident neutron combines with a target nucleus; it quickly decays by emitting some combination of neutrons, charged particles, fission fragments, and gamma rays.

Condenser. A component of a power plant in which heat energy is removed from steam, changing water from the vapor phase to the liquid phase.

Constant dollars. A unit used to compare costs in different years by correcting for the effects of inflation. (See Section 15.5.1.)

Containment building (or containment). A heavy structure housing the reactor vessel and associated equipment; it is designed to isolate the reactor from the outside environment.

Continuum region. The region, either of nuclear excitation energies or of neutron energies, in which the neutron absorption cross section varies only slowly with energy because the compound nuclear states formed by neutron absorption overlap in energy. (See Section 3.3.)

Control rod. A rod composed of a *control substance*; used to regulate the power output of a reactor and, if necessary, to provide for rapid shutdown of the reactor.

Control substance (or control material). A material in solid or liquid form, with a high thermal-neutron absorption cross section, used to control the reactivity of a nuclear reactor. (See Section 5.5.2.)

Conversion. The production of fissile material from fertile material in a nuclear reactor.

Conversion ratio (C). The ratio of the rate of production of fissile nuclei in a reactor to the rate of consumption of fissile nuclei; for newly inserted uranium fuel, C is the ratio of the rate of production of ^{239}Pu to the rate of consumption of ^{235}U. (See Section 5.4.)

Converter reactor (or converter). A reactor in which there is substantial conversion of fertile material to fissile material; sometimes restricted to reactors with conversion ratios between 0.7 and 1.0.

Coolant. Liquid or gas used to transfer heat away from the reactor core.

Cooling pool. A water tank near the reactor, in which spent fuel is cooled and kept isolated from the environment for periods ranging from several months to several decades.

Core. *See* **reactor core**.

Cosmic rays. Atomic and sub-atomic particles impinging upon the earth, originating from the sun or from regions of space within and beyond our galaxy.

Coulomb force. The force between electrically charged particles; described by Coulomb's Law.

Critical. An arrangement of fissile material is critical when the effective multiplication factor is unity.

Criticality. The state of being critical; this is the necessary condition for sustaining a chain reaction at a constant power level.

Criticality accident. A nuclear reactor accident caused by some or all of the reactor fuel becoming super-critical.

Criticality factor. *See* **effective multiplication factor**.

Critical mass. The minimum mass of fissile material required for a sustained chain reaction in a given nuclear device.

Cross section, neutron (σ). A measure of the probability that an incident neutron will interact with a particular nuclide; cross sections are separately specified for different target nuclides and different reactions; the cross section has the units of area and can be (loosely) thought of as an effective target area for a specific process. (See Section 3.1.2.)

Crust. For the Earth, the solid outer layer composed mostly of rock.

Curie (Ci). A unit used to specify the rate of decay of a radioactive sample, equal to 3.7×10^{10} disintegrations per second (3.7×10^{10} Bq). (See Section 2.2.2.)

Current dollars. The unit for expressing actual costs at a given time, with no correction for inflation. (See Section 15.5.1.)

Daughter nucleus. Alternative term for *decay product* or *progeny*.

Decay chain (or decay series). *See* **radioactive decay series**.

Decay constant (λ). The probability per unit time for the radioactive decay of an atom of a given radioactive species. (See Section A.7.1.)

Decay product. The residual nucleus produced in radioactive decay; also referred to as *daughter nucleus*.

De-excitation. A transformation in which an excited nucleus gives up some or all of its excitation energy, typically through gamma-ray emission.

Delayed critical reactor. A nuclear reactor in which the fissions caused by delayed neutrons are essential to achieving criticality. (See Section 5.3.3.)

Delayed neutron. A neutron which is emitted from a fission fragment at an appreciable time after the fission event, typically more than 0.01 seconds later. (See Section 4.3.2.)

Design basis accident. A hypothetical reactor accident that defines reference conditions which the safety systems of the reactor must be able to handle successfully.

Deterministic. (a) In radiation protection, a deterministic effect is one for which the magnitude of the effect depends upon the magnitude of the dose and is approximately the same for similar individuals; *see, in contrast,* **stochastic**. (b) In nuclear accident analysis, a deterministic safety assessment focuses on the meeting of specifications that are designed to prevent failures or enable the reactor to withstand them without serious harmful consequences; *see, in contrast,* **probabilistic safety assessment**.

Deuterium (^{2}H or D). The hydrogen isotope with $A=2$, where A is the atomic mass number.

Deuteron. The nucleus of the deuterium atom; it consists of one proton and one neutron.

Diffusion. *See* **gaseous diffusion**.

Doppler broadening. In reactors, the increase in the effective width of resonance peaks for neutron-induced reactions, caused by thermal motion of the target nuclei. (See Section 3.2.3.)

Dose. In radiation physics, a brief form equivalent to *radiation dose*.

Dose commitment. The cumulative radiation dose produced by inhaled or ingested radioactive material, summed over the years following intake (typically up to a limit of 50 y).

Dose equivalent (*H*). A measure of the estimated biological effect of exposure to ionizing radiation, equal to the product of the quality factor for the radiation and the physical dose; the common units for the dose equivalent are the *sievert* and the *rem*. (See Section 2.4.1.)

Dose rate effectiveness factor (DREF). A factor sometimes applied in estimating radiation effects, corresponding to a presumed reduction in risk if the dose per unit time is small. (See Section 2.5.4.)

Dry storage. The storage of nuclear wastes under conditions where cooling is provided by the natural or forced circulation of air.

Effective dose equivalent (*H_E*). An overall measure of the estimated biological effects of radiation exposure, taking into account both the type of radiation and the region of the body exposed. (See Section 2.4.1.)

Effective multiplication factor (*k*). The ratio of the rate at which neutrons are produced in a reactor by fission to the rate of neutron loss through absorption and leakage; equivalently, the ratio of the number of neutrons produced by fission in one generation of a chain reaction to the number produced in the preceding generation; also known as the *criticality factor*. (See Section 5.1.1.)

Efficiency. *See* **thermal conversion efficiency**.

Elastic scattering. For neutrons, a reaction in which the incident neutron and the target nucleus remain intact; their total kinetic energy is unchanged while the neutron kinetic energy and direction are both changed.

Electrons. Stable elementary particles of small mass and negative charge that surround the nucleus of an atom and balance its positive charge; sometimes the term is used to include the "positive electron" or *positron*. *See also* **beta particle**.

Electron volt (eV). The kinetic energy gained by an electron in being accelerated through a potential difference of 1 volt; $1 \text{ eV} = 1.6022 \times 10^{-19}$ J. (See Section A.2.4.)

Elemental abundance. The relative abundance of an element in a sample of matter, expressed as the ratio of mass of that element to the total mass of the sample. (See Section A.3.2.)

Energy state (or energy level). For nuclei or atoms, one of the possible configurations of the constituent particles, characterized by its energy.

Enrichment. (a) The relative abundance of an isotope, often applied to cases where the isotopic abundance is greater than the level found in nature [for uranium the enrichment is usually expressed as a ratio by mass, not by number of atoms]. (b) The process of increasing the isotopic abundance of an element above the level found in nature.

Epidemiological study. For radiation, a study of the correlation between the radiation exposure of a population and selected illnesses or deaths in that population.

Excited state. A state of the nucleus in which the nucleus has more than its minimum possible energy; usually this energy is released by the emission of one or more gamma rays. (See Section A.4.)

Exposure. In radiation physics, a brief form equivalent to *radiation exposure*.

External costs. Costs of electrical power generation attributable to indirect factors, such as environmental pollution; also referred to as "social costs." (See Section 15.1.1.)

Fast breeder reactor (or fast breeder). A breeder reactor in which most fission events are initiated by fast neutrons.

Fast neutrons. Neutrons with energies comparable to the energies of neutrons emitted in fission, typically greater than 0.1 MeV.

Fertile nucleus. A non-fissile nucleus, which can be converted into a fissile nucleus through neutron capture, typically followed by beta decay. (See Section 4.2.2.)

Fissile nucleus. A nucleus with an appreciable fission cross section for thermal neutrons. (See Section 4.2.2.)

Fission. *See* **nuclear fission.**

Fission fragment. One of the two nuclei of intermediate mass produced in fission.

Fossil fuels. Fuels derived from organic materials which decayed many millions of years ago, namely coal, oil, and natural gas.

Fractionation. The process of changing the elemental or isotopic abundances of material in a region due to preferential transport of certain elements or isotopes out of or into the region.

Fracture. A break in the underground rock. (See Section 9.2.2, footnote 8.)

Fragment. *See* **fission fragment**.

Front end (of fuel cycle). In the nuclear fuel cycle, the steps before the fuel is placed in the reactor. (See Section 7.1.)

Fuel cycle. *See* **nuclear fuel cycle**.

Fuel rod (or fuel pin). A structure which contains a number of pellets of fuel; typically a long cylindrical rod with a thin metal wall. (See Section 6.2.3.)

Fusion power. Power derived from nuclear reactions in which energy is released in the combining of light atoms into heavier ones (e.g., of hydrogen into helium).

Gamma ray. A photon emitted in the transition of a nucleus from an excited state to a state of lower excitation energy. (See Section A.6.4.)

Gas-cooled reactor (GCR). A nuclear reactor in which the coolant is a gas, typically carbon dioxide or helium.

Gaseous diffusion. An isotopic enrichment process based upon the dependence on mass of the average velocity of the molecules in a gas at a given temperature. (See Section 7.2.2.)

Generation costs (or generating costs). The total costs of generating electrical power, including operations, maintenance, fuel, and carrying charges that cover capital costs of construction; sometimes termed "busbar cost." (See Section 15.2.1.)

Geologic waste disposal. Disposal of nuclear wastes in an underground site, excavated from the surrounding rock formation.

Geometric cross section. The cross section equal to the projected area of the nucleus. (See Section 3.1.2.)

Giga (G). 10^9, e.g., 1 gigawatt$=10^9$ W.

Gigawatt electric (GWe). Electrical power produced or consumed at the rate of 1 gigawatt.

Gigawatt thermal (GWt). Thermal power produced or consumed at the rate of 1 gigawatt.

Gigawatt-year (GWyr). The electrical energy corresponding to a power of 1 GWe maintained for a period of one year; equal to 3.15×10^{16} J.

Graphite. A crystalline form of carbon; the carbon used for reactor moderators is in this form.

Gray (Gy). The SI unit of absorbed dose for ionizing radiation; equal to 1 joule per kilogram. (See Section 2.4.1.)

Gross capacity. The capacity of a generating facility as measured at the output of the generator unit (also known as "busbar capacity"); equal to the sum of the net capacity and the power consumed within the plant.

Ground state. The state of a nucleus in which the nucleus has its minimum possible energy and therefore cannot decay by the emission of gamma rays.

Ground water. Water present in an underground zone.

Half-life (*T*). The average time required for the decay of one-half of the atoms in a radioactive sample.

Heavy metal. An element used as a nuclear fuel or produced by neutron capture in nuclear fuel; for example, thorium, uranium, and plutonium.

Heavy water. Water (H_2O) in which the hydrogen atoms are primarily atoms of ^2H (deuterium).

Heavy water reactor. A reactor in which heavy water is used as the coolant and moderator.

High-level waste. Highly radioactive material discharged from a nuclear reactor, including spent fuel and liquid or solid products of reprocessing.

Hormesis. A theory that for small radiation doses the incidence of damage, e.g., cancer induction, is reduced by small increases in the magnitude of the dose.

Implosion bomb. A nuclear bomb in which criticality is achieved by the rapid compression of uranium or plutonium to high density.

Inelastic scattering. For neutrons, a neutron-induced reaction in which the two final particles (neutron and nucleus) are the same as the initial particles, but with the final nucleus in an excited state and the total kinetic energy of the neutron and nucleus reduced.

Initial heavy metal. The inventory of heavy metal (typically uranium) present in reactor fuel before irradiation.

Interim storage. Temporary storage of spent fuel or reprocessed wastes, in anticipation of further treatment or transfer to a long-term storage facility.

International System of Units (SI units). The internationally adopted system of physical units, based upon the metric system; designated in French as the Système International d'Unités.

Ionizing radiation. Radiation in which individual particles are energetic enough to ionize atoms of the material through which they pass, either directly for charged particles (e.g., alpha particles and beta particles) or indirectly for neutral particles (e.g., x rays, gamma rays, and neutrons) through the production of charged particles.

Irradiation. The act of exposing to radiation.

Isobar. One of a set of nuclides that have the same atomic mass number (*A*) and differing atomic numbers (*Z*).

Isomer. One of a set of nuclides that have the same atomic number (*Z*) and atomic mass number (*A*), but differing excitation energies; a state is considered to be an isomeric state only if a significant time elapses before its decay.

Isotope. One of a set of nuclides that have the same atomic number (*Z*) and differing atomic mass numbers (*A*).

Isotopic abundance (or fractional isotopic abundance). The relative abundance of an isotope of an element in a mixture of isotopes, usually expressed as the ratio of the number of atoms of that isotope to the total number of atoms of the element. (*See also* **enrichment**, for exception in the case of uranium.)

Joule (J). The SI unit of energy.

Kilo. 10^3, e.g., 1 kilowatt=10^3 W.

Kiloton. Used as a unit of energy to describe the explosive yield of a nuclear weapon; equal to the energy released in the detonation of 1 million kg of TNT, nominally taken to be 10^{12} calories or 4.18×10^{12} joules.

Kinetic energy. Energy attributable to the motion of a particle or system of particles.

Level. *See* **energy level**.

Level width (Γ). The energy interval between points on either side of a resonance maximum where the absorption cross section is one-half the cross section at the maximum. (See Section 3.2.2.)

Light water. Water (H_2O) in which the hydrogen isotopes have their natural (or "ordinary") isotopic abundances (99.985% ^1H).

Light water reactor (LWR). A reactor which uses ordinary water as both coolant and moderator. (See Section 6.1.4.)

Linear energy transfer (LET). The rate of energy delivery to a medium by an ionizing particle passing through it, expressed as energy per unit distance traversed. (See Section 2.4.1.)

Linear hypothesis (or linearity hypothesis). The hypothesis that the quantitative impact of exposure to ionizing radiation is linearly proportional to the magnitude of the dose, remaining so even at low doses. (See Section 2.5.3.)

Liquid metal reactor. A reactor in which the coolant is a liquid metal, typically molten sodium.

Low-level wastes. A classification of nuclear wastes that excludes high-level and transuranic wastes. (See Section 8.1.3.)

Low-power license. An authorization granted by the Nuclear Regulatory Commission for the limited operation of a nuclear reactor.

Mean free path (λ). For neutrons, the mean distance a neutron travels before undergoing a nuclear reaction in the medium through which it is moving.

Mean life (τ). The mean time for nuclear decay, equal to the reciprocal of the decay constant. (See Section A.7.2.)

Mega (M). 10^6, e.g., 1 megawatt = 10^6 W.

Megawatt electric (MWe). Electrical power produced or consumed at the rate of 1 megawatt.

Megawatt thermal (MWt). Thermal power produced or consumed at the rate of 1 megawatt.

Metallic fuel. Nuclear fuel in metallic form, as distinct from oxide form; e.g., uranium rather than uranium oxide.

Milli (m). 10^{-3}, e.g., 1 mrem = 10^{-3} rem.

Millibarn (mb). A cross section unit equal to 10^{-3} barn.

Milling. The process by which uranium oxides are extracted from uranium ore and concentrated.

Mill tailings. The residues of milling, containing the materials remaining after uranium oxides are extracted from uranium ore; the tailings include radionuclides in the ^{238}U series, such as ^{230}Th. (See Section 7.2.1.)

Mixed oxide fuel (MOX). Reactor fuel composed of uranium and plutonium oxides (UO_2 and PuO_2).

Moderating ratio. A figure of merit for comparing moderators. (See Section 5.2.2.)

Moderator. A material of low atomic mass number in which the neutron kinetic energy is reduced to very low (thermal) energies, mainly through repeated elastic scattering.

Mole. The amount of a substance for which the mass in grams is numerically equal to the atomic (or molecular) mass of an atom (or molecule) of the substance expressed in atomic mass units. *See also* **Avogadro's number**.

Multiplication factor. *See* **effective multiplication factor**.

Natural radionuclide. A radionuclide produced by natural processes such as stellar nucleosynthesis or the radioactive decay of a natural radioactive precursor.

Net capacity. The capacity of a generating facility measured at the input to the transmission lines carrying power away from the plant; equal to the difference between the gross capacity and the power consumed within the plant.

Neutrino. A neutral particle, with zero or small mass, emitted together with a positron in β^+ decay; sometimes the term *neutrino* is used generically to include the *antineutrino*. (See Section A.6.3.)

Neutron. A neutral particle with a mass of approximately one atomic mass unit; it is important as a constituent of the nucleus and as the particle which initiates fission in nuclear reactors.

Neutron capture. A process in which a neutron combines with a target nucleus to form a product from which no nucleons are emitted and that therefore has the same atomic number as the target nucleus and an atomic mass number greater by unity. (See Section 3.1.1.)

Noble gas. One of the elements that are chemically inert and therefore are in gaseous form at ordinary temperatures and pressures; the noble gases are helium, neon, argon, krypton, xenon, and radon.

Nuclear binding energy. The total energy required to dissociate a nucleus in its ground state into its constituent nucleons.

Nuclear energy. Energy derived from nuclear fission or fusion, usually converted to electricity but sometimes used directly in the form of heat; it is sometimes loosely used as a synonym for *nuclear power*.

Nuclear fission. A process in which a nucleus separates into two main fragments, usually accompanied by the emission of other particles, particularly neutrons. (See Section 3.1.1.)

Nuclear force. The force that exists between every pair of nucleons, in addition to and independent of the Coulomb force between protons.

Nuclear fuel cycle. The succession of steps involved in the generation of nuclear power, starting with the mining of the ore containing the fuel and ending with the final disposition of the spent fuel and wastes produced in a nuclear reactor. (See Section 7.1.)

Nuclear power. Electrical power derived from a nuclear reactor; it is sometimes loosely used as a synonym for *nuclear energy*.

Nuclear power plant. A facility for producing electricity using energy from fission or fusion processes.

Nuclear reaction. For neutrons, an event in which a neutron interacts with an atomic nucleus to give products which differ in composition or kinetic energy from the initial neutron and nucleus. (See Section 3.1.1.)

Nuclear reactor. A device in which heat is produced at a controlled rate, by nuclear fission or nuclear fusion.

Nuclear waste package. The nuclear wastes, their container and its contents, and the immediately surrounding materials used to isolate the container.

Nuclear wastes. Radioactive materials remaining as a residue from nuclear power production or from the use of radioisotopes.

Nuclear weapon (or nuclear bomb). A weapon which derives its explosive force from nuclear fission or nuclear fusion.

Nucleon. A generic term for neutrons and protons, reflecting similarities in some of their properties.

Nucleosynthesis. The series of nuclear reactions by which the elements are formed in nature; these reactions started in the big bang and continued in stars and the interstellar medium. (See Section 2.3.1.)

Nucleus. The very dense central core of an atom, composed of neutrons and protons. (See Section A.1.1.)

Nuclide. An atomic species defined by its atomic number and atomic mass number.

Ocean sediment. Material on the ocean floor deposited by the settling of insoluble solid material from the ocean.

Once-through cycle. A nuclear fuel cycle in which the spent fuel is stored or disposed of without extraction of any part of the fuel for further use in a reactor.

Operating costs. The costs of operating and maintaining a power plant, sometimes including the costs of capital additions and usually excluding fuel costs. (See Section 15.2.1.)

Operating license. For nuclear reactors, a legally required prerequisite for the operation of the reactor; in the United States the license is granted by the Nuclear Regulatory Commission.

Overpack. A container placed around a nuclear waste canister to provide additional protection against corrosion and physical damage.

Oxide fuel. Nuclear fuel in oxide form, as distinct from metallic form; e.g., uranium oxide rather than uranium.

Parent nucleus. A preceding radioactive nucleus in a chain of one or more radioactive decays.

Passive safety. Reactor safety achieved without dependence on human intervention or the performance of specialized equipment; e.g., achieved by the action of unquestionable physical effects such as the force of gravity or thermal expansion.

Photon. A discrete unit or quantum of electromagnetic radiation, which carries a prescribed amount of energy proportional to the frequency of the radiation. (See Section A.4.)

Physical dose. An alternative term for **absorbed dose**.

Pitchblende. An ore rich in uranium oxides.

Plutonium (Pu). The 94th element ($Z=94$); does not occur in nature because of its short half life, but is produced in nuclear reactors for use as a fissile reactor fuel or for nuclear weapons—most importantly the isotope ^{239}Pu.

Poison. *See* **reactor poison**.

Porosity (ϵ). The fraction of the volume of a rock formation which is open, and may be filled with air or water.

Positron. A positively charged particle which is identical to the electron in mass and the magnitude of its charge; sometimes called a "positive electron." *See also* **beta particle**.

Potassium (K). The 19th element ($Z=19$); the natural radionuclide ^{40}K in the body contributes substantially to the total human radiation exposure. (See Section 2.4.2.)

Potential energy. The energy associated with a particular configuration of a system; configuration changes that reduce the potential energy may result in a release of kinetic energy as, for example, in fission.

Precursor. In radioactive decay, the parent nucleus or an earlier nucleus in a radioactive decay series.

Predetonation. In a nuclear weapon, the initiation of a fission chain reaction before the fissile material has been sufficiently compacted for fission to take place in a large fraction of the material. (See Section 14.2.4.)

Pressure vessel. *See* **reactor vessel**.

Pressurized heavy water reactor (PHWR). *See* **CANDU reactor**.

Pressurized water reactor (or pressurized light water reactor) (PWR). A light water reactor in which the water is at a high enough pressure to prevent boiling. *See also* **steam generator**.

Primary cooling system. The system in a reactor which provides cooling for the reactor core, carrying heat from the core to a heat exchanger or turbine generator.

Probabilistic safety (or risk) assessment (PSA or PRA). A method for estimating accident probabilities by considering the failure probabilities for individual components and determining the overall probability of combinations of individual failures that result in harmful consequences; used interchangeably with "probabilistic safety analysis." (See Section 11.4.2.)

Production costs. In electricity generation, the sum of operating costs and fuel costs. (See Section 15.2.1.)

Progeny. Residual nuclei following radioactive decay; often called *daughter nuclei*.

Prompt critical reactor. A nuclear reactor in which the fissions caused by delayed neutrons are not essential to achieving criticality. (See Section 5.3.3.)

Prompt neutron. A neutron emitted at the time of nuclear fission, with no significant delay.

Proton. A particle with a mass of approximately one atomic mass unit and a charge equal and opposite to that of an electron; a basic constituent of all atomic nuclei.

Quality factor (Q). A measure of the relative effectiveness of different types of ionizing radiation in producing biological damage, taken to be unity for x rays and gamma rays. (See Section 2.4.1.)

Rad. A traditional unit for the absorbed dose of ionizing radiation; equal to 100 ergs per gram or 0.01 gray. (See Section 2.4.1.)

Radiation. In nuclear physics and engineering, often used as shorthand for *ionizing radiation*.

Radiation dose. A measure of the magnitude of exposure to radiation, expressed more specifically as the *absorbed dose* or as the *dose equivalent*.

Radiation exposure. The incidence of radiation upon an object, e.g., upon a person; the term is also used in a specific sense for the dose received from incident x rays. (See Section 2.4.1.)

Radioactive. Possessing the property of *radioactivity*.

Radioactive decay. The spontaneous emission by a nucleus of one or more particles, as in *alpha-particle decay* or *beta-particle decay*. (See Section A.6.1.)

Radioactive decay series (or chain). A sequence of radioactive nuclei, connected by the successive emission of alpha particles or beta particles and terminating in a stable "final" nucleus. (See Section 2.3.2.)

Radioactivity. (a) The phenomenon, observed for some nuclides, of spontaneous emission of one or more particles, leaving a residual nucleus of different atomic number. (See Section A.6.1.) (b) The activity of a sample, commonly in units of becquerels or curies.

Radionuclide (or radioisotope). A nuclear species that is radioactive.

Radium (Ra). The 88th element ($Z=88$); a natural radionuclide in the ^{238}U radioactive decay series.

Radon (Rn). The 86th element ($Z=86$); the isotope ^{222}Rn, a product of the decay of ^{226}Ra, is responsible for a large fraction of the radiation dose received by most individuals. (See Section 2.4.2.)

Reactivity (ρ). A measure of the extent to which a reactor is supercritical ($\rho>0$) or subcritical ($\rho<0$). (See Section 5.3.1.)

Reactor. *See* **nuclear reactor**.

Reactor core. The region of a reactor in which the nuclear chain reaction proceeds.

Reactor grade plutonium. Plutonium with the isotopic mixture characteristic of spent fuel, typically 20% or more ^{240}Pu. (See Section 14.2.4.)

Reactor period. The time during which the neutron flux and reactor power level change by a factor of e, where e (equal to 2.718) is the base of the natural logarithms. (See Section 5.3.2.)

Reactor poison. A material, produced in fission or otherwise present in the reactor, with a high cross section for absorption of thermal neutrons. (See Section 5.5.1.)

Reactor vessel. The tank containing the reactor core and coolant; when, as in water-cooled reactors, a high pressure is maintained within the vessel, it is alternatively known as the *pressure vessel*. (See Section 6.2.2.)

Rem. A unit of radiation dose equivalent, equal to 0.01 sievert; historically an abbreviation for roentgen-equivalent-man. (See Section 2.4.1.)

Renewable energy. Energy that is derived from sources that will remain undiminished into the very distant future, for example, hydroelectric energy, wind energy, direct solar energy, and energy from ocean tides.

Reprocessing. The processing of spent fuel from a nuclear reactor in order to extract and separate plutonium and other selected components.

Resonance (or resonance peak). For neutrons, a local maximum in the neutron absorption cross section, arising when the kinetic energy of the neutron corresponds to the energy of an excited state of the composite nucleus.

Resonance escape probability (p). The probability that a neutron emitted in fission in a reactor will be moderated to thermal energies, avoiding resonance capture in the fuel. (See Section 5.1.2.)

Resonance region. The region of incident neutron energies within which the reaction cross section varies rapidly with energy due to discrete resonance peaks.

Retardation factor (R). The ratio of the velocity of water through a medium to the average velocity with which a given ion travels through the medium. (See Section 9.2.3.)

Retrievable storage. The storage of nuclear wastes in a repository in a manner that permits the later removal of the wastes, for example for reasons of safety or to extract selected nuclides.

Roentgen (R). A measure of radiation exposure, usually restricted to x ray exposures. (See Section 2.4.1.)

Runaway chain reaction. A chain reaction in which the reactivity increases at a rapid, uncontrolled rate.

Saturated zone. A zone in the ground in which the voids in the rock are filled with water; this is the zone lying below the water table.

Scattering. *See* **elastic scattering** and **inelastic scattering**.

Seabed. Top layer of solid material under the ocean.

Secular equilibrium. A condition where the number of decays per unit time is virtually the same for all members of a radionuclide decay series in a sample of radioactive material. (See Section 2.3.2.)

Seismic. Relating to vibrations in the earth, particularly due to earthquakes.

Separation energy. For nucleons, the energy required to remove the nucleon from a nucleus in its ground state. (See Section 4.2.2.)

Separative work. For uranium, a measure of the difficulty of enriching the uranium to a higher ^{235}U fraction (defined in Section 7.2.2, footnote 11).

Separative work unit (*SWU*). A unit, with dimensions of mass, used to specify the magnitude of separative work. (See Section 7.2.2.)

Sievert (Sv). The SI unit of radiation dose equivalent. (See Section 2.4.1.)

SI unit. A unit in the *International System of Units*, widely used as the standard system of units in physics and engineering.

Slow neutrons. Neutrons of low kinetic energy, often taken to be less than 1 eV. (See Section 3.4.1.)

Sorption. Processes that retard the transport of nuclides in ground water by binding them (usually temporarily) to the rock through which the ground water flows. (See Section 9.2.3.)

Source term. In a nuclear reactor accident, the inventory of radionuclides that escapes from the containment into the environment.

Specific activity. The rate of decay of a radionuclide sample divided by its mass. (See Section 2.2.3.)

Spent fuel. Nuclear fuel that has been used in a reactor and has been removed, or is ready to be removed, from the reactor.

Stable nucleus. A nucleus which, in its ground state, does not undergo radioactive decay.

State. *See* **energy state**.

Steam generator. A heat exchanger used to produce steam in a pressurized water reactor. (See Section 6.2.1.)

Stochastic. In radiation protection, a stochastic effect impacts individuals randomly; the probability of a given impact on an individual (e.g., cancer induction) depends upon the magnitude of the dose received. (See Section 2.4.1.)

Subcritical. For nuclear fuel, having an effective multiplication factor that is less than unity.

Supercritical. For nuclear fuel, having an effective multiplication factor that is greater than unity.

Symmetric fission. Fission events in which the masses of the two fission fragments are approximately equal. (See Section 4.3.1.)

Thermal conversion efficiency (or thermal efficiency). For an electrical power plant, the ratio of the electrical energy produced to the total heat energy produced in the consumption of the fuel.

Thermal energy. (a) The energy, kT, characteristic of particles in thermal equilibrium in a system at temperature T; k is here the Boltzmann constant. (See Section A.2.4.) (b) The total heat energy produced in a power plant (as distinct from the electrical energy produced).

Thermal equilibrium. A system is in thermal equilibrium when there is no net flow of heat from one part to another; in a reactor at temperature T, neutrons reach thermal equilibrium when their average kinetic energy is the same as that of the molecules of a gas at temperature T.

Thermalization. The reduction, primarily by elastic scattering in the moderator, of the energies of fission neutrons from their initial values (typically near 1 MeV) to thermal energies (below 0.1 eV).

Thermal neutrons. Neutrons with a kinetic energy distribution characteristic of a gas at thermal equilibrium, conventionally at a temperature of 293 K; this temperature corresponds to a most probable kinetic energy of 0.025 eV. (See Section 3.4.2.)

Thermal reactor. A reactor in which most fission events are initiated by thermal neutrons.

Thermal utilization factor (f). The fraction of thermal neutrons that are captured in the fuel (rather than in the coolant or structural materials). (See Section 5.1.2.)

Ton. The unit of mass in the English system, equal to 2000 lbs (907 kg).

Tonne (or metric ton). The unit of mass in the metric system, equal to 1000 kg.

Total cross section (σ_T). The sum of all reaction cross sections, for a given pair of particles at a given energy. (See Section 3.1.2.)

Transmutation. The conversion of a nuclide into another nuclide in a reactor or accelerator.

Transuranic element. An element of atomic number greater than 92.

Transuranic wastes. Wastes that contain more than 0.1 μC of long-lived alpha-particle emitting transuranic nuclides per gram of material, but which are not *high-level wastes*. (See Section 8.1.3.)

Tuff. A rock formed by the compaction of the ashes of material emitted in volcanic explosions.

Uranium (U). The 92nd element ($Z=92$); important in nuclear reactors and nuclear weapons, particularly for its fissile isotope ^{235}U.

Unsaturated zone. A zone in the ground in which voids in the rock are at least partly filled with air, rather than completely filled with water; this is the zone lying above the water table.

Vitrification. For nuclear wastes, incorporation of liquid wastes into a molten glass, which solidifies on cooling.

Void. In reactors, a region (or bubble) of vapor in a coolant that normally is a liquid; in rock formations, gaps in the rock matrix that can be filled by air or water.

Void coefficient. The rate of change of the reactivity with change in the void volume; a positive void coefficient corresponds to an increase in reactivity when the void volume increases. (See Section 11.2.2.)

Waste canister. A container in which solid nuclear wastes, including spent fuel, are placed for storage or transportation.

Wastes. *See* **nuclear wastes**.

Water dilution volume (WDV). The volume of water required to dilute a sample of nuclear wastes to the maximum permissible concentration for drinking water; it serves as a measure of the relative hazard posed by individual radionuclides in the wastes. (See Section 8.3.2.)

Water table. The top of the region of the ground where voids in the rock are filled with water; i.e., the top of the *saturated zone*.

Watt (W). The SI unit of power; 1 W = 1 joule per second.

Weapons-grade plutonium. Plutonium with isotopic concentrations that make it effective in nuclear weapons; commonly this means a ^{240}Pu concentration of less than 6%. (See Section 14.2.4.)

Welded tuff. *Tuff* which has been welded together by heat and pressure.

Xenon (Xe). The 54th element ($Z=54$); the fission product ^{135}Xe has a very high absorption cross section for thermal neutrons.

Xenon poisoning. The reduction of reactivity in the core due to a build-up of the ^{135}Xe concentration. (See Section 5.5.3.)

X rays. Photons emitted by excited atoms or in the rapid acceleration (or deceleration) of electrons (commonly reserved for wavelengths below the ultraviolet region).

Zircaloy. An alloy of zirconium, commonly used as a cladding material in nuclear fuel rods.

Zirconium (Zr). The 40th element ($Z=40$); a major component of the fuel cladding alloy, zircaloy.

Subject Index